AF805730

Multicopter Design and Control Practice

Quan Quan · Xunhua Dai · Shuai Wang

Multicopter Design and Control Practice

A Series Experiments based on MATLAB and Pixhawk

電子工業出版社
PUBLISHING HOUSE OF ELECTRONICS INDUSTRY
http://www.phei.com.cn

Springer

Quan Quan
School of Automation Science
and Electrical Engineering
Beihang University
Beijing, China

Xunhua Dai
School of Computer Science
and Engineering
Central South University
Changsha, China

Shuai Wang
School of Automation Science
and Electrical Engineering
Beihang University
Beijing, China

ISBN 978-981-15-3137-8 ISBN 978-981-15-3138-5 (eBook)
https://doi.org/10.1007/978-981-15-3138-5

Jointly published with Publishing House of Electronics Industry
The print edition is not for sale in China (Mainland). Customers from China (Mainland) please order the print book from: Publishing House of Electronics Industry.

© Publishing House of Electronics Industry 2020, corrected publication 2021
This work is subject to copyright. All rights are reserved by the Publishers, whether the whole or part of the material is concerned, specifically the rights of translation, reprinting, reuse of illustrations, recitation, broadcasting, reproduction on microfilms or in any other physical way, and transmission or information storage and retrieval, electronic adaptation, computer software, or by similar or dissimilar methodology now known or hereafter developed.
The use of general descriptive names, registered names, trademarks, service marks, etc. in this publication does not imply, even in the absence of a specific statement, that such names are exempt from the relevant protective laws and regulations and therefore free for general use.
The publishers, the authors, and the editors are safe to assume that the advice and information in this book are believed to be true and accurate at the date of publication. Neither the publishers nor the authors or the editors give a warranty, express or implied, with respect to the material contained herein or for any errors or omissions that may have been made. The publishers remain neutral with regard to jurisdictional claims in published maps and institutional affiliations.

This Springer imprint is published by the registered company Springer Nature Singapore Pte Ltd.
The registered company address is: 152 Beach Road, #21-01/04 Gateway East, Singapore 189721, Singapore

Foreword

Drones and new types of aerial vehicles are conquering the skies at an impressive pace—enabled by rapid progress in sensors, electric components and embedded computing power. Whereas 20 years ago, development, integration and testing of automated functions for aerospace application were restricted to people working for large-scale transport or military aircraft; everybody can nowadays develop, build and test novel algorithms, leveraging the power of freely available consumer Unmanned Aircraft System (UAS) and open-source projects.

This development is nothing else than a revolution—it empowers any interested person to not only engage in the fascinating fields of modeling, simulation, guidance, navigation and control of aerial vehicles but also turn own ideas into reality and actually fly and test them.

So the big question is now—How to get started?!?—How to find the right balance between theoretical depth and practical pragmatism?—How to be guided through the jungle of drivers, settings and wirings of embedded hardware?

Multicopter Design and Control Practice—A Series Experiments Based on MATLAB and Pixhawk, the new book by Quan Quan, Xunhua Dai and Shuai Wang of Beihang University, Beijing, China, is the appropriate answer. After their successful book *Introduction to Multicopter Design and Control* where the interesting reader can learn about the theory of designing, simulating and controlling multicopter vehicles, this new book builds on the powerful resources provided by the MathWorks toolchains (MATLAB, Simulink, Stateflow, …) as the backbone for model-based system engineering and the Pixhawk with PX4 as the most distributed open-source hardware and software solution for guidance, navigation and control of low-cost aerial vehicles in a multitude of configurations.

The book provides a detailed step-by-step tutorial describing the Pixhawk system, setting up hardware, software, simulation and hardware in the loop simulation. It details the experimental process and addresses every involved component in an intuitive and tangible way. Sensor calibration, state-estimation and filter design, attitude control, position control and supervisory logic are presented in detailed experiments step-by-step that all follow the scheme of introducing the experiment, presenting preliminaries, theory, analysis and finally the design task and a summary.

After successfully accomplishing the book by executing the practical experiments by oneself, the reader is capable of developing and more important deploying and testing algorithms for multicopters and other types of drones.

The fascinating world of guiding, navigation and controlling aerial vehicles is no longer a privilege of a small group—it is accessible to everyone. Even if you never interacted with embedded hardware and software before, the book empowers you, to become a part of it. It gives you the hard skills required to make things happen.

November 2019

Florian Holzapfel
Institute of Flight System Dynamics
Technical University of Munich
Munich, Germany

Preface

Multicopter development began to flourish in the year 2013 because of the high demand for aerial photography and the emergence of low-cost multicopter products. New ideas, technologies, products, applications, and investments associated with multicopters emerged one after another. During the years 2016 and 2017, owing to an increase in market bubbles and frequent incidents wherein civil aviation flights were threatened by Unmanned Aerial Vehicles (UAVs), UAV-related industries suffered a small decline. Meanwhile, many countries began to establish laws, regulations, and standards for micro-small UAVs, open test bases, organize the study of air traffic control systems for UAVs, and start the processes of education and training. It can be said that the micro-small UAV industries represented by multicopters have entered a new, orderly phase with more opportunities for development. With the rise of 5G, micro-small UAVs may become big data collection platforms at low altitudes, which is expected to bring seven to ten times more business opportunities for industry.[1]

The rapid development of multicopters is inseparable from the support of open-source flight control systems. Whether based on open-source flight control systems or fully independent development, developers need to have a comprehensive grasp of multicopter design, modeling, perception, control and decision-making to produce multicopters that deliver outstanding performance. However, much micro-small multicopter research and development started with micro-small enterprises. Unlike traditional aviation institutes, micro-small enterprises have disadvantages such as fewer engineers, less experience, and fewer resources; thus, chief and full-stack engineers are urgently needed. An important reason for the shortage of such engineers is that textbooks and curriculum have failed to keep up with the rapid development of multicopters. Motivated by this, Beihang Reliable Flight Control Group (http://rfly.buaa.edu.cn) published a book in 2017 titled *Introduction to Multicopter Design and Control* in Springer.

[1]Connected drones - a new perspective on the digital economy. Available on https://cdn.microdrones.com/fileadmin/web/_downloads/papers/Huawei_whitepaper.pdf

Although we recognize that *Introduction to Multicopter Design and Control* can explain some theory problems for readers, the lack of practical exercises means that it cannot help readers in deepening their practical understanding. There is a certain gap that still exists between the requirements of the industry and current engineering education, which has motivated us to develop new tools and tutorials based on multicopters for allowing readers to apply the theory to practice.

Based on these considerations, the experiment of this book adopts the following:

(1) the most widely-used flight platform, multicopters, as a flight platform;
(2) the most widely-used open-source autopilot systems of aerial vehicles, Pixhawk/PX4, as a control platform;
(3) one of the most widely-used programming languages in the field of control engineering, MATLAB/Simulink, as a programming language.

Based on the current advanced development of the *Model-Based Design* (MBD) process, the above three are closely linked. In addition to the advancement of software, hardware, and development concepts, the experiments consider coverage and differentiation. The book covers eight tasks that include multicopter propulsion system design, dynamic modeling, sensor calibration, state estimation and filter design, attitude controller design, set-point controller design, semi-autonomous control mode design, and failsafe logic design. These tasks cover a comprehensive level of knowledge, and they can be completed by following a progressive route. Each task consists of three step-by-step experiments from shallow to deep, namely a basic experiment, an analysis experiment, and a design experiment so that readers with different backgrounds can benefit from them. Each experiment must be implemented in MATLAB/Simulink, and simulation tests are carried out in the released software-in-the-loop simulation platform. Furthermore, readers can upload the control algorithms to the Pixhawk autopilot through the automatic code generation technology and get a closed-loop control system with a given real-time simulator for Hardware-In-the-Loop (HIL) simulation tests. Throughout the above process, readers can familiarize themselves with the basic process of MBD, including the composition, mathematical model, and control of a multicopter. Furthermore, readers can master a variety of modern tools such as MATLAB/Simulink and FlightGear in the development and computer simulation, a HIL simulator, Pixhawk autopilot, and a remote control transmitter in the HIL simulation test. The design task also includes outdoor flight experiments to allow readers to experience the full development of a multicopter. Based on this, the platform designed in this book won the gold award of National Experimental Equipment Design Contest for College Automation Education in China. Advanced development tools and processes allow micro-small enterprises to realize new ideas rapidly, thereby overcoming the disadvantages they face, which include fewer people, less experience, and fewer resources. Furthermore, by lowering the learning threshold, more people with other related backgrounds have opportunities to enter the aviation field.

This book can be considered a companion to *Introduction to Multicopter Design and Control*. To make this book self-contained, we have included some theoretical parts of *Introduction to Multicopter Design and Control* that are required for the experiments in this book. The process of preparing this book, which commenced in 2018, spanned nearly 2 years. During this time, we first designed and implemented all the experiments and then continually revised the book to present our ideas as clearly as possible. Quan Quan designed the structure, experimental process, and experiments of the entire book; Xunhua Dai designed the experimental platform and development process; Shuai Wang conducted all the experiments according to the experimental process. The successful completion of this book is inseparable from the enthusiastic support and help of the students of Beihang Reliable Flight Control Group. First of all, we thank Lanjiang Yang, Hao Liu, and Junqing Ning for participating in the experiments and assisting in the revision of this book carefully and repeatedly. Secondly, we appreciate Fanning Liu for the basic development of the experimental process and some experiments that she conducted during her graduation project with our group. This book was used for undergraduate students' curriculum design (B3I034140) and graduate students' course (031574) partly, as well as in the courses opened by a company, Tang Dynasty Robot (https://www.chinarobot.com). We thank all the students who participated in the courses for their helpful feedback.

The development of the experiments and tools in this book has been expensive; therefore, we reserve the copyright for them. Experimental codes and the RflySim platform that we developed are available at https://rflysim.com for personal use. No company or individual is allowed to sell the codes and tools included in this book as educational products without the authorization of Beihang Reliable Flight Control Group. Otherwise, a legal response shall be pursued. For questions and licenses related to this book, please contact us via https://rflysim.com.

Beijing, China
July 2019

Quan Quan

The original version of the book was revised based on the corrections provided by the author. The correction to the book is available at https://doi.org/10.1007/978-981-15-3138-5_13

Contents

Chapter 1
Introduction

Considering the multicopter as a typical research object, this book introduces its salient features and main differences between the multicopter and other types of aerial vehicles. The multicopter is one of the most important types of aerial vehicles; it has a very broad development potential in various fields owing to its advantages such as ease of use and Vertical Take-Off and Landing (VTOL) capability. For new task requirements in specific fields, the airframe configuration, aerodynamic configuration, propulsion system, and control system of a multicopter usually need to be redesigned. This entails a large number of flight tests to verify whether the desired performance requirements are satisfied. Traditional development and testing methods for small aerial vehicles are typically based on outdoor flight experiments directly; this leads to two major difficulties in cultivating professional and technical engineers required by the industry. The first difficulty is that a multicopter system is composed of various components involving a great deal of interdisciplinary knowledge and skill that leads to a high threshold for beginners. The second difficulty is that outdoor flight tests are typically dangerous and are often limited by the local airspace management policies; thus, the Verification and Validation (V&V) [1] of the development process are hard to perform. With traditional development methods, it is difficult for beginners to implement flight control algorithms to real aerial vehicles, so the current education of aerial vehicles often focuses on theoretical learning and numerical simulation. To change this situation, this book presents a series of experiments including Software-In-the-Loop (SIL) simulation, automatic code generation, Hardware-In-the-Loop (HIL) simulation, indoor flight experiments, and outdoor flight experiments based on the idea of *Model-Based Design* (MBD).[1] The introduction to the latest MBD tools and methods ensures that readers can efficiently focus on studying algorithms for multicopters with no need to access low-level source code and hardware. This book also releases an experimental platform for rapidly implementing and verifying ideas involving multicopters. This chapter introduces the following: (1) the pre-knowledge for the study of this book, (2) the experimental courses with basic learning routes,

[1] https://www.mathworks.com/solutions/model-based-design.html.

© Publishing House of Electronics Industry 2020, corrected publication 2021
Q. Quan et al., *Multicopter Design and Control Practice*,
https://doi.org/10.1007/978-981-15-3138-5_1

and (3) the expected benefits that this book can offer. It should be noted that the tools and methods presented in this book are not limited to multicopters; they can also be applied to other types of aerial vehicles.

1.1 What Are Multicopters

1.1.1 Classification of Common Small Aerial Vehicles

As shown in Fig. 1.1, commonly-used small aerial vehicles are mainly classified into three types.

(1) Fixed-wing aircraft
As shown in Fig. 1.1a, wings of a fixed-wing aircraft are permanently attached to the airframe. Most civil aerial vehicles and fighters are fixed-wing aircraft. Their propulsion systems generate a forward airspeed so that the wings can produce lift to balance the weight of these vehicles. Based on this principle, fixed-wing aircraft must maintain a certain forward airspeed; therefore, they cannot take off or land vertically. In comparison with traditional helicopters, a fixed-wing aircraft has a much simpler structure and can carry a heavier payload over a longer distance while consuming less power. The disadvantage of fixed-wing aircraft is the requirement of a runway or launcher for takeoff and landing.

(2) Single-rotor helicopter
As shown in Fig. 1.1b, a helicopter is a type of rotorcraft in which lift is directly supplied by rotors. A single-rotor helicopter has four flight control inputs that include the cyclic, the collective, the anti-torque pedals, and the throttle. The collective is used to control the angle of attack of the rotor. Although the lift of a helicopter is primarily controlled by the collective and the throttle, the fast dynamic response of the lift is adjusted by the collective control. From the introduction above, it is understood that the single-rotor helicopter has the ability of VTOL because the lift is not induced by the velocity (also called linear velocity) of the airframe. Therefore, no runway or launcher is required for takeoff and landing. Compared with a fixed-wing aerial vehicle, it does not have the advantage in terms of the time of endurance. Moreover, its complex structure incurs a high maintenance cost.

(a) Fixed-wing aircraft (b) Helicopter (c) Multicopter

Fig. 1.1 Common aerial vehicles

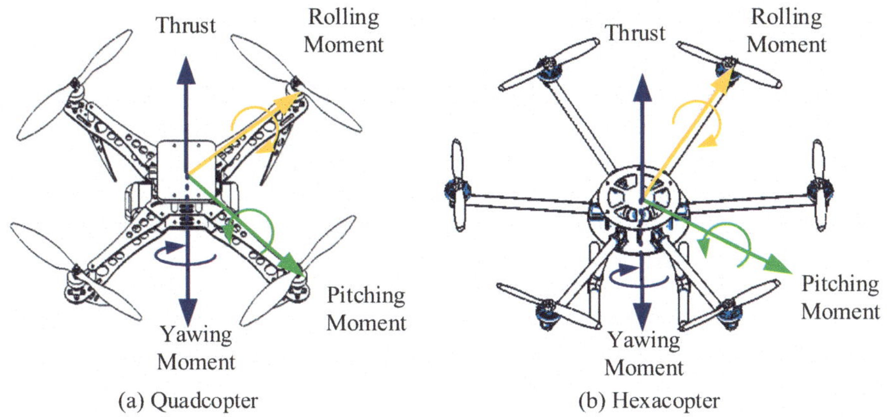

Fig. 1.2 Thrust and moments of quadcopter and hexacopter

(3) Multicopter

A multicopter is also called *multirotor*.[2] It can be considered as a type of helicopter that has three or more propellers (rotors). It also has the ability of VTOL. The most popular multicopter is the quadcopter whose typical configuration is presented in Fig. 1.1c. A quadcopter has four control inputs that include the four motor throttle signals. Unlike a single-rotor helicopter, rapid lift adjustment is realized by controlling propeller angular speeds. Because of the multicopter structure, the anti-torque moments can cancel each other out. Because of its simple structure, a multicopter is easy to use and offers advantages including high reliability and low maintenance cost. However, its payload capacity and the time of endurance are limited. The differences between a multicopter and a quadcopter are summarized as follows. As shown in Fig. 1.2, a multicopter, such as a hexacopter, has multiple propellers to generate the thrust, pitching moment, rolling moment, and yawing moment, whereas a quadcopter only has four propellers to generate the thrust and the three-axis moments. This means that there are hardly any fundamental differences except the allocation of thrust and moments to these propellers.

[2] In fact, a rotor is different from a propeller. Airplanes often use propellers to produce thrust, whereas helicopters use rotors to produce lift. Unlike airplane propeller (often with a fixed blade-pitch angle), the blade-pitch angle of a rotor of a helicopter is controlled by a swashplate. By definition, multicopters are often mounted with propellers rather than rotors [2, pp. 79–85], because the propellers of multicopters often have fixed blade-pitch angles. However, the terms "quadrotor" and "multirotor" are widely-used. Therefore, the two terms "rotor" and "propeller" are not distinguished in this book strictly. However, the term "propeller" is preferred.

1.1.2 Unmanned Aerial Vehicles and Model Aircraft

Unmanned Aerial Vehicle or Uninhabited Aerial Vehicle (UAV), i.e., an aircraft without pilots on board. The flight of UAVs may either be autonomously controlled by onboard computers or remotely controlled by a pilot on the ground. UAVs are also called *drones*. In this book, small UAVs, namely small drones, are mainly considered.

Model Aircraft. "An aircraft of limited dimensions, with or without a propulsion device, not able to carry a human being and to be used for aerial competition, sport or recreational purposes" is called a model aircraft [3]. It is also referred to as a Radio Control (RC) model aircraft or an RC aircraft. For the entire flight, an RC model aircraft must be within the Visual Line of Sight (VLoS) of the remote pilot. The statutory parameters of a model aircraft operation are outlined in [4].

As shown in Table 1.1, differences between drones and model aircraft are summarized below.

(1) *Composition*
A small drone is more complex than a model aircraft in terms of its composition. A drone consists of an airframe, a propulsion system, an autopilot, a task system, a communication link system, and a Ground Control Station (GCS), etc. A model aircraft generally consists of an airframe, a propulsion system, a simple stabilizing control system, an RC system (includes a pair of transmitter and receiver), etc.

(2) *Operation*
Drones are controlled either autonomously by onboard computers or by remote pilots on the ground or in another vehicle, whereas model aircraft are only controlled by remote pilots.

(3) *Function*
Drones are often used for military or special civil applications. They are expected to carry out particular missions. Model aircraft are more like toys.

Most multicopters have two high-level control modes: Semi-Autonomous Control (SAC) and Fully-Autonomous Control (FAC). Many open-source autopilots support both modes. The SAC mode implies that autopilots can be used to stabilize the attitude of multicopters, and they can help multicopters in holding their altitude and position. Considering the open-source autopilot ArduPilot Mega (APM)[3] as an

Table 1.1 Differences between drones and model aircraft

	Drones	Model aircraft
Composition	Complex	Simple
Operation	Autonomous control and remote control	Remote control
Function	Military or special civil applications	Entertainment

[3]Refer to http://ardupilot.org for details.

example, under the SAC mode, users are allowed to choose one of the following modes: "stabilize mode", "altitude hold mode", or "loiter mode". In such a high-level mode, a multicopter is still under the control of remote pilots. Therefore, it is more like a model aircraft. On the other hand, the FAC mode implies that the multicopter can follow a pre-programmed mission script stored in the autopilot that is made up of navigation commands and can take off and land automatically. In this mode, remote pilots on the ground only need to schedule tasks. The multicopter is then similar to a drone. Some multicopter autopilots support both modes, which can be switched by remote pilots, with each mode corresponding to some particular applications. In this book, multirotor drones and multirotor model aircraft are both called multicopters for simplicity.

1.2 Why Multicopters

The structures of multicopter systems are simple and yet complex. They are integrated systems that require interdisciplinary knowledge, which makes them a perfect research object and touchstone for the practice of control methods. The characteristics and future research requirements of multicopter systems are summarized below.

(1) Multicopters can be controlled autonomously by onboard computers or remotely through wireless communication (involving knowledge in *Information and Communication Engineering*) by the GCS or the RC system. Therefore, the reliability and security of the communication link are important concerns by users to avoid the threat of hackers. Researchers are also investigating methods to detect illegal flights by monitoring wireless communication links and tracking operators of remotely controlled multicopters.
(2) Multicopter systems are composed of many electronic components (involving knowledge in *Electronics Science and Technology*). The electronic circuit must be reliable in the presence of outside electromagnetic radiation. In addition, an onboard embedded processor is typically required to provide more computing resources with less power consumption and less weight.
(3) Multicopter systems require operating systems to run flight control algorithms (involving knowledge in *Computer Science*). Thus, a Real-Time Operating System (RTOS) plays an important role in providing communication interfaces with onboard devices. For example, the open-source autopilot software PX4[4] is based on a lightweight RTOS—Nuttx.[5]
(4) For multicopter designs, the selection of materials, configuration, and structure (involving knowledge in *Mechanics and Mechanical Engineering*) must be considered; additionally, the propulsion system (involving knowledge in *Mechanics and Electrical Engineering*) must be taken into account. For example, a lighter

[4]https://dev.px4.io/master/en/concept/architecture.html.
[5]https://nuttx.org/.

and more stable structure is always expected for multicopters; a streamline fuselage design is expected for high-speed flight; a propeller and a motor must be well-matched to achieve maximum efficiency.

(5) For the state estimation of multicopters, it is necessary to consider the problems of signal non-synchronization, sampling time difference, data delay, and sensor failures (e.g., GPS, gyroscope, accelerometer, magnetometer, barometer, ultrasonic rangefinder, and photoelectric sensor), which make the state estimation of the multicopter robust and high-performance (involving knowledge in *Instruments Science and Technology*).

(6) The multicopter system is a typical closed-loop control system (involving knowledge in *Control Science and Engineering*) with many interesting features that are unstable, nonlinear, underactuated and control-direction-limited (the propeller can only generate positive thrust perpendicular to the plane of the fuselage). In comparison with fixed-wing aircraft and helicopters, multicopters with less aerodynamic design requirements reduce the difficulty in modeling, analysis, and control. As a result, multicopter systems allow beginners or engineers from non-aeronautical fields to use multicopters to realize their ideas quickly and study the control methods of aerial vehicles.

(7) Multicopters are also cheap and easy to pilot, thereby making it easy to obtain flight data, which provides a basis for the health assessment of multicopters.

Based on the above characteristics, the multicopter is an excellent object for the education and research of many interdisciplinary subjects, especially *Control Science and Engineering* for universities or research institutes. Moreover, system engineering for the entire development process of a multicopter is a challenge for learners and engineers. Full-size aerial vehicles are often developed by traditional and large aerospace institutes or companies, where there is no lack of engineers, financial support, experience, resources, etc. However, in the expanding and increasingly fierce competitive market environment, labs in universities and start-up companies face problems such as limited personnel, lack of experience, and fewer resources for the development of micro-small aerial vehicles. Thus, a core team of engineers is required to have independent development capability in airframe configuration and structure design, propulsion system design, vehicle modeling and system identification, state estimation, control system design, path planning, decision-making logic, health evaluation, failsafe design, etc. Furthermore, the team must have practical experience in operating-system development, software debugging, and flight tests. Through studying and training based on the multicopter platform, it is possible to cultivate industrial demand interdisciplinary engineers to improve their capabilities related to software development, analysis, algorithm design, management, and presentation. To say the least, the multicopter itself is very practical in the industry. It can be a bridge to connect education and practical application, which is also very attractive to university students. These considerations motivated us to write this book.

1.3 What This Book Includes

This book involves knowledge offered in university professional courses such as *Model Control Engineering*, *Circuit*, *Computer Control System*, and *Optimal Estimation*. Readers can take the related professional courses in advance or acquire the necessary required knowledge during the experimental process. Because flight control related courses have strong requirements for engineering practice, it is difficult to master flight control technology only through theoretical learning. Motivated by this, we have developed a multicopter experimental platform for the rapid control algorithm development based on the Pixhawk[6] autopilot and MATLAB/Simulink,[7] with a series of experimental courses. These benefit readers to use comprehensive theoretical knowledge, thereby solving practical problems for multicopter systems.

1.3.1 Experimental Platform

The RflySim platform released with this book mainly consists of five parts: the Simulink-based controller design and simulation platform, HIL simulation platform, Pixhawk autopilot system, multicopter system, and instructional package. Each part is described in detail below.

(1) *Simulink-Based Controller Design and Simulation Platform*
It includes a high-fidelity multicopter Simulink simulation model that readers can use to simulate various dynamic characteristics of multicopters by changing the model parameters in Simulink. During simulations, the real-time attitude and trajectory of the multicopter can be observed through a three-dimensional (3D) visualization environment. Readers can also design multicopter control algorithms in Simulink and perform SIL simulation to verify the control performance. After the SIL simulation verification process, the control algorithms can be complied and uploaded to the Pixhawk autopilot through the code generation toolbox.

(2) *HIL Simulation Platform*
The core components of the HIL simulation platform include a Real-time Motion Simulation Software—*CopterSim* and a 3D Visual Display Software—*3DDisplay* which we developed. To ensure model consistency in different simulation phases, the simulation model of CopterSim is directly obtained from the previously mentioned Simulink multicopter model through the code generation technique. CopterSim and 3DDisplay run on a computer that connects with the Pixhawk autopilot via a USB cable to exchange sensor data and control signals, thereby constituting a closed-loop control system for the HIL simulation.

[6]http://pixhawk.org/.

[7]https://www.mathworks.com/products/simulink.html.

(3) *Pixhawk Autopilot System*
The Pixhawk autopilot system is a control system to sense the multicopter states and compute the desired control signals for motors, and it usually includes a Pixhawk autopilot, an RC system, a GCS, etc.
(4) *Multicopter Hardware System*
Multicopters, especially quadcopters, are currently the most widely-used aerial vehicles or aerial robot platforms. The multicopter hardware system (including the fuselage, propulsion system, and landing gear) with a Pixhawk autopilot comprises an integrated multicopter flight platform for autonomous or semi-autonomous flight tests.
(5) *Instructional Package*
It includes this book, online related courses, experimental instruction, and source codes.

1.3.2 Experimental Courses

This book provides eight specific experimental courses:

(1) Propulsion system design experiment
(2) Dynamic modeling experiment
(3) Sensor calibration experiment
(4) State estimation and filter design experiment
(5) Attitude controller design experiment
(6) Set-point controller design experiment
(7) Semi-autonomous control mode design experiment
(8) Failsafe logic design experiment.

The above eight experiments can be classified into four groups: design and modeling experiments, estimation experiments, control experiments, and decision-making experiments. Their relationships are shown in Fig. 1.3.

The code examples provided by this book ensure that each experiment or each part of an experiment can be finished independently. To make the task objectives different for different readers, our experiments can be completed by following different progressive routes. The progressive studying routes have:

(a) Design and modeling experiments → Control experiments
(b) Design and modeling experiments → Control experiments → Decision-making experiments
(c) Design and modeling experiments → Estimation experiments → Control experiments → Decision-making experiments

Route (a) is classic, and route (c) is comprehensive and challenging. In the design and modeling experiments, readers can design various aerial vehicles such as quadcopters or hexacopters. In addition, readers can use various methods such as the Euler angle method, the rotation matrix method, and the quaternion method [5] in the multicopter modeling. With these settings, various results can be obtained in the subsequent

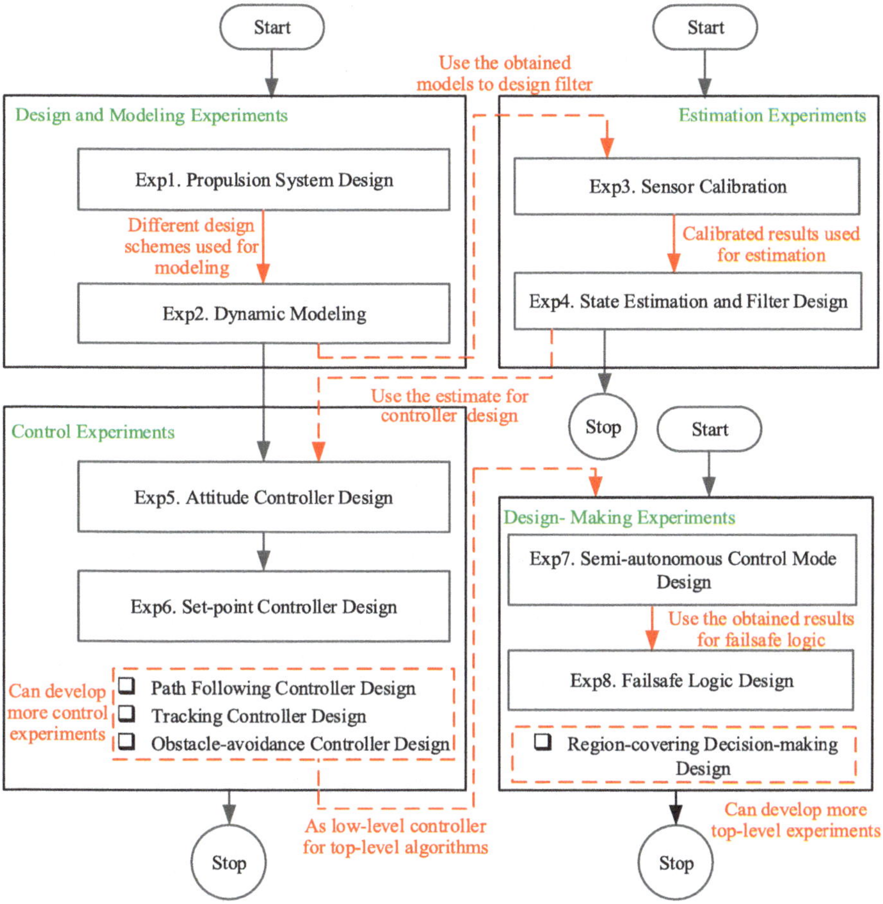

Fig. 1.3 Progressive routes for eight experiments

experiments that can help readers in increasing their independent learning ability, and they can share their design experiences with others. Readers can also design experiments by themselves (e.g., path-following controller design, tracking controller design, obstacle-avoidance controller design, and region-covering decision-making design).

Inspired by [6], each of the above eight experiments includes three step-by-step experiments in a progressive way: a basic experiment, an analysis experiment, and a design experiment.

(1) *Basic experiment.* Open the given code example. Then, read and run its source code directly to observe and record the results.
(2) *Analysis experiment.* Modify the given code example. Then, run the modified example program to collect and analyze the data.

Table 1.2 Experimental types, objectives and content

Objective	Basic experiment	Analysis experiment	Design experiment
Development platform	✓	✓	✓
Analysis process	×	✓	✓
Design method	×	×	✓
SIL simulation	✓	✓	✓
HIL simulation	✓	✓	✓
Flight Test	×	×	✓

(3) *Design experiment.* Based on the above two experiments, complete the given design task independently.

Through the basic experiment and the analysis experiment, readers can gain a deep understanding of the architecture, model, and algorithm performance of the whole system associated with the corresponding experiment. In the design experiment, the readers design and develop algorithms by referring to the code examples offered in the basic experiments and analysis experiments. The three step-by-step experiments constitute a learning ladder from shallow to deep, which makes it convenient for readers to reach the final experimental goal. As listed in Table 1.2, the contents and objectives of step-by-step experiments are different from each other.

After the algorithm design is completed in each experiment, they are realized in MATLAB/Simulink and then tested on the *Simulink-Based SIL Simulation Platform*. After the SIL simulation tests are successfully conducted, the obtained controller must be compiled through the automatic code generation technology and uploaded into the *Pixhawk Autopilot System*. Then, the *Pixhawk Autopilot System* can connect with the *HIL Simulation Platform* to constitute a closed-loop HIL simulation for further tests. Finally, the Pixhawk autopilot running the designed controller is installed on the provided *Multicopter Hardware System*, and the preliminary verification experiments are carried out on the indoor testing platform with safety protection, which is followed by outdoor flight tests. Considering the multicopter attitude controller design experiment as an example, the simulations and experiments shown in Fig. 1.4 are required in step-by-step experiments.

1.3.3 Features

(1) *Comprehensiveness*

The entire set of experiments include knowledge of aerodynamic, theoretical mechanics, sensor measurement, digital signal processing, and automation engineering. For example, the design of the multicopter propulsion system requires circuit knowledge for the calculations of the current and voltage, fluid knowledge for the computation and measurement of aerodynamic parameters. The modeling

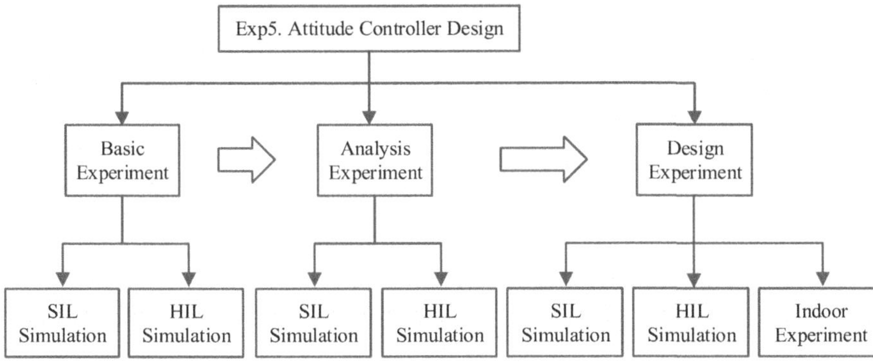

Fig. 1.4 Relationship among step-by-step experiments

of multicopter involves knowledge of theoretical mechanics; the calibration of multicopter sensors requires to collect sensor data from practical embedded control systems; the control experiments involve knowledge of the transfer function, Bode plot, and frequency domain compensation in control theory, whereas the decision-making experiments involve logic switch algorithms of state machines.

(2) *Adaptability*
Each experiment consists of three step-by-step experiments, namely a basic experiment, an analysis experiment, and a design experiment. The purpose is to consider the practical knowledge background of different readers from different fields. Readers with strong abilities can conduct all the experiments in a progressive route or try new experiments based on the offered platform.

(3) *Engineering*
The experimental courses require readers to flexibly use their mathematical, natural science, and engineering knowledge to realize modeling, filtering, control, and decision-making algorithm design independently. The ability to conduct algorithm programming, debugging, and overall system testing is required throughout the experiments, which exercise the ability of readers to solve practical engineering problems. Meanwhile, in the development process of a multicopter system, readers get the opportunity to gain engineering practice that includes physical parameter measurement, development, testing, and piloting of a real aerial vehicle. This practice can enable readers to considerably improve their practical skill and consolidate their theoretical knowledge.

(4) *Innovation*
Readers only need basic programming knowledge in MATLAB/Simulink to develop algorithms for multicopters and use the obtained algorithms in practical flight tests. In the experimental platform presented in this book, readers do not need to conduct low-level programming tasks such as C/C++ algorithm writing, sensor driver development, and communication protocols; instead, they can purely focus on the development of multicopter design and control algorithms. Most of the development and verification work presented in this book

is conducted indoor, which greatly accelerates the development process of the flight control algorithms. Furthermore, the entire design and development process is based on the latest MBD framework that can help readers in improving efficiency. In summary, the innovations of the platform designed for this book include lowering the programming requirements, shortening the development cycle, improving test efficiency, and having high extensibility and standard compatibility.

(5) *Interestingness*

1) Design and build a multicopter. In the propulsion system design phase, the readers can design their own multicopters and perform performance calculations to create a purchasing list of products for assembling a real multicopter.
2) Pilot a multicopter in the simulator. During the provided HIL simulation phase, readers can control their designed multicopter with an RC transmitter on our HIL simulation platform.
3) Make a story for autonomous flight tasks. Using the experimental platform, readers can design their tasks to create various scenarios (e.g., package delivering, plant protection, area monitoring).

(6) *Practicality*

1) The practicality of multicopters. The multicopter is very practical in the industry, and it can effectively connect education and practical application. At present, many universities have opened a new major in *Robotics Engineering* in China, and the multicopter, as a type of aerial robotics platform, is a perfect research object.
2) The practicability of the experimental platform. The hardware devices of the entire platform related to this book are relatively simple and inexpensive, which ensures that most readers can buy and build the hardware platform by themselves. The MATLAB/Simulink Personal Edition is economical and is also purchased by many universities, which makes it easy to open related experimental courses for students. This also enables readers to build their experimental platform for future design and verification of multicopter algorithms.
3) The practicality of the development process. The development process is based on the current MBD development framework in MATLAB/Simulink, wherein an automatic code generation technology is used to generate C/C++ codes for uploading to the target embedded control systems. The MBD framework is widely adopted by many famous companies to use MATLAB/Simulink as an efficient development tool for aerial vehicle designs, such as the Lockheed Martin F-35 fighter and NASA's Mars rovers. MATLAB/Simulink provides a unified software environment for system design in a variety of industrial fields and can automatically and efficiently generate embedded code in compliance with the DO-178B and MISRA-C standards. Therefore, the entire development process is simple, advanced, efficient, reliable, and practical.

4) The practicality of content. This book covers key techniques for multicopter system development that include propulsion system design, modeling, state estimation, control, and decision-making. These contents can increase the comprehensive abilities of readers and can also be applied to other embedded control systems such as driverless cars or robotics.

1.4 Engineering Education Certification Standards

Because the design and control tasks for multicopter systems are complex engineering issues, this book completely covers all aspects of *engineering education certification standards* [7]. The following are the main points to be explained.

(1) Engineering knowledge. Modeling, filter design, controller design, and decision-making logic design based on various multicopter design requirements can reflect the educational objective that trains the ability to "apply knowledge of mathematics, natural science, engineering fundamentals and an engineering specialization to the solution of complex engineering problems".
(2) Problem analysis. Multicopter modeling, algorithm design, and code debugging techniques can reflect the educational objective that trains the ability to "identify, formulate, research literature and analyze complex engineering problems reaching substantiated conclusions using first principles of mathematics, natural sciences, and engineering sciences".
(3) Design/development of solutions. The multicopter system design and flight test can reflect the educational objective that trains the ability to "design solutions for complex engineering problems and design systems, components or processes that meet specified needs with appropriate consideration for public health, and safety, cultural, societal and environmental considerations".
(4) Investigation. Multicopter modeling and algorithm design can reflect the educational objective that trains the ability to "conduct investigations of complex problems using research-based knowledge and research methods including design of experiments, analysis and interpretation of data, and synthesis of the information to provide valid conclusions".
(5) Modern tool usage. In the development process of multicopter systems, many tools are needed to improve the efficiency of the prototype development. The commonly-used tools include airframe design analysis tools, fluid analysis software, mechanical testing tools, operating systems, algorithm design tools, HIL simulation platforms, etc. Learning to use appropriate technologies, resources, modern engineering tools, and information technology tools for complex engineering problems can reflect the educational objective that trains the ability to "create, select and apply appropriate techniques, resources and modern engineering and IT tools, including prediction and modeling, to complex engineering problems, with an understanding of the limitations".

(6) The engineer and society. Designers need to have some innovative ability to explore the application scenarios for multicopter. This can reflect the educational objective that trains the ability to "apply reasoning informed by contextual knowledge to assess societal, health, safety, legal and cultural issues and the consequent responsibilities relevant to professional engineering practice and solutions to complex engineering problems".
(7) Individual and teamwork. Multicopter development generally requires a small team. This can reflect the educational objective that trains the ability to "function effectively as an individual, and as a member or leader in diverse teams and multidisciplinary settings".
(8) Communication. For the presentation of multicopter research, students must have the corresponding communication ability. This can reflect the educational objective that trains the ability to "communicate effectively on complex engineering activities with the engineering community and society at large, such as being able to comprehend and write effective reports and design documentation, make effective presentations and give and receive clear instructions".

If you have any question, please go to https://rflysim.com for your information.

Chapter 2
Experimental Platform Configuration

This chapter is mainly for readers or experimental course teachers who need to deploy a code generation environment and prepare a platform for practical flight experiments. If the experimental platform has already been configured, readers can skip this chapter and directly start the experimental courses. The experimental platform can be divided into two parts: the hardware platform and the software platform. This chapter shall introduce the basic components of each platform and present the deployment procedure in detail.

2.1 Overall Introduction

2.1.1 Hardware Platform

Because all control algorithms are eventually deployed in a real aerial vehicle to perform flight tests, a hardware platform must be prepared for the basic flight test requirements. As shown in Fig. 2.1, the experimental hardware platform recommended in this book is composed of five main parts.

(1) **Ground computer**: it is a high-performance Personal Computer (PC) with an operating system that performs two main tasks.

   1) Providing the software operating environment for the simulation software tools to perform functions such as control algorithm development, SIL simulation, automatic code generation, and HIL simulation;
   2) Working as a ground control station in outdoor flight tests to achieve functions such as sensor calibration, parameter tuning, real-time control, and communication.

© Publishing House of Electronics Industry 2020, corrected publication 2021

Q. Quan et al., *Multicopter Design and Control Practice*,
https://doi.org/10.1007/978-981-15-3138-5_2

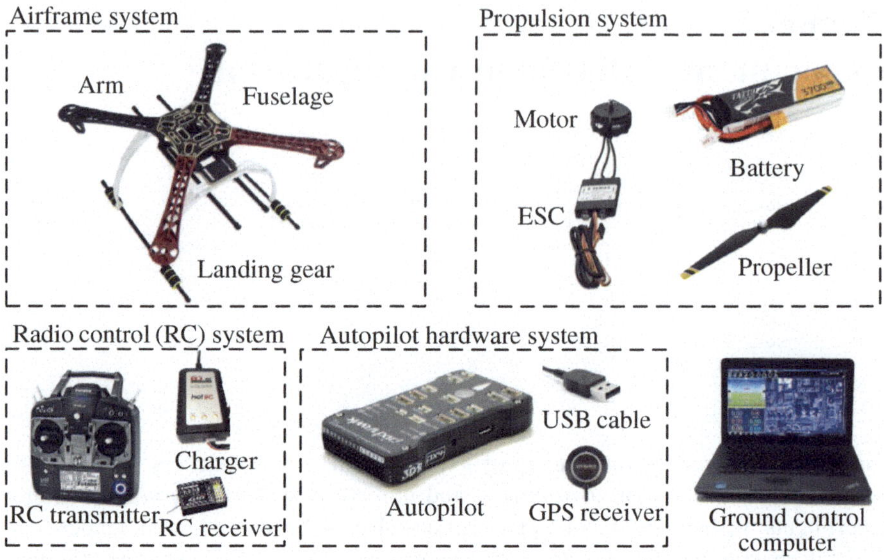

Fig. 2.1 Composition of experimental hardware platform

To ensure all the software tools run smoothly on the ground computer, the following basic configuration requirements must be satisfied.

- Operating System (OS): Windows 7–10, 64-bit system
- Processor: Intel I5 series or above
- Memory: 8G or above
- Graphics: discrete graphics, memory 2G or more
- Storage: 30G available space (solid-state drives are recommended)
- Interface: at least one USB Type-A
- Monitor: screen resolution 1080P or above

Noteworthy, for higher development efficiency, computer performance must be as high as possible.

(2) **Autopilot system (also called flight control system)**: as the operating platform of control algorithms, it has many sensors and powerful computing processor to estimate flight states and calculate the control signals for the propulsion system to realize the flight control of multicopters. For this book, we selected a widely-used open-source autopilot—Pixhawk[1] as the development and experimental autopilot system. Pixhawk is an independent open-hardware project that aims to provide standard readily-available, high-quality, and low-cost autopilot hardware for education, amateurs, and developers. For different flight mission, performance, and cost requirements, the Pixhawk provides a series of autopilot hardware products that highly promote the development of multicopters.

[1] http://pixhawk.org/.

(3) **Radio Control (RC) system**: it mainly includes an RC transmitter, an RC receiver, and a battery charger. The RC system is used to send control commands from the pilot on the ground to the autopilot system on the multicopter to realize remote flight control.
(4) **Propulsion system**: it mainly includes a battery pack and several propellers, Electronic Speed Controllers (ESCs), and motors. The propulsion system is used to receive the Pulse Width Modulation (PWM) control signals from the autopilot system, and control the movement of a multicopter with thrust and torque generated by the rotation of propellers and motors.
(5) **Airframe system**: it includes a fuselage, landing gear, and several arms. The airframe is used to support the propulsion system and the autopilot system and carry a payload; thus, it is required to have excellent aerodynamic performance and structural strength to ensure that flight missions are successfully and reliably accomplished.

2.1.2 Software Platform

This experimental platform relies on many software tools to realize controller design, code generation, autopilot code compilation, HIL simulation, and other functions. The *Simulation Software Package* published along with this book has a one-click installation script. Readers can click the script to finish all the installation and configuration process of the required software environment. The MATLAB/Simulink and the Simulation Software Package comprise the software platform, which contains the following.

(1) **MATLAB/Simulink**: it is a visual simulation tool developed by Mathworks,[2] which is widely-used in aerial vehicles, cars, and other applications. It can be easily applied to develop simulation systems for dynamic system modeling, controller design, hardware and software simulation, and performance analysis through a modular programming language. The simulation software package and source code published along with this book support MATLAB R2017b and above. The required MATLAB toolboxes include the following.

- MATLAB/Simulink
- Control System Toolbox
- Curve Fitting Toolbox
- Aerospace Blockset
- Aerospace Toolbox
- MATLAB Coder
- Simulink Coder
- Stateflow

[2]https://www.mathworks.com/.

This book does not provide the installation package or installation process for MATLAB, so please purchase and install the above-required MATLAB toolboxes by yourself. If conditions are permitted, it is recommended to install MATLAB R2017b version with all toolboxes.

(2) **Pixhawk Support Package (PSP) Toolbox**[3]: it is a Simulink toolbox officially released by Mathworks for controller design, code generation, and firmware upload of the Pixhawk autopilot. We have made some updates and optimizations based on the official PSP toolbox to ensure compatibility with the latest Pixhawk and MATLAB versions.

(3) **FlightGear—Flight Simulator**[4]: it is a popular open-source flight simulator that can be used to easily observe the flight states of a simulated aerial vehicle in Simulink by receiving flight data from Simulink via a User Datagram Protocol (UDP) interface.

(4) **PX4 Software—Source Code**[5]: PX4[6] is an open-source flight control software system that runs on the Pixhawk hardware platform. The Pixhawk hardware + PX4 software constitutes an integrated autopilot system, which is one of the most widely-used autopilot systems for aerial vehicles.

(5) **PX4 Toolchain—Compiling Environment**: it is used to compile the PX4 source code along with the controller algorithms generated by the PSP toolbox into a ".px4" format firmware file. Then, the firmware file is uploaded into the Pixhawk autopilot hardware (similar to the process of reinstalling an operating system on a PC). The control algorithm generated by the PSP toolbox will automatically run after Pixhawk restarts.

(6) **Eclipse C/C++—Integrated Development Environment (IDE)**[7]: it is used to read and modify the PX4 source code. Eclipse C/C++ is a compact C/C++ IDE with functions similar to those of Microsoft Visual Studio.

(7) **QGroundControl (QGC)—Ground Control Station**[8]: it is used to perform the pre-flight tasks (e.g., sensor calibration and parameter tuning) for the Pixhawk autopilot before the multicopter takes off. The QGC is also used to receive the flight states and send the control commands of the multicopter through wireless radio telemetry during flight tests.

(8) **CopterSim—Real-Time Motion Simulation Software**: it is a real-time motion simulation software developed for the Pixhawk/PX4 autopilot system. Readers can configure multicopter models in CopterSim, and connect it to the Pixhawk autopilot via the USB serial port to perform indoor HIL simulations.

---

[3]https://ww2.mathworks.cn/hardware-support/forms/pixhawk-downloads-conf.html.

[4]http://home.flightgear.org/.

[5]https://github.com/PX4/Firmware.

[6]https://px4.io/.

[7]https://www.eclipse.org/downloads/packages/.

[8]http://qgroundcontrol.com/.

(9) **3DDisplay—3D Visual Display Software**: it is a real-time 3D visual display software. It receives the flight data of CopterSim through UDP to display the attitude and position of a multicopter in real-time. CopterSim and 3DDisplay together constitute an integrated HIL simulation platform. The distributed independent operation mechanism of CopterSim and 3DDisplay provides future compatibility for swarm simulations.

2.1.3 Relationship Between Software and Hardware Platforms

The previous subsection introduced the hardware and software components of the required experimental platform. These components seem to be diverse and complex, but they are necessary for the development and practical flight experiments of multicopters. Familiarization with these tools can reduce the development difficulty and

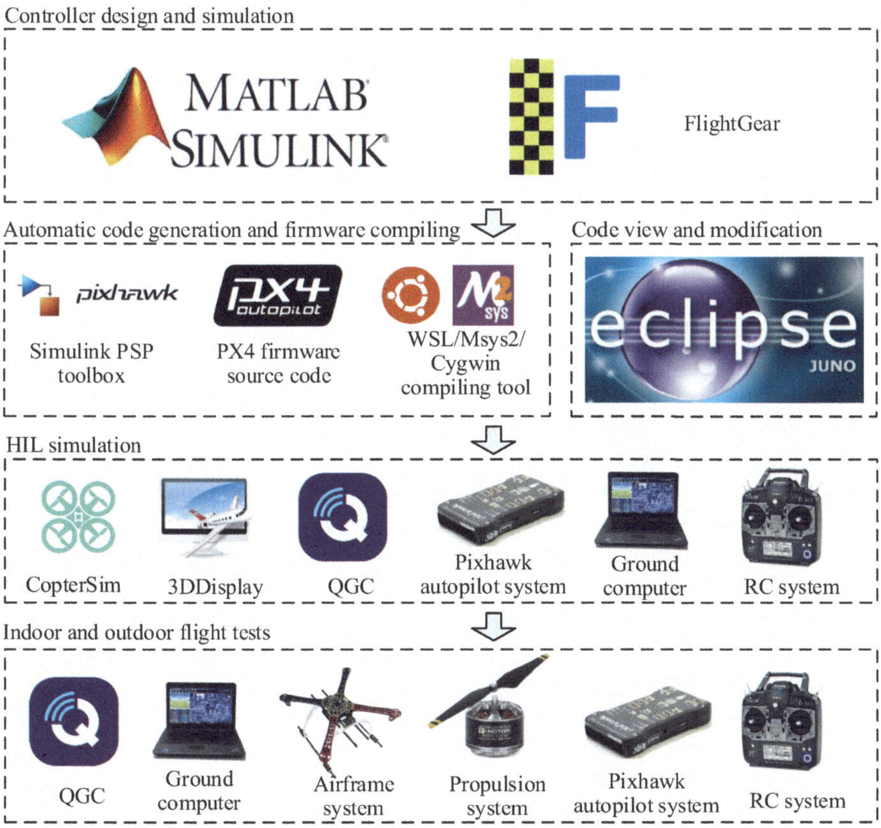

Fig. 2.2 Hardware and software components used in different phases

significantly improved efficiency, which can save a lot of time during the learning process. Figure 2.2 shows the relationships among the various hardware and software components and the overall process of the experimental platform. Most of the software tools play important roles in all the phases of multicopter development. The roles and application methods of the components presented in Fig. 2.2 are introduced in detail in the following sections.

2.2 Software Package Installation

The installation process of the Simulation Software Package is highly automated and readers can run a one-click installation script to achieve rapid and automatic deployment.

2.2.1 Installation Steps

The installation steps are summarized below.

(1) Download the installation image file "RFLYExpPspSoftwarePack.iso" of the Simulation Software Package from the official website: https://rflysim.com, and then unzip it or mount it to a virtual drive folder.
(2) Open MATLAB and click the "Browse for Folder" button (see Fig. 2.3). On the folder selection window, select the folder "RFLYExpPspSoftwarePack" obtained in the previous step.
(3) As shown in Fig. 2.3, tap the command "PX4InstallScript" in the "Command Window" of MATLAB and press the "Enter" key on the keyboard to run the one-key installation script. Note that another convenient way to run the one-key script is to select the "PX4InstallScript.p" file (see Fig. 2.3) with the mouse right key, and click the "Run" button on the pop-up menu.
(4) In the pop-up configuration window shown in Fig. 2.4, select the required configuration according to the actual hardware and software requirements (the default configuration is recommended for beginners, where the compiling command is px4fmu-v3_default, the PX4 firmware version is 1.7.3, and the installation directory is the C disk, which may occupy around 6G storage), and click the "OK" button in Fig. 2.4. Note, if you are using Pixhawk 4 (fmu-v5), please select firmware version 1.8.2 (or above) because the LED and RC will not work properly in old firmware versions.
(5) Wait patiently for the package to be successfully installed and deployed, which may take around 30 min.

Noteworthy:

(1) Antivirus software may prevent this script from generating desktop shortcuts. If the script prompts that the shortcut generation has failed, please close the

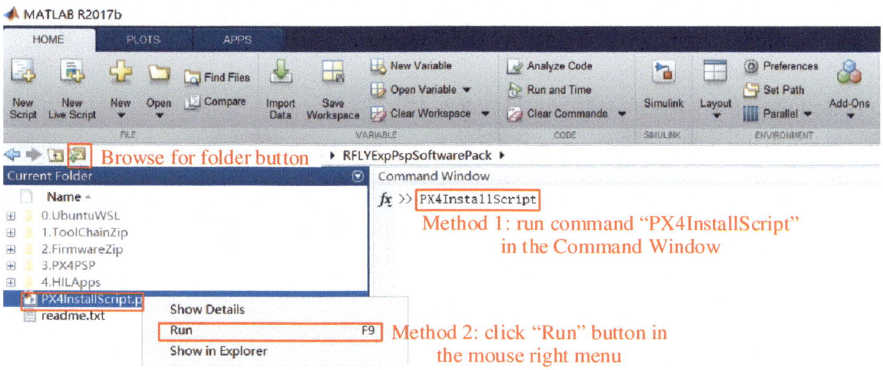

Fig. 2.3 Installing multicopter simulation software package with one-click installation script

antivirus software (Windows 10 should also turn off the "Real-time protection" option in the Settings page) and manually click the "GenerateShortcutCMD.bat" script in the installation directory (the default directory is C:\PX4PSP) to automatically generate all the software shortcuts.

(2) If readers want to change the firmware configurations or restore the compiling environment, just run the "PX4InstallScript" command again and select the required options.

(3) Readers can check the document "readme.txt" in the folder "RFLYExpPspSoftwarePack" for more detailed notes.

2.2.2 *Advanced Settings*

For advanced independent developers, Fig. 2.4 provides options to select the installation directory, Pixhawk hardware version, PX4 firmware version, compiling command, compiling environment, etc. The options in Fig. 2.4 are explained in detail below.

(1) **Software package installation directory**. All dependent files on the software package are installed in this directory, which requires around 6G storage. The default installation directory is "C: \PX4PSP". If the C disk space is not sufficient, readers should choose a directory in other disks; the directory name must be correct and only in English to prevent compilation failures.

(2) **PX4 firmware compiling command**. The default compiling command for PX4 is "px4fmu-v3_default". By selecting this compiling command, the compiling toolchain is automatically called to compile the PX4 source code to a firmware file "px4fmu-v3_default.px4" after the PSP generates the controller code. Then, the file ".px4" is uploaded to the supported hardware to realize the deployment of the control algorithms. Different Pixhawk hardware products must select

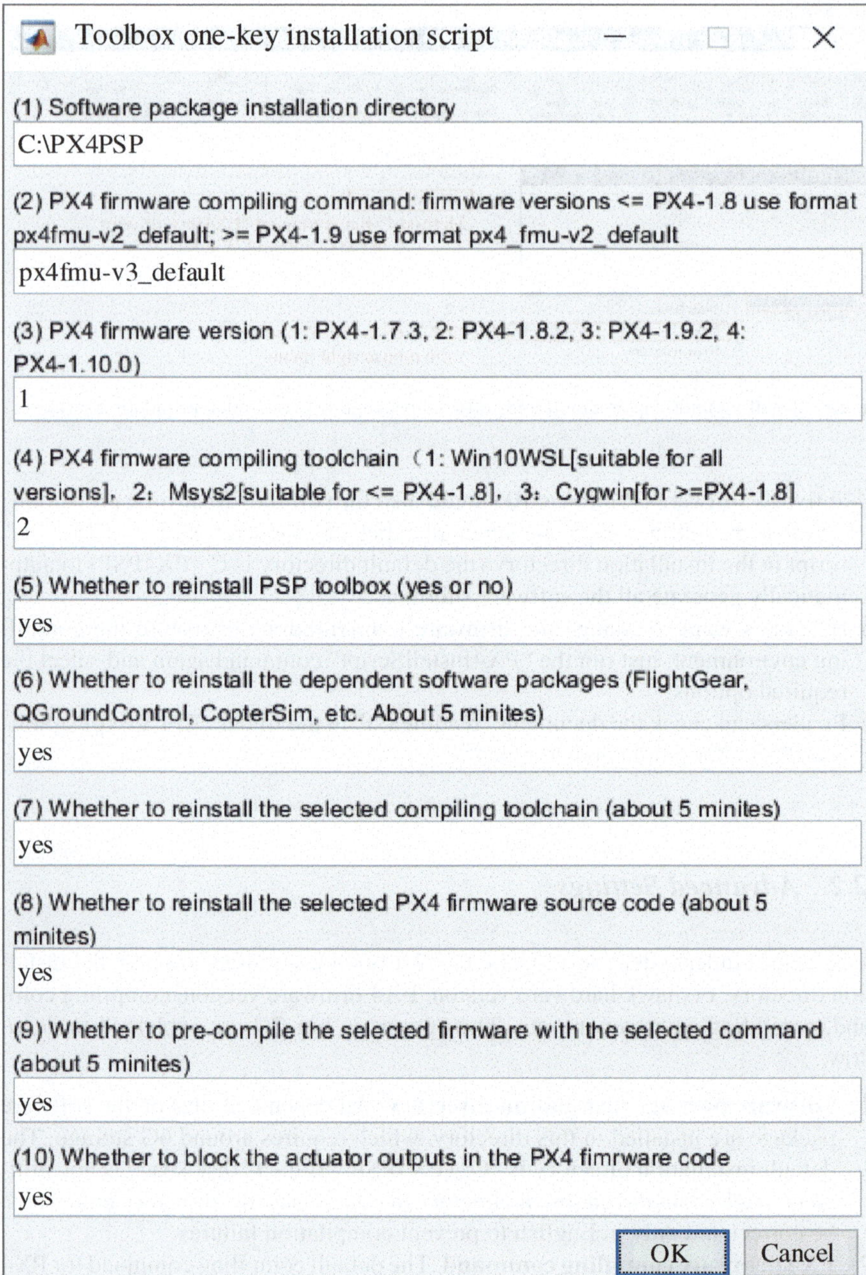

Fig. 2.4 Options of PX4InstallScript

Fig. 2.5 Pixhawk hardware series products with compiling commands

different PX4 firmware compiling commands. Figure 2.5 shows some Pixhawk hardware products, where "px4fmu-v3_default" can be used for three popular products: Pixhawk 1 (2MB flash version), mRo Pixhawk[9] and Cube (Pixhawk 2).[10] The command "px4fmu-v2_default" corresponds to the most famous Pixhawk 1.[11] PX4 also supports other hardware (for example, Intel Aero,[12] Crazyfile,[13] and so on). The corresponding compiling commands[14] are listed below.

- Pixhawk 1: px4fmu-v2_default.px4
- Pixhawk 1 (2MB flash version): px4fmu-v3_default.px4
- Pixhawk 4: px4fmu-v5_default
- Pixracer: px4fmu-v4_default
- Pixhawk 3 Pro: px4fmu-v4pro_default
- Pixhawk Mini: px4fmu-v3_default
- Pixhawk 2: px4fmu-v3_default
- mRo Pixhawk: px4fmu-v3_default
- HKPilot32: px4fmu-v2_default
- Pixfalcon: px4fmu-v2_default
- Dropix: px4fmu-v2_default
- MindPX/MindRacer: mindpx-v2_default
- mRo X-2.1: auav-x21_default
- Crazyflie 2.0: crazyflie_default
- Intel Aero Ready to Fly Drone: aerofc-v1_default.

[9]https://docs.px4.io/master/en/flight_controller/mro_pixhawk.html.
[10]https://docs.px4.io/master/en/flight_controller/pixhawk-2.html.
[11]https://docs.px4.io/master/en/flight_controller/pixhawk.html.
[12]https://software.intel.com/en-us/aero/drone-kit.
[13]https://www.bitcraze.io/crazyflie-2/.
[14]http://dev.px4.io/master/en/setup/building_px4.html.

(3) **PX4 firmware version**. The version of the PX4 source code is updated constantly, and the latest firmware version was 1.10 when this book was written. As the firmware version is upgraded, new features may be introduced, and more new products will be supported, but the compatibility with some old autopilot hardware will be affected. Because the Pixhawk 1 (2MB flash version, or mRo Pixhawk) hardware selected in this book is a classic Pixhawk product with LED for better experimental observation effect, the older PX4 firmware version 1.7.3 was selected with the compiling command "px4fmu-v3_default" to achieve better-using effect.

(4) **PX4 firmware compiling toolchain**. Because the compilation of PX4 source code depends on the Linux compiling environment, the software package provides three sets of compiling toolchains to realize the simulation of the Linux compiling environment under the Windows environment.

1) *Win10WSL* based on the Windows Subsystem compiler environment for Linux (WSL)[15];
2) the *Msys2Toolchain* based on Msys2[16] toolchain;
3) the *CygwinToolchain* based on the Cygwin[17] toolchain.

Note that: the CygwinToolchain only supports PX4 firmware with version 1.8 or above; the Msys2Toolchain only supports PX4 firmware with version 1.8 or below. Both the CygwinToolchain and the Msys2Toolchain support Windows 7 and above, which are easy to deploy, but the compiling speed is slow. In RflySim advanced platform, for Windows10 1809 and above, readers can follow the tutorial in "0.UbuntuWSL\Win10UbuntuInstallationStep.docx" to install an Ubuntu subsystem in Windows and then choose the Win10WSL toolchain shown in Fig. 2.4. The Win10WSL toolchain can greatly accelerate the compiling speed and support all versions of PX4 firmware.

(5) **Whether to reinstall the PSP Toolbox (yes or no)**. If this option is set to "yes", the PSP Toolbox is installed on MATLAB/Simulink. If the PSP toolbox has already been installed, a new installation of the PSP Toolbox is performed. If this option is set to "no", the script does not do anything on the existing PSP toolbox (it will not uninstall the PSP toolbox or carry out other actions).

(6) **Whether to reinstall the dependent software packages**. If this option is set to "yes", software tools (such as FlightGear, QGC, CopterSim, and 3DDisplay) are deployed to the selected installation directory and shortcuts for them are generated on the desktop. The related drivers for Pixhawk hardware are also installed. If the software tools have already been installed, selecting "yes" will remove the old installation files and reinstall them. If this option is set to "no", then no change will be made.

(7) **Whether to reinstall the selected compiling toolchain**. If this option is set to "yes", the selected compiling toolchain (Win10WSL, CygwinToolchain, or

[15]https://en.wikipedia.org/wiki/Windows_Subsystem_for_Linux.

[16]https://baike.baidu.com/item/MSYS2.

[17]https://www.cygwin.com/.

Msys2Toolchain) will be deployed to the selected installation directory. If the toolchain already installed, the script will remove the old toolchain files and reinstall it. In contrast, if this option is set to "no", then no change will be made.

(8) **Whether to reinstall the selected PX4 firmware source code**. If this option is set to "yes", the selected PX4 firmware source code will be deployed to the selected installation directory. If the firmware files already exist, the old firmware folder will be deleted, and a new copy of the source code will be deployed. If this option is set to "no", then no change will be made.

(9) **Whether to pre-compile the selected firmware**. If this option is set to "yes", the PX4 source code will be pre-compiled. This can greatly save the compiling time of the subsequent code generation process; whether the compiling environment is installed properly can also be checked. If this option is set to "no", then no change will be made.

(10) **Whether to block the actuator outputs of the PX4 original controller**. If this option is set to "yes", the control signals of the PX4 original controller will be blocked to prevent them from conflicting with the generated controller in Simulink. This option must be set to "yes" for the simulations and experiments in this book. If this option is set to "no", the PX4 outputs will not be blocked, and this mode can be used to test the PX4 original controller.

For the firmware versions 1.9 and above, the PX4 starts to use new compiling commands with the form "px4_fmu-v3_default.px4" instead of "px4fmu-v3_default.px4". Because the simulation software package of this book will be continuously updated for the latest PX4 firmware when using the latest version of the firmware (1.9 and above), readers need to modify the compiling command in Fig. 2.4 to the correct format.

2.2.3 Installation Completion

When the above one-key installation scripted is successfully executed, readers can check the installed content with the following procedure.

(1) As shown in Fig. 2.6, the shortcuts for the core software tools will be generated on the desktop.

(2) As shown in Fig. 2.7, the folders of all software tools are stored in the selected installation directory (the default is "C:\PX4PSP"). Note that all the software tools are completely portable and independent of the original software (e.g., official versions of QGC and FlightGear) on Windows. In Fig. 2.7, the folder "Firmware" stores the PX4 source code. The folder "examples" stores Simulink source code examples of the PSP toolbox; the folder "drivers" stores Pixhawk drivers. The folder "Python27" stores a Python environment for the automatic firmware uploading of the PSP toolbox. The names of other folders are the same as the software names, whose detailed introduction can be found in Sect. 2.1.2.

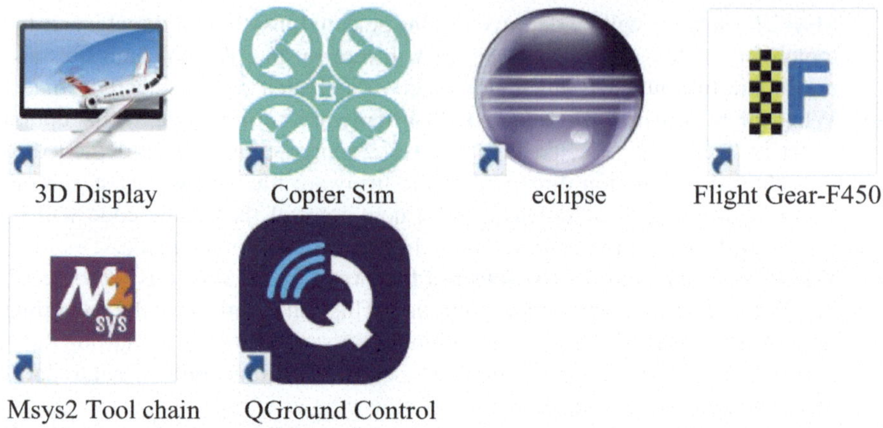

Fig. 2.6 Desktop shortcuts of simulation software package

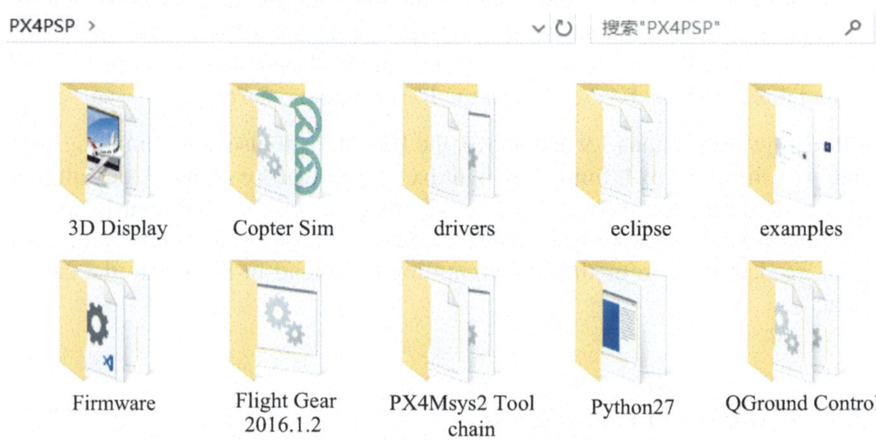

Fig. 2.7 All files in installation directory of simulation software package

(3) As shown in Fig. 2.8, the installed PSP toolbox can be found on the "Add-Ons" - "Manage Add-Ons" page of MATLAB. On this page, some management operations can be performed for the PSP toolbox that includes disabling, uninstalling, and viewing the installation directory. Note that the PSP toolbox can be installed once for all the MATLAB applications on a computer whose versions are higher than or equal to MATLAB R2017b.
(4) As shown in Fig. 2.9, readers can open any Simulink file and click the "Simulink Library Brower" button to open the Simulink library browser, and then find the "Pixhawk Target Blocks" library generated by the PSP toolbox.

If readers want to uninstall the simulation software package, they can carry out the following steps.

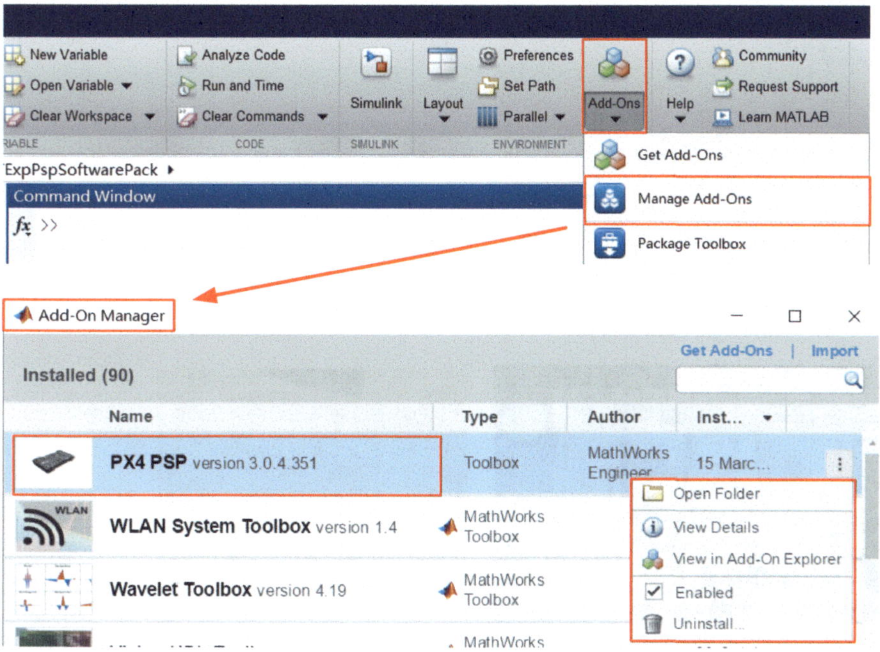

Fig. 2.8 PSP Toolbox management page in MATLAB

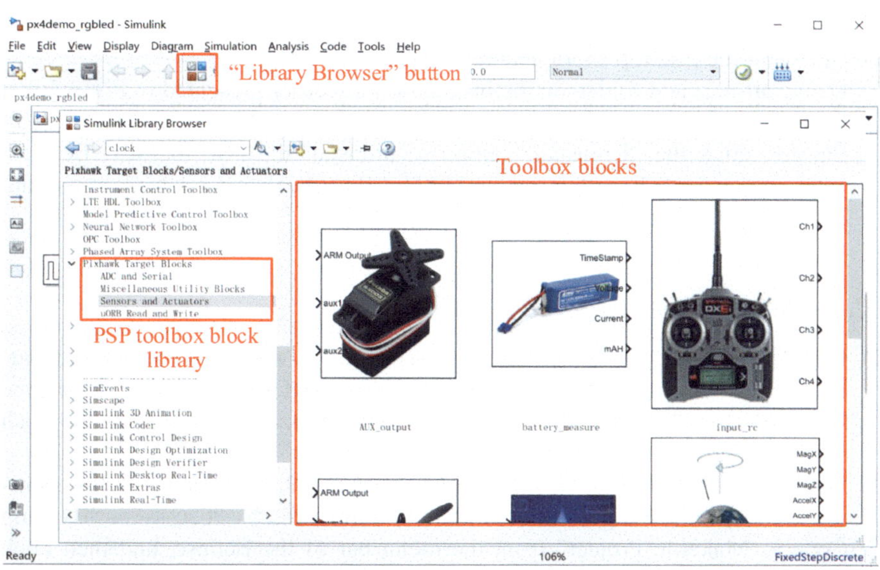

Fig. 2.9 PSP toolbox in Simulink library browser

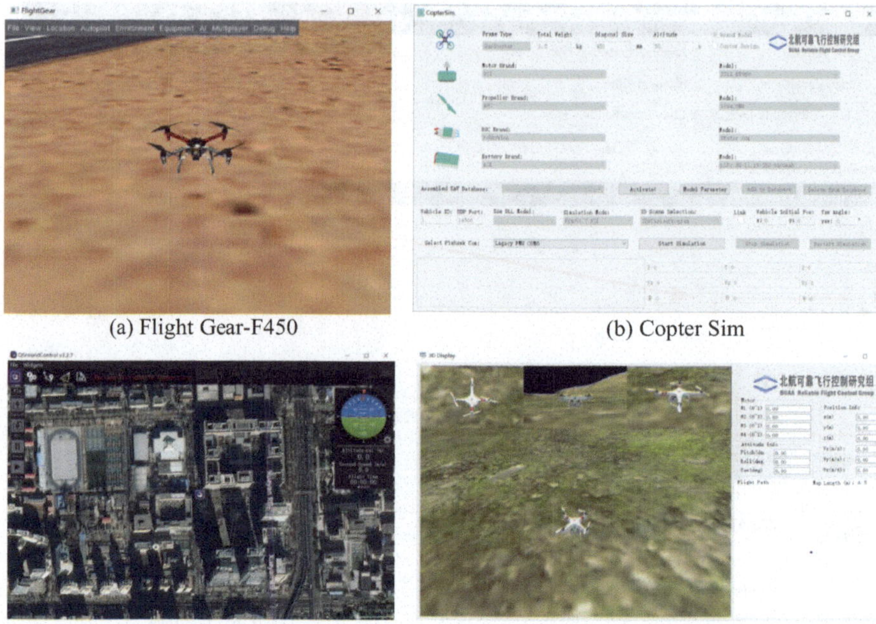

(a) Flight Gear-F450 (b) Copter Sim

(c) QGround Control (d) 3D Display

Fig. 2.10 Basic software UIs for simulation tools

(1) Delete all the desktop shortcuts presented in Fig. 2.6;
(2) Delete all files and folders in the installation directory presented in Fig. 2.7;
(3) In the "Management Additional Functions" page of MATLAB presented in Fig. 2.8, click the "Uninstall" button to uninstall the PSP toolbox. Besides, enter command "edit pathdef.m" in MATLAB to check and delete all remaining PX4PSP path items.

2.2.4 Brief Introduction to Software

(1) Double-click the desktop shortcuts shown in Fig. 2.6, which include "FlightGear-F450", "CopterSim", "QGroundControl" and "3DDisplay". Then, check the software User Interface (UI) one by one with Fig. 2.10 to confirm that each software can operate correctly.
(2) Double-click the desktop shortcut "Eclipse" in Fig. 2.6 to open the Eclipse software. As shown in Fig. 2.11, click "File" – "Import..." – "C/C++"—"Existing Code as Makefile Project" from the menu bar of the Eclipse, and then click "Next". In the "Existing Code Location" section of the pop-up window, click the "Browse" button to locate the "Firmware" folder in the installation directory (default is "C:\PX4PSP"), then choose "Cross GCC" and click the "Finish" button.

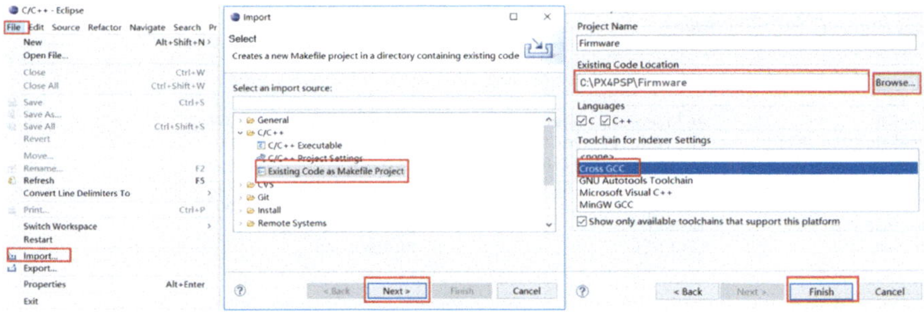

Fig. 2.11 Importing PX4 Firmware source code into Eclipse

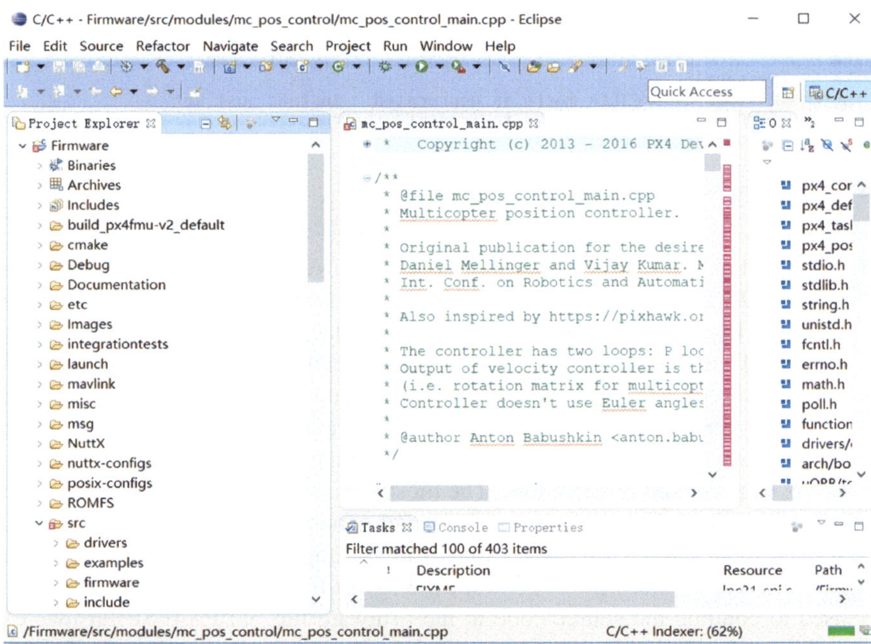

Fig. 2.12 Eclipse code reading interface

After completing the above steps, as shown in Fig. 2.12, the source code and directory structure of the "Firmware" can be found in the "Project Explorer" window, where readers can read the PX4 source code and try to modify it. Readers can also view the PX4 developer documentation website[18] to clearly understand the architecture and implementation principles of the PX4 algorithms and deepen the understanding of an actual flight control system. Note that: a "Welcome" tab will cover the content shown in Fig. 2.12 when readers first open Eclipse, it must be closed manually.

[18]http://dev.px4.io/master/en/index.html.

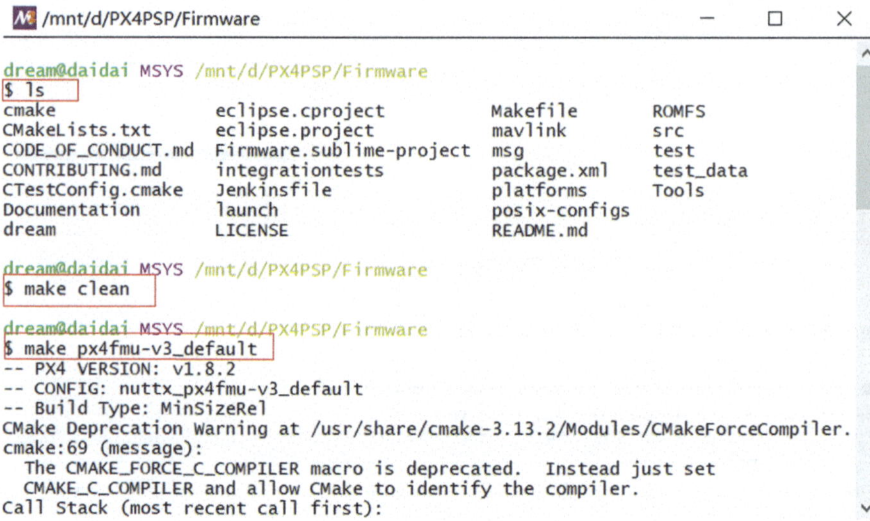

Fig. 2.13 Commands tapped in compiling toolchains

(3) Double-click one of the three shortcuts "Win10WSL", "Msys2Toolchain" or "CygwinToolchain" on the desktop to pop up the command window interface shown in Fig. 2.13 (the original UI has a pure black background, and the image color has been reversed for reading). Because the compiling toolchains are essentially Linux emulation software, the basic Linux commands (such as "ls", "pwd", and "gcc–version") can be tapped on the command line. For readers who are unfamiliar with Linux operations, this compiling toolchain can also be used as a Linux learning and practice tool. The most important function of this toolchain is to compile the source code of PX4 and generate the ".px4" firmware file. As shown in Fig. 2.13, "make clean" can be tapped on the command line to clear the old compiling files, and the "make px4fmu-v3_default" command is used to compile the source code to the firmware file "C:\PX4PSP\Firmware\build\px4fmu-v3_default\px4fmu-v3_default.px4" for Pixhawk 1 (2MB flash version). Because the PSP toolbox will automatically call this compiling command after the code is generated, readers do not need to know how to use it.

2.3 Hardware Platform Configuration

This chapter will introduce the RC system configuration, Pixhawk autopilot system configuration, airframe, and propulsion system configuration.

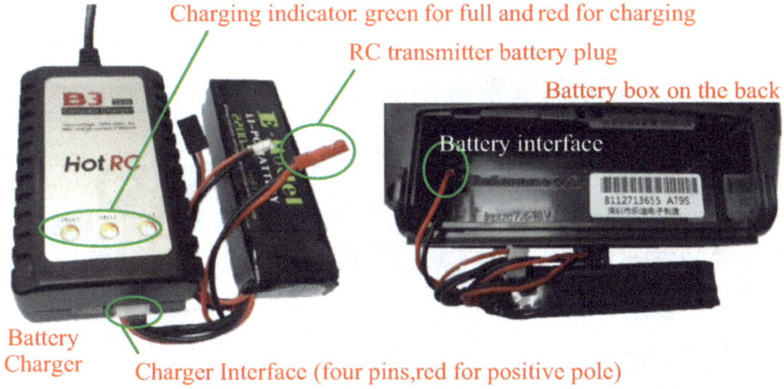

Fig. 2.14 RC transmitter, battery, and charger

2.3.1 RC System Configuration

There are two RC system products presented in this book, which are RadioLink AT9S and Futaba T14SG. The receivers of these RC systems have the S.BUS output function that can transmit the PWM signals of all channels to the flight control through one data line. Radio Link AT9S is relatively inexpensive, and it is suitable for indoor experiments; Futaba T14SG is relatively expensive, but it offers better performance and reliability, which makes it more suitable for actual outdoor flight tests. RC transmitters with "Left-hand throttle (Mode 2)" configuration are selected in this book, whose left stick is the throttle lever without the auto-return function. RC transmitters with "Right-hand throttle (Mode 1)" and "Left-hand throttle" configurations have different hardware structures that cannot be modified through the software setting page; thus, readers need to pay attention to this. The following subsections detail the configuration steps of the two RC systems. Other RC systems can be configured in a similar way.

2.3.1.1 RadioLink AT9S Configuration Method

RadioLink AT9S includes an RC transmitter and an R9DS RC receiver. Other necessary accessories include a battery (LiPo lithium polymer battery, 3S, 11.1V), a battery charger, a JR line (or DuPont line) for connecting the RC receiver to the Pixhawk autopilot, and a MicroUSB cable for connecting the Pixhawk autopilot with the computer.

(1) Battery and charger instructions

The left side of Fig. 2.14 shows the battery and the charger. The battery begins to charge when the four-port charging head of the battery is inserted into the socket on the charger. Red and green indicator colors represent the "charging" and "fully charged" status, respectively. The RC transmitter battery is installed

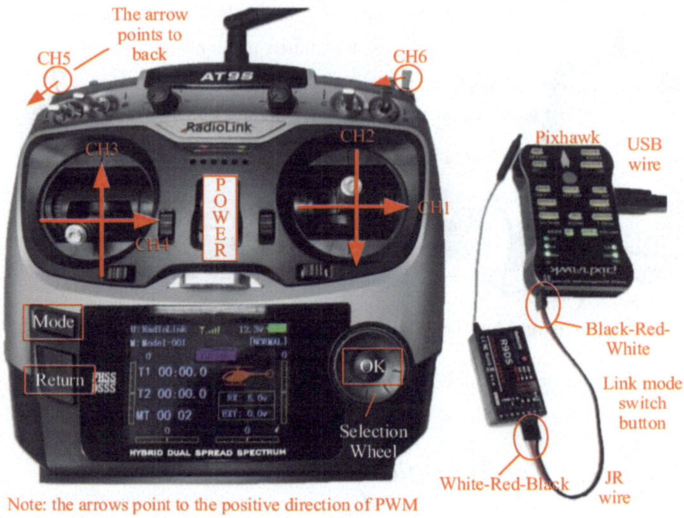

Fig. 2.15 RC transmitter and receiver configuration

as follows. Open the battery slot on the back of the RC transmitter and insert the two-line battery power supply interface (red line for the battery positive pole) into the battery power slot.

(2) RC receiver initial setting

1) Connect the RC receiver and the Pixhawk according to Fig. 2.15. The horizontal pin on the downside of the tail face of the receiver must be connected to the left-most RC pin on the tail face of the Pixhawk with a three-line JR line, and the Pixhawk MicroUSB port must be connected to the computer USB Type-A interface to supply power to the receiver and the Pixhawk.
2) How to rematch the RC transmitter with an RC receiver (the connection has been completed by default, and this step needs to be performed only when problems occur in the connection of the receiver and the transmitter). Turn on the power of the RC transmitter (all other RC transmitters should be turned off), and correctly connect the receiver with the Pixhawk and the computer. Then, press the matching switch on the right side of the receiver with a pen tip or needle (see Fig. 2.15) for more than one second. At this time, the LED of the receiver starts to flash, which indicates that it is searching for the nearest RC transmitter. When the receiver LED flashes seven or eight times and then remains constant, it means that the matching process is finished and a connection has been successfully established between the RC transmitter and receiver.
3) S.BUS signal mode selection (the receiver is in this mode by default; thus, this step is typically not performed). The S.BUS mode allows the Pixhawk to transmit all channel PWM signals through one JR line. If the Pixhawk

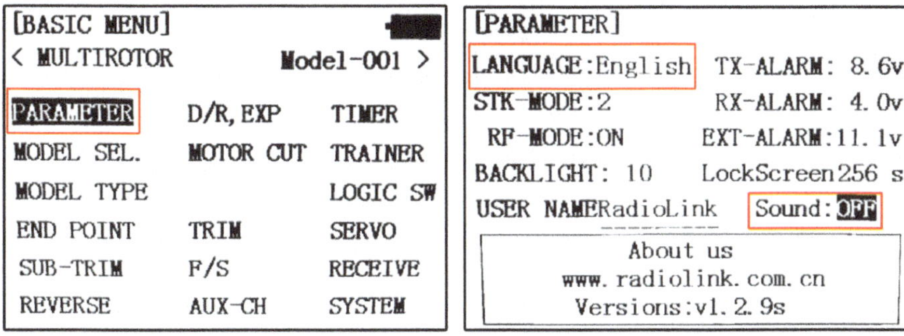

Fig. 2.16 RC transmitter parameter setting page

is powered up and connected to the receiver, the LED on the receiver is blue-white, which indicates that it is already in S.BUS mode and no setup is required. If the receiver LED is red, readers need to double-press (press twice within one second) the matching switch on the right side of the receiver. If the receiver LED turns blue-white, then the S.BUS mode has been successfully set.

(3) RC transmitter setting

1) Pull up the "POWER" switch shown in Fig. 2.15 to open the RC transmitter.
2) Setting the Language and Turning off the Sound
   - Press the "Mode" button on the RC transmitter shown in Fig. 2.15 for several seconds to enter the model setting page shown in Fig. 2.16. Roll the "Selection Wheel" on the RC transmitter shown in Fig. 2.15, move the cursor to "PARAMETER", and press the "OK" button on the RC transmitter shown in Fig. 2.15 to enter the RC transmitter parameter setting page.

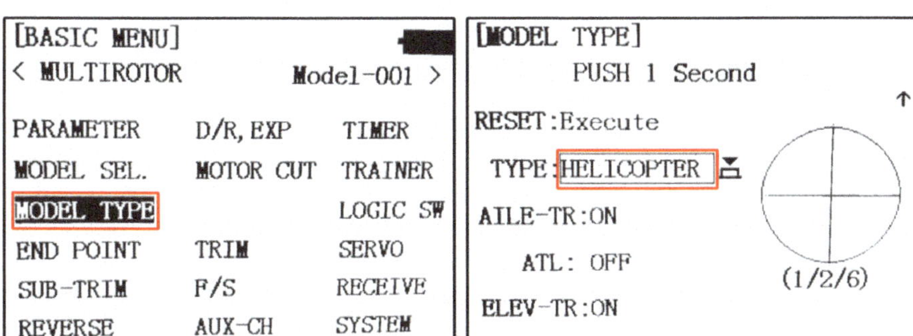

Fig. 2.17 Multicopter control mode switching of RC transmitter

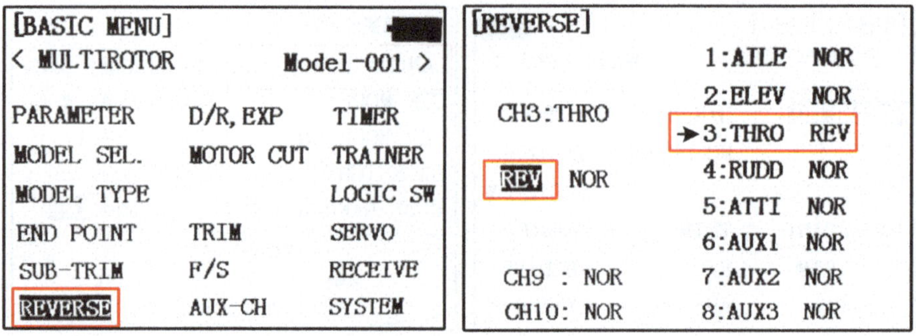

Fig. 2.18 Throttle channel reverse setting

- Scroll the "Selection Wheel" to select the "English" item, click the "OK" button, and then scroll the "Direction Wheel" again to select the desired display language. Then click the "OK" button to confirm the selection.
- Because this experiment is mainly performed indoor, it is recommended to turn off the speakers of the RC transmitter to prevent disturbing people nearby. As shown in Fig. 2.16, modify the "Sound" option from "ON" to "OFF".

3) Multicopter mode setting
   - Press the "Mode" button for several seconds to enter the "BASIC MENU" page, and click the "MODEL TYPE" item to enter the model type selection page shown in Fig. 2.17.
   - Change the "TYPE" item from "HELICOPTER" to "MULTICOPTER", and then press the "OK" button for several seconds to set the control mode to "Multicopter".
4) Throttle channel reverse setting
   - The throttle channel of the RadioLink transmitter is opposite to normal RC transmitters, and the throttle channel reverse needs to be set. Press the "Mode" button for several seconds to enter the "REVERSE" setting page shown in Fig. 2.18, and change the throttle channel from "NOR" to "REV".
5) CH5-CH6 mode switching channel setting
   - Because of experimental requirements, CH5 shown in Fig. 2.15 of the RC transmitter needs to be mapped to a three-position switch for mode switching of the Pixhawk. Press the "Mode" button for several seconds, and click the "AUX-CH" item next to the "REVERSE" item shown in Fig. 2.18.
   - As shown in Fig. 2.19, on the "AUX-CH" setting page, click the "CH5" item to enter the channel setting page, and map CH5 to a three-position switch "SwE" on the RC transmitter (switch "E" is located in the top-left corner of the RC transmitter in Fig. 2.15).
   - Similarly, the "CH6" item in Fig. 2.19 must be modified to a three-position switch "SwG" of the RC transmitter (switch "G" is located in the upper-right corner of the RC transmitter in Fig. 2.15).

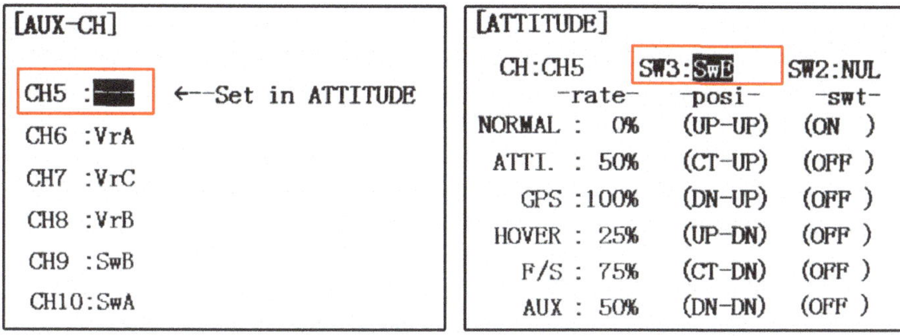

Fig. 2.19 CH5-CH6 mode switching channel setting

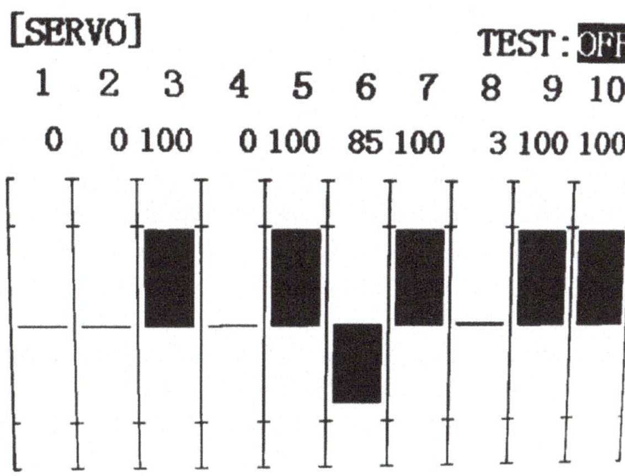

Fig. 2.20 RC transmitter stick position and direction

6) Channel confirmation

- Restart the RC transmitter, and press the "Return" button (see Fig. 2.15) on the RC transmitter to enter the "SERVO" page (see Fig. 2.20). In this page, the PWM value of each channel can be verified by moving sticks and switches on the RC transmitter. Note that, as shown in Fig. 2.20, the channel value reaches an upper limit of 100 corresponding to the desired PWM value of 1100 μs; the channel value reaches the lower limit 100, corresponding to the desired output PWM value of 1900 μs. Note that the actual PWM value range received by the RC receiver may not equal to 1100–1900 due to various errors. So RC calibration is important for autopilots to correctly recognize the control commands from the ground pilot. For example, in Fig. 2.20, the third channel is located at an upper limit of 100, which indicates that the PWM value is about 1100 μs; the other three channels are located at 0 positions, which indicates that the corresponding PWM value is about 1500 μs.

- It is important to understand the correct relationship between the stick position and the PWM value of each channel. Move each channel stick in Fig. 2.15 to confirm that each channel corresponds correctly to the following rules.
  - CH1: this corresponds to the horizontal movement of the right-hand stick of the RC transmitter. The right-hand stick moves from left to right, corresponding to a PWM value that changes from 1100 to 1900.
  - CH2: this corresponds to the vertical movement of the right-hand stick of the RC transmitter. The right-hand stick moves from top to bottom, corresponding to a PWM value that changes from 1100 to 1900.
  - CH3: this corresponds to the vertical movement of the left-hand stick of the RC transmitter. The left-hand stick moves from top to bottom, corresponding to a PWM value that changes from 1900 to 1100 (opposite to CH2).
  - CH4: this corresponds to the horizontal movement of the left-hand stick of the RC transmitter. The left-hand stick moves from left to right, corresponding to a PWM value that changes from 1100 to 1900.
  - CH5: this corresponds to the three-position switch on the upper-left side of the RC transmitter. The switch moves from the top position (the farthest position from the user), middle position, and bottom position (the closest position from the user), corresponding to PWM values of 1100, 1500, and 1900, respectively.
  - CH6: this corresponds to the three-position switch on the upper-right side of the RC transmitter. The switch moves to the top position (the farthest position from the user), middle position, and bottom position (the closest position from the user), corresponding PWM values of 1100, 1500, and 1900, respectively.

2.3.1.2 Configuration for Futaba T14SG

The connection between the Futaba receiver and the Pixhawk autopilot is slightly different from that of the RadioLink receiver. The specific connection is shown in Fig. 2.21. In the following paragraphs, the setup process for the Futaba RC transmitter is introduced.

As shown in Fig. 2.22, the Futaba T14SG RC transmitter needs to use six channels: the J1 stick (CH1, roll channel), J2 stick (CH2, pitch channel), J3 stick (CH3, throttle channel), J4 stick (CH4, yaw channel), SE three-position switch (upper-left switch, CH5 mode channel), and SG three-position switch (upper-right switch, CH6 mode channel). The basic process of setting the Futaba T14SG RC transmitter is summarized as follows.

(1) Double-click the "LINK" button on the RC transmitter in Fig. 2.22 to enter the "LINKAGE MENU" link setting page. As shown in Fig. 2.23a, enter the "MODEL TYPE" page and change the "TYPE" to "MULTICOPTER";

(2) Go back to the "LINKAGE MENU" page, and enter the "FUNCTION" page to confirm the channel mapping is the same as that shown in Fig. 2.23b, wherein the first to fourth channels of the RC transmitter correspond to the J1–J4 sticks;
(3) Go back to the "LINKAGE MENU" page, and enter the "REVERSE" page to confirm that the reverse direction of the channel is as shown in Fig. 2.23c, i.e., only the third channel (throttle) is reversed;
(4) Go back to the "LINKAGE MENU" page, and enter the "FUNCTION", and scroll to the second page for the setting of CH5 to CH8. As shown in Fig. 2.23d, set the "CTRL" option of the "5 MODE" channel to "SE" stick (the upper-left stick of the RC transmitter).
(5) As in the previous step, set the "6 AUX1" channel shown in Fig. 2.23d to the "SG" stick (the upper-right stick of the RC transmitter).

After the above settings, similar to the RadioLink AT9S in Fig. 2.20, it is also necessary to verify that the PWM output of each stick follows the correct definition required by this book.

2.3.2 *Pixhawk Autopilot System Configuration*

Several basic firmware uploading and configuration operations are required for the brand-new Pixhawk to ensure that the Pixhawk autopilot meets the experimental requirements and ensure that the operation and configuration of Pixhawk are correct. The configuration method is summarized below.

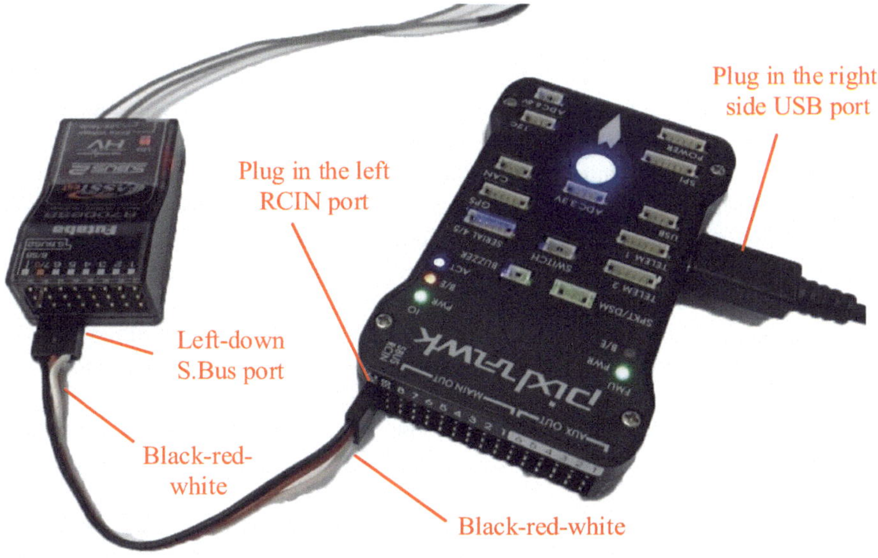

Fig. 2.21 Pixhawk and Futaba receiver connection diagram

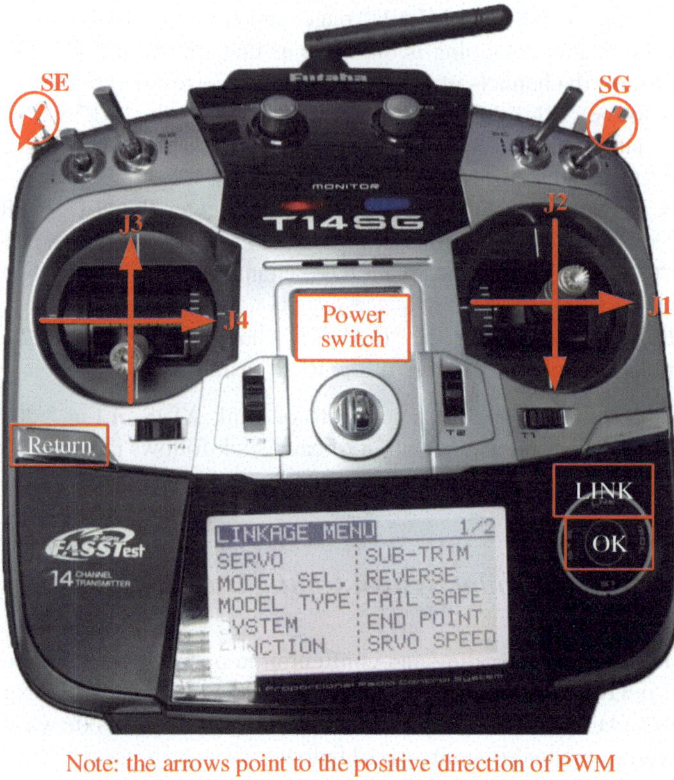

Fig. 2.22 Futaba T14SG RC transmitter

(1) Open the QGC software.
(2) As shown in Fig. 2.24a, click the "gear" icon to enter the setting page; then, click the "Firmware" button to enter the firmware burning page.
(3) Connect the Pixhawk autopilot and the computer using a USB cable. At this time, the software will automatically recognize the Pixhawk hardware. As shown in Fig. 2.24b, a firmware configuration window will pop up on the right side of the UI. Select the "PX4 Flight Stack" item, and click "OK"; then, QGC begins to automatically download (see Fig. 3.44 if the computer is not connected to the Internet) and burn the latest PX4 firmware into the Pixhawk autopilot hardware.
(4) After the firmware is burned, the Pixhawk will automatically restart and reconnect to the QGC software. Then, as shown in Fig. 2.25, enter the "Airframe" tab, select "HIL Quadcopter X" airframe type , and click the "Apply and Restart" button. Then the autopilot will automatically restart to finish the configuration for HIL simulation.
(5) After rebooting, QGC will automatically reconnect to Pixhawk. Check each configuration page to ensure that the Pixhawk autopilot has been in the HIL simulation mode.

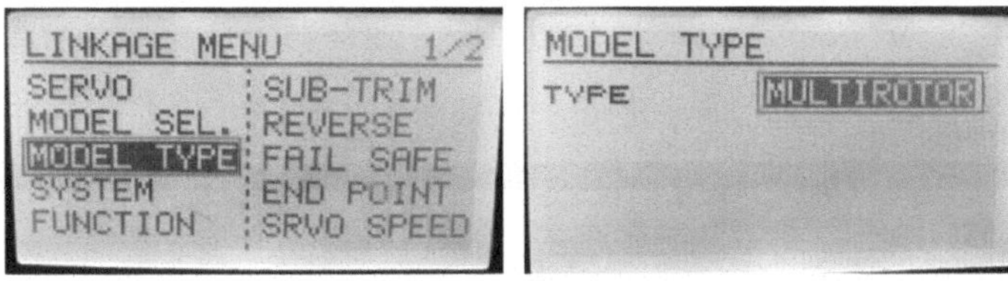

(a) Model type for multicopter

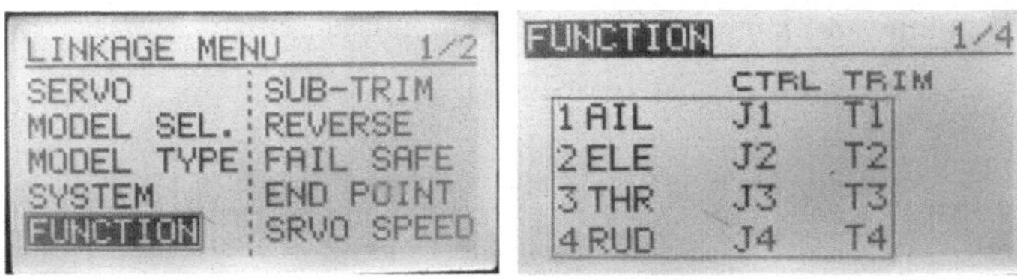

(b) Check channel definitions

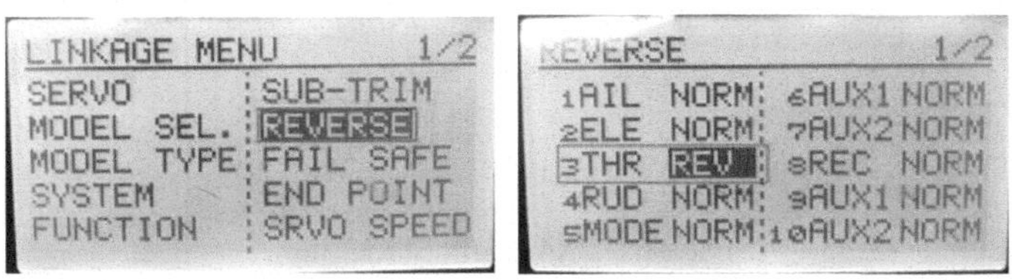

(c) Reverse throttle channel

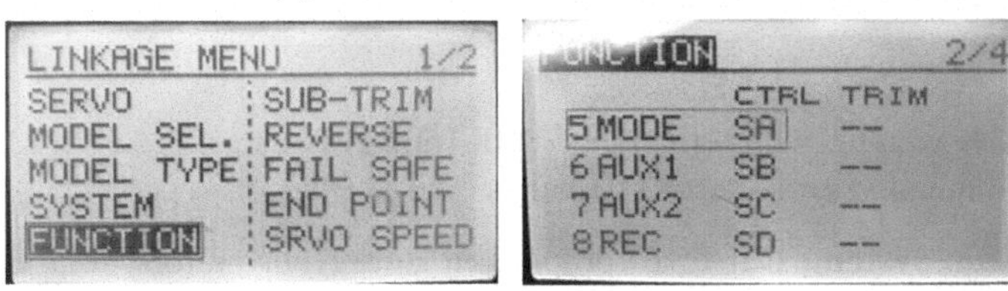

(d) Mode switch channel selection

Fig. 2.23 Futaba T14SG RC transmitter

2.3.3 Airframe and Propulsion System Configuration

The parameters of the multicopter simulation model in both SIL and HIL simulations are from a quadcopter with a diagonal size (opposite motor axis distance) of 450 mm and a weight of 1.4 kg. For the subsequent flight experiments, it is necessary to ensure that the configuration of the multicopter is as close as possible to the simulation model. The experiments presented in this book select the most popular F450 multicopter (see Fig. 2.26) with the following configuration.

(1) **Airframe:** DJI Flame Wheel F450 airframe

- Airframe weight (fuselage + arm + landing gear): 282 g
- Protection airframe: weight: 4×32 g
- Diagonal size: 450 mm

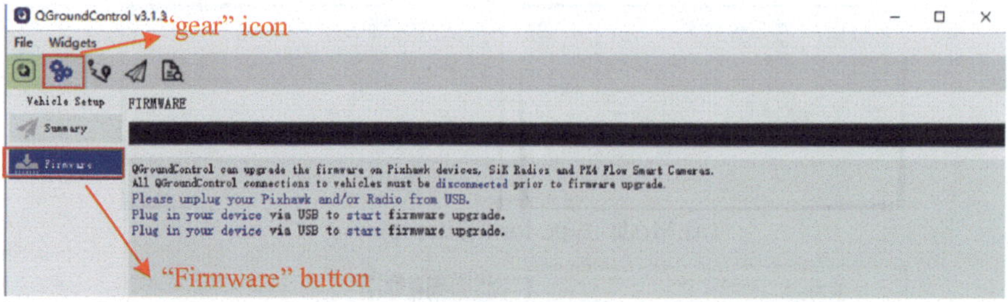

(a) Setting page

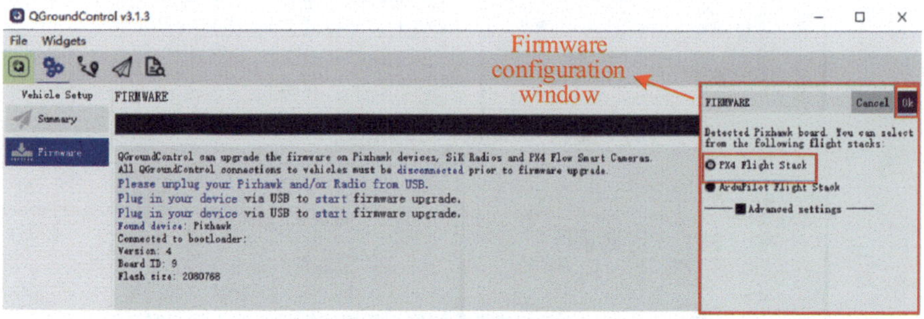

(b) Select "PX4 Flight Stack" item

Fig. 2.24 Pixhawk autopilot configuration on QGC

- Take-off weight: within the range 800–1600 g
- Recommended propeller: 8–10 inches

(2) **Propulsion system:** DJI E310 propulsion suite (four motors, four ESCs, and four propellers)

- Motor size: 23×12 mm, KV value: 960 RPM/V, weight: 60 g
- Propeller size: 24×12.7 cm (9.4×5.0 in), weight: 13 g
- ESC size: 74×32×10 mm, maximum continuous current: 20 A, weight: 43 g

(3) **Battery**: GENS ACE LiPo battery

- Capacity: 4000 mAh
- Voltage: 3S (11.1 V)
- Discharge rate: 25 C
- Weight: 300 g

(4) **Autopilot:** Pixhawk 1 (2MB flash version)

- Compiling command: px4fmu-v3_default, size: 81×47×16 mm, weight: 36 g
- GPS module: UBlox NEO-M8N GPS, module weight: 14 g, weight: 24 g
- Other accessories: power module, buzzer, safety switch, connector, and anti-vibration damper weighing a total of 60 g

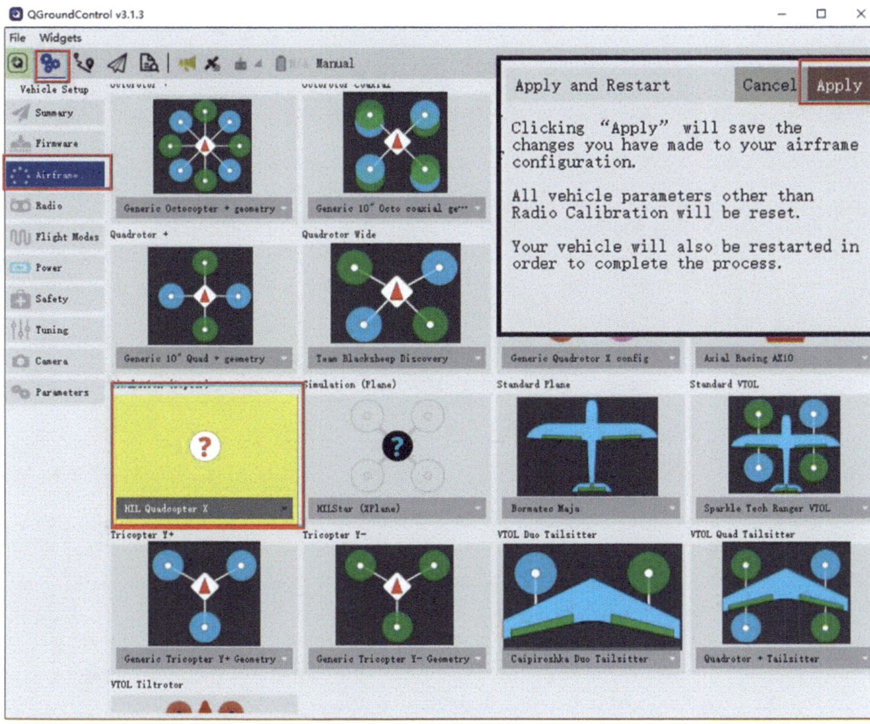

Fig. 2.25 Selecting "HIL Quadcopter X" airframe type

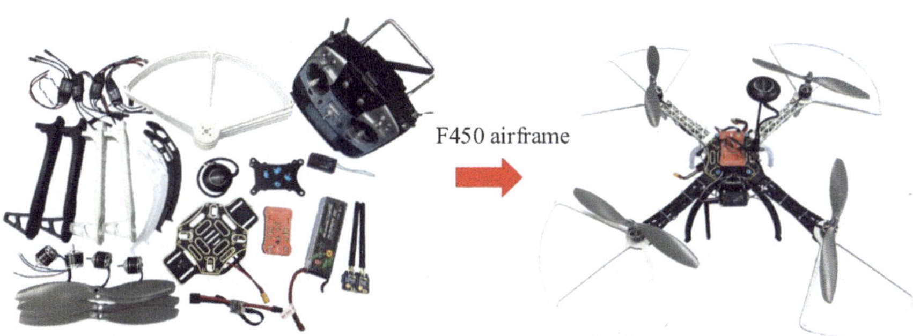

Fig. 2.26 F450 airframe and its components

After completing the assembly of the aerial vehicle, readers can follow the PX4 official website tutorial[19] to conduct preliminary flight tests to ensure that all functions are normal.

If you have any question, please go to https://rflysim.com for your information.

[19]https://docs.px4.io.

Chapter 3
Experimental Platform Usage

This chapter introduces the composition of the experimental platform released with this book and provides a thorough explanation of the role of each platform component. Examples with detailed experimental procedure are also presented in this chapter to familiarize readers with the basic functions and usage of the offered software and hardware, thereby laying a foundation for subsequent experiments to improve the learning efficiency.

3.1 Brief Introduction to Experimental Platforms

3.1.1 Platform Composition

The experimental platform is a rapid development platform for multicopter control algorithm design based on MATLAB/Simulink and Pixhawk. It is mainly composed of the following five parts: the *Simulink-based Controller Design and Simulation Platform*, the *HIL Simulation Platform*, the *Pixhawk Autopilot System*,[1] the *Multicopter Hardware System*, and the *Instructional Package*.

(1) *Simulink-based Controller Design and Simulation Platform*
As shown in Fig. 3.1, this platform has a high-fidelity nonlinear model for simulating various multicopter dynamics. Furthermore, there is an interface for communicating with the FlightGear simulator to provide a real-time 3D display for the flight status (e.g., trajectory and attitude) of the simulated multicopter. Multicopter control algorithms can be conveniently designed on this Simulink-based simulation platform and then verified with SIL simulations. Furthermore, the Pixhawk Support Package (PSP) toolbox can be used to generate C/C++ code

[1] https://docs.px4.io/master/en/flight_controller/pixhawk_series.html.

© Publishing House of Electronics Industry 2020, corrected publication 2021

Q. Quan et al., *Multicopter Design and Control Practice*,
https://doi.org/10.1007/978-981-15-3138-5_3

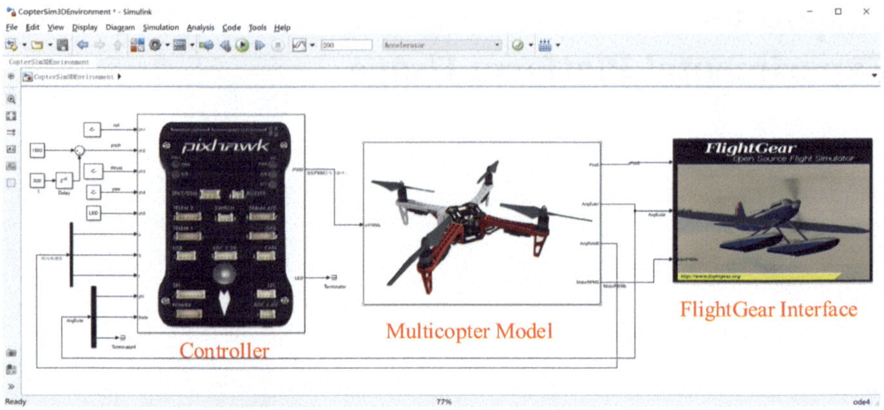

Fig. 3.1 Simulink-based controller design and simulation platform

of the control algorithms, which is then compiled and uploaded to a Pixhawk autopilot.

(2) *HIL Simulation Platform*
The HIL simulation platform includes a real-time motion simulation software—*CopterSim* (see Fig. 3.2a) and a 3D visual display software—*3DDisplay* (see Fig. 3.2b). The simulation model of CopterSim is obtained by importing parameters from the Simulink-based simulation platform mentioned earlier. Both CopterSim and 3DDisplay must run on a computer with Windows OS (Win7 or higher, x64). They are connected with the Pixhawk autopilot through a USB cable, thereby establishing a closed-loop control system for HIL simulations.

(3) *Pixhawk Autopilot System*
Figure 3.3 illustrates the entire Pixhawk autopilot system that includes a Pixhawk autopilot, an RC transmitter, an RC receiver, and a Ground Control Station (GCS). The Pixhawk autopilot used in this book is composed of Pixhawk 1 (2MB flash version)[2] autopilot hardware and PX4 autopilot software. The PX4 autopilot software[3] has a built-in Real-Time Operating System (RTOS) to ensure basic functions, such as multi-thread scheduling, low-level drivers, and algorithm execution. The Pixhawk autopilot uses an RC receiver to wirelessly connect with the corresponding RC transmitter for receiving the control commands from the remote pilot on the ground. Further, the Pixhawk autopilot can also wirelessly connect with GCS through a pair of radio telemetry to exchange flight data and mission commands.

(4) *Multicopter Hardware System*
As shown in Fig. 3.4, a multicopter hardware system is usually composed of an airframe system, a propulsion system, external sensors, a test stand, etc.

---

[2]corresponding to the circuit diagram version Pixhawk 2.4.6; for more details, please visit this website: https://docs.px4.io/master/en/flight_controller/pixhawk.html.

[3]In addition to supporting the PX4 autopilot software used in this book, the Pixhawk series autopilots also support Ardupilot open-source autopilot software; see: http://ardupilot.org/dev/index.html.

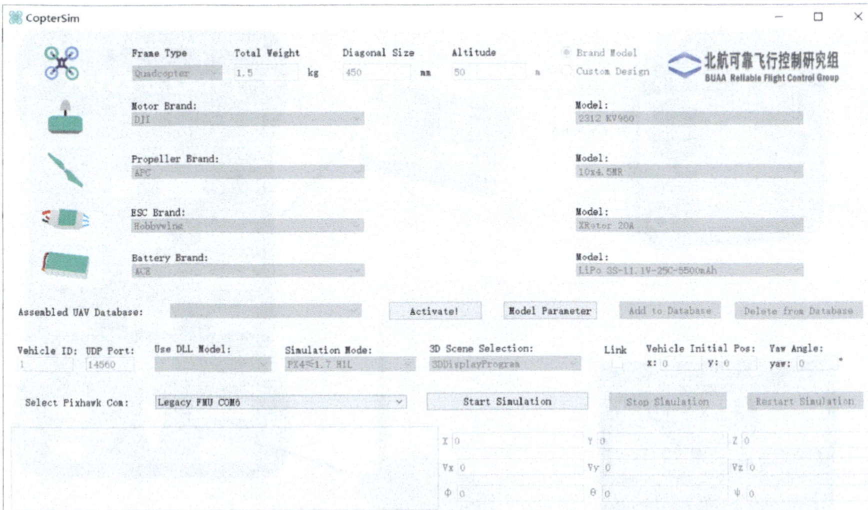

(a) CopterSim: real-time motion simulation software

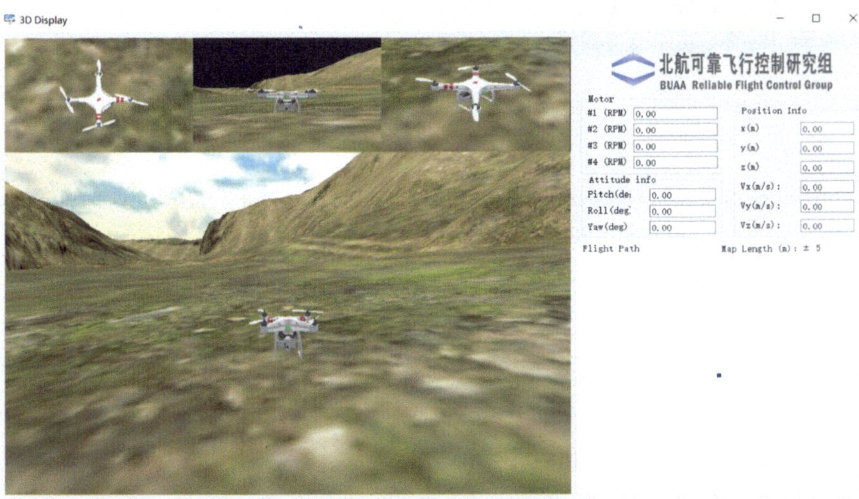

(b) 3DDisplay: 3D visual display software

Fig. 3.2 HIL simulation platform

Combining the Pixhawk autopilot and a multicopter hardware system requires an integrated multicopter flight platform that can be used to perform specific flight tasks. According to the actual flight performance requirements, the multicopter airframe system can be designed with different configurations, such as quadcopters, hexacopters, and coaxial octocopters. In this book, a small quadcopter is chosen as an example to study its hardware system design, such as

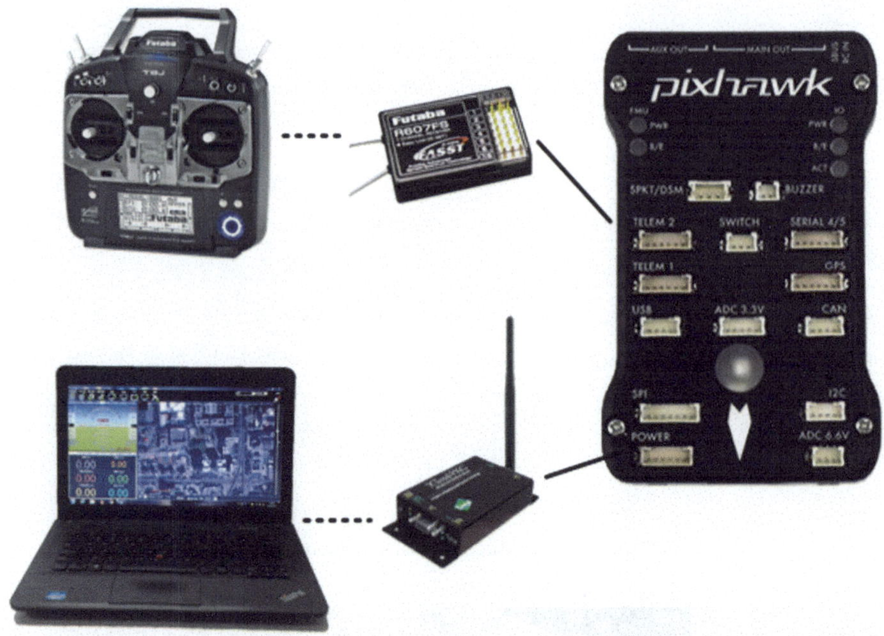

Fig. 3.3 Pixhawk autopilot system

airframe design, propulsion system selection. Then, the control algorithms are designed for the resulting multicopter hardware system.

(5) *Instructional Package*
This book encompasses video tutorials, PowerPoint files, and source code examples. The topic is closely related to our previous book [8].

3.1.2 Platform Advantage

This experimental platform provides interfaces for multicopter controller design in MATLAB/Simulink. Readers (beginners, students, or engineers) can develop a rapid design and verify the control algorithms by SIL simulation. After the control algorithms are well designed and verified, the platform also provides a code generation function to generate Simulink controllers to C/C++ code. Then, the code can be compiled into the autopilot software firmware file; finally, it is automatically uploaded to the Pixhawk hardware. The platform also provides a HIL simulation platform for preliminary simulation tests on a Pixhawk autopilot system that may help in eliminating potential problems that may exist in flight tests. After all the tests are completed, indoor and outdoor flight tests can be carried out by assembling the Pix-

Fig. 3.4 Multicopter hardware system

hawk autopilot onto a real multicopter hardware system. The performance of the designed controllers can be evaluated through experimental tests.

3.2 Simulink-Based Controller Design and Simulation Platform

To improve the design efficiency of multicopter controllers, as shown in Fig. 3.5, this book provides a high-fidelity simulation environment based on Simulink/FlightGear. The main source code file is presented in "e0\1.SoftwareSimExps\CopterSim3D Environment.slx".

The following steps describe the correct way to open Simulink files (those with the file suffix ".slx").

(1) Open MATLAB via the Windows desktop shortcut or the Windows start menu.
(2) As shown in Fig. 3.6, click the "Browse Folder" button in the MATLAB User Interface (UI) to set the current directory to the folder of the ".slx" file to be opened.
(3) Double-click the ".slx" file in the "Current Folder" window (the lower-left side in Fig. 3.6) to open it.

All ".slx" files should be opened in this way to ensure that the working directory is correct and that the initialization scripts are successfully loaded.

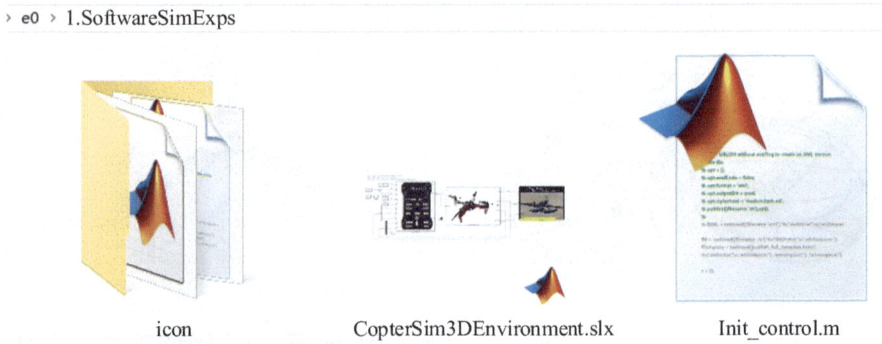

Fig. 3.5 Files in "SoftwareSimExps" folder

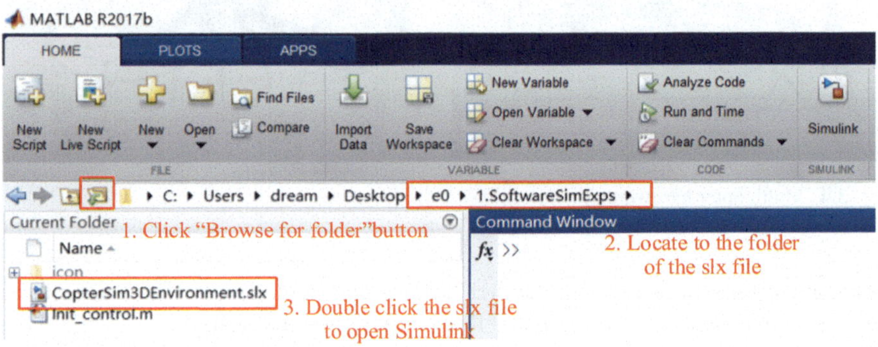

Fig. 3.6 Method to correctly open a Simulink slx file

Open the "CopterSim3DEnvironment.slx" file according to the above procedure; then, a SIL simulation example will appear as shown in Fig. 3.7. The SIL simulation example contains three subsystems: the "Controller" subsystem, the "Multicopter Model" subsystem, and the "FlightGear Interface" subsystem. Their key features are summarized below.

(1) The input, output, and feedback signals of the "Controller" subsystem are consistent with the available signals in a real autopilot system. For example, the input signals of the controller in Fig. 3.7 are the control commands for pitch, roll, yaw, and altitude control from a simulated RC transmitter, and the output signals are the motor PWM signals for a multicopter model.
(2) The controller uses the sensor estimated states (e.g., attitude, angular velocity, position, velocity, and other state information) to achieve stable attitude control.
(3) The input and output signals of the "Multicopter Model" subsystem are consistent with those of a real multicopter. For example, the input signals of the "Multicopter Model" subsystem are the PWM control signals of eight motors (the multicopter model will choose the actual number of motors according to the selected model) defined by the Pixhawk autopilot (the data range is 1000–2000 corresponding

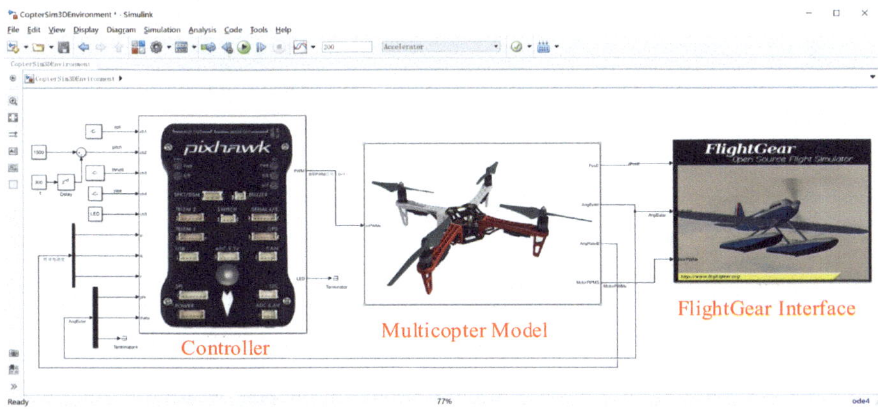

Fig. 3.7 Simulink SIL simulation example

to a throttle command of 0–1), and the output signals are the data from various sensors.

(4) "FlightGear Interface" subsystem provides a communication interface to send the flight data to FlightGear, where the real-time vehicle attitude and flight trajectory can be observed in a realistic 3D scene.

3.2.1 Controller

Double-click the "Controller" subsystem in Fig. 3.7 yields the internal structure of the controller, as presented in Fig. 3.8. This example shows a simple attitude controller for pitch and roll angles. The controller receives the control signals from the RC transmitter and controls the multicopter to achieve the desired pitch and roll angles.

As shown in Fig. 3.8, the 1st–5th input ports are five input channels from the RC transmitter ("ch1"–"ch5"); the 6th–8th input ports are the angular velocity ("p", "q", "r") from the gyroscope sensor; the 9th–10th input ports are the roll angle and pitch angles ("phi", "theta") estimated from the inertial sensor. As shown in Fig. 3.8, the computing process of the entire "Controller" subsystem is roughly divided into five steps.

(1) The "Input Interfaces" module receives the RC transmitter signals and the multicopter state estimation signals.[4]

[4]In a real autopilot system, these signals should be obtained from the modules related to state estimation (e.g., raw sensor data, Kalman filter, and complementary filter). For simplicity, in the controller design during the SIL simulation the true values of the multicopter model can be used first.

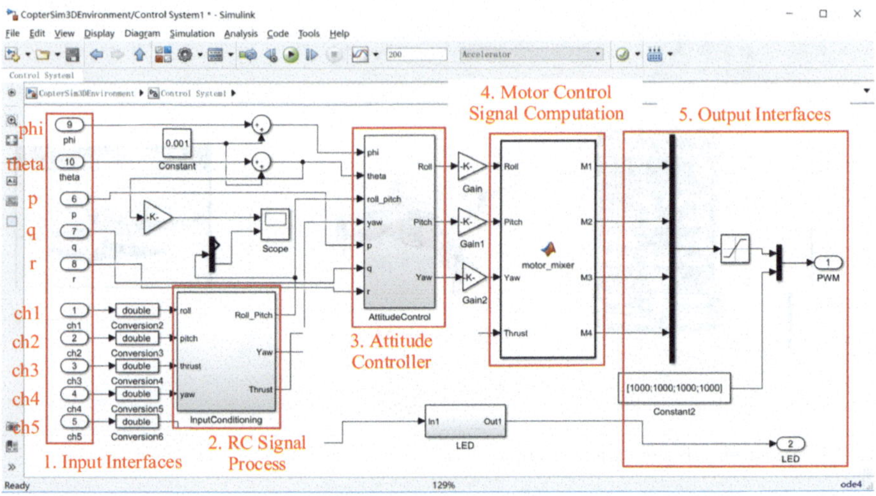

Fig. 3.8 Internal structure of "Controller" subsystem

(2) The "RC Signal Process" module maps the five-channel signals of the RC transmitter to the desired roll and pitch angle values.
(3) The "Attitude Controller" module computes the desired force and torque values to control the multicopter to the desired attitude.
(4) The "Motor Control Signal Computation" module maps the force and torque values to the control signals (ranging from 1000 to 2000) for the four motors.
(5) The "Output Interfaces" module fills the remaining 4-dimensional control signals and generates an 8-dimensional PWM signal (there are eight PWM output ports on Pixhawk) ranging from 1000 to 2000 μs.[5]

3.2.2 *Multicopter Model*

Double-click the "Multicopter Model" subsystem in Fig. 3.7, and the internal structure of the multicopter model is presented in Fig. 3.9. The multicopter model simulates a real multicopter system to output the flight state and sensor signals based on the motor PWM controls from the control system.

[5] A value within the range from 1000 to 2000 corresponds to a higher level duration (in microseconds) of PWM signals. Given that the period of an RC PWM signal is usually 20 ms (50 Hz), the duty ratio of the PWM signal measured by a multimeter usually ranges from 0.05 to 0.1 instead of from 0 to 1.

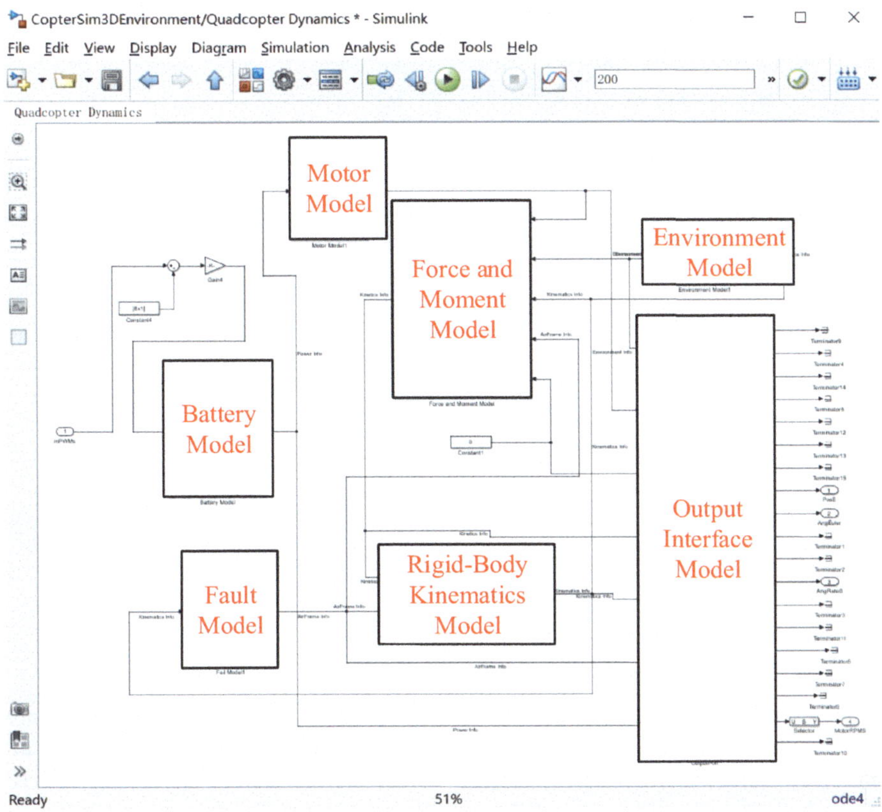

Fig. 3.9 Internal structure of "Multicopter Model" subsystem

As shown in Fig. 3.9, the "Multicopter Model" subsystem contains the following seven main modules.

(1) "Motor Model" module: it simulates the motor dynamics.
(2) "Force and Moment Model" module: it simulates all external forces and moments acting on the body, such as the propeller thrust, fuselage aerodynamics, gravity, and ground supporting force.
(3) "Rigid Body Kinematics Model" module: it calculates the vehicle kinematics of the multicopter, such as speed, position, and attitude.
(4) "Environmental Model" module: it calculates the environmental data, such as gravitational acceleration, air density, wind disturbances, and geomagnetic field.
(5) "Fault Model" module: it is mainly used to inject model uncertainties (related to mass and moment of inertia) as well as faults.
(6) "Battery Model" module: it simulates the discharge process of the battery.
(7) "Output Interface Model" module: it packs the output signals in the desired format.

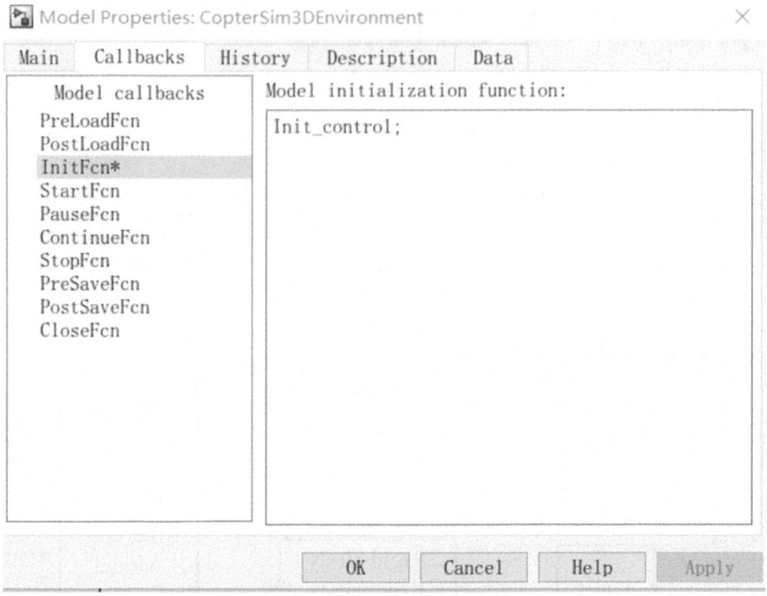

Fig. 3.10 Initialization interface for the "Init_control.m" script

The controller parameters are stored in an initialization script "e0\1.Software SimExps\Init_control.m". This script will be automatically called to import all parameters into the Simulink workspace when the simulation starts. Figure 3.10 depicts how to proceed to autorun the initialization script. Readers can open the UI in Fig. 3.10 by clicking "File"—"Model Properties"—"Callbacks"—"InitFcn" in the Simulink menu within the "CopterSim3DEnvironment.slx" project. The source code of the "Init_control.m" initialization script is listed in Table 3.1. Note, for MATLAB R2019b and above, the "Callbacks" option can be found at "MODELING"—"Model Settings"—"Model properties" on the Simulink toolbar.

All the parameters required by the "Multicopter Model" are stored in the script "e0\1.SoftwareSimExps\icon\Init.m. This script will be automatically called when the second line (see Table 3.1) of the "Init_control.m" script is executed. The key model parameters are listed in Tables 3.2, 3.3, 3.4, 3.5, and 3.6. By modifying the above model parameters, multicopters with different sizes and configurations (see Table 3.5) can be obtained, and flight simulations under different environments (see Table 3.6) can be performed.

3.2.3 FlightGear Interface

As shown in Fig. 3.7, the "FlightGear Interface" subsystem has three input ports representing the multicopter position, Euler angles, and motor PWM signals, respectively. This subsystem sends multicopter flight state information to FlightGear to

Table 3.1 Init_control.m: script for initialization of the control parameters

```
path(path,'./icon/'); %add folder 'icon' to MATLAB path
call
Init; %run the model parameter initialization script

%PID parameter
Kp_RP_ANGLE =6.5;
Kp_RP_AgngleRate=0.55;
Ki_RP_AgngleRate=0.01;
Kd_RP_AgngleRate=0.005;
Kp_YAW_AngleRate=3.2;
Ki_YAW_AngleRate=0.8;
Kd_YAW_AngleRate=0.05;

%maximum control angle, unit: degree
MAX_CONTROL_ANGLE_RP=45;
MAX_CONTROL_ANGLE_Y=180;

%maximum control angle rate, unit: degree per second
MAX_CONTROL_ANGLE_RATE_RP=180;
MAX_CONTROL_ANGLE_RATE_Y=90;
```

Table 3.2 Init.m: initial state of the multicopter

```
%aerial vehicle initial
ModelInit_PosE=[0,0,0]; %initial position in earth frame(unit:m)
ModelInit_VelB=[0,0,0]; %initial velocity in body frame(unit:m/s)
ModelInit_AngEuler=[0,0,0];%initial angle(unit:rad)
ModelInit_RateB=[0,0,0];%initial angular velocity(unit:rad/s)
ModelInit_RPM=0;%initial motor speed(unit: RPM- Revolutions per minute)
```

Table 3.3 Init.m: multicopter's motor parameters

```
%Motor parameters
ModelParam_motorMinThr=0.05;%motor minimum throttle dead zone
ModelParam_motorCr=1148; %motor speed-throttle curve slope(unit:rad/s)
ModelParam_motorWb=-141.4; %motor speed-throttle curve constant term(unit:rad/s)
ModelParam_motorT=0.02; %motor inertia time constant(unit:s)
ModelParam_motorJm=0.0001287; %the moment of inertia of motor-propeller(unit:kg.
    m^2)
```

Table 3.4 Init.m: multicopter's propeller parameters

```
% propeller parameters
% propeller moment coefficient Cm(unit: N.m/(rad/s)^2)
% definition: moment M (unit:N.m), propeller speed w(unit:rad/s), M=Cm*w^2
ModelParam_rotorCm=2.783e-07;

% propeller thrust coefficient Ct(unit:N/(rad/s)^2)
% Definition: Thrust T (unit:N), T =Ct*w^2
ModelParam_rotorCt=1.681e-05;
```

Table 3.5 Init.m: multicopter configuration parameters

```
%multicopter parameters
ModelParam_uavType = int8(3); % multicopter configuration, 3 means X-type
     quadcopter
ModelParam_uavMotNumbs = int8(4); % motor count
% drag coefficient Cd
% definition: resistance D(N), forward flight speed V(m/s), D = Cd*V^2
ModelParam_uavCd =0.1365;% drag coefficient (unit: N/(m/s)^2)
ModelParam_uavCCm = [0.0035 0.0039 0.0034]; % Rotational Damping Coefficient
ModelParam_uavMass=1.515; % multicopter mass (unit: kg)
ModelParam_uavDearo =0.05; % deviation between aerodynamic center and mass
     center
ModelParam_uavR=0.225; % multicopter frame radius (unit: m)
ModelParam_uavJxx =0.0211; %x axis moment of inertia (unit: kg.m^2)
ModelParam_uavJyy =0.0219; %y axis moment of inertia (unit: kg.m^2)
ModelParam_uavJzz =0.0366;%z axis moment of inertia (unit: kg.m^2)
```

Table 3.6 Init.m: multicopter environmental parameters

```
%Environmental parameters
ModelParam_envLongitude = 116.2593683; % initial longitude
ModelParam_envLatitude = 47.397742; % initial latitude
ModelParam_envAltitude = -50;       % reference altitude, negative value for
     pointing up
```

observe the flight attitude and trajectory of the multicopter in a 3D scene. The steps to follow are described next.

(1) Double-click the FlightGear-F450 shortcut on the desktop to open FlightGear; Note that if FlightGear cannot run properly on your computer, you can open 3DDisplay to observe the result.
(2) Click the "Run" button on the Simulink toolbar (see Fig. 3.11) to run the "CopterSim3DEnvironment.slx" file;
(3) Then, as shown in Fig. 3.12, the multicopter takes off vertically from the ground and starts flying forward at a certain pitch angle after 5 s.

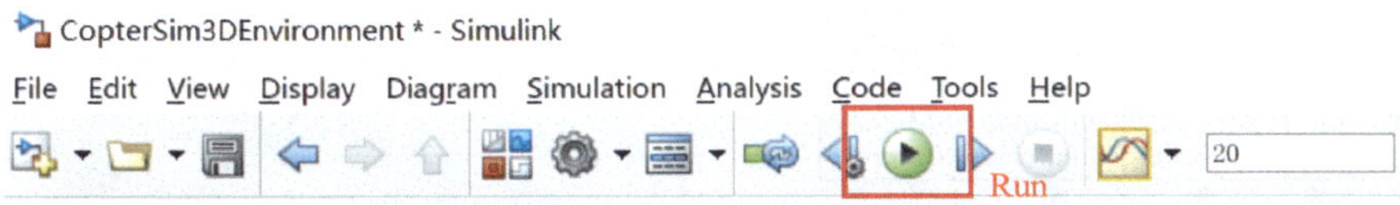

(a) Simulink "Run" button on MATLAB 2017b ~ 2019a

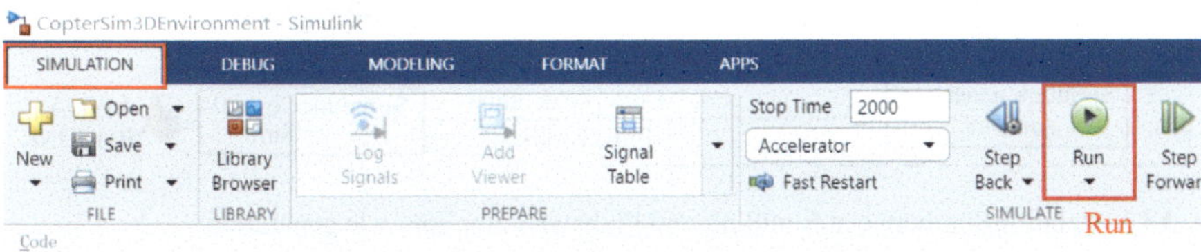

(b) Simulink "Run" button on MATLAB 2019b and above

Fig. 3.11 Simulink "Run" button for different MATLAB versions

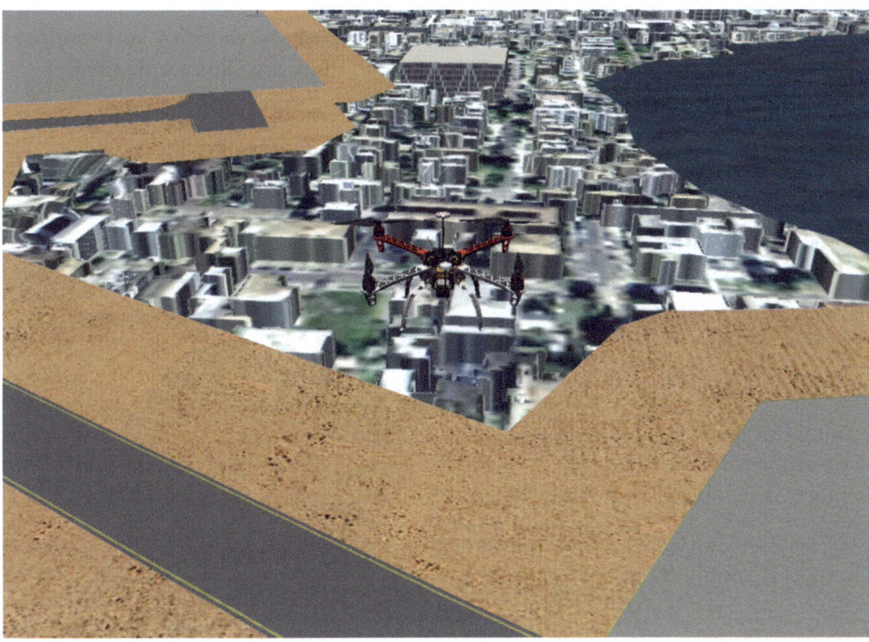

Fig. 3.12 A quadcopter in FlightGear

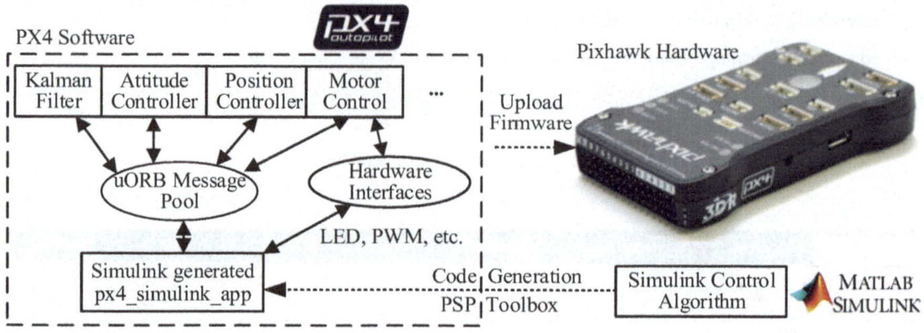

Fig. 3.13 Relationship between Simulink and Pixhawk autopilot code generation

3.3 PSP Toolbox

Figure 3.13 shows the relationship among the PSP toolbox, the PX4 software, and the Pixhawk hardware. The main features of the toolbox are summarized below.

(1) The toolbox can simulate and test different multicopter models and flight control algorithms in Simulink and then automatically deploy the algorithms to the Pixhawk autopilot.
(2) The toolbox provides many practical examples, including LED control, RC data process, and attitude controller.
(3) The toolbox provides many interface modules to access the Pixhawk hardware and software components.
(4) It automatically records flight data from sensors, actuators, and controllers deployed by themselves.
(5) It can subscribe and publish uORB topic messages. All messages in the PX4 autopilot software are temporarily stored in a uORB message pool. The subscription function can read topics of interest from the message pool, and the publishing function can publish specific topics to the message pool for other modules.

The relationship between the code generated by Simulink and the Pixhawk autopilot system is summarized below.

(1) The structure of a Pixhawk autopilot system includes two parts, namely the Pixhawk hardware (similar to the computer hardware) and the PX4 software (similar to the operating system and applications running on a computer).
(2) The PX4 software system can be further divided into several small modules, which run independently in parallel multi-thread. Each module exchanges data with other modules through the subscription and publication of uORB messages.
(3) After the algorithm code is generated by the PSP toolbox, it is embedded into the PX4 software system. This will not affect the operation of the native control modules in the PX4 software. Instead, a new independent module (with an independent thread) named "px4_simulink_app" will be created to run in parallel with other modules.

(4) As shown in Fig. 3.13, the whole code generation and deployment procedures are presented as follows.

1) The PSP toolbox generates the C/C++ code from the control algorithm designed in Simulink.
2) The obtained algorithm code is imported into the PX4 source code to generate a "px4_simulink_app" independent of other modules.
3) The PSP toolbox calls the compiling toolchain (Win10WSL, Msys2, or Cygwin) to compile all the code into a ".px4" PX4 firmware file (similar to a software installation package).
4) Upload the obtained firmware file to the Pixhawk hardware; then, the Pixhawk autopilot can execute the PX4 software with the generated algorithm code.

(5) The native modules in the PX4 software may assess the same hardware outputs as the generated "px4_simulink_app" module, which may cause read and write conflicts. Therefore, in the one-key installation script, the hardware output accessing codes of the PX4 native modules have been blocked in the last option shown in Fig. 2.4. This will ensure that only the "px4_simulink_app" module can send motor control signals.

(6) The generated Simulink code can also be used to replace some of the native modules (sensors, filters, attitude controllers, etc.) of the PX4 software shown in Fig. 3.13. However, the PX4 software code needs to be manually modified to block the output interface of the corresponding native module (Another feasible way is to block the module in the startup script "Firmware\ROMFS\px4fmu_common\init.d\rcS"). For example, if readers want to use Simulink to replace the filter module (input sensor data, output filter data) of the PX4 software, they should manually block uORB publishing code of the PX4 "Position and Attitude Estimator" module in Fig. 3.14 to prevent it from publishing estimate data. The detailed steps are described next.

1) Open the "Firmware\src\modules\ekf2\ekf2_main.cpp" file (corresponding to the code of the extended Kalman filter module).
2) Search for the text "ORB_ID(vehicle_attitude)" and comment out the related code. Note: for PX4 firmware version 1.8 and above, you can search text "_att_pub" to find the lines with words "publish" and "att", and replace these lines with "UNUSED(att);".
3) Write attitude filter in Simulink to publish "vehicle_attitude" message with the "uORB Write" block.

3.3.1 Simulink Pixhawk Target Blocks Library of PSP Toolbox

As shown in Fig. 3.15, after installing the PSP toolbox, a "Pixhawk Target Blocks" interface module library can be found in the Simulink library browser. These modules provide interfaces to access the Pixhawk hardware I/Os and the PX4 internal messages. The "Pixhawk Target Blocks" library consists of four sub-libraries, namely

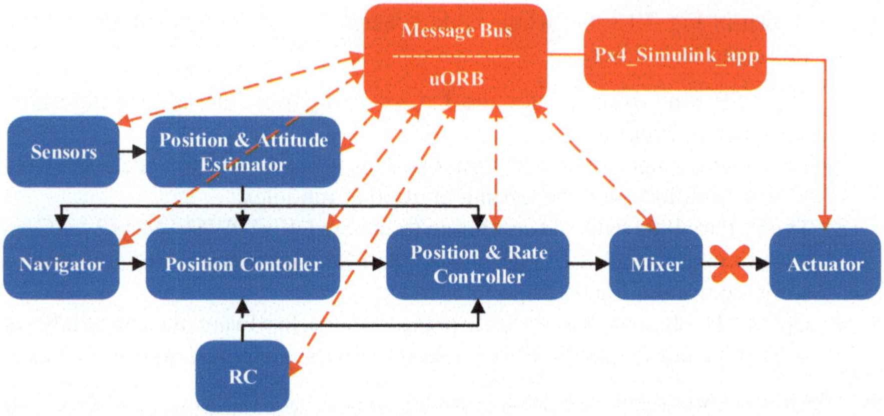

Fig. 3.14 Relationship between PX4 native modules and module generated by Simulink

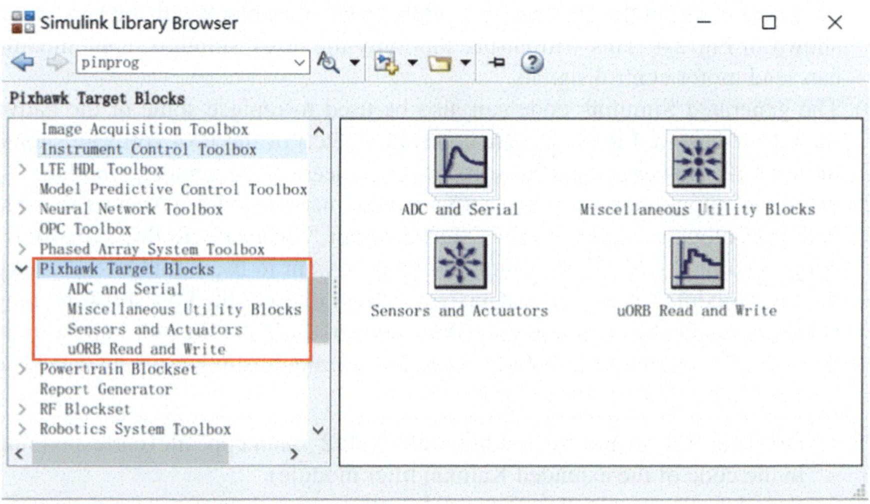

Fig. 3.15 Simulink PSP module library

"ADC and Serial library", "Miscellaneous Utility Blocks" library, "Sensors and Actuators" library, and "uORB Read and Write" library.

Figure 3.16 presents several key I/O interface modules in the "Sensors and Actuators" library. These modules make it easy to acquire the sensor data or estimate data for designing flight controllers that compute output signals for the motors, LEDs, and buzzers.

As shown in Fig. 3.17, the PSP toolbox also provides many examples (see folder "e0\2.PSPOfficialExps") with an official manual (see document "e0\2.PSPOfficialExps\Pixhawk_Pilot_Support_Package.pdf" for details) for readers to be quickly familiar with functions and usage methods of PSP toolbox.

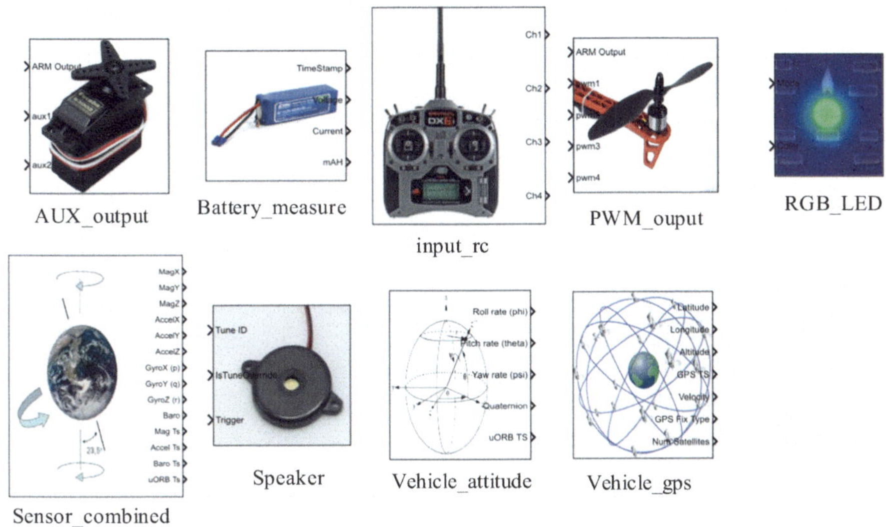

Fig. 3.16 Schematic diagram of PSP toolbox sensor and actuator interface library

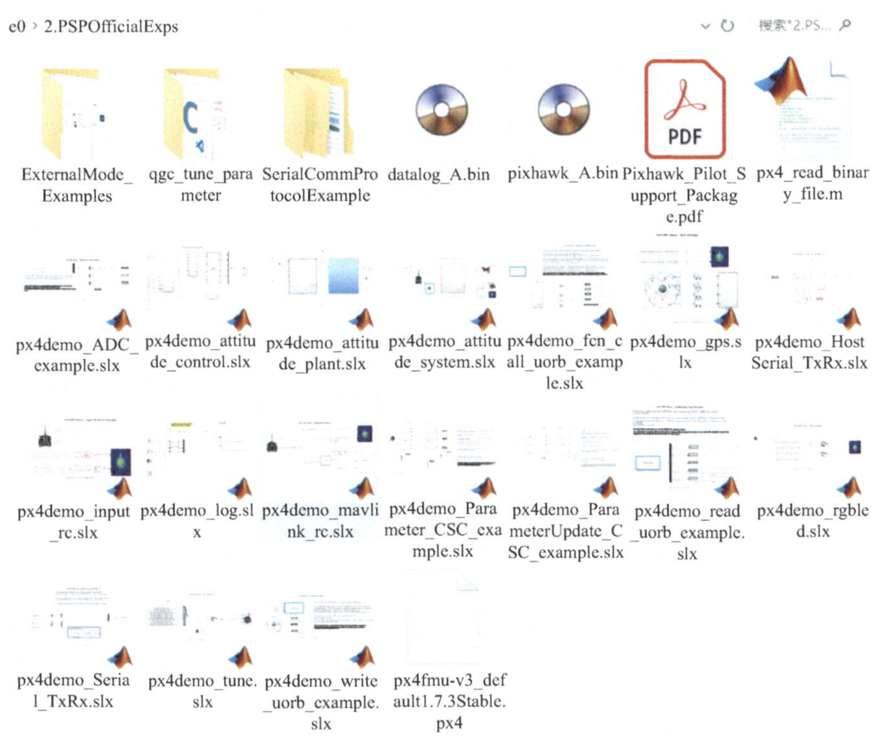

Fig. 3.17 Official examples and manuals for PSP toolbox

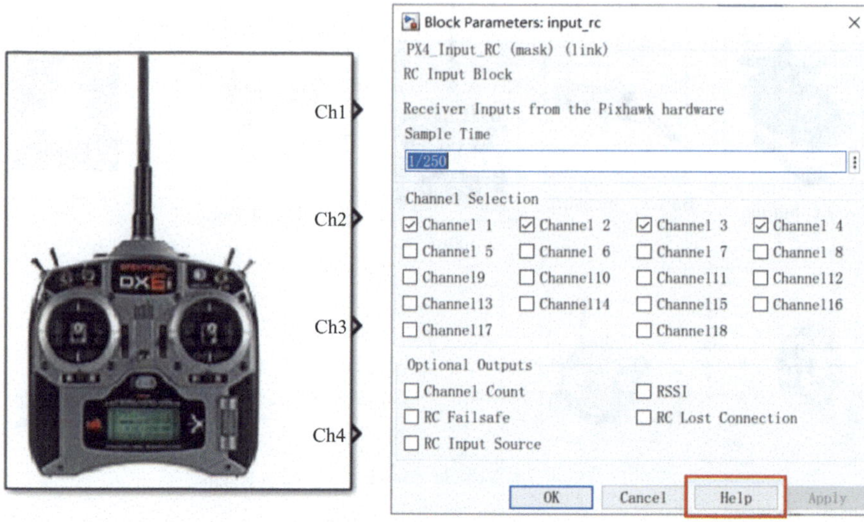

Fig. 3.18 RC input module and its parameter setting box

3.3.2 Instructions for Modules in PSP Toolbox

(1) RC input module
Figure 3.18 presents the RC input module and its parameter setting box. It is convenient to select RC channels and other information to be used by Simulink. The definition and application of each option can be viewed by clicking the "help" button of the box or by consulting the official PDF document. The PSP toolbox also provides an example (see file"e0\2.PSPOfficialExps\px4demo_input_rc.slx") to show how to use this module.

(2) PWM output module
Figure 3.19 depicts the PWM output module, which is used to send PWM signals to PX4IO ports to control the motor. The PWM update frequency and the number of output channels can be configured in the setting box.

(3) AUX output module
Figure 3.20 presents the AUX_output module, which is used to send PWM signals to PX4FMU ports to control the servo deflection. The PWM update frequency and the number of output channels can be configured in the setting box.

(4) Buzzer module
Figure 3.21 presents the Buzzer module, which is used when the buzzer is required to make a warning sound. There is an example (see file "C:\PX4PSP\examples\px4demo_tune.slx", where "C:\PX4PSP" should be modified to your installation directory) for detailed information.

(5) RGB_LED module
This module can control the blink mode and color of the LED on Pixhawk. As shown in Fig. 3.22, the module receives two inputs, namely "Mode" and "Color"

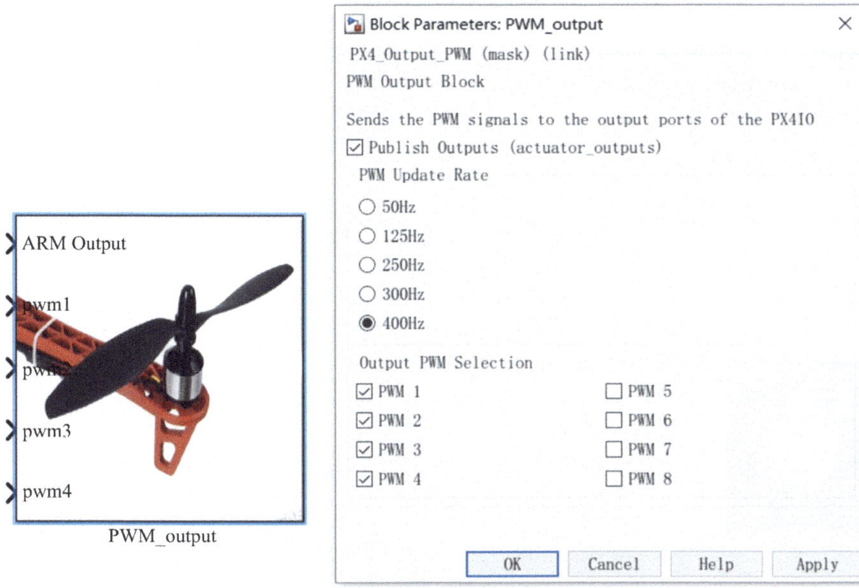

Fig. 3.19 PWM output module and its parameter setting box

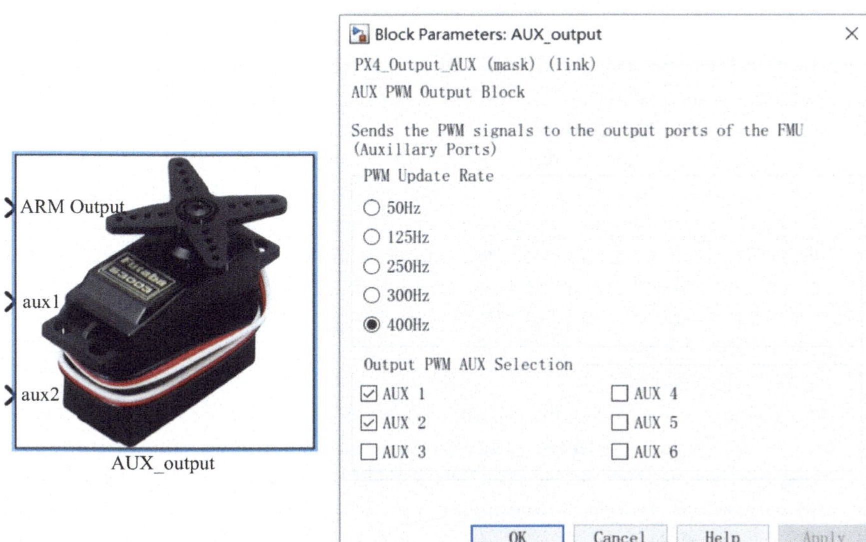

Fig. 3.20 AUX_output module and its parameter setting box

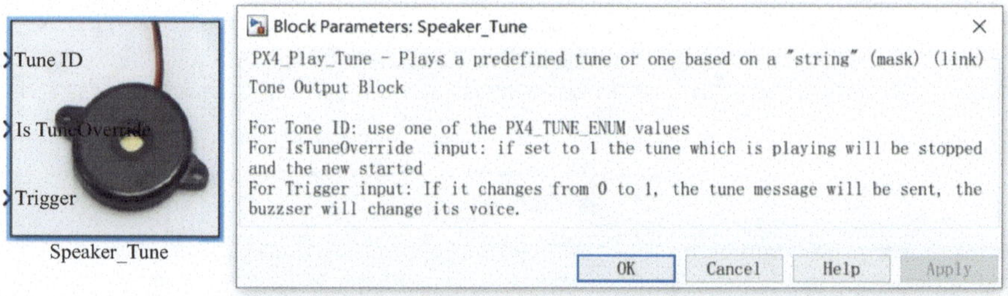

Fig. 3.21 Buzzer module and its parameter setting box

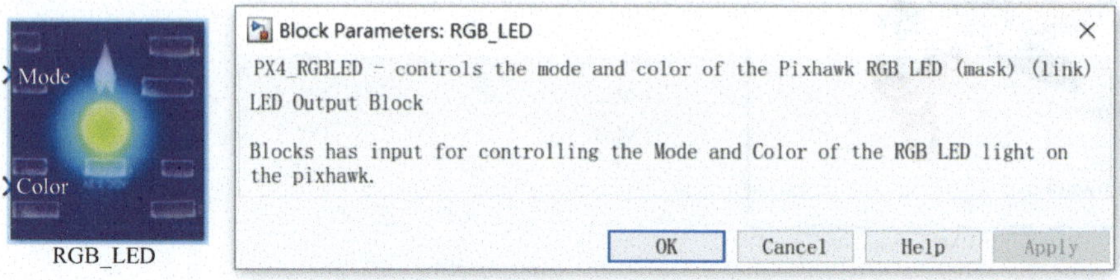

Fig. 3.22 LED light module and its parameter setting box

representing the mode and color of the LED. The PSP toolbox provides an example (see file "e0\2.PSPOfficialExps\px4demo_rgbled.slx") to study this module.

(6) Sensor combination module
This module can access the sensor data available in the Pixhawk autopilot, which can then be used for controller design in Simulink. Available sensor data include magnetometers, accelerometers, gyroscopes, barometers, and timestamps. As shown in Fig. 3.23, the sample rate and the required sensor data can be configured in the parameter setting box. The PSP toolbox also provides an example (see file "e0\2.PSPOfficialExps\px4demo_attitude_system.slx") to study this module.

(7) Attitude data module
As shown in Fig. 3.24, the attitude data module provides an interface to access the attitude estimate (angular rate and quaternion). The PSP toolbox also provides an example (see file "e0\2.PSPOfficialExps\ px4demo_attitude_control.slx") to study this module.

(8) GPS data module
This module, shown in Fig. 3.25, can be used to access the Pixhawk GPS data, which are achieved by subscribing to the uORB topic "vehicle_gps". Therefore, in practical operation, it is necessary to ensure that the GPS module is inserted into the Pixhawk hardware and then works. The PSP toolbox also provides an example (see file "e0\2.PSPOfficialExps\px4demo_gps.slx") to study this module.

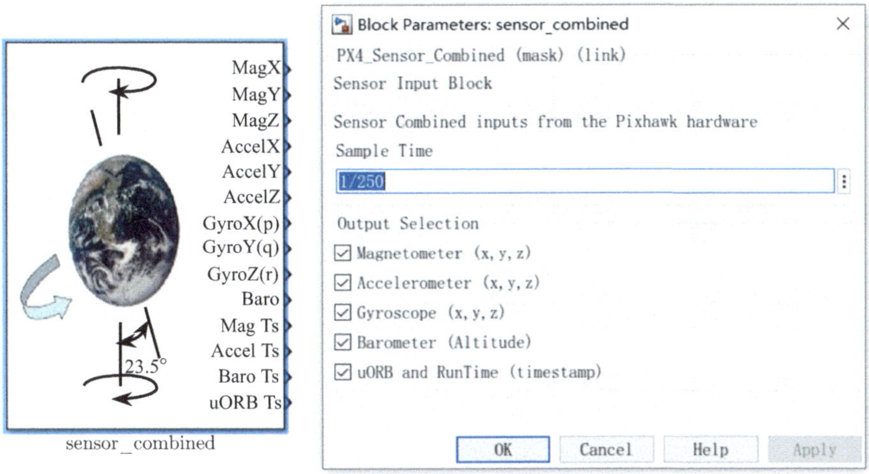

Fig. 3.23 Sensor combination module and its parameter setting box

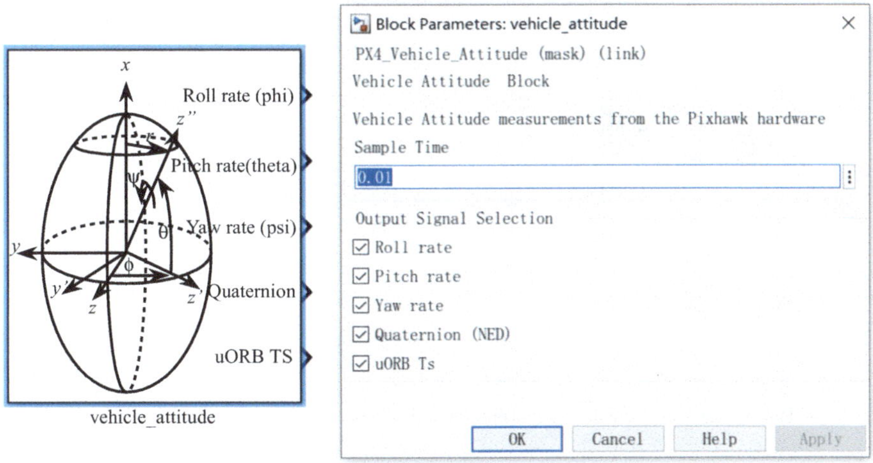

Fig. 3.24 Attitude data module and its parameter setting box

(9) Battery data module
This module, shown in 3.26, can be used to obtain the real-time status of the battery. It is implemented by subscribing to the uORB topic "battery_status". Therefore, in practical operation, it is necessary to ensure that the power module is inserted into the Pixhawk hardware and then works correctly.

(10) uORB modules
These modules, presented in Fig. 3.27, are used to read or write uORB messages from the PX4 autopilot software. All the uORB message topics supported by

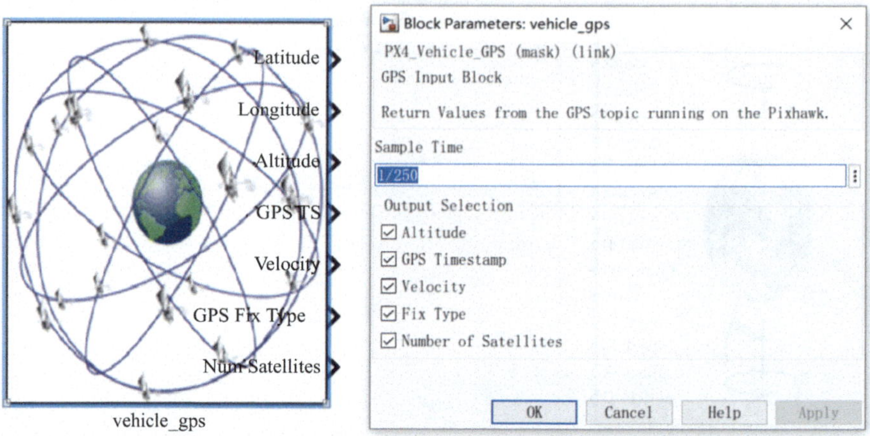

Fig. 3.25 GPS data module and its parameter setting box

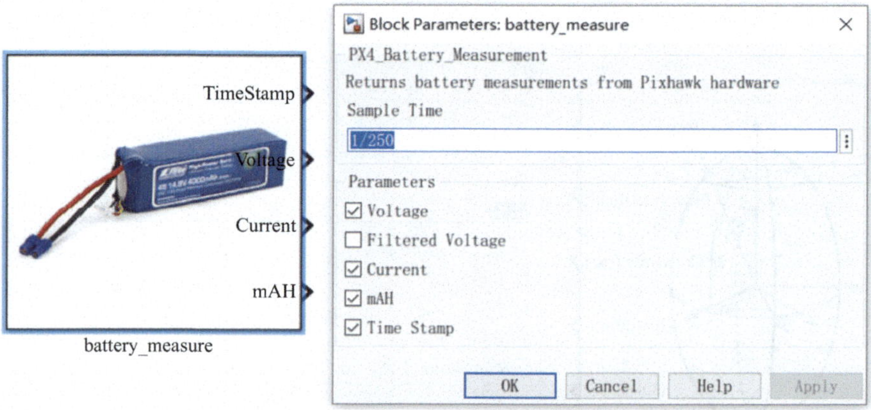

Fig. 3.26 Battery data module and its parameter setting box

the PX4 autopilot are listed in the directory "Firmware\msg" of the software package installation directory (configured as in Fig. 2.4; the default directory is "C:\PX4PSP"). Double-click the "uORB Write" module in Fig. 3.27, then the obtained parameter setting box of the "uORB Write" module is presented in Fig. 3.28, where the uORB topic name and the message variables to be sent can be configured. Clicking the "Open .msg file" button in Fig. 3.28 yields the content of the select ".msg" file (see Fig. 3.29), and clicking the "Open .msg folder" button yields the list of all supported uORB messages (See Fig. 3.30). There are two advanced "uORB Write" modules presented in Fig. 3.31, which provide more convenient ways to send uORB messages.

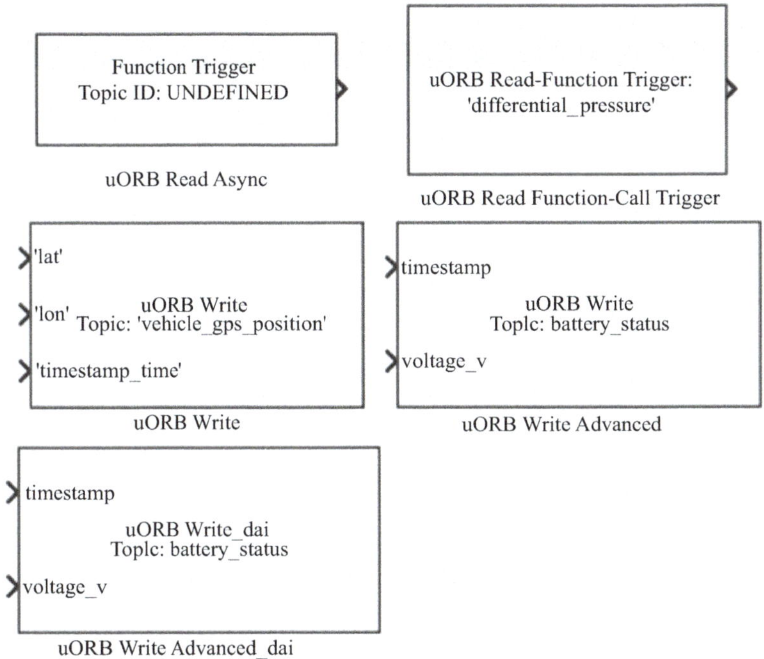

Fig. 3.27 uORB modules for message reading and writing

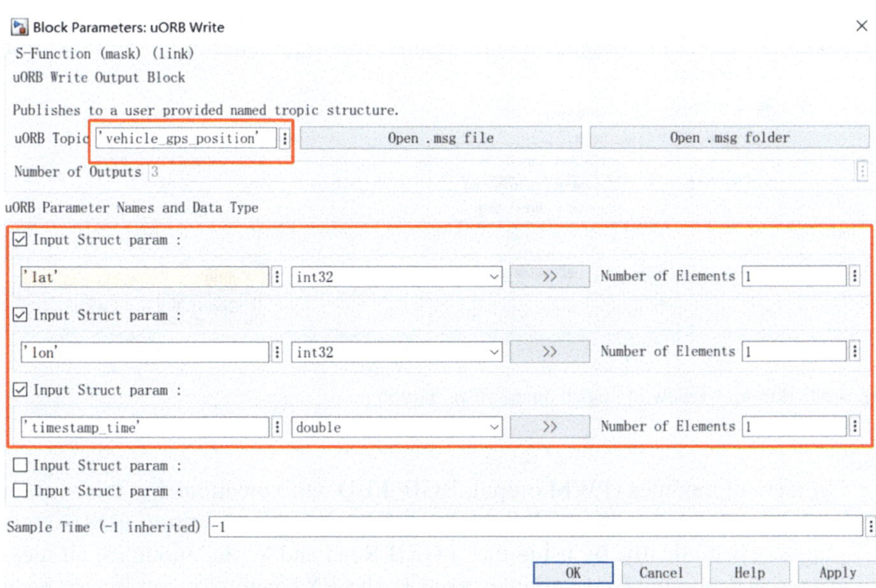

Fig. 3.28 "uORB Write" module parameter setting box

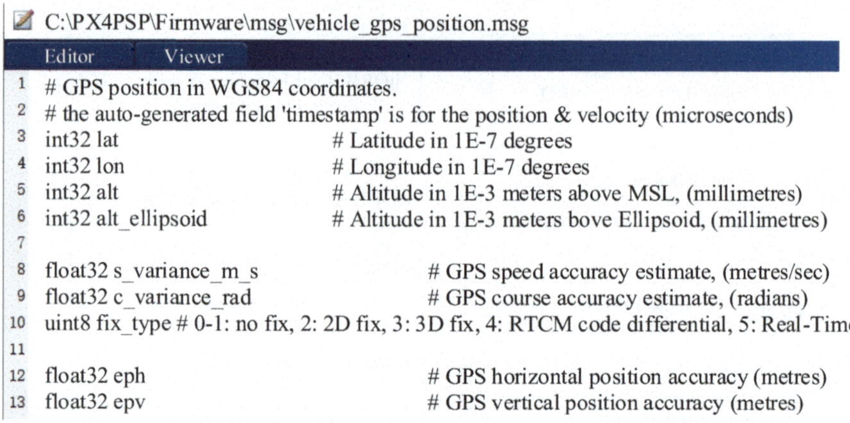

Fig. 3.29 uORB message file

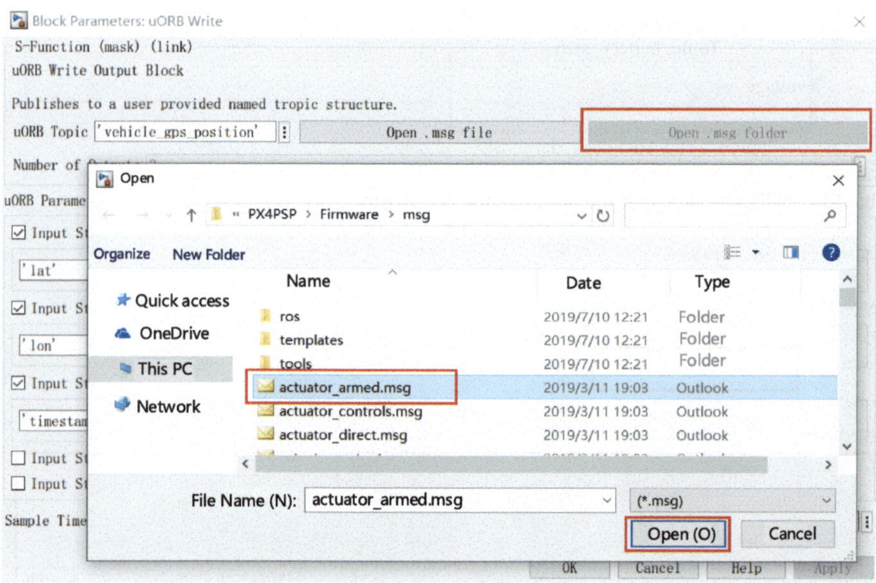

Fig. 3.30 Pop-up window of "Open .msg folder" button

In fact, all modules (PWM output, RGB_LED, etc.) mentioned in this section are implemented at the underlying code by reading and writing uORB messages. Theoretically, by using the "uORB Read and Write" modules, all messages and intermediate variables used in the PX4 autopilot can be accessed by Simulink. This simplifies the implementation of more advanced functions

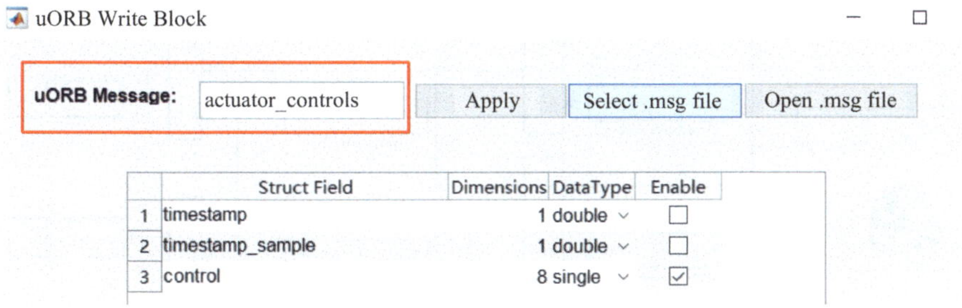

(a) uORB Write Advanced module. uORB MsgID is the name of ".msg" file.

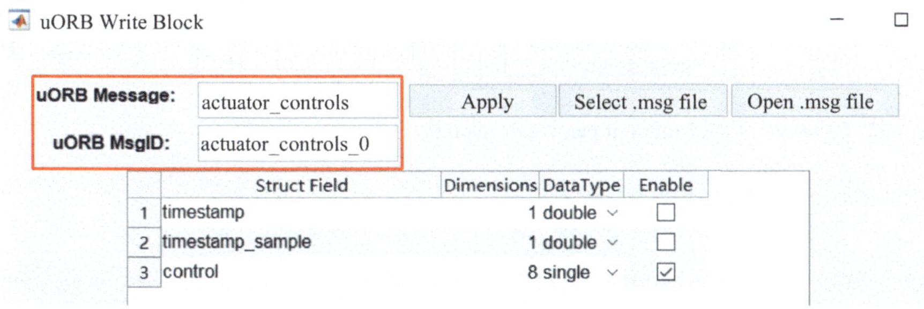

(b) uORB Write Advanced_dai module. uORB MsgID can be set different from the name of the ".msg" file.

Fig. 3.31 Advanced "uORB Write" modules and difference between them

for controller design. The PSP toolbox also provides two examples (see file "e0\2.PSPOfficialExps\px4demo_fcn_call_uorb_example.slx", and file "e0\2.PSPOfficialExps\px4demo_write_uorb_example.slx") to study this module.

In the PX4 development website, there are detailed documents for creating a new uORB message[6] and receiving a new MAVLink message[7] to communicate with external devices. In addition to the uORB modules presented in Fig. 3.27, it is convenient for the Simulink controller "px4_simulink_app" to exchange data with external devices, such as cameras, sensors, and host computers.

(11) Accessing PX4 internal parameters

For the sake of convenience for controller parameter tuning in flight tests, the PSP Toolbox also provides interfaces to access the PX4 internal parameters. In this way, the parameters of the controller generated by Simulink can be tuned online in the GCS software, instead of modifying the controller parameters in Simulink, generating code, and uploading the firmware file again. As shown in Fig. 3.32, an example of how to access the PX4 internal parameters is presented in file "e0\2.PSPOfficialExps\px4demo_Parameter_CSC_example.slx".

---

[6] http://dev.px4.io/master/en/middleware/uorb.html.

[7] http://dev.px4.io/master/en/middleware/mavlink.html.

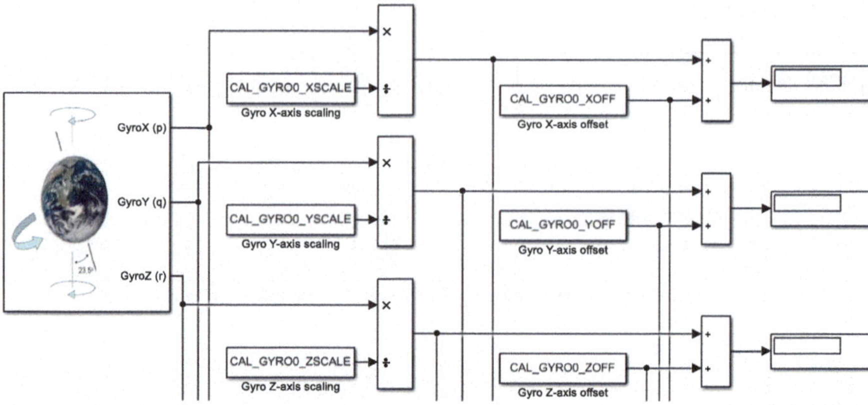

Fig. 3.32 Example of PX4 internal parameter reading

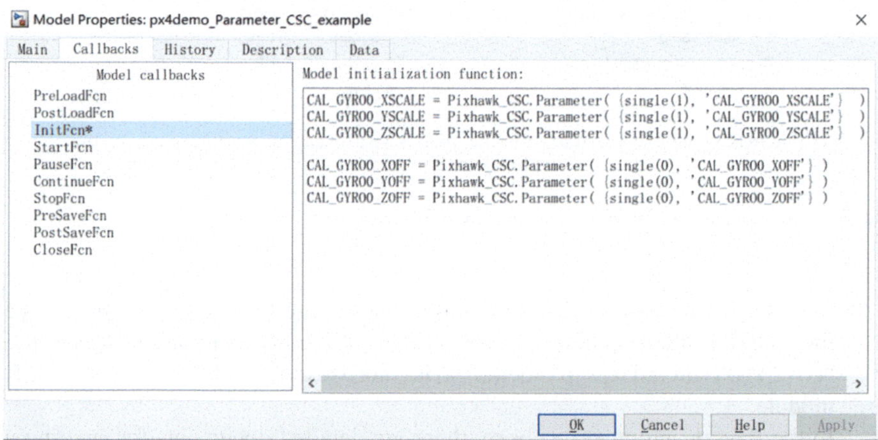

Fig. 3.33 Simulink initialization script for accessing PX4 parameters

PX4 internal parameter access is realized by using the function "Pixhawk_CSC.Parameter({*, *})", which needs to be called in the Simulink initialization function (click "Simple"—"Model Properties"—"Callbacks" - "InitFcn" in the Simulink menu bar). For the example shown in Fig. 3.32, the corresponding parameter initialization script is shown in Fig. 3.33.

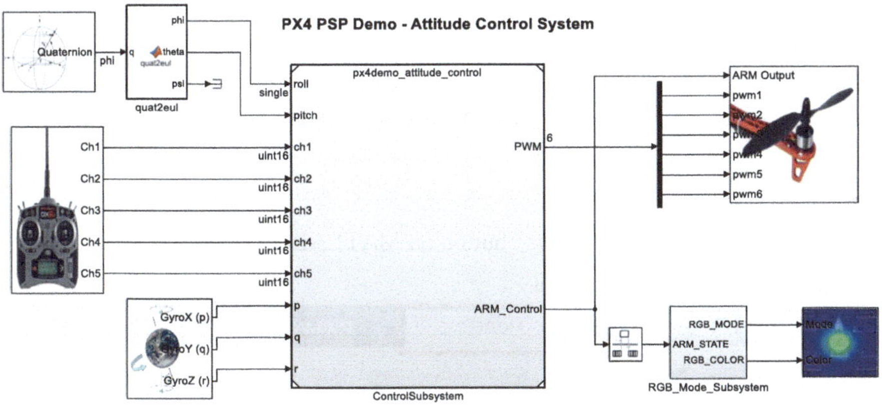

Fig. 3.34 Example of Simulink controller connecting with PSP modules

3.3.3 Simulink Configuration for Code Generation of PSP Toolbox

(1) Preparation of the Simulink controller for code generation
The preparation procedure is described below.

1) As shown in Fig. 3.7, design a controller in Simulink and verify it with SIL simulations.
2) Copy the verified controller to a new Simulink file.
3) Connect the input and output ports of the controller subsystem with the input (e.g., combined sensor module and RC input module) and output (e.g., PWM module and uORB modules) interface modules in the PSP module library presented in Fig. 3.15.
4) An example of the obtained Simulink controller file is presented in Fig. 3.34. The example file is available in "e0\2.PSPOfficialExps\px4demo_attitude _system.slx".

(2) Open the Simulink setting panel
The new created Simulink file must be configured to support the code generation function of the PSP toolbox. First of all, as shown in Fig. 3.35, the Simulink setting panel can be opened by clicking "Simulation"—"Model Configuration Parameters" in the Simulink menu bar.

(3) Setting for PSP code generation
As indicated in Fig. 3.36, go to the "Hardware Implementation" tab and select the "Pixhawk PX4" item in the pull-down menu of the "Hardware board" option. Then, all necessary parameter setting for PSP code generation is automatically configured.

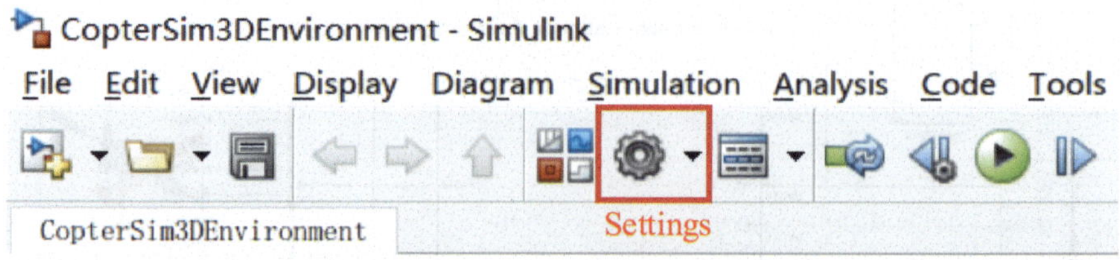

(a) Simulink "Settings" button on MATLAB 2017b - 2019a

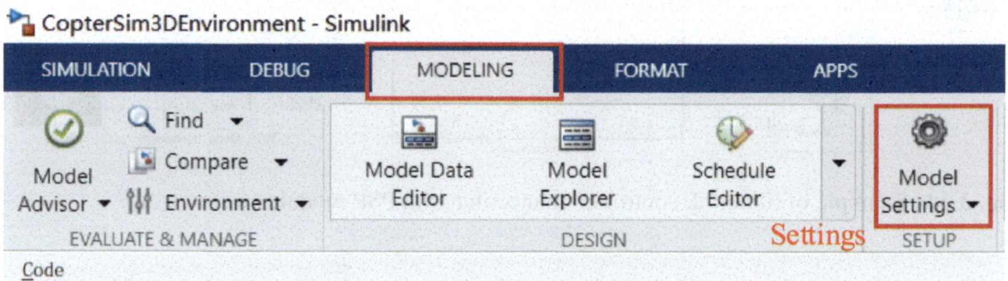

(b) Simulink "Settings" button on MATLAB 2019b and above

Fig. 3.35 Simulink "Settings" button for different MATLAB versions

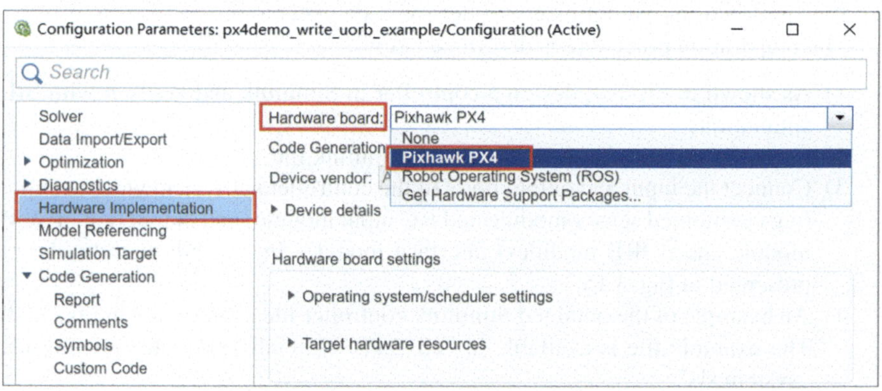

Fig. 3.36 Selecting target hardware

(4) Source code compilation and firmware generation
Click the "Build" button in Fig. 3.37 to convert the Simulink controller into C/C++ code and then compile it into the PX4 firmware. As shown in Fig. 3.38, the code generation and compiling process can also be observed by clicking the "Diagnostics" button on Simulink. A successful compiling process in the "Diagnostic Viewer" dialog is shown in Fig. 3.39, where the compiling process is finished with the following text "Successfully generated all binary outputs". It can also be observed in Fig. 3.39 that a "Code Generation Report" document will pop up after the compiling process is finished.

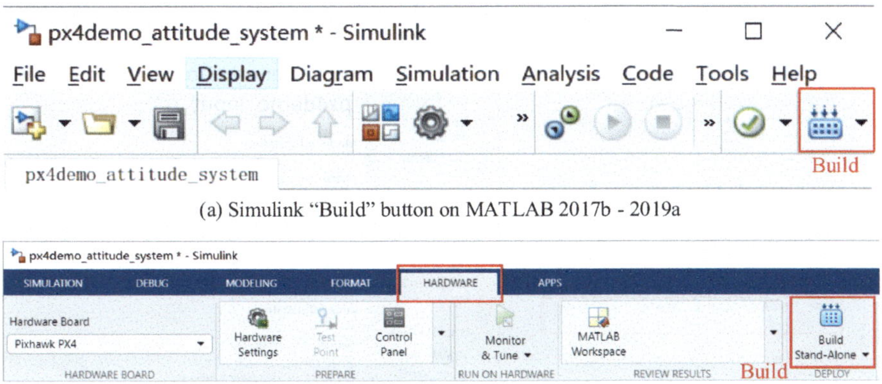

(a) Simulink “Build” button on MATLAB 2017b - 2019a

(b) Simulink “Build” button on MATLAB 2019b and above

Fig. 3.37 Simulink “Build” button for different MATLAB versions

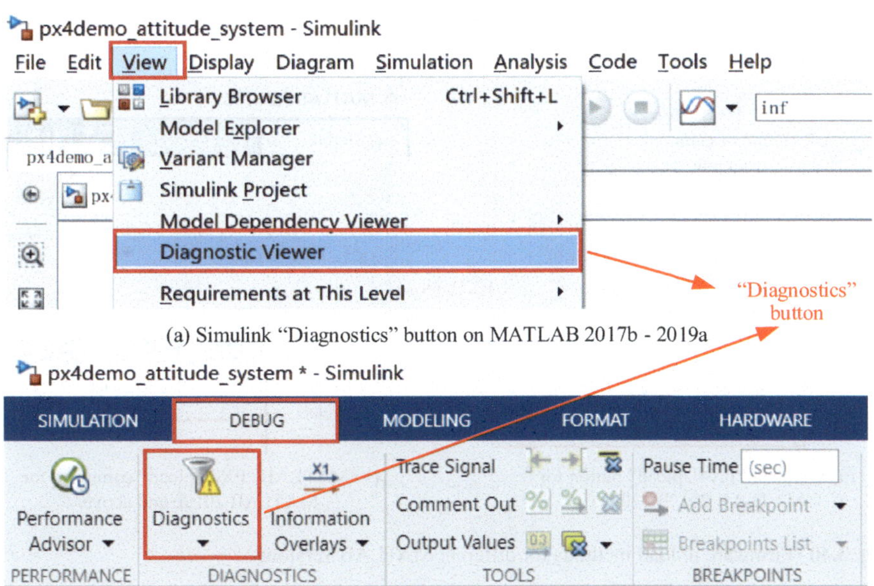

(a) Simulink “Diagnostics” button on MATLAB 2017b - 2019a

(b) Simulink “Diagnostics” button on MATLAB 2019b and above

Fig. 3.38 Simulink “Diagnostics” button for different MATLAB versions

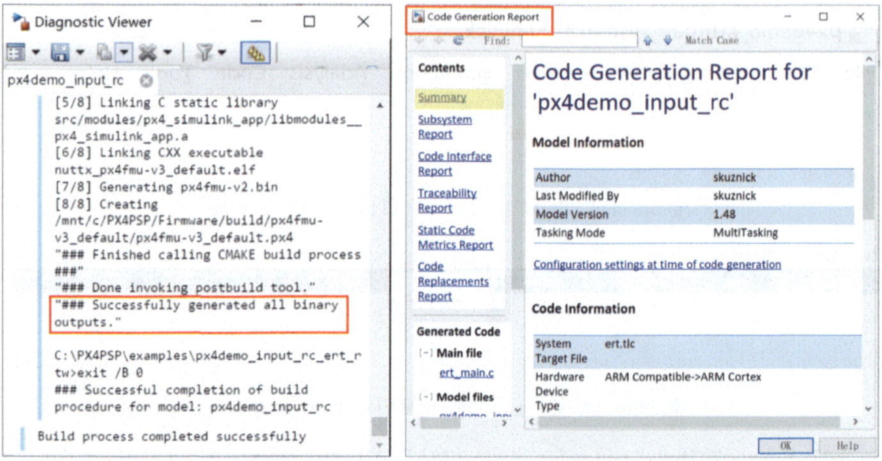

(a) Compilation progress (b) Code generation report

Fig. 3.39 Display dialog of code generation and firmware compiling

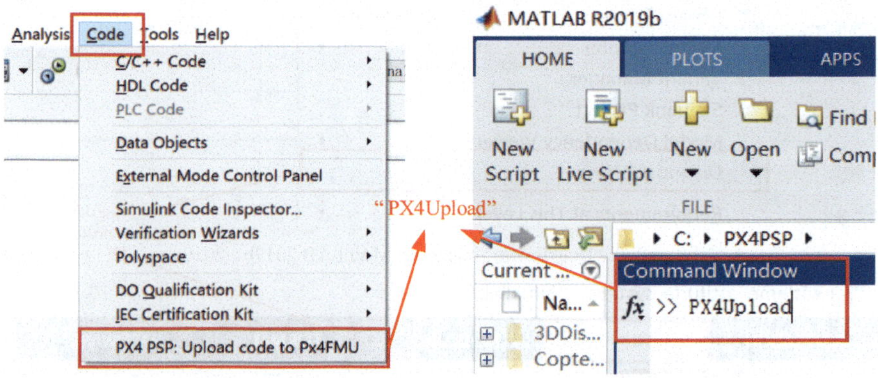

(a) Simulink "PX4Upload" button for MATLAB 2017b - 2019a

(b) MATLAB "PX4Upload" command for MATLAB 2019b and above

Fig. 3.40 Firmware upload methods for different MATLAB versions

(5) Upload PX4 firmware to Pixhawk hardware
Use the one-key upload function provided by the PSP toolbox to upload and burn the firmware to the Pixhawk hardware. The specific steps are described below.

1) Use a USB cable to connect the MicroUSB port (on the side of the Pixhawk hardware) with the USB port on the computer.
2) As shown in Fig. 3.40, for MATLAB 2017b–2019a, click "Code"—"PX4 PSP: Upload code to Px4FMU" on the Simulink menu bar, then the firmware will be automatically uploaded to the Pixhawk autopilot; for MATLAB 2019b and above, since the Simulink menu is deprecated, readers can input the

"PX4Upload" command in the "Command Window" of the MATLAB interface to upload the firmware .

3) Check the pop-up window carefully; sometimes, the Pixhawk autopilot has to be re-plugged to start the firmware uploading process.

After completing the above steps, the controller designed in Simulink has been run on the Pixhawk autopilot.

3.4 Pixhawk Hardware System

3.4.1 Hardware System Composition and Connection

As shown in Fig. 3.41, the hardware components required by this book include an RC transmitter, an RC receiver, a JR signal cable (connecting the Pixhawk autopilot and the RC receiver), a Pixhawk autopilot (Pixhawk 1 is recommended for studying, and higher hardware versions are recommended for outdoor flight tests), and a MicroUSB cable (connecting the computer and the Pixhawk hardware for power supply and data transmission). The connection relationships among the above components are presented in Fig. 3.42.

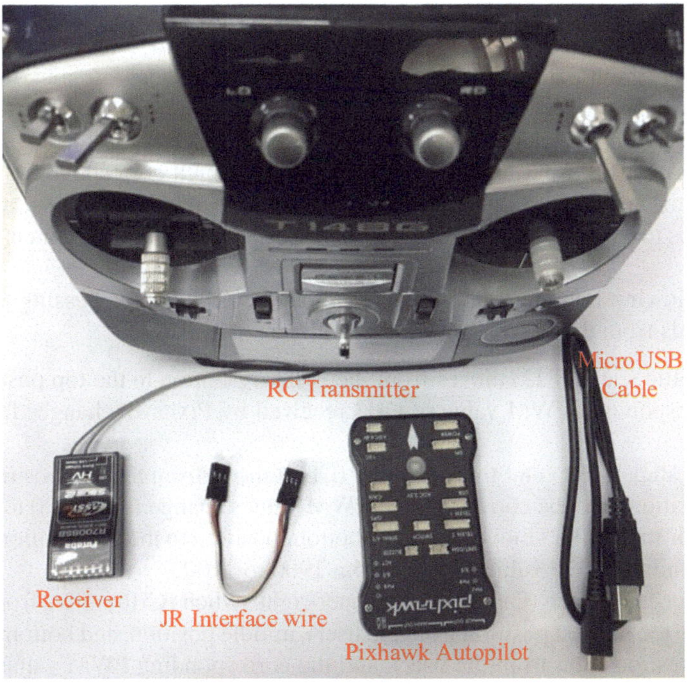

Fig. 3.41 Pixhawk autopilot, RC transmitter, and RC receiver

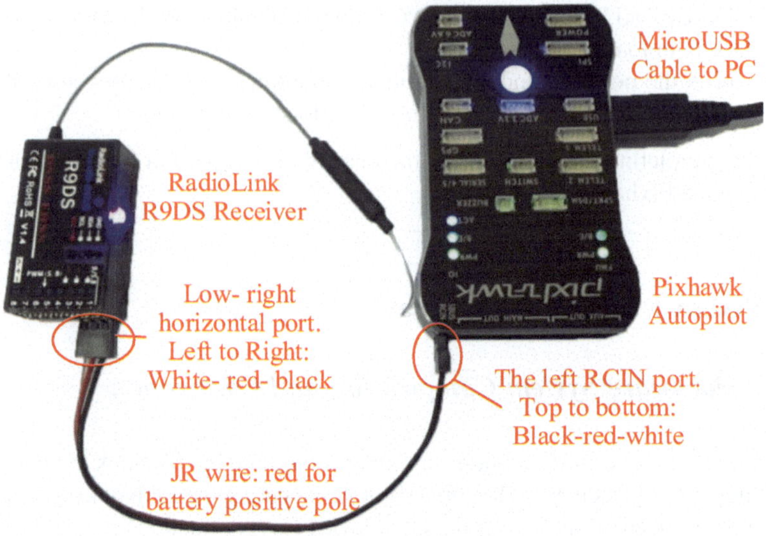

Fig. 3.42 Connection between Pixhawk hardware and RadioLink receiver

3.4.2 Basic Operation Method for RC Transmitter

The RC transmitter used in this book should be set to "Mode 1". As illustrated in Fig. 3.43, the throttle and yaw channels are controlled by the left stick on the RC transmitter, whereas the roll and pitch channels are controlled by the right stick. The roll, pitch, throttle, and yaw channels correspond to the CH1 to CH4 of the RC receiver respectively; the upper-left three-position switch corresponds to CH5 for triggering the autopilot to switch flight modes; the upper-right three-position switch corresponds to CH6 for triggering the autopilot to switch flight modes or enable other functions.

The following relationships should be remembered when processing the PWM input signals from the RC system for designing a controller.

- The throttle stick (CH3) moves from the bottom position to the top position, and the corresponding PWM value of CH3 received by Pixhawk changes from 1100 to 1900;
- The roll stick (CH1) and the yaw stick (CH4) move from the left position to the right position, and the corresponding PWM values change from 1100 to 1900;
- The pitch stick (CH2) moves from the bottom position to the top position, and the corresponding PWM value changes from 1900 to 1100;
- The upper-left switch (CH5) and the upper-right switch (CH6) move from the top position (the farthest position from the user), middle position, and bottom position (the closest position from the user), and the corresponding PWM values change from 1100, 1500, to 1900.

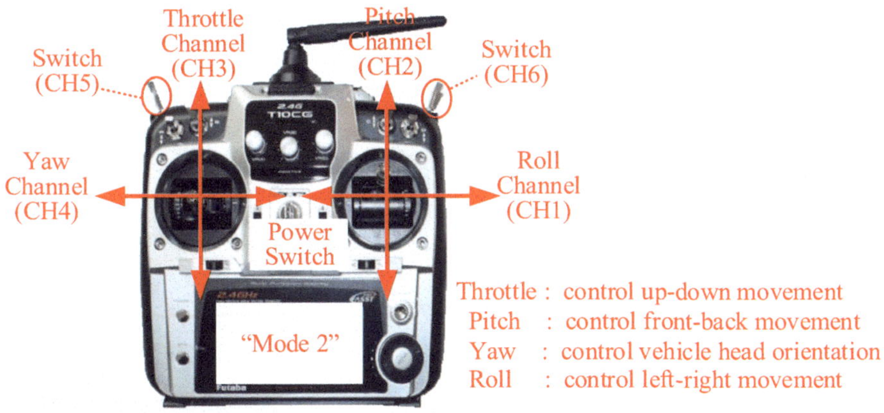

Fig. 3.43 RC transmitter channels

3.4.3 Method for Uploading Firmware Through QGC

Find the QGroundControl icon on the desktop and open the QGC software,[8] the software UI is presented in Fig. 3.44. Then, the following procedure presents the way to upload a PX4 firmware file to the Pixhawk hardware through QGC.

(1) Click the "Settings" button (the gear icon on the toolbar in Fig. 3.44) to enter the QGC setting page.
(2) Click the "Firmware" tab, and then connect the Pixhawk hardware with a USB cable. The QGC will automatically detect the Pixhawk autopilot and pop up the right tap, as shown in Fig. 3.44.
(3) Click the "Advanced settings" checkbox.
(4) Click on the "Standard Version (stable)" tab.
(5) Select the "Custom firmware file .." option in the pop-up menu.
(6) Click the "OK" button.
(7) As shown in Fig. 3.45, select file "e0\2.PSPOfficialExps\px4fmu-v3_default1.7.3Stable.px4" in the pop-up file selection window, and click the "Open" button. Then, the QGC will upload and burn the firmware into the Pixhawk hardware. Note that, the .px4 firmware files for other Pixhawk hardware can be found in the folder "e0\4.PX4Firmwares".

3.4.4 Pixhawk Setting for HIL Simulation Mode

Open the QGC software, and connect the Pixhawk autopilot with a USB cable. Then, QGC will automatically recognize the Pixhawk autopilot and create a connection for

[8]http://qgroundcontrol.com.

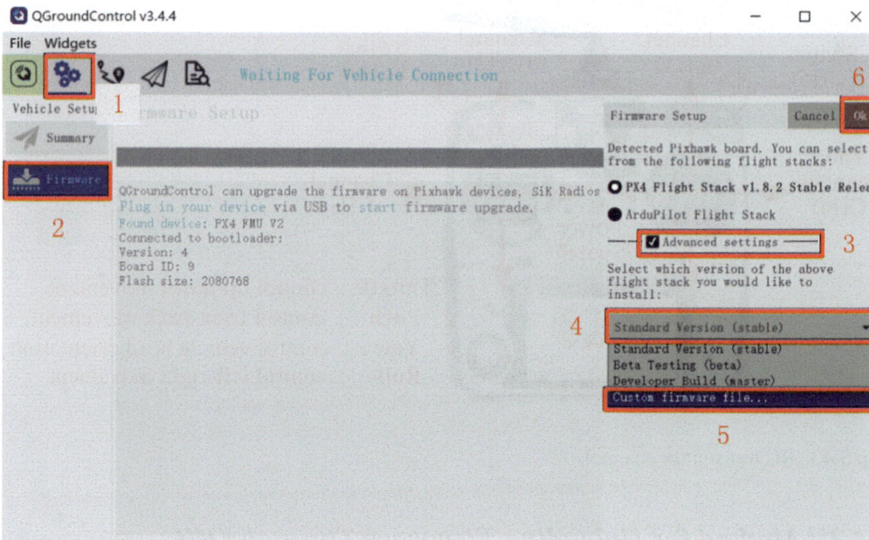

Fig. 3.44 QGC firmware upload interface

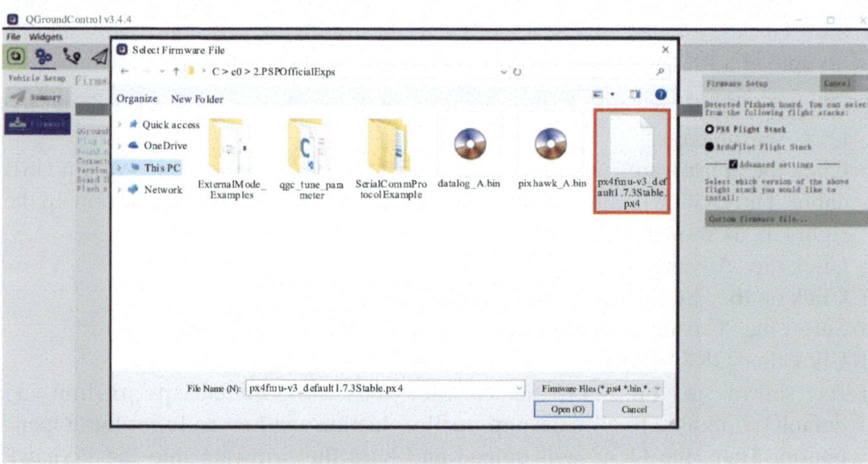

Fig. 3.45 QGC custom firmware selection page

parameter setting and data transmission. Fig. 3.46 presents the UI of QGC after connecting with the Pixhawk autopilot.

Click the "Airframe" tab (the third item on the left side in Fig. 3.46) on the QGC setting page to confirm that the "HIL Quadcopter X" airframe mode (see Fig. 3.47) is selected by default. This setting is critical for subsequent HIL simulation. Otherwise, a manual setup will be required. The airframe setting method is presented below.

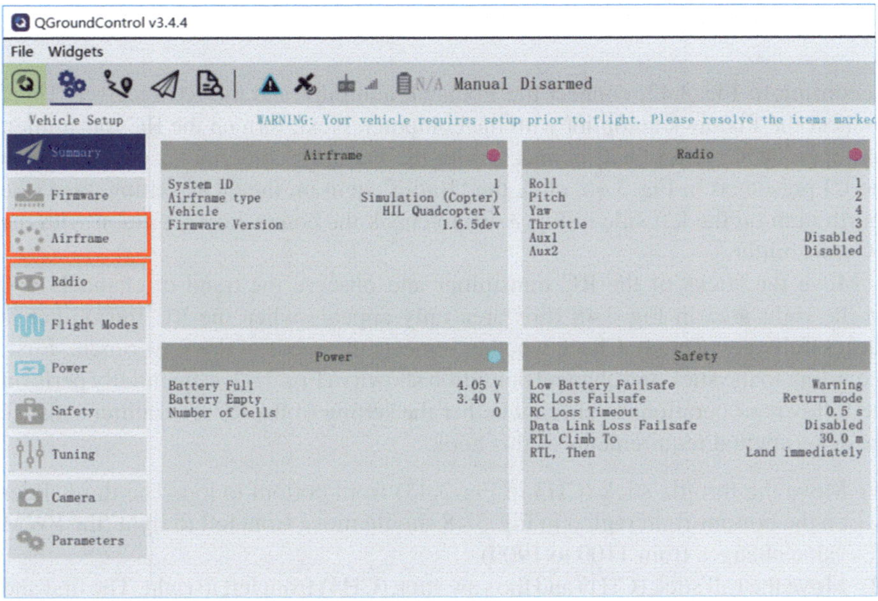

Fig. 3.46 Pixhawk autopilot connected to QGC

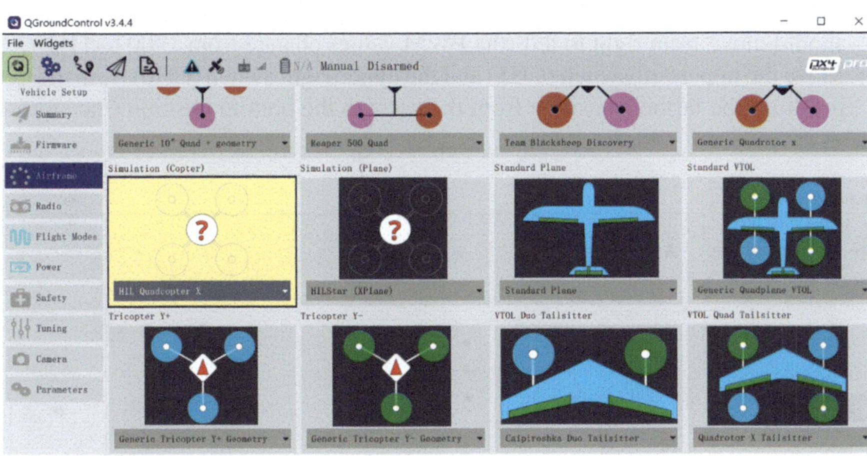

Fig. 3.47 QGC Airframe setting page

(1) Select the "HIL Quadcopter X" airframe icon in Fig. 3.47.
(2) Click the "Apply and Restart" button in the upper right corner of the UI. Then, the autopilot will restart to make the new airframe available.
(3) Wait for a few seconds; QGC will connect to the autopilot again and will check whether the setting in Fig. 3.47 is correctly established.

3.4.5 RC Transmitter Configuration and Calibration

According to Fig. 3.42, connect the Pixhawk autopilot and the RC receiver. Then, connect the Pixhawk autopilot with the computer. Next, turn on the RC transmitter, and open QGC. After QGC connects with the Pixhawk autopilot successfully with the UI presented in Fig. 3.46, click the "Radio" item on the QGC setting page (the fourth item on the left side in Fig. 3.46) to check the connection condition with the RC transmitter.

Move the sticks of the RC transmitter and observe the trend of channels 1–6 on the right area in Fig. 3.48 (this area only appears when the RC transmitter is successfully connected). Check whether it meets the needs of this experiment. First, according to the stick and channel definition shown in Fig. 3.43, sequentially perform the following operations to check whether the setting of the RC transmitter satisfies the experimental requirements of this book.

(1) Move the throttle stick (CH3 in Fig. 3.43) from bottom to top. The third slider on the bottom-right region in Fig. 3.48 should move from left to right (the PWM value changes from 1100 to 1900).
(2) Move the roll stick (CH1) and the yaw stick (CH4) from left to right. The first and the fourth sliders in Fig. 3.48 should move from left to right (the PWM values change from 1100 to 1900).
(3) Move the pitch stick (CH2) from bottom to top. The second slider in Fig. 3.48 should move from right to left (the PWM values change from 1900 to 1100).
(4) Move the upper-left switch (CH5) and the upper-right switch (CH6) from the top position (the farthest position from the user) to the bottom position (the closest

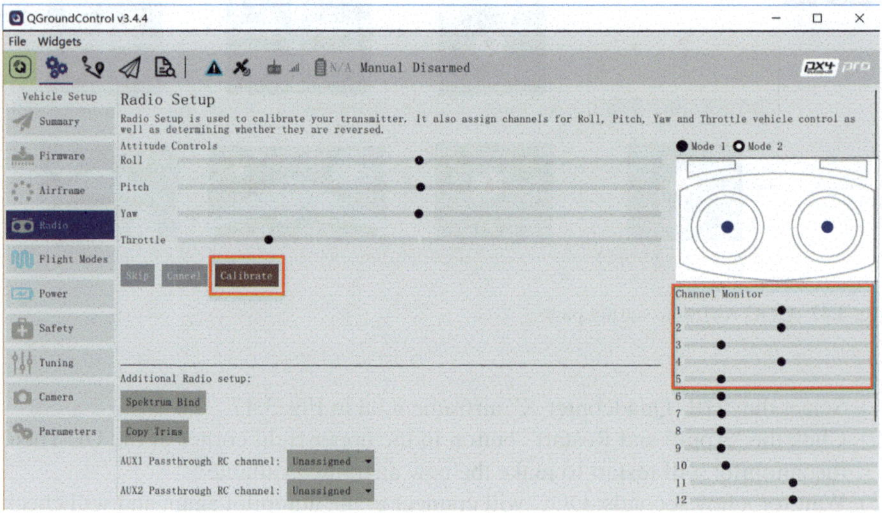

Fig. 3.48 QGC RC transmitter setting and calibration page

position from the user). The fifth and the sixth sliders in Fig. 3.48 should move from left to right (the PWM values change from 1100 to 1900).

If the above rules are not satisfied, it means that the RC transmitter is not set correctly, so the RC transmitter should be re-configured by the methods presented in Sect. 2.3.1. Besides, if QGC prompts a warning message saying that the RC transmitter is not calibrated, click the "Calibrate" button in the middle in Fig. 3.48 and complete the RC calibration by moving the sticks according to the instructions on QGC.

3.4.6 *Flight Mode Settings*

After the RC transmitter is successfully calibrated, enter the "Flight Modes" setting page (see Fig. 3.49) and select "Mode Channel" as the previously tested CH6 channel. Since the CH6 channel is a three-position switch, the top position (the farthest position from the user), middle position, and bottom position (the closest position from the user) of the switch correspond to "Flight Mode 1, 4, 6" in Fig. 3.49, respectively. Also, according to Fig. 3.49, associate these three modes to "Stabilized" (the stabilized mode, only including attitude control), "Altitude" (the altitude hold mode, including attitude and altitude control), and "Position" (the loiter mode, including attitude and position control).

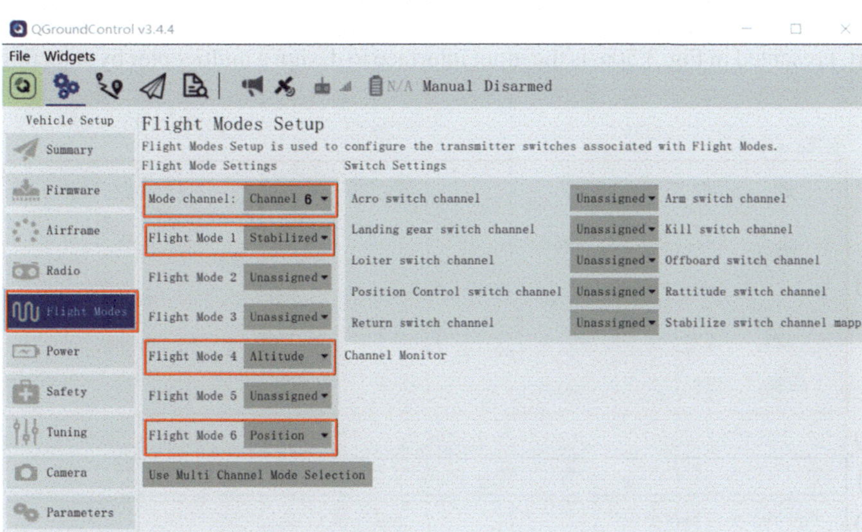

Fig. 3.49 Flight mode setting page in QGC

3.5 HIL Simulation Platform

The HIL simulation platform includes a Real-time Motion Simulation Software—*CopterSim* and a 3D Visual Display Software—*3DDisplay*.

3.5.1 CopterSim

Double-click the CopterSim shortcut on the Windows desktop to open the CopterSim software, whose UI is presented in Fig. 3.50. The default simulation model and parameters are the same as for the Simulink multicopter model used in the SIL simulation system (see Fig. 3.1). This is because the CopterSim is developed based on the code generation technique with the Simulink multicopter model. CopterSim needs to run on a 64-bit Windows computer platform with a serial port and a MicroUSB cable to communicate with the Pixhawk autopilot (see Fig. 3.42).

CopterSim sends sensor data to the Pixhawk autopilot, and then the autopilot solves the motor PWM control signal and returns it to CopterSim. As a result, the Pixhawk autopilot can perform real-time control on the simulated multicopter in CopterSim, as well as control a real multicopter. Meanwhile, CopterSim will send the attitude and position information of the multicopter to the local network through the UDP protocol, and the 3DDisplay receives the multicopter flight information to complete the corresponding real-time 3D scene display.

As shown in Fig. 3.50, the UI of CopterSim is divided into two parts. The upper part, presented in Fig. 3.50a, is the input interface to design a multicopter by selecting

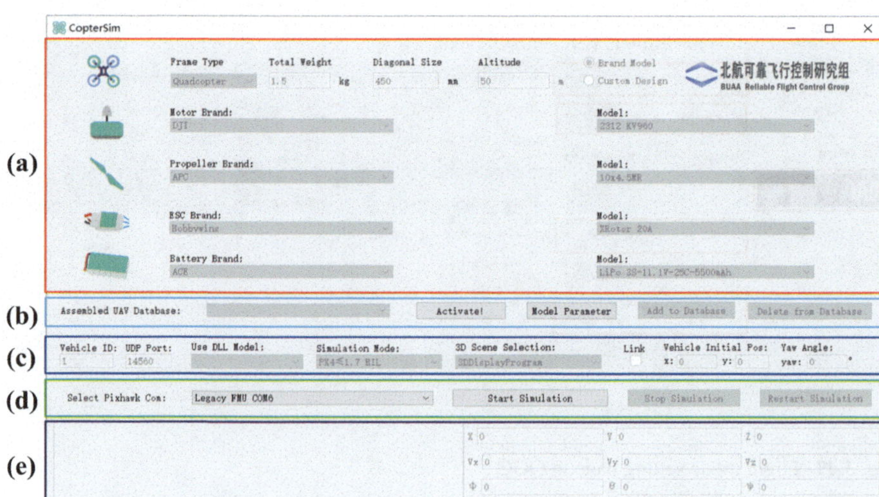

Fig. 3.50 CopterSim main UI

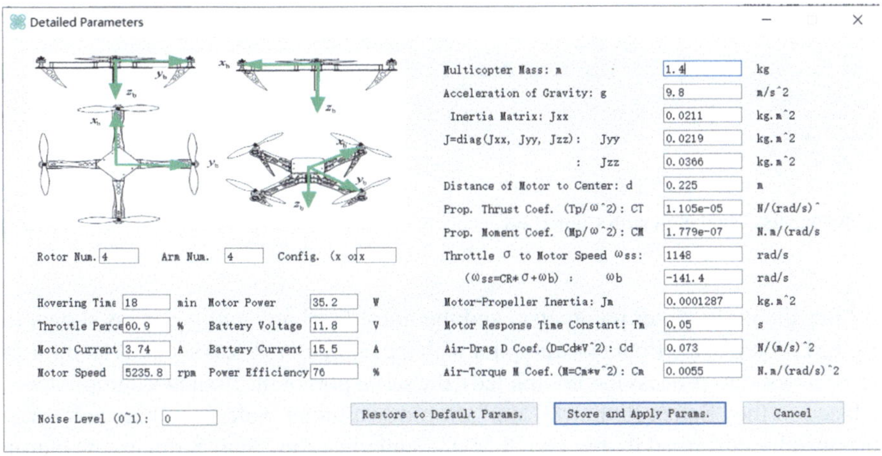

Fig. 3.51 Model parameter configuration dialog

popular components on the market. The lower part presented in Figs. 3.50b–e is the interface to connect with the autopilot for HIL simulation. Note that CopterSim enables by default only the basic functions required by this book. Registration is required to use many other practical functions, such as swarm simulation, high-fidelity UE4 scenes, and HIL simulations for other aerial vehicles (e.g., fixed-wing aircraft). Please, see Appendix A for more information.

Click the "Model Parameter" button in the middle of the CopterSim UI in Fig. 3.50b. The model parameter configuration dialog in Fig. 3.51 will pop up; the model parameters stored in the previous simulation will be displayed here. The parameter dialog in Fig. 3.51 mainly includes two parts: the hover information (hover endurance, throttle, output power, motor speed, etc.) and the basic multicopter parameters (total mass, the moment of inertia, size, thrust coefficient, and drag coefficient). Clicking the "Restore to Default Params" button on the dialog in Fig. 3.51 will restore the model parameters to the default values; clicking the "Save and Apply Params" button will store the current parameters to the database for subsequent HIL simulations.

CopterSim also allows readers to directly modify the model parameters on the right page of Fig. 3.51. For example, enter the same parameters as the multicopter model used in Simulink SIL simulations (the parameters are stored in file "e0\1.SoftwareSimExps\icon\Init.m"). Then, click the "Store and Apply parameters" button in Fig. 3.51 to store and apply the model parameters. The "noise level (0–1)" in Fig. 3.51 allows selecting the noise level of the simulated sensors, where "0" denotes that the sensor noise is not enabled, and "1" denotes that the noise level is consistent with the real Pixhawk autopilot. A noise level between 0–1 or larger than one can also be selected to represent the noise level of actual sensors. This enables the possibility of testing the anti-interference ability of the designed control algorithms.

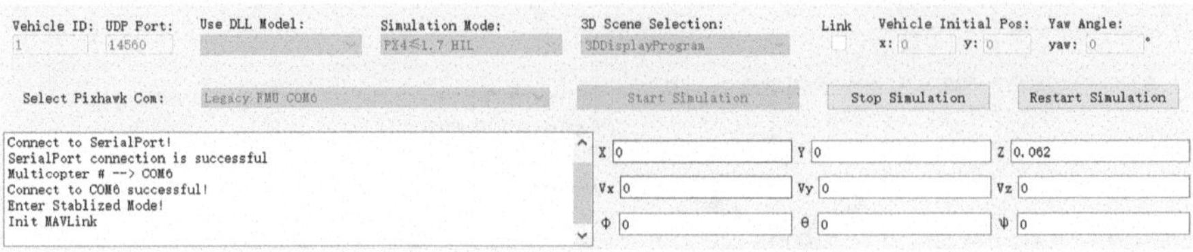

Fig. 3.52 HIL simulation with CopterSim

After the multicopter parameters and the noise level are configured, as shown in Fig. 3.51, connect the Pixhawk autopilot with the computer. A few seconds later (after the Pixhawk start process has completed), the serial port of the Pixhawk autopilot will be listed in the "Select Pixhawk Com" drop-down menu. Select the Pixhawk serial port (usually described by the text "FMU"), and click the "Start Simulation" button to start the HIL simulation. As shown in Fig. 3.52, the messages from the Pixhawk are printed on the CopterSim UI, which indicates that the HIL simulation is running correctly. When the Pixhawk LED turns green or the text "** EKF initialization finished" is observed on CopterSim (the estimation system is initialized), you can use RC to arm the drone and takeoff. During the HIL simulation process, clicking the "Stop Simulation" button will stop the HIL simulation, and clicking the "Restart Simulation" will re-initialize the multicopter position and states to their initial values.

3.5.2 3DDisplay

Double-click the 3DDisplay shortcut on the Windows desktop to open the 3DDisplay software. As shown in Fig. 3.53, the "3D Scene Viewer" on the left side of the 3DDisplay UI presents the current flight status of the multicopter in the 3D scene. The basic flight parameters are displayed in the upper right window of the 3DDisplay UI, including motor speed, position, and attitude information. The flight trajectory of the multicopter is displayed on the lower right window of the 3DDisplay UI.

3.5.3 Flight Tests with HIL Simulation Platform

In the HIL simulation platform, when controlling a real multicopter, it is convenient to control the simulated multicopter with a real RC transmitter to perform basic actions, such as arming, taking off, manual flight, landing, etc. The detailed steps are described next.

(1) Push up the POWER switch to turn on the RC transmitter.
(2) Correctly connect the computer with the Pixhawk hardware system (including the Pixhawk autopilot and the RC receiver) and start the HIL simulation in CopterSim according to the procedure mentioned above.

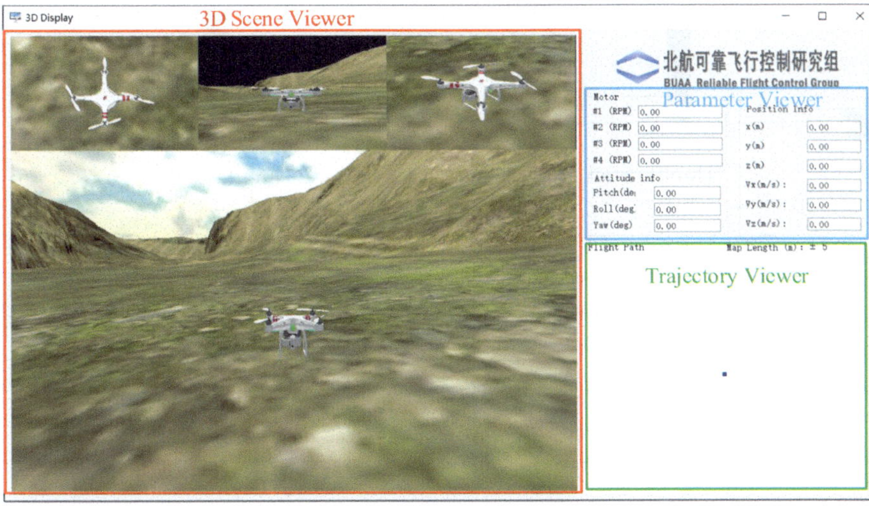

Fig. 3.53 User interface of 3DDisplay

(3) As shown in Fig. 3.54a, arm the Pixhawk autopilot by moving the left-hand stick on the RC transmitter (CH3) to the lower-right corner for 2–3 s.

(4) Pixhawk is successfully armed when its LED turns from slow flashing to always on,[9] and the CopterSim print message "Detect Px4 Armed" is received from Pixhawk. If arming Pixhawk fails, please disconnect all hardware and software and repeat the above steps.

(5) Pull up the left-hand stick on the RC transmitter (CH3) for the multicopter to take off and fly up to a certain altitude. Next, vertically move the left-hand stick to verify the vertical motion control of the multicopter.

(6) Horizontally move the left-hand stick on the RC transmitter (CH4) to verify the yaw angle motion control of the multicopter.

(7) Vertically move the right-hand stick on the RC transmitter (CH2) to verify the pitch angle control as well as the forward and backward motion control of the multicopter.

(8) Horizontally move the right-hand stick on the RC transmitter (CH1) to verify the roll angle control as well as the left and right motion control of the multicopter.

(9) Change the position of the top-right switch on the RC transmitter (CH6) to verify the mode switching control of the multicopter.

(10) Pull down the left-hand stick on the RC transmitter (CH3) to land the multicopter to ground.

(11) As shown in Fig. 3.54b, move the left-hand stick on the RC transmitter (CH3) to the lower-left corner for 2–3 s to disarm the Pixhawk.

[9] Higher Pixhawk hardware (e.g., Pixhawk 2/3/4/5) starts to discard LED module, so an external I2C LED module is required to observe the lighting effect.

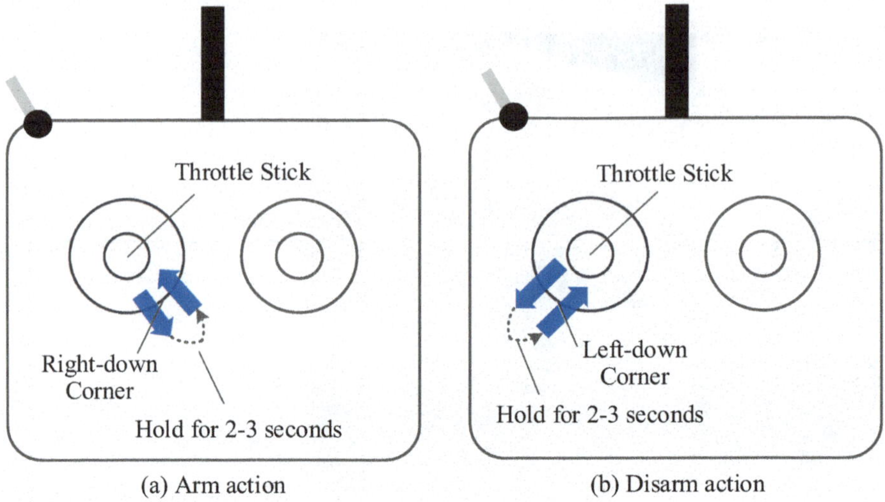

Fig. 3.54 Arm and disarm of Pixhawk autopilot through RC transmitter

(12) Click the "Stop Simulation" button on the CopterSim UI to stop the HIL simulation. Then, disconnect all software and hardware connections between the computer and Pixhawk.

If you have any question, please go to https://rflysim.com for your information.

Chapter 4
Experimental Process

This chapter first introduces the experimental process from a general perspective and then describes two demonstrative experiments to introduce the usage method of the experimental platform as well as the basic operation process of the experiments. These demonstrative experiments consist of LED light control and multicopter attitude control. The objective of the LED light experiment is to make readers familiar with the basic code generation process of the experimental platform, including designing light control algorithms in Simulink, generating code and firmware for the Pixhawk hardware, and finally observing the lighting control effects of LED light on Pixhawk. After the LED light control experiment, the attitude control experiment shows how to perform the HIL simulation and outdoor flight.

4.1 Experimental Process

The experimental courses covered in the following chapters contain theory and techniques for multicopter design and control practice. They include propulsion system design, modeling, filtering, control, and decision-making. To ensure a progressive learning process, each experimental course includes three step-by-step experiments, i.e., a basic experiment, an analysis experiment, and a design experiment.

(1) *Basic experiment.* Open the given code example, read, and run its source code to observe, record, and analyze the results.
(2) *Analysis experiment.* Modify the given code example and then run it to collect and analyze the data.
(3) *Design experiment.* Based on the above two experiments, complete the given design task independently.

© Publishing House of Electronics Industry 2020, corrected publication 2021

Q. Quan et al., *Multicopter Design and Control Practice*,
https://doi.org/10.1007/978-981-15-3138-5_4

For the basic experiments and the analysis experiments, this book provides source code examples to ensure that all readers can complete them easily. Through the above two experiments, readers will acquire a deep understanding of the theoretical and practical methods. In design experiments, readers will independently design and verify algorithms by referring to the code examples offered in basic experiments and analysis experiments. The three step-by-step experiments constitute a learning ladder from shallow to deep, which is convenient for readers to reach the final experimental goal.

Each experiment generally includes the following three phases: the *Simulink-based Algorithm Design and SIL Simulation Phase*, the *HIL Simulation Phase*, and the *Flight Test Phase*. The overall experimental process is illustrated in Fig. 4.1.

(1) *Simulink-based Algorithm Design and SIL Simulation Phase*
The entire phase is performed in the MATLAB environment. Readers can design control algorithms in Simulink based on the given multicopter simulation model. The input and output ports of the designed controller and the multicopter simulation model should be correctly connected to ensure that the data transmission is consistent with the real multicopter control system. Similar to a real multicopter system, the multicopter simulation model sends sensor data or estimated states (e.g., attitude, angular rate, position, and speed) to the controller, and the controller sends back the motor PWM control signals to the multicopter simulation model. This results in a closed-loop HIL simulation system. In this phase, readers can analyze the control performance by observing the simulation results, and then modify the controller or re-design it according to the desired performance requirement.

(2) *HIL Simulation Phase*
In this phase, readers will import the multicopter model from Simulink to CopterSim, upload the control algorithms from Simulink to a real Pixhawk with the code generation technology, and replace the Simulink virtual data connection with a real USB cable. CopterSim sends sensor data (e.g., accelerometer, gyroscope, GPS, barometer and electronic compass) to the Pixhawk system via the USB cable; the PX4 autopilot software in the Pixhawk system filters the sensor data and estimates the multicopter states; next, it sends the estimated multicopter states to the designed controller via the internal uORB message bus; the controller computes the motor PWM control signals and sends them back to the CopterSim via the USB cable to establish a closed-loop HIL simulation system. Compared with a SIL simulation system, the multicopter simulation model in CopterSim is running with the same clock period as the real world to ensure the real-time requirement, and the control algorithms are running in a real embedded system, namely the Pixhawk autopilot, which is closer to a practical multicopter system. Note that the data transmission delay is inevitable in a real communication system; and the operating environments of the simulation model, as well as the control algorithms in the HIL simulation system, are difficult to be consistent with the SIL simulation system. Therefore, the controller parameters may need

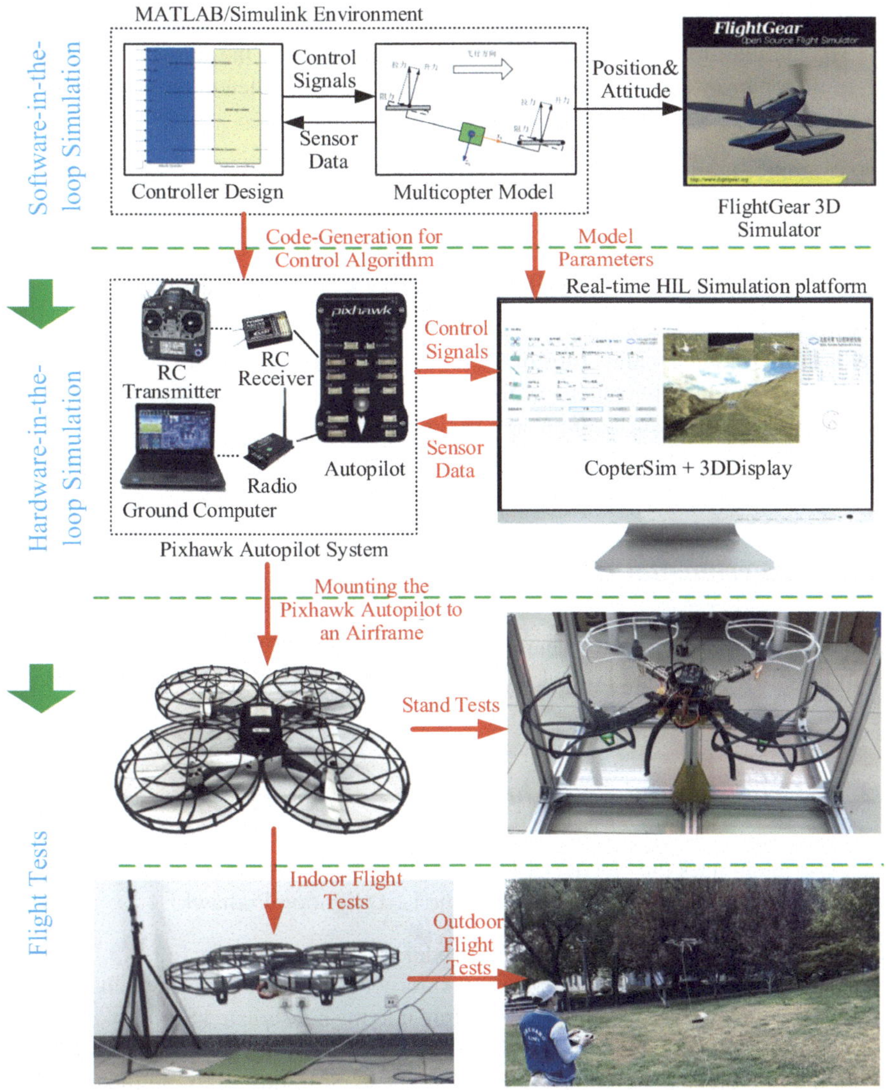

Fig. 4.1 Full experimental process

to be slightly tuned based on the simulation results. In any case, this is consistent with the system development process of a real multicopter system.

(3) *Flight Test Phase*
In this phase, the multicopter simulation model in CopterSim is further replaced with a real multicopter hardware system; the sensor data are replaced with signals from real sensor chips by sensing the multicopter motion and states and the control signals are directly output to the Pixhawk I/O ports to control the motors.

Note that it is difficult to ensure that the simulation model used in both the SIL simulation and the HIL simulation is identical to the dynamics of a real multicopter system. Consequently, the controller parameters may need further slight tuning according to the experimental results.

Two examples are in the following to familiarize the reader with the entire process.

4.2 Experimental Procedure for LED Control Experiment

The flashing frequency and color of the LED on the Pixhawk hardware can be controlled by designing controller in Simulink. This section takes a simple LED light control experiment[1] as an example to introduce the operation process of the hardware and software components of the experimental platform released with this book.

4.2.1 Experimental Objective

As shown in Fig. 4.2, use two channels from CH1 to CH5 of the RC transmitter to control the LED light on Pixhawk in two different colors and two different modes.

4.2.2 Experimental Procedure

The experiment uses the "RGB_LED" module in the PSP toolbox introduced in the previous chapter (see Fig. 4.3) to control the LED light on Pixhawk.

(1) Create and open a new Simulink model file. As shown in Fig. 4.3, find the "RGB_LED" module in the "Pixhawk Target Blocks" toolbox, which is in the "Library Browser" of Simulink, and drag it to the new created Simulink file. The RC transmitter module "input_rc" in Fig. 4.3 is also dragged into the Simulink as the input signals to control the LED light. A Simulink example file completing the whole process of configuration is available in "e0\2.PSPOfficialExps\px4demo_input_rc.slx".
(2) The RBG_LED module can be used to control the mode and color of the LED light on Pixhawk. Its mode enumeration variable "RGBLED_MODE_ENUM" includes the following seven options

- SL_MODE_OFF
- SL_MODE_ON

[1]Higher Pixhawk hardware (e.g., Pixhawk 2/3/4/5) starts to discard LED module, so an external I2C LED module is required to observe the experiment result.

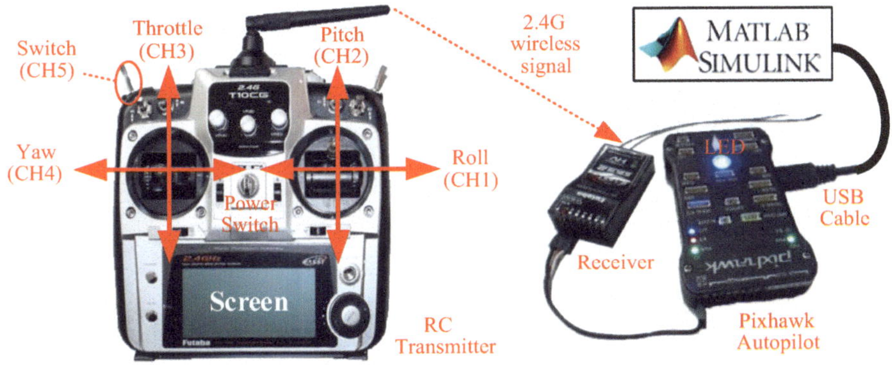

Fig. 4.2 Hardware connection of LED control experiment

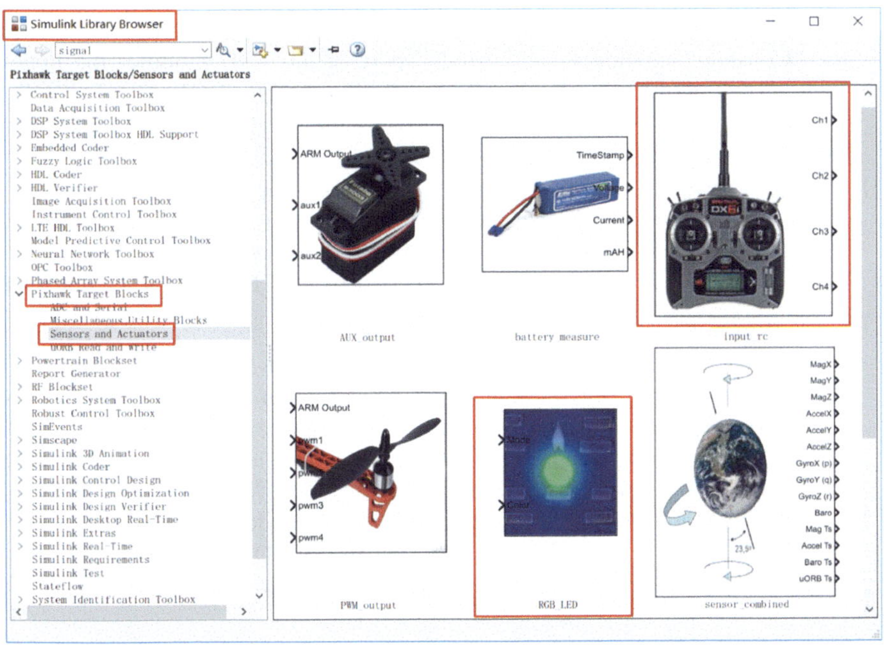

Fig. 4.3 LED output interface model for PSP toolbox

- SL_MODE_DISABLED
- SL_MODE_BLINK_SLOW
- SL_MODE_BLINK_NORMAL
- SL_MODE_BLINK_FAST
- SL_MODE_BREATHE.

The color variable "RGBLED_COLOR_ENUM" has the following options

- SL_COLOR_OFF
- SL_COLOR_RED
- SL_COLOR_GREEN
- SL_COLOR_BLUE
- SL_COLOR_YELLOW
- SL_COLOR_PURPLE
- SL_COLOR_AMBER
- SL_COLOR_CYAN
- SL_COLOR_WHITE.

The above variables have been registered as MATLAB global parameters when installing the PSP toolbox so that they can be directly applied. For example, an LED light with blue color and fast blinking mode can be obtained by sending the variable "RGBLED_MODE_ENUM.SL_MODE_BLINK_FAST" with the "Constant" module to the "Mode" port of the LED module (the upper input port of the LED module), and sending the variable "RGBLED_COLOR_ENUM.SL_COLOR_BLUE" with the "Constant" module to the "Color" port (the lower input port of the LED module).

(3) Controller design. According to the introduction to the RC system provided in the Sect. 2.3.1, the data range of the PWM signals received by Pixhawk is 1100–1900. As shown in Fig. 4.4, we will use two channels of the RC transmitter in this experiment to control the mode and color of the LED light. The controller design procedure is described next.

1) Use the CH3 channel of the RC transmitter (see the throttle channel in Fig. 4.2) to change the blink mode of the LED light. When the throttle stick is turned to the upper side, i.e., when the PWM value of the CH3 channel is higher than 1500 microseconds (abbreviated as CH3>1500), then the "Mode" port receives an "RGBLED_MODE_ENUM.SL_MODE_BLINK_FAST" variable corresponding to a fast blinking mode; when CH3≤1500, the "Mode" port receives the "RGBLED_MODE_ENUM.SL_MODE_BLINK_NORMAL" variable corresponding to a slow blinking mode.

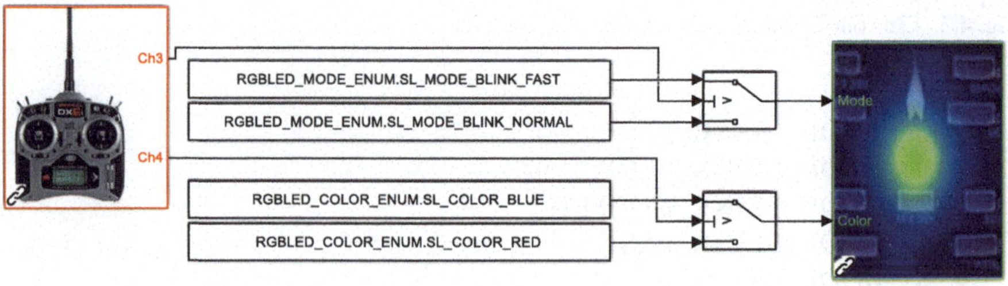

Fig. 4.4 LED controller with an RC transmitter

2) Use the CH4 channel of the RC transmitter to change the color of the LED light. When CH4>1500, the "Color" port receives an "RGBLED_COLOR_ENUM.COLOR_RED" variable corresponding to red color; when CH4≤1500, the "Color" port receives an "RGBLED_COLOR_ENUM.COLOR_BLUE" variable corresponding to blue color.

4.2.3 Controller Code Generation and Firmware Uploading

(1) For MATLAB 2017b–2019a, as shown in Fig. 4.5, click the "Simulation"—"Model Configuration Parameters" option on the Simulink menu bar to enter the Simulink setting dialog; for MATLAB 2019b and above, click the "Settings" button. The obtained Simulink setting window is presented in Fig. 4.6.
(2) Select the target hardware. As shown in Fig. 4.7, set the option "Hardware Implementation"—"Hardware Board" to "Pixhawk PX4".
(3) Compile the model. As shown in Fig. 4.8, compile the designed controller into the PX4 firmware file by clicking the "Build" button on the Simulink toolbar. The code generation and firmware compiling process can be observed in the "Diagnostic Viewer" window of Simulink. For MATLAB 2017b–2019a, it can be opened by clicking on the "View diagnostics" button in the status bar below Simulink (see the lower box in Fig. 4.9) or by clicking the "View" - "Diagnostic Viewer" (see Fig. 4.9) button on the Simulink menu. For MATLAB 2019b and

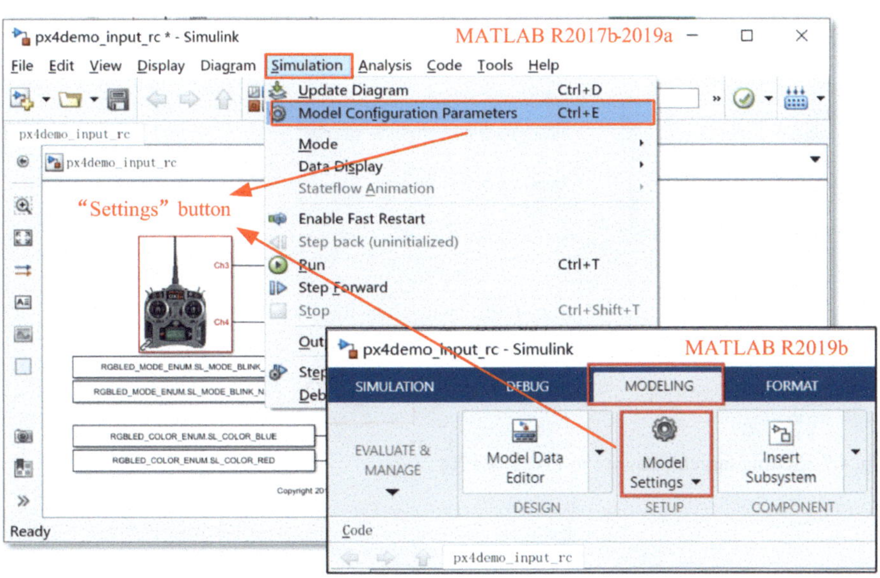

Fig. 4.5 "Model Configuration parameters" option on Simulink

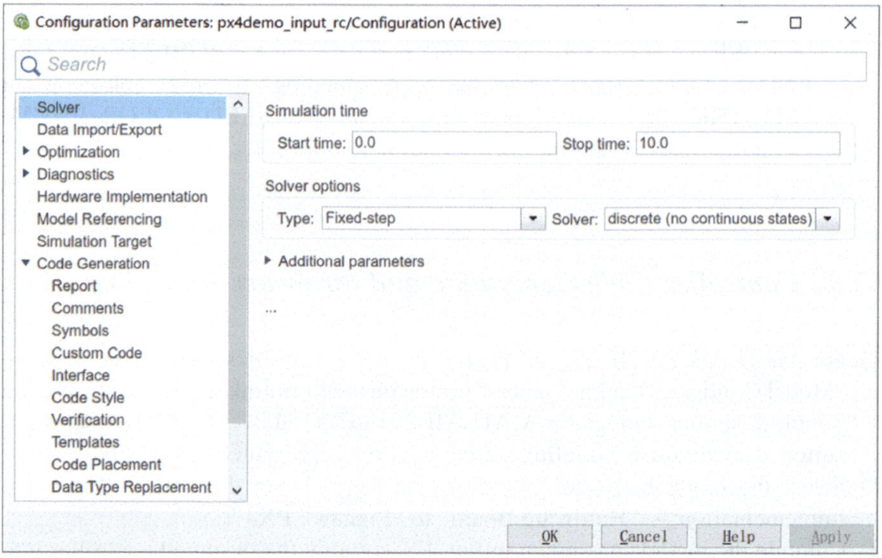

Fig. 4.6 Simulink setting dialog

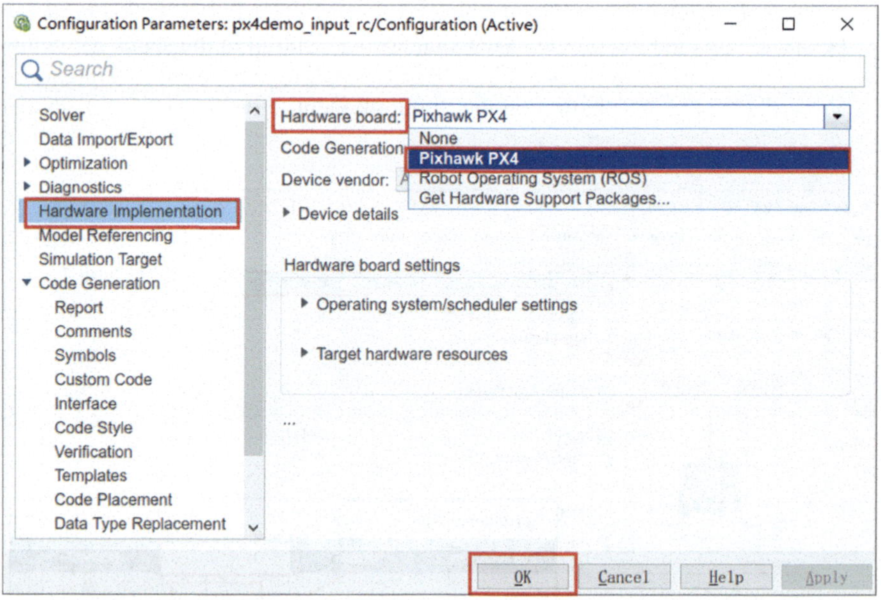

Fig. 4.7 Setting target hardware to Pixhawk PX4

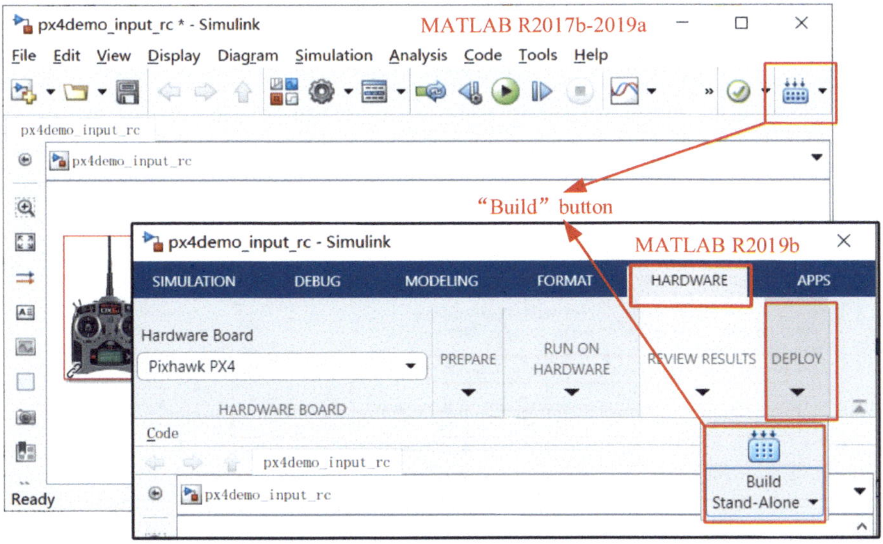

Fig. 4.8 Button for compiling code in Simulink

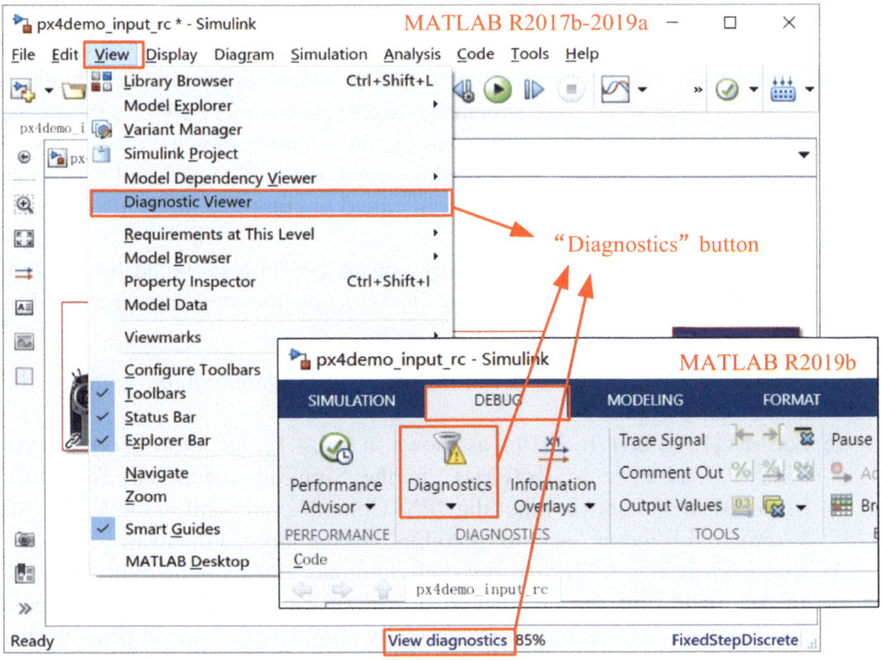

Fig. 4.9 "Diagnostics" option in different Simulink versions

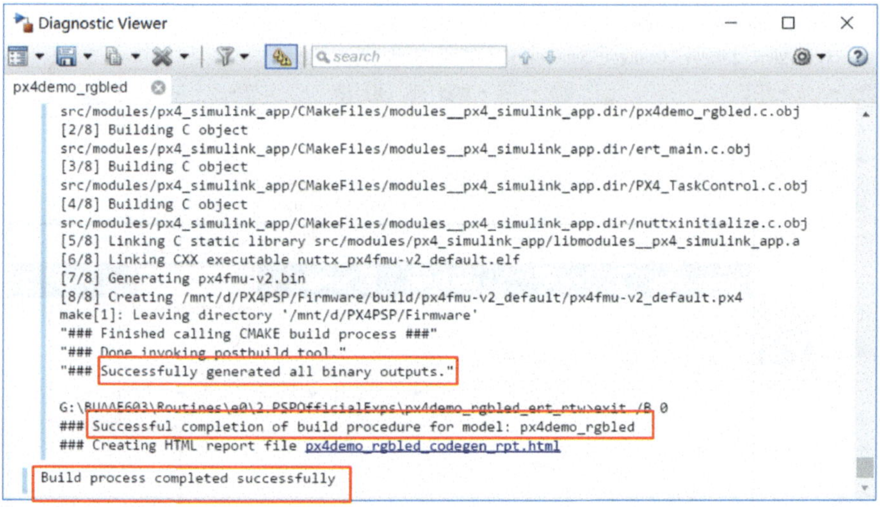

Fig. 4.10 Diagnostic viewer showing "build process completed successfully"

above, click the "Diagnostics" button to open the "Diagnostic Viewer" window. As shown in Fig. 4.10, the code and firmware are compiled successfully when the "Build process completed successfully" message appears in the "Diagnostic Viewer" window. A code generation report shown in Fig. 4.11 can also be obtained. At this point, the corresponding C/C++ code files have been generated in the folder "Firmware\src\modules\px4_simulink_app", and the "make px4fmu-v3_default" command has been called to complete the firmware compilation process.

(4) Upload the firmware. Use the one-key upload function provided by the PSP toolbox to upload the PX4 firmware file. The specific steps are presented as follows.

1) Connect the MicroUSB port of the Pixhawk hardware with the computer by using a USB cable.
2) For MATLAB 2017b–2019a, as shown in Fig. 4.12, click the "Code"-"PX4 PSP: Upload code to Px4FMU" in the Simulink menu bar; for MATLAB 2019b and above, input the "PX4Upload" command in the MATLAB "Command Window" according to Fig. 3.40b to upload the firmware.
3) As shown in Fig. 4.13, Simulink will automatically recognize the Pixhawk autopilot and start to upload and deploy the PX4 firmware file. The uploading and deploying process is successfully completed when the progress bar reaches 100%. Note that Pixhawk may need to be re-plugged in some cases to start the upload progress.

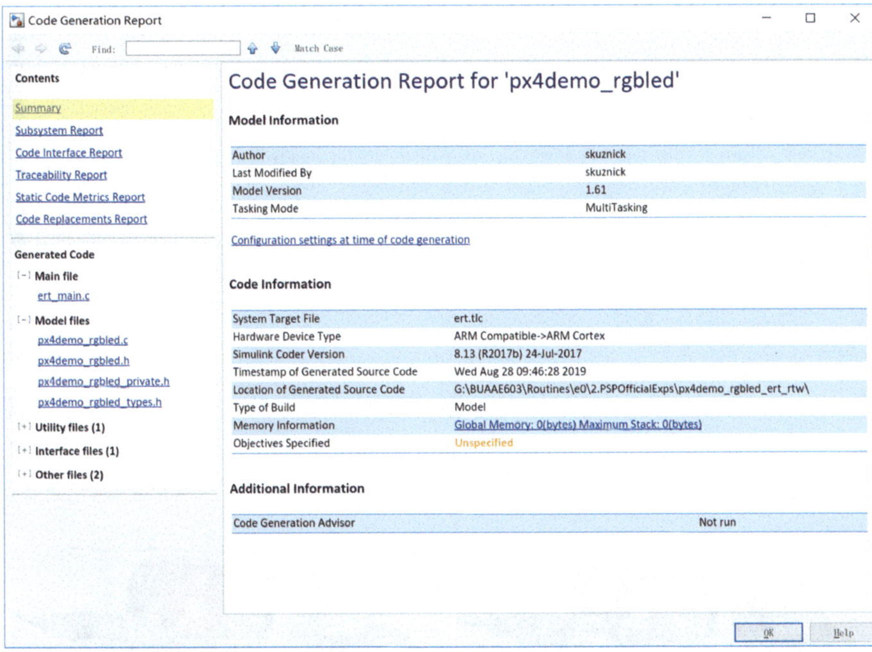

Fig. 4.11 Generated report after compiling

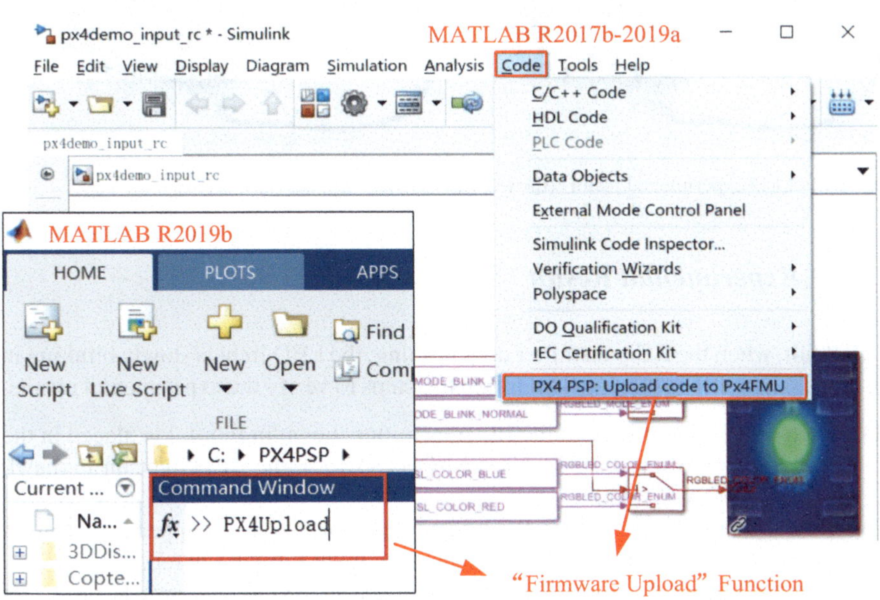

Fig. 4.12 Simulink firmware upload menu

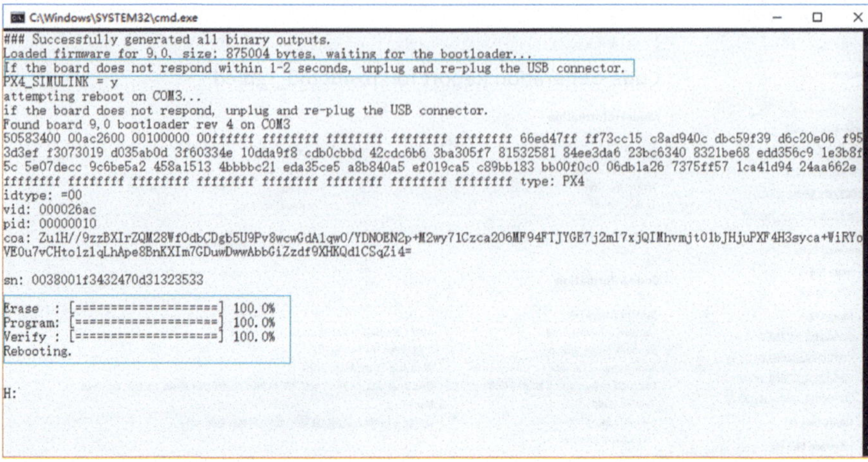

Fig. 4.13 Firmware is successfully uploaded

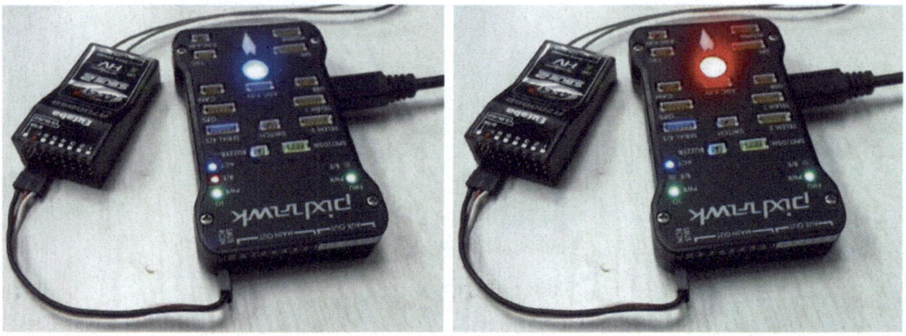

Fig. 4.14 LED experimental results (the left LED is blue, and the right is red)

4.2.4 Experimental Result

By default, when the RC transmitter does nothing, the LED light is slowly blinking in blue. As shown in Fig. 4.14, do the following steps to verify the experimental results.

(1) When the left-hand stick of the RC transmitter shown in Fig. 4.2 is placed in the upper-right position (CH3 > 1500 and CH4 > 1500), the LED light on Pixhawk is quickly blinking in blue.
(2) When the throttle stick of the RC transmitter is placed in the upper-left position (CH3 > 1500 and CH4 < 1500), the LED is quickly blinking in red.
(3) When the throttle stick of the RC transmitter is placed in the lower-left position (CH3 < 1500 and CH4 < 1500), the LED is slowly blinking in red.
(4) When the throttle stick of the RC transmitter is placed in the lower-right position (CH3 < 1500 and CH4 > 1500), the LED is slowly blinking in blue.

4.3 Experimental Procedure of Attitude Control Experiment

This section uses a well-designed attitude control system as an example to introduce the basic operation process of all the controller design experiments. This example is certainly complicated than the previous LED light control experiment.

4.3.1 Simulink-Based Algorithm Design and SIL Simulation

(1) **Step 1: controller design**
Create a new Simulink file and design a multicopter attitude controller in it. For simplicity, an example of a well-designed attitude controller is available in "e0\3.DesignExp\Exp1_AttitudeController.slx". Open it, and the controller details are presented in Fig. 4.15. The design requirements for the controller are in the following.

1) Input data
- CH1-CH5 channel signals of the RC transmitter (see Fig. 4.15), which correspond to the "ch1"-"ch5" input ports in Fig. 4.15. The actual data approximately range from 1100 to 1900, so calibration or dead zones are required in processing the RC data.
- Multicopter angular velocity (corresponding to the "p", "q", "r" input ports in Fig. 4.15, unit: rad/s). The above three inputs represent the velocity rotating around the x-axis of the body, the velocity rotating around the y-axis of the body, and the velocity rotating along the z-axis of the body.
- Multicopter Euler angles (unit: rad). Here, the roll angle and the pitch angle (corresponding to the "roll" and "pitch" input ports in Fig. 4.15) are

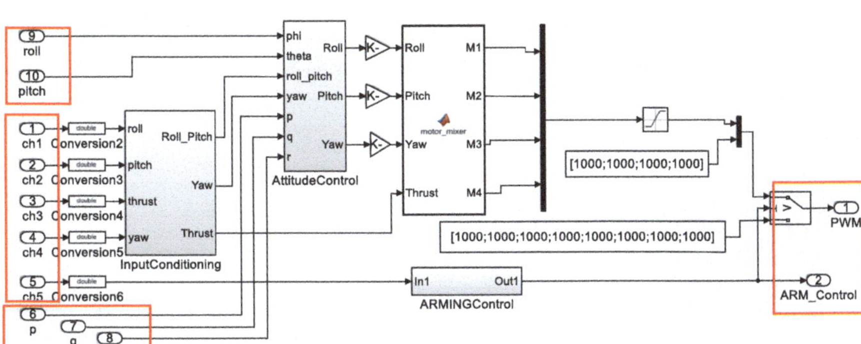

Fig. 4.15 Attitude controller example

mainly considered, and the yaw control is temporarily not considered in this experiment.

2) Output data
   - PWM control signals of four motors (corresponding to the “PWM” output port in Fig. 4.15). The data range is 1000–2000.
   - Identifier for the armed state (corresponding to the “ARM_Control” output port in Fig. 4.15). The data type is Boolean.
3) Expected effects
   - The throttle stick (CH3 on the RC transmitter) controls multicopters to perform the upward-and-downward movement.
   - Push up the pitch stick (CH2 < 1500) to control the multicopter flying forward.
   - Move the roll stick leftward (CH1 < 1500) to control the multicopter flying leftward.
   - Pull back (or down) the three-position switch (CH5 > 1500) to disarm the multicopter.

(2) **Step 2: generate the controller subsystem**
Select all the components in Fig. 4.15 with the mouse (or simultaneously press the keys CTRL + A), and right-click the mouse, choosing “Create Subsystem From Selection” to pack the controller to a subsystem in Simulink. Right-click the obtained subsystem and click “Mask”—“Create Mask” to open the mask setting box in Fig. 4.16. Then, enter text “image(‘./icon/Pixhawk.png’);” in the

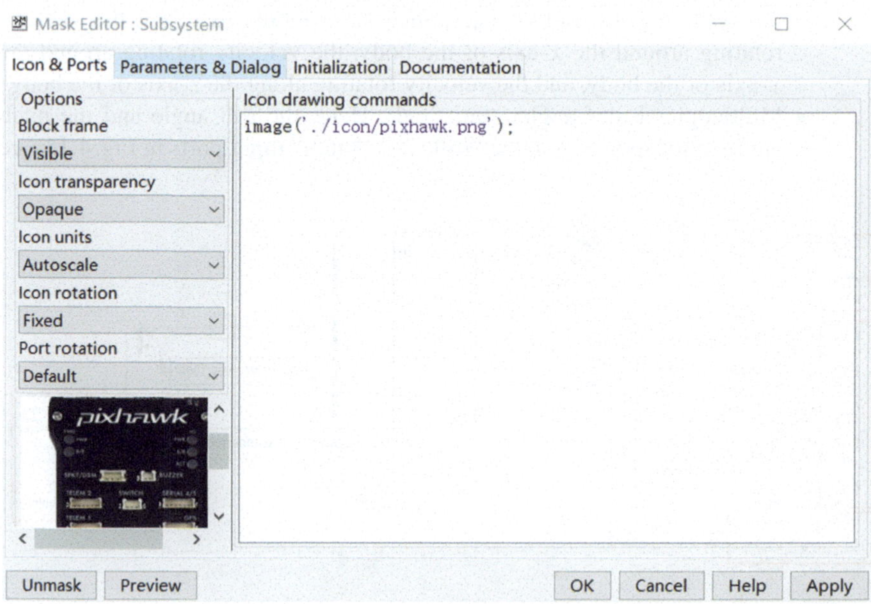

Fig. 4.16 Mask setting dialog for Simulink subsystem

"Icon drawing commands" input box in Fig. 4.16. Finally, click the "OK" button and adjust the positions of the input and output ports of the obtained subsystem to get a subsystem as presented in Fig. 4.17.

(3) **Step 3: integrate the controller with the model**
Open the Simulink file for SIL simulation, i.e., file "e0\1.SoftwareSimExps\CopterSim3DEnvironment.slx" used in the previous chapter; delete its original controller subsystem (remember to create a backup); then, copy the new controller subsystem obtained in **Step 2** to replace it.

(4) **Step 4: connect and configure the inputs and outputs**
As shown in Fig. 4.18, reconnect the controller to the multicopter model, where the output port "PosE" denotes the multicopter position vector in the earth frame, "AngEuler" denotes the Euler angle vector of the multicopter attitude, and "AngRateB" denotes the angular velocity vector of the multicopter. Given that the RC transmitter signals cannot be obtained in the SIL simulation phase, readers can use constant values to replace them or use the MATLAB functions to simulate the corresponding RC transmitter actions. The angular velocity input ports "p", "q", and "r" of the controller in Fig. 4.17 can be obtained from the "AngRateB" vector of the multicopter model; the angle "roll" and "pitch" can be obtained from the "AngEuler" vector. An example is also available in "e0\3.DesignExps\Exp2_ControlSystemDemo.slx" whose controller and practical RC transmitter signals have been connected as presented in Fig. 4.18.

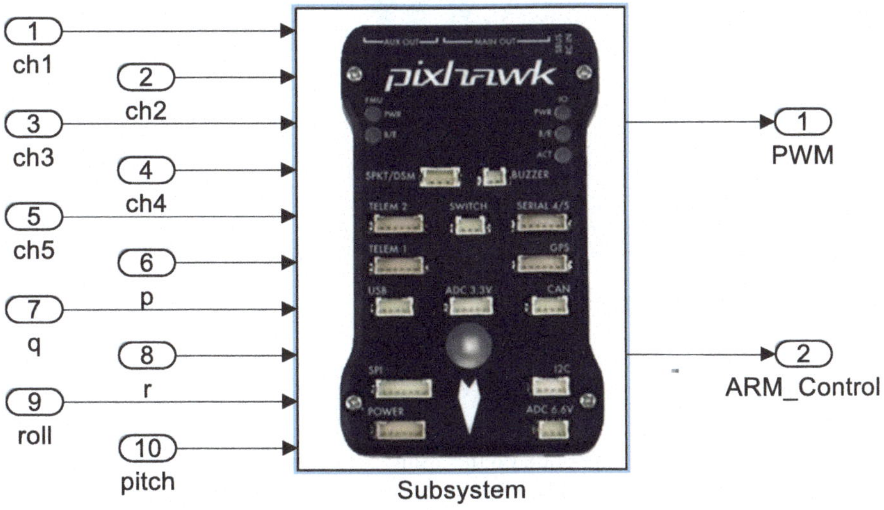

Fig. 4.17 Obtained attitude controller subsystem

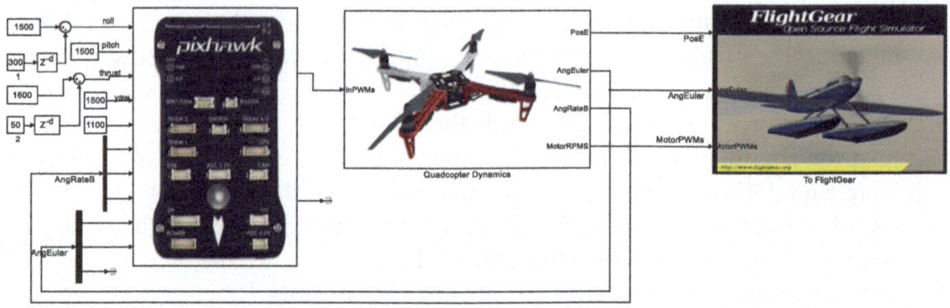

Fig. 4.18 Controller connected with multicopter model

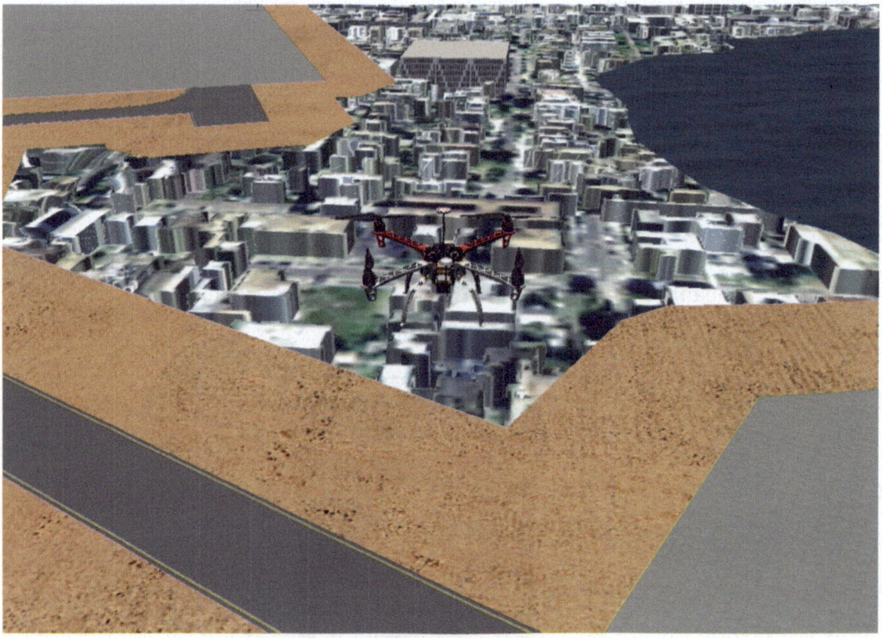

Fig. 4.19 A quadcopter in FlightGear

(5) **Step 5: start a joint simulation**
Click the FlightGear-F450 shortcut on the Windows desktop to open FlightGear (or open 3DDisplay), and click on the "Start Simulation" button on the Simulink UI to start the simulation. Then, it can be observed in FlightGear (see Fig. 4.19) that a quadcopter climbs up for some time and then rolls left and flies leftward, indicating that the controller has achieved the expected requirements.

4.3.2 Code Generation and Configuration

(6) **Step 6: configuration of the code generation environment**
After finishing the SIL simulation in Simulink, copy the obtained controller subsystem to file "e0\3.DesignExps\Exp3_BlankTemp.slx" (this file has been configured with all the settings required for code generation). Readers can also create a blank Simulink file and configure it according to Fig. 4.7.

(7) **Step 7: connect the controller to the PSP modules**
Extract the corresponding I/O interfaces from the Simulink PSP module library (see Fig. 4.3) and connect it to the controller obtained in **Step 6**. As shown in Fig. 4.20, a complete example is available in "e0\3.DesignExps\Exp4_AttitudeSystemCodeGen_old.slx". Note that the motor control signals should be sent the uORB message of "actuator_outputs" to the "uORB Write" module instead of the PWM output module. This is because the controller is currently used for HIL simulation instead of actual flight tests.

(8) **Step 8: generate code and compile the firmware**
Click the "Build" button (see Fig. 3.37) on the Simulink toolbar to automatically generate code and PX4 firmware file. The result in Fig. 4.21 shows a successful compilation process.

(9) **Step 9: upload the firmware**
Connect the computer to the Pixhawk autopilot with a USB cable, then use the "PX4Upload" function (see Fig. 3.40) to upload the firmware to the Pixhawk. A successful uploading result is shown in Fig. 4.22.

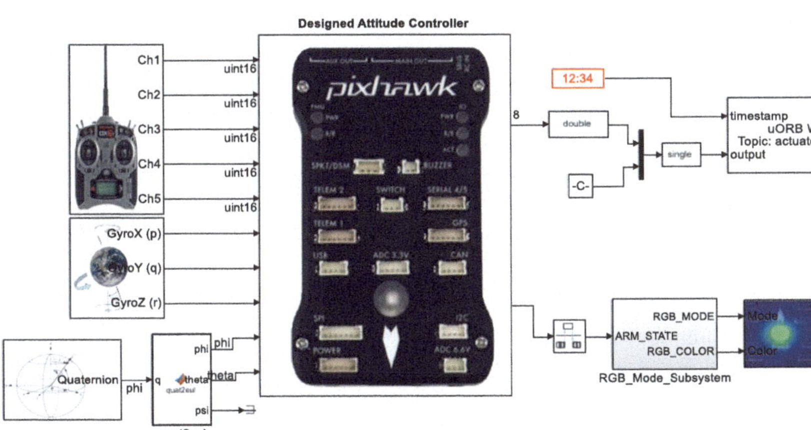

Fig. 4.20 Attitude controller in Simulink for code generation

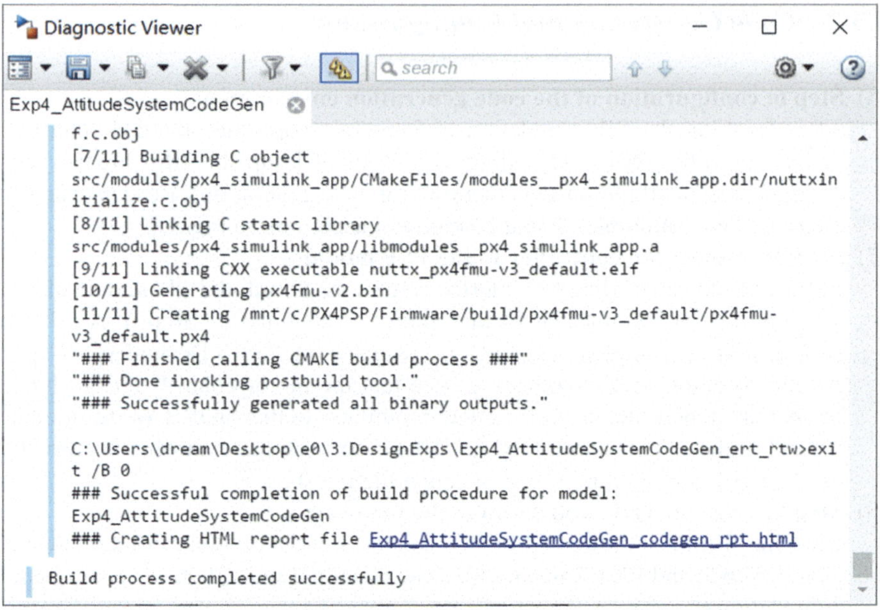

Fig. 4.21 Simulink controller compiling successfully

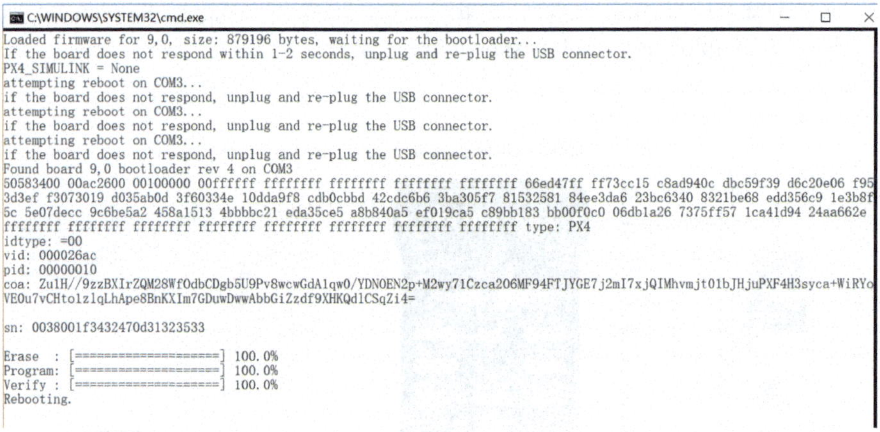

Fig. 4.22 Upload firmware successfully

4.3.3 HIL Simulation

(10) **Step 10: hardware system connection**
As shown in Fig. 4.2, connect the RC receiver to Pixhawk with a three-color JR cable, then connect Pixhawk to the computer via a USB cable. At this point, readers can observe that the LED on Pixhawk lights up and blinks slowly,[2] the LED on the RC receiver is blue and white (this is for RadioLink and green light for Futaba receiver). Then, turn on the RC transmitter, readers can observe that the LED light on the Pixhawk blinks quickly for a few seconds, indicating that the RC transmitter data has been successfully received. If there is no change for the Pixhawk LED light, indicating that the connection between the RC transmitter and the receiver is not correct, readers should check and confirm the hardware connection.

(11) **Step 11: CopterSim configuration**
Double-click the CopterSim shortcut on the Windows desktop to open it. There is no need to change the model parameters (or click "Model Parameters" - "Restore Default Parameters" - "Storage and Use Parameters" to restore aerial vehicle parameters to default values). Select the serial port of the Pixhawk autopilot in the "Select Pixhawk Com" drop-down box (usually in the format "*** FMU COM*"), click the "Start Simulation" button to start HIL simulation. As shown in Fig. 4.23, the message returned by the Pixhawk autopilot will be printed on the lower-left box of the CopterSim UI.

(12) **Step 12: 3DDisplay configuration**
Double-click the 3DDisplay shortcut on the desktop to open it. This software does not require any configuration; it passively receives the flight attitude and trajectory information of multicopters sent by CopterSim and displays it in real-time. The multicopter can be controlled by the RC transmitter to verify the designed attitude control algorithm. Move the throttle stick on the RC transmitter to the lower-right corner for three seconds to disarm the Pixhawk, and pull down (or back) the CH5 stick to the rearming position to disarm the designed controller. Then, the RC transmitter is able to control the multicopter to complete the corresponding action. As shown in Fig. 4.24, the multicopter attitude and position can be observed on the left side of the 3DDisplay interface. The real-time flight data are observed in the upper-right region, whereas the multicopter trajectory is observed in the lower-right region of the 3DDisplay UI.

[2]Higher Pixhawk hardware (e.g., Pixhawk 2/3/4/5) starts to discard LED module, so an external I2C LED module is required to observe the lighting effect.

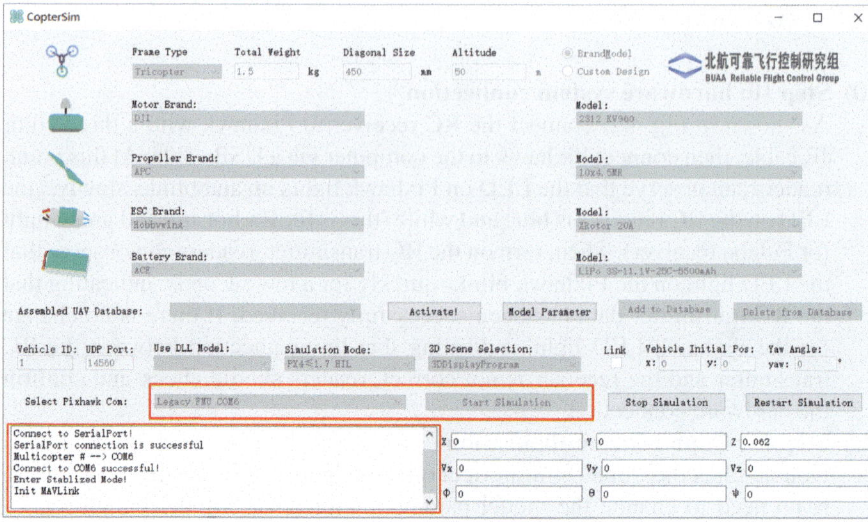

Fig. 4.23 Model simulator software configuration in CopterSim

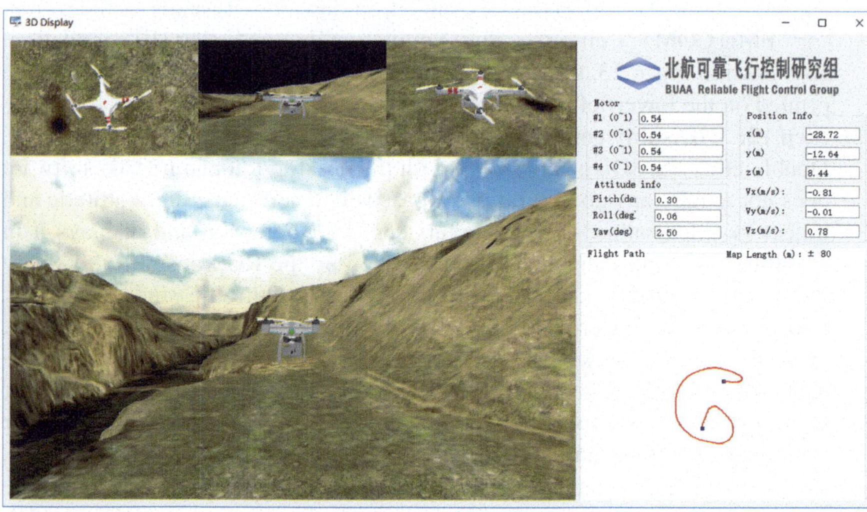

Fig. 4.24 Multicopter 3DDisplay interface

4.3.4 *Flight Test*

(13) **Step 13: mount Pixhawk onto a multicopter airframe**
The multicopter used in the outdoor flight tests is an F450 quadcopter (see Fig. 4.25). The parameters of the multicopter are accurately measured and identified by the system identification methods to ensure that the multicopter simulation model is consistent with the dynamics of the real multicopter system. For outdoor flight tests, the airframe of Pixhawk should be changed from "HIL Quadcopter X" to "DJI Flame Wheel F450" in QGC, which is presented in Fig. 4.26. All sensors should also be calibrated in QGC.

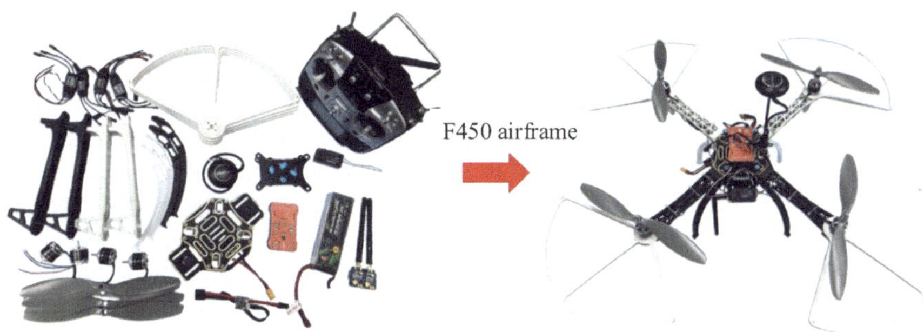

Fig. 4.25 F450 airframe and its components

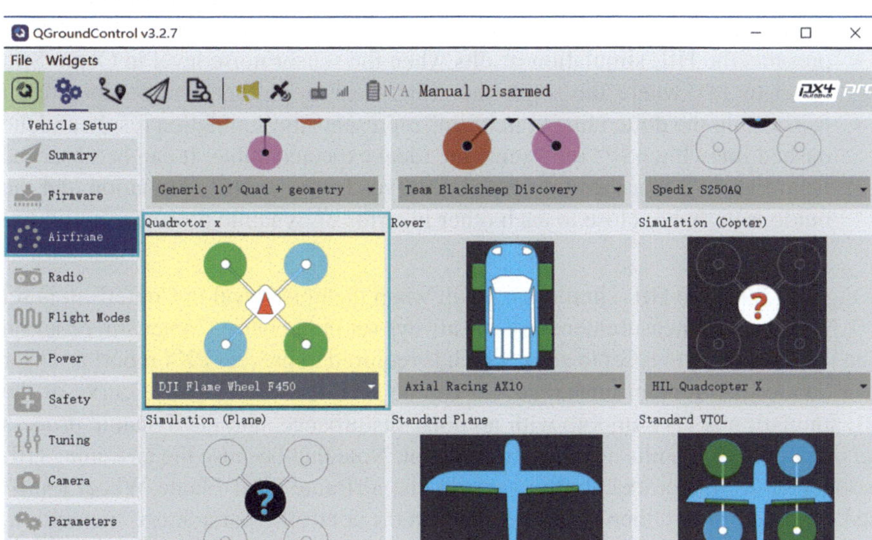

Fig. 4.26 Flight test airframe in QGC

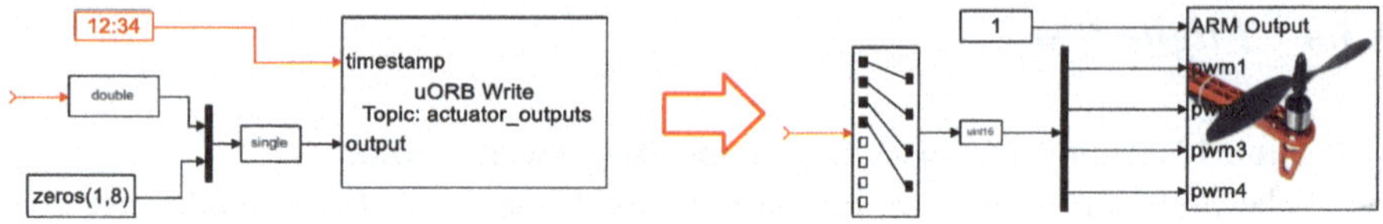

Fig. 4.27 Changing "uORB Write" module to PWM_output

(14) **Step 14: modify the Simulink controller**
Open the Simulink file in Fig. 4.20 and change the "uORB Write" module to the "PWM_out" module provided by the PSP toolbox, according to Fig. 4.27. An example with the modified motor output is presented in "e0\3.DesignExps\Exp5_AttitudeSystemCodeGenRealFlight_old.slx". Then, click the Simulink "compile" button to compile the controller into PX4 firmware and upload it to Pixhawk.

(15) **Step 15: preparation for flight tests**
Owing to the lack of complete failsafe logic for the generated control algorithm, safety should be fully considered to prevent accidents in outdoor flight tests. For example, the tests should be carried out in a relatively open area (such as a lawn) and on a nice day with good weather and low wind speed. When the above conditions are met, connect Pixhawk with the battery, and press the safety switch[3] on Pixhawk for more than three seconds. Then, control the multicopter with an RC transmitter to verify the actual performance of the controller.

(16) **Step 16: test results and analysis**
Next, we will present the results from the HIL simulation and outdoor flight tests to verify the accuracy of the multicopter simulation model. Figure 4.28 presents the HIL simulation results when the sensor noise level in CopterSim is set to 1.0, where the solid line "PitchReal" represents the outdoor flight tests result, the dotted line "PitchSim" represents the simulation result, and the dashed line "PitchSP" represents the ideal expected value. It can be observed from Fig. 4.28 that the step response curves from the HIL simulation and the outdoor flight are close to each other in terms of dynamic processes and noise levels.

Figure 4.29 is the HIL simulation result when the noise level in CopterSim is set to 0. Note that the measurement noise is always set in the latest CopterSim program even if the noise level is set to zero, which is required to prevent PX4 report a sensor error due to data remain unchanged. It can be seen that the response curve of the HIL simulation is very smooth with no noise disturbance, and the dynamic process and outdoor flight results are slightly different. Note that because the airframe "HIL Quadcopter X" is not exactly the same as the airframe "DJI Flame Wheel F450" used in outdoor flight, there are differences in the controller parameters. In addition, the data transmission speed of the Pixhawk serial port may fluctuate during the HIL

[3]https://docs.px4.io/master/en/config/airframe.html#safety_switch.

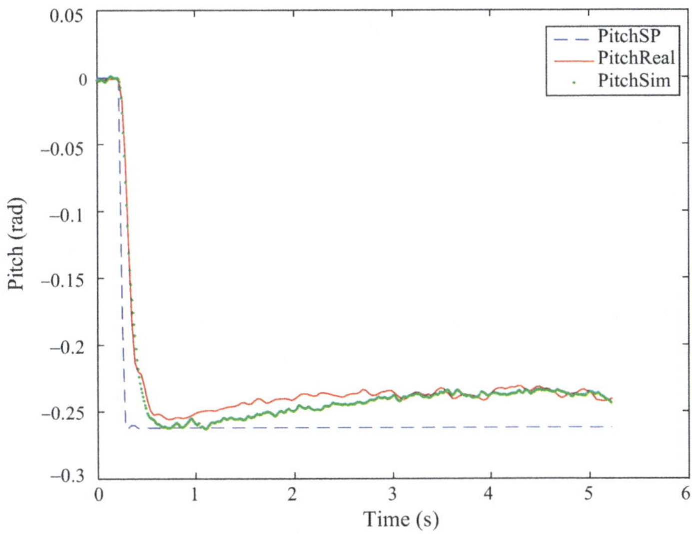

Fig. 4.28 HIL simulation result when noise level is set to 1.0

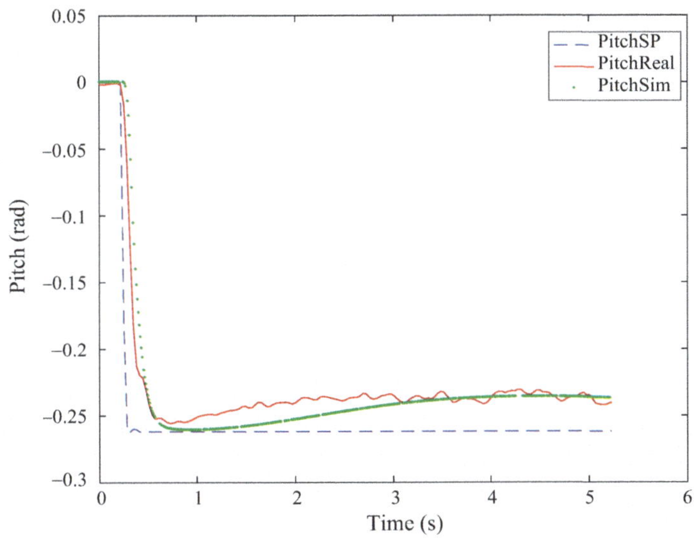

Fig. 4.29 HIL simulation result when noise level is set to 0

simulation, thereby affecting the real-time performance. Finally, the aerodynamics of the multicopter in the outdoor flight is very complicated, but a simplified aerodynamic model is used in the multicopter simulation model. Therefore, the error between their response curves is acceptable.

If you have any question, please go to https://rflysim.com for your information.

Chapter 5
Propulsion System Design Experiment

The composition of a multicopter system is both simple and complex. This chapter aims to introduce the composition of multicopter systems and performance modeling of a propulsion system by using three step-by-step experiments from shallow to deep, namely a basic experiment, an analysis experiment, and a design experiment. In the basic experiment, readers can configure propulsion systems and understand the function of a propulsion system on flight performance. In the analysis experiment, readers can calculate the hover endurance, and analyze the hover endurance with respect to temperature, altitude, propeller size, and the number of propellers. Finally, in the design experiment, readers can design a multicopter that meets a given requirement.

5.1 Preliminary

In order to make this chapter self-contained, the experiment preliminary is from Chaps. 3 and 4 of *Introduction to Multicopter Design and Control* [8].

5.1.1 Propulsion System

One of the most important components of a multicopter is the propulsion system, which decides the performance requirements of the multicopter, such as hover endurance, speed, payload, and flying range. Propulsion systems are composed of propellers, motors, Electronic Speed Controller (ESCs), and batteries. The components of the propulsion system must be matched and compatible with each other. Otherwise, the multicopter may not work properly, and in some extreme cases, may have a sudden failure causing an accident.

© Publishing House of Electronics Industry 2020, corrected publication 2021

Q. Quan et al., *Multicopter Design and Control Practice*,
https://doi.org/10.1007/978-981-15-3138-5_5

5.1.1.1 Propeller

(1) Function

A propeller is a component that produces thrust and torque to control a multicopter. The motor efficiency varies with the output torque (depending on the type, size, speed, and other factors of the propeller). Therefore, a good match should ensure that the motor operates in a high-efficiency condition, which will guarantee less power consumption for the same thrust, and extend the time of endurance of the multicopter. Accordingly, choosing appropriate propellers is a very direct approach to improving the performance and efficiency of multicopters.

(2) Parameters

1) Type

Generally, a propeller model is described by a four-digit number, such as a 1045 (or 10×45) propeller. In the number, the first two digits represent the diameter of the propeller (unit: inch), and the latter two digits represent the propeller pitch (also referred to as the screw pitch, blade pitch, or simply pitch; unit: inch). Therefore, a model "APC1045" propeller implies that the propeller belongs to the APC series, and the diameter and pitch of the propeller are 10 inches and 4.5 inches, respectively. The propeller pitch is defined as "the distance a propeller would move in one revolution if it were moving through a soft solid, like a screw through wood". For example, a 21-inches-pitch propeller would move forward 21 inches per revolution.

2) Chord length

The definition of the chord length of a propeller is shown in Fig. 5.1, and varies along the radius. Generally, the chord located at 2/3 of the radius of the propeller is chosen as the nominal chord length.

3) Moment of inertia

The moment of inertia is a quantity expressing the tendency of a body to resist angular acceleration, and is the sum of the products of the mass of each particle in the body with the square of its distance from the rotation axis. A smaller moment of inertia of the propeller can improve the response speed of the motor, which is important for control effects and performance.

4) Number of blades

Some typical propellers with different numbers of blades are shown in Fig. 5.2. Experiments have indicated that the efficiency of a two-blade propeller is better than that of a three-blade propeller. As shown in Fig. 5.3, the maximum thrust of the propeller increases with an increased number of blades (Fig. 5.3a), whereas the efficiency is decreased (Fig. 5.3b). It can also be found that to obtain the same thrust, a three-blade propeller requires less diameter than a corresponding two-blade propeller. Although the efficiency is reduced, the endurance time may be improved to a certain extent by the reduction of the size and weight of the airframe.

5) Safe rotation rate

Generally, the materials of the propellers used in multicopters are flexible.

As such, when the rotation rate exceeds a certain value, the propellers may deform, reducing their efficiency. Therefore, when calculating a safe rotation rate limit, all possible conditions should be considered. The APC Web site[1] gives an empirical formula for the maximum speed of its multicopters' propellers, which is 105,000 Revolution Per Minute (RPM)/propeller diameter (unit: inch). By contrast, the maximum speed of Slow Flyer (SF) propellers is only 65,000 RPM/propeller diameter.

6) Propeller specific thrust
The specific thrust, also referred to as the efficiency (unit: g/W), is a very important parameter for measuring the efficiency of energy transformation. The propeller specific thrust is defined as follows

$$\begin{aligned}\text{Mechanical power (uint: W)} &= \text{Torque (unit: N} \cdot \text{m)} \times \text{Propeller speed (unit: rad/s)}\\ \text{Propeller speciffic thrust (uint: g/W)} &= \frac{\text{Thrust (uint: g)}}{\text{Mechanical (uint:W)}}.\end{aligned}$$

7) Material
Materials for propellers include carbon fiber, plastic, and wood. Although propellers made of carbon fiber cost approximately twice as much as those made of plastic, they are more popular, because of the following advantages: 1) less vibration and noise, owing to its high rigidity; 2) lighter and stronger; and 3) more suitable for a motor with a high KV rating. However, because of the high rigidity, the motor will absorb most of the impact when a crash occurs, and the fiber blade can be treated as a high-speed rotating razor which is dangerous to nearby humans. Propellers made of wood are much heavier and more expensive, and are suitable for multicopters with a large payload capacity.

(3) Static Balance and Dynamic Balance
The goal of static balance and dynamic balance is to reduce vibration caused by the asymmetric centrifugal force and by the asymmetrical distribution of mass and shape. The imbalance of the propeller is a major source of vibration for a multicopter. A propeller static imbalance occurs when the Center of Gravity (CoG) of the propeller does not coincide with the axis of rotation. A dynamic imbalance occurs when the CoG of the propeller does not coincide with the center of inertia. The imbalanced force not only affects the sensor measurement, but also makes the motor bearing wear down more quickly and increases power consumption. These characteristics will shorten the lifetime of the multicopter, and increase the possibility of failure. Moreover, an imbalanced propeller is far noisier than a balanced one. A balancer as shown in Fig. 5.4 can be used to test the static balance of a propeller. The test of dynamic balance is often difficult, as it requires sensors to record the data. If an imbalance exists, some measures can be taken, such as pasting scotch tape on the lighter blade, or grinding the heavier one (but not the edge) using sandpapers.

---

[1] http://www.apcprop.com/Articles.asp?ID=255.

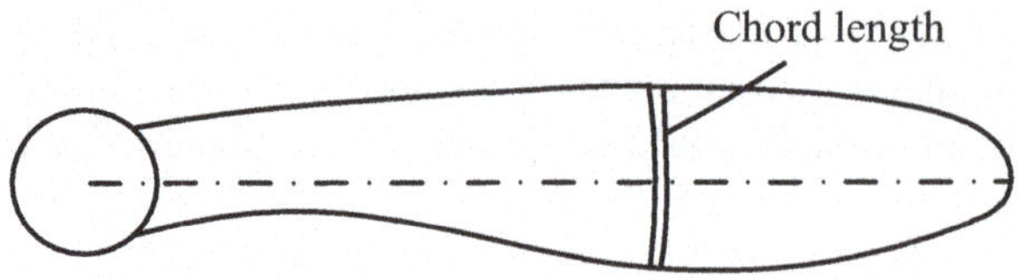

Fig. 5.1 Chord length of a propeller

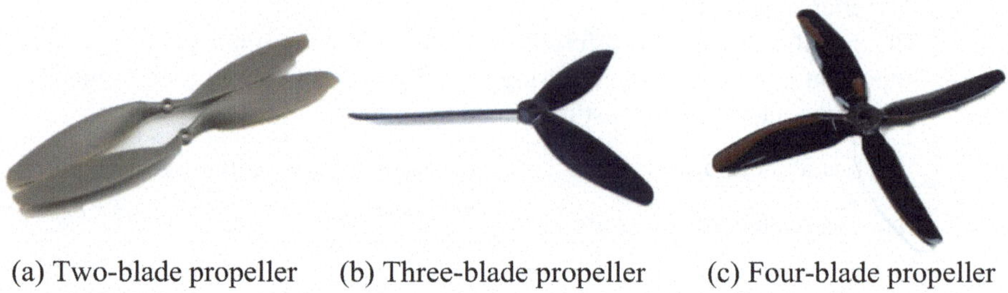

Fig. 5.2 Propellers with different number of blades

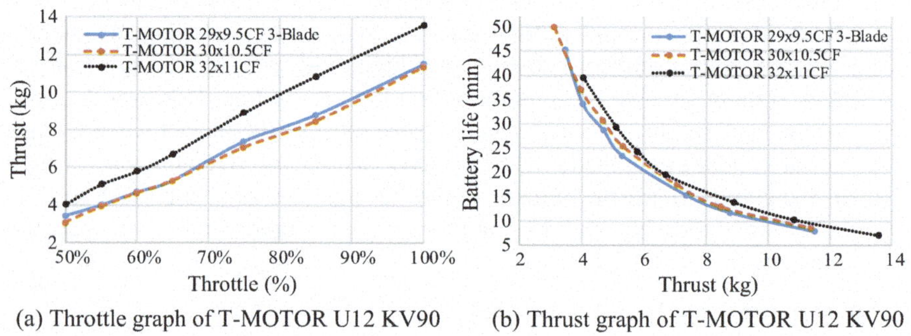

Fig. 5.3 Thrust and efficiency of two-blade propeller and three-blade propeller. Taking T-MOTOR 29 × 9.5CF 3-Blade as an example, it indicates that the propeller has three blades, and it is made of Carbon Fiber (CF). The default number of blades is two

5.1.1.2 Motor

(1) Function
The motors of multicopters are mainly brushless Direct Current (DC) motors, owing to various advantages such as high efficiency, the potential to downsize, and low manufacturing costs. Brushless DC motors are used to convert electrical energy (stored in a battery) into mechanical energy for the propeller. Concretely, and based on the position of the rotors, brushless DC motors can be classified into outer rotor type and inner rotor type, as shown in Fig. 5.5. Considering that the motor of a multicopter is supposed to drive larger propellers to improve efficiency, the outer rotor type outperforms the inner rotor type as it can provide

Fig. 5.4 A Du-Bro propeller balancer

larger torques. Moreover, as compared with that of the inner rotor type, the speed of the outer rotor type is more stable. Therefore, the outer rotor type is more popular in multicopters and most other aircraft.

(2) Working Principle

As shown in Fig. 5.6, the control circuit generates a Pulse Width Modulated (PWM) signal to the ESC, where the signal is amplified by a driving circuit and sent to the power switch of the inverter. Then, the control circuit controls the motor winding to operate in a certain sequence and generate a jump-type rotating magnetic field in the air gap of the motor. Common types of main circuits of brushless DC motors include the star-type three-phase half-bridge, star-type three-phase bridge, and angle-type three-phase bridge. Among them, the star-type three-phase bridge is the most widely-used. Three output signals of a position detector are controlled by a logic circuit to control the on and off states of a switch. There are two control modes: two-two conduction mode and three-three conduction mode. As shown in Fig. 5.7, for every 60° the rotor rotates, the switches of the inverters commutate once and the field of the state of the stator changes once. There are six magnetic states and three phases for the motors, and each phase conducts 120°.

(3) Parameters

1) Size

The size of a motor is generally represented by its stator size with a four-digit number, such as motor 2212 (also written as 22×12), among which the first two digits indicate its stator diameter (mm), and the latter two digits indicate its stator height (mm). For example, the motor number "2212" indicates that the stator diameter of the motor is 22 mm and that the stator height is 12 mm. That means, the larger the former two digits are, the wider the motor is; the larger the latter two digits are, the higher the motor is. A wide and high motor has high power, which is more suitable for large multicopters.

2) KV value for motors

The KV value for brushless DC motors is the number of RPM that the motor will revolve when 1 V (Volt) is applied with no load attached to the motor. For example, 1000 KV simply means that when 1 V is applied, the no-load motor speed will be 1000 RPM. A low-KV motor has more windings of thinner wires, meaning that it will carry more power, produce higher torque, and drive a bigger propeller. In contrast, a high KV motor can produce a low torque so that it can only drive a small propeller.

3) No-load current and voltage

In the no-load test, the current passing through the three-phase winding of the stator after applying a nominal voltage (generally 10 or 24 V) is defined as the nominal no-load current.

4) Maximum current/power

This is the maximum current or power the motor can undertake. For example, a maximum continuous current of "25 A/30s" indicates that the motor can work safely with continuous current up to 25 A, but after 30s, the motor may be burnt out. The same definition can be applied for the maximum continuous power.

5) Resistance

There is resistance in all motor armatures. It is very small but cannot be ignored, as the current flowing through the resistance is tremendously large and sometimes reaches tens of amperes. The existence of the resistance generates heat during the running of the motor, which may overheat the motor and reduce efficiency.

6) Motor efficiency

Motor efficiency is an important parameter for measuring performance. It is defined as follows

$$\begin{aligned}\text{Electrical power (unit: W)} &= \text{Input Voltage (unit: V)} \times \text{Effective current (unit: A)}\\ \text{Motor efficiency} &= \frac{\text{Mechanical power (unit: W)}}{\text{Electrical power (unit: W)}}.\end{aligned}$$

The motor efficiency is not constant. In general, it varies with input voltage (throttle) and load (propeller). For the same propeller, the efficiency of the motor may be reduced as the input voltage (current) is increased. That is because the larger the current is, the more the heat (caused by the resistance) and other losses will be, which reduces the ratio of the effective mechanical power.

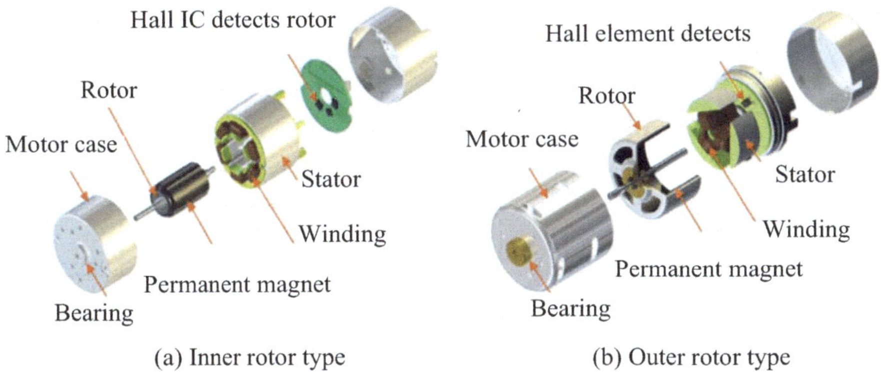

Fig. 5.5 Outer rotor type and inner rotor type. Photo courtesy of www.nidec.com

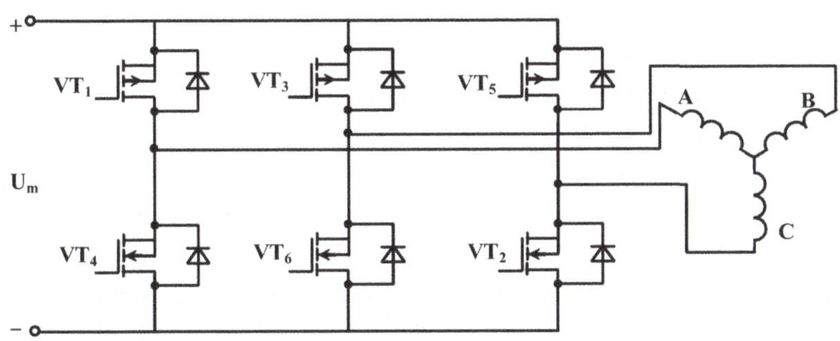

Fig. 5.6 Main circuit of star-type three-phase bridge

7) Overall specific thrust
The overall performance of the propulsion system depends largely on a well-matched combination of motor and propeller. To evaluate the efficiency of the motor and propeller together, the overall specific thrust is calculated as

$$\begin{aligned}\text{Overall specific thrust (unit: g/W)} &= \frac{\text{Thrust (unit: g)}}{\text{Electrical power (unit: W)}}\\ &= \text{Propeller specific thrust} \times \text{Motor efficiency}.\end{aligned}$$

As both the propeller specific thrust and motor efficiency are not constant, the overall specific thrust changes with the working condition. The overall specific thrust is often given by the motor producers. Taking a motor as an example, the overall specific thrust under different states is displayed in Fig. 5.8, where "Efficiency (G/W)" is the overall specific thrust. This will help designers in choosing combinations of motor and propeller according to their requirements.

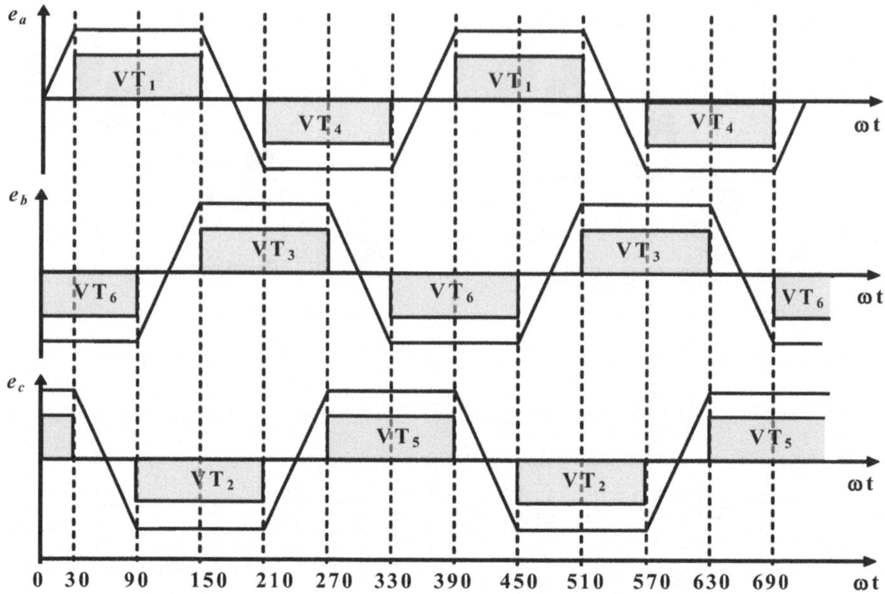

Fig. 5.7 Counter electromotive force waveform of three-phase winding and its two-two conduction mode

Motor model	Voltage (V)	Propeller Model	Throttle	Current (A)	Power (W)	Thrust (g)	Speed (RPM)	Efficiency (G/W)	Torque (N·m)	Temperature ℃
T-MOTOR MN5212 KV340	24	T-MOTOR 15x5CF	50%	3.3	79	745	3821	9.44	0.142	38
			55%	4.2	99.8	910	4220	9.11	0.172	
			60%	5.2	123.6	1075	4576	8.7	0.198	
			65%	6.3	150.7	1254	4925	8.32	0.232	
			75%	9.1	217.2	1681	5663	7.74	0.31	
			85%	12.2	292.1	2115	6315	7.24	0.382	
			100%	17.8	426.7	2746	7167	6.44	0.498	
		T-MOTOR 18x6.1CF	50%	5.7	137.5	1318	3596	9.58	0.29	74
			55%	7.4	178.1	1612	3958	9.05	0.344	
			60%	9.3	222	1901	4310	8.56	0.411	
			65%	11.6	278.2	2259	4622	8.12	0.472	
			75%	16.5	395.5	2835	5226	7.17	0.605	
			85%	22.1	531.1	3477	5751	6.55	0.737	
			100%	31	744.7	4355	6358	5.85	0.918	

Fig. 5.8 Overall specific thrust of motor MN5212 KV420 (data from http://uav-en.tmotor.com/)

5.1.1.3 Electronic Speed Controller

(1) Function

The basic function of ESCs is to control the speeds of motors based on PWM signals that autopilots send, which are too weak to directly drive brushless DC motors. Furthermore, some ESCs act as dynamic brakes, or power supplies (battery elimination circuit module) for RC receiver or servo motors. Unlike a general ESC, a brushless ESC has a new function, i.e., it can act as an inverter to trans-

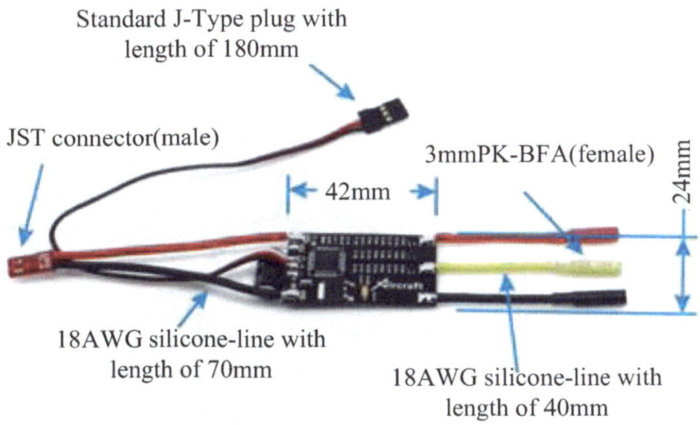

Fig. 5.9 XAircraft ESC-S20A for multicopters

form an onboard DC power input into a three-phase Alternating Current (AC) power for brushless DC motors. Undoubtedly, there are some other auxiliary functions, such as battery protection and starting protection. Figure 5.9 shows the X-aircraft ESC-S20 A for multicopters.

(2) Parameters

1) Maximum continuous/peak current
The most important parameter for brushless ESCs is the current, which is usually represented by Ampere (A), such as 10, 20 and 30 A. Different motors need to be equipped with different ESCs. An inappropriate matching will burn out ESCs, or even cause motor failure. Concretely, there are two important parameters of the brushless ESC: the maximum continuous current, and the peak current. The former is the maximum continuous current in a normal working condition, whereas the latter is the maximum instantaneous current that the ESC can withstand. Each ESC is labeled with a specified value, such as Hobbywing "XRotor 15 A", indicating the maximum continuous current allowed. When choosing the type of ESC, attention should be paid to the maximum continuous current determine whether it leaves a safety margin (20% for example), so as to efficiently avoid burning out the power tube. Taking a 50 A ESC as an example, a 10 A current margin is often left as a safety margin.

2) Voltage range
The range of voltage that allows the ESC to work properly is also an important parameter. Usually, an index such as "3–4 S LiPo" can be found on an ESC specification, and mean that the voltage range of this ESC is 3–4 cells of a LiPo battery, i.e., 11.1–14.8 V.

Fig. 5.10 Hobbywing brushless ESC programmable card

3) Resistance
   All ESCs have resistance, and the resultant heating power cannot be ignored, as the current flowing through them can sometimes reach 10 A. Considering the heat dissipation, the resistance of ESCs with high current is always designed to be small.
4) Refresh rate
   The motor response has a significant relationship with the refresh rate of ESCs. Before the development of multicopters, ESCs were designed specifically for model airplanes or cars. The maximum operating frequency of the servo motors was 50 Hz, and therefore the refresh rate of the ESCs was 50 Hz. Theoretically, the higher the refresh rate is, the faster the response will be. As multicopters differ from other types of model airplanes in that rapid thrust adjustment is realized by rapid control of the propeller angular speed, the refresh rate of multicopter ESCs is often faster. To ensure smooth outputs, low-pass filtering is often applied to the input or output of ESCs at the cost of reducing their response rate. This also implies a reduced control frequency.
5) Programmability
   The performance of ESCs can be optimized by tuning internal parameters. There are three ways to set the parameters of ESCs, i.e., programmable cards as shown in Fig. 5.10, computer software via a USB, and RC transmitters. The parameters that can be set up include throttle range calibration, low voltage protection, power outage value, current limitation, brakes mode, throttle control mode, switch timing setting, starting mode, and PWM mode setting.

6) Compatibility
   If the ESC and motor are incompatible, the motor is likely to be jammed, which may result in a fall and crash for a multicopter in the air. Sometimes motors may get jammed in extreme cases, such as when the control command for mode transitions changes sharply, generating large instantaneous currents, which are not easy to detect.

(3) Square-Wave Driver and Sinusoidal Driver

1) Square-wave driver
   A square-wave driver type ESC outputs a square wave. Its control elements work in the switch state, which makes it simpler, cheaper, and easier to control.
2) Sinusoidal driver
   A sinusoidal driver-type ESC outputs a sinusoidal wave, which uses field-oriented control. Therefore, the sinusoidal driver performs better in the aspects of operation stability, speed range, efficiency, and reduction of vibration-noise. Methods for measuring the angle of the rotor include an optical encoder, a Hall sensor, and observer-based methods. As multicopter's motor rotors are working in a high-speed state, the field-oriented control can be applied based on an observer of the rotor electrical angle with information including the motor model, current, and voltage. This is a good way to reduce costs.

5.1.1.4 Battery

(1) Function
   The battery is used to provide energy. A battery for small multicopters is shown in Fig. 5.11. A problem often encountered with current multicopters is the time of endurance, which heavily depends on the capacity of the batteries. Nowadays, there are many types of batteries. Lithium Polymer (LiPo) battery and Nickel Metal Hydride (NiMH) battery are the most commonly-used, because of their superior performance and cheap price.

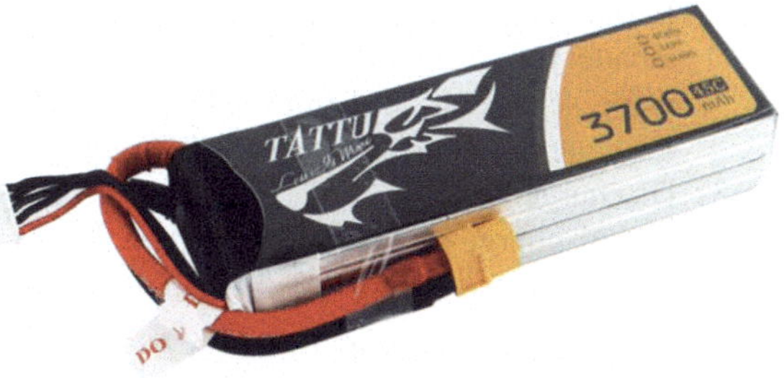

Fig. 5.11 GENS ACE Tattu UAV battery

(2) Parameters

The basic parameters of a battery include its voltage, discharge capacity, internal resistance, and discharge rate. The nominal voltage of a single cell of LiPo battery is 3.7 V. When it is fully charged, the voltage can reach 4.2 V. To ensure that the total battery capacity and voltage are sufficient, several cells can be assembled together. The remaining voltage is decreased gradually with the discharge of the battery. Most research shows that in a certain range, the remaining voltage is in a linear relationship with the remaining capacity of the battery. However, in the late stages of discharge, the voltage may drop sharply, which may result in a rapid thrust loss in a multicopter. To ensure that a multicopter has enough power or capacity to return home before the carried battery runs out, it is necessary to set a safe voltage threshold for the battery. In addition, the output voltage will drop as the current of discharge is increased, as more voltage is allocated to the internal resistance. It should be noted that the battery should not be completely discharged; otherwise, it may have irreversible damage.

1) Connection

By combining battery cells in series, a higher voltage can be obtained, while the capacity remains unchanged. In contrast, by combining battery cells in parallel, a larger capacity can be obtained, while the voltage remains unchanged. In some cases, there exist different combinations of both series and parallel arrangements in battery packs. The letters S and P are used to represent the series connection and parallel connection, respectively. For example, as shown in Fig. 5.12a, assuming that the voltage of one cell is 3.7 V and its capacity is 100 mAh, then 3S1P represents three cells in series connection (total voltage is 11.1 V, capacity is 100 mAh). As shown in Fig. 5.12b, for a 2S2P battery, its total voltage is 7.4 V and its total capacity is 200 mAh.

2) Capacity

The milliAmpere-hour (mAh) or Ampere-hour (Ah) is a technical index that how much electrical charge a particular battery has. A capacity of 5000 mAh for a LiPo battery means that the discharge of the battery will last for an hour with a current of 5000 mA when the voltage of a single cell is decreased from 4.2 to 3.0 V. However, the discharge ability will be decreased along with the process of discharge, and its output voltage will also be slowly decreased. As a result, the remaining capacity is not a linear function of the discharge time. In practice, there are two ways to detect whether the remaining capacity of a battery can support a flight. One way is to detect the voltage of the battery using sensors in real-time, which is commonly used. The other way is to estimate the State of Charge (SoC) value of the batteries.

3) Discharge Rate

The discharge rate is represented by

$$\text{Discharge Rate (unit: C)} = \frac{\text{Current of Discharge (uint: mA)}}{\text{Capacity (uint: mAh)}}.$$

Fig. 5.12 Battery pack connection

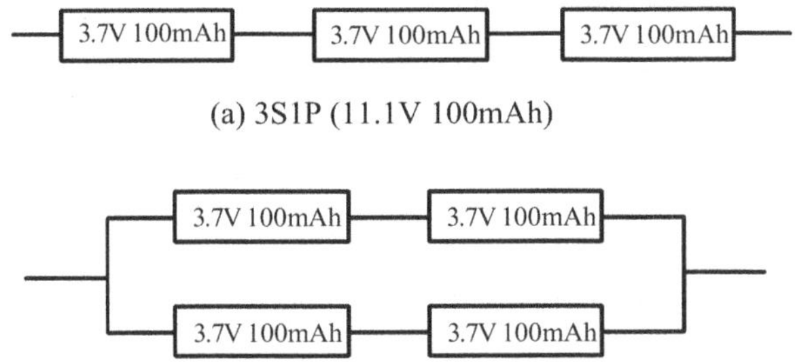

For example, the discharge rate of a battery will be 0.2 C when its nominal capacity is 100 mAh and the discharge current is 20 mA. Evidently, the discharge rate measures the rate of discharge. When the maximum discharge rate of a battery with a nominal capacity of 5000 mAh is 20 C, the maximum current of discharge is calculated as $5000\ \text{mA} \times 20\ \text{C} = 100\ \text{A}$. The total current of a multicopter cannot exceed the maximum current limit of the battery; otherwise, the battery may be burnt out. A battery with a higher discharge rate can generate more current, and can be applied to multicopters demanding higher current.

4) Resistance
The resistance of a battery is not a constant value, and varies with the power status and service life. The resistance of a rechargeable battery is relatively small in its initial state. However, after a long period of use, and because of the exhaustion of electrolytes and decrease in chemical substance activity of the battery, the internal resistance will be gradually increased, until a certain degree where power in the battery cannot be released. Then, the battery can be regarded as being run out of.

5) Energy Density
The energy density is the amount of energy stored in a given system or region of space per unit volume or mass, and the latter is more accurately termed "specific energy". In general, the units for energy density and specific energy are (Watt × hour)/kg and (Watt × hour)/L, i.e., Wh/kg and Wh/L, respectively. Batteries with higher energy density are more popular owing to the contradiction between volume (weight) and endurance for a product. A Lithium-ion battery, as a type of clean energy, is getting increased attention, and is widely-used in many applications. The energy density of Lithium-ion batteries varies by chemistry, and the energy density can range from 240 to 300 Wh/L (double that of the NiCd, 1.5 times that of NiMH).

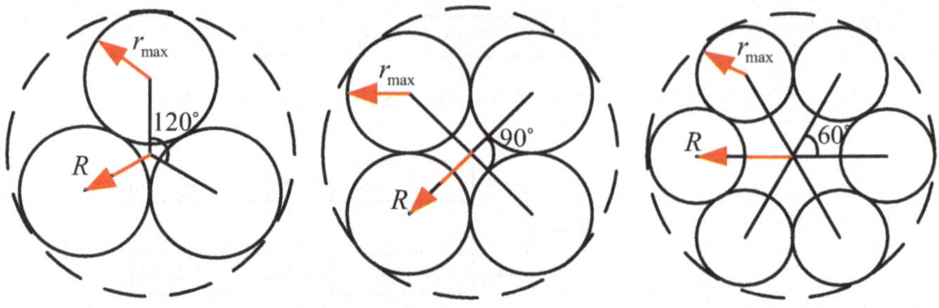

(a) Y6-configuration hexacopter (b) Quadcopter (c) Hexacopter

Fig. 5.13 Multicopters with different airframe configurations and their geometry parameters

5.1.2 Propeller Radius and Airframe Radius

The size of a multicopter is closely related to the radius of the propeller. Assuming that the angle between the two arms is θ, then for a Y6-configuration hexacopter (Fig. 5.13a), $\theta = 120°$; for a traditional quadcopter (Fig. 5.13b), $\theta = 90°$; and for a traditional hexacopter (Fig. 5.13c), $\theta = 60°$. According to Fig. 5.13, if the number of arms of the multicopter is n, and $\theta = 360°/n$, then the relationship between the airframe radius R and the maximum radius of a propeller $r_{\max}$ is

$$R = r_{\max}/\sin\frac{\theta}{2} = r_{\max}/\sin\frac{180°}{n}. \tag{5.1}$$

Numerical simulation results show that the thrust is decreased because of aerodynamic interference between propeller flow fields, and that a cyclic fluctuation of the thrust appears. However, this influence is negligible when the rotors are not too close to each other. Additionally, by using smoke plume to visualize airflow, it can be clearly observed that when the space between propellers varies from $0.1r_{\max}$ to $r_{\max}$, each propeller has its independent vortex without affecting propeller performance. Hence, to make a multicopter more compact without losing much efficiency via aerodynamic interference, the following rule of thumb can be considered where the propeller radius r_{p} is determined by the weight and the maximum payload of the multicopter. Thus, once the propeller radius is determined, the airframe radius is determined according to Eqs. (5.1) and (5.2) as

$$r_{\max} = 1.05r_{\mathrm{p}} \sim 1.2r_{\mathrm{p}}. \tag{5.2}$$

The rotor radius r_{p} is selected based on the total weight and maximum load of the multicopter. Therefore, once the propeller model size is determined, the frame radius according to Eqs. (5.1) and (5.2) can also be determined.

Table 5.1 Propulsion system parameters and symbols

Component	Parameters
Propeller	$\Theta_{\rm p}$ = {Diameter $D_{\rm p}$, Pitch $H_{\rm p}$, Blade Number $B_{\rm p}$, Propeller Weight $G_{\rm p}$}
Motor	$\Theta_{\rm m}$ = {KV Value $K_{\rm V0,}$ Maximum Continuous Current $I_{\rm eMax}$, Nominal No-load Current $I_{\rm m0}$, Nominal No-load Voltage $U_{\rm m0}$, Resistance $R_{\rm m}$, Motor Weight $G_{\rm m}$}
ESC	$\Theta_{\rm e}$ = {Maximum Current $I_{\rm eMax}$, Resistance $R_{\rm e}$, ESC Weight $G_{\rm e}$}
Battery	$\Theta_{\rm b}$ = {Capacity $C_{\rm b}$, Resistance $R_{\rm b}$, Total Voltage $U_{\rm b}$, Maximum Discharge Rate $K_{\rm b}$, Battery Weight $G_{\rm b}$}

5.1.3 Propulsion System Modeling

The propulsion system modeling is divided into four parts: propeller modeling, motor modeling, ESC modeling, and battery modeling. To ensure the practicability of the method, all input parameters of these models should be basic parameters that can be easily found in the product descriptions, as shown in Table 5.1. The outputs are a series of performance indices. For simplicity, they are formulated into the following four problems.

(1) Propeller Modeling
Fixed-pitch propellers in multicopters are often used. The propeller performance depends on the thrust T (unit: N) and torque M (unit: N·m). They are expressed as

$$T = C_{\rm T}\rho \cdot \left(\frac{N}{60}\right)^2 D_{\rm p}^4 \tag{5.3}$$

$$M = C_{\rm M}\rho \cdot \left(\frac{N}{60}\right)^2 D_{\rm p}^5 \tag{5.4}$$

where N (unit: RPM) denotes the propeller speed, $D_{\rm p}$ (unit: m) denotes the diameter of the propeller, $C_{\rm T}$ and $C_{\rm M}$ denote the dimensionless thrust coefficient and torque coefficient, respectively. The air density ρ (unit: kg/m^3), which varies with respect to the local altitude h (unit: m) and the temperature $T_{\rm t}$ (unit: $^\circ$ C), is written as

$$\rho = \frac{273 P_{\rm a}}{101325(273 + T_{\rm t})}\rho_0 \tag{5.5}$$

where the standard air density $\rho_0 = 1.293$ (kg/m^3) (0 °C, 273 K). The atmospheric pressure $P_{\rm a}$ (unit: Pa) is obtained as

$$P_{\rm a} = 101325\left(1 - 0.0065\frac{h}{273 + T_t}\right)^{5.2561}. \tag{5.6}$$

The height of a multicopter varies slightly while performing a task. Thus, h and T_t are treated as constants in the following performance evaluation.

(2) Motor Modeling

The electric motors used in current multicopters are often brushless DC motors. A brushless DC motor can be modeled as a permanent magnet DC motor. Its equivalent circuit is shown in Fig. 5.14, where U_m (unit: V) is the equivalent motor voltage, I_m (unit: A) is the equivalent motor current, E_a (unit: V) is the back-electromotive force, R_m (unit: Ω) is the armature resistance, and L_m (unit: H) is the armature inductance of the motor. $\hat{I}_0$ (unit: A) is the no-load current required to overcome mechanical friction and air friction in the motor, as well as magnetic hysteresis and eddy current losses in the motor, and is approximately a constant at certain motor speed. In addition, $I_a = I_m - \hat{I}_0$ helps to produce the electromagnetic torque. Here, the armature inductance L_m and the transient processes caused by switching elements are ignored.

The motor modeling aims to obtain U_m (unit: V) and I_m (unit: A) from the motor speed N (equal to the propeller speed), the motor load torque M (equal to the propeller torque), and the motor parameter class Θ_m. For clarity, they are expressed in functions as

$$\begin{aligned} U_m &= f_{U_m}(\Theta_m, M, N) \\ I_m &= f_{I_m}(\Theta_m, M, N) \end{aligned} \tag{5.7}$$

where N and M are determined based on Eqs. (5.3) and (5.4) in the above section, and Θ_m is shown in Table 5.1. The detailed procedure to obtain $f_{U_m}(\Theta_m, M, N)$ and $f_{I_m}(\Theta_m, M, N)$ is shown

$$\begin{aligned} U_m &= f_{U_m}(\Theta_m, M, N) \triangleq \left(\frac{M K_{V0} U_{m0}}{9.55(U_{m0} - I_{m0} R_m)} + I_{m0}\right) R_m + \frac{U_{m0} - I_{m0} R_m}{K_{V0} U_{m0}} N \\ I_m &= f_{I_m}(\Theta_m, M, N) \triangleq \frac{M K_{V0} U_{m0}}{9.55(U_{m0} - I_{m0} R_m)} + I_{m0}. \end{aligned} \tag{5.8}$$

(3) ESC Modeling

An ESC is a device external to the motor that electronically performs the commutation achieved mechanically in brushed motors. The ESC converts the DC voltage of the battery to a three-phase alternating current electrical signal, which is synchronized with the rotation of the rotor and applied to the armature windings. ESCs regulate motor speed within a range depending on the motor load torque and battery voltage. An ESC equivalent circuit is shown in Fig. 5.15. Given the motor modeling results U_m and I_m, the ESC parameter set Θ_e, and the battery parameter set Θ_b (see Table 5.1 for details), the ESC modeling aims to obtain the input throttle command σ (scale value, range from 0 to 1, no unit), the input voltage U_e (unit: V), and the input current I_e (unit: A). For clarity, they are expressed in functions as

$$\begin{aligned}\sigma &= f_\sigma\left(\Theta_{\rm e}, U_{\rm m}, I_{\rm m}, U_{\rm b}\right)\\ I_{\rm e} &= f_{I_{\rm e}}\left(\sigma, I_{\rm m}\right)\\ U_{\rm e} &= f_{U_{\rm e}}\left(\Theta_{\rm b}, I_{\rm e}\right).\end{aligned} \tag{5.9}$$

As shown in Fig. 5.15, $U_{\rm eo}$ denotes the equivalent DC voltage, given by

$$U_{\rm eo} = U_{\rm m} + I_{\rm m} R_{\rm e}. \tag{5.10}$$

Based on Eq. (5.10), the throttle command σ is obtained as

$$\sigma = \frac{U_{\rm eo}}{U_{\rm e}} \approx \frac{U_{\rm eo}}{U_{\rm b}}. \tag{5.11}$$

Furthermore, since the input power of ESC is equal to the output power, the ESC input current is as follows

$$I_{\rm e} = \sigma I_{\rm m} \tag{5.12}$$

which is limited by

$$I_{\rm e} \leq I_{\rm eMax} \tag{5.13}$$

where $I_{\rm eMax}$ (unit: A, from $\Theta_{\rm e}$) denotes the maximum current of ESC. The ESC voltage $U_{\rm e}$ is supplied by the battery voltage $U_{\rm b}$ (unit: V, from $\Theta_{\rm b}$), given by

$$U_{\rm e} = U_{\rm b} - I_{\rm b} R_{\rm b} \tag{5.14}$$

where $I_{\rm b}$ (unit: A) denotes the battery current and $R_{\rm b}$ (unit: Ω, from $\Theta_{\rm b}$) denotes the battery resistance. For a multicopter, the number of ESCs is equal to the number of propulsors $n_{\rm r}$, so

$$I_{\rm b} = n_{\rm r} I_{\rm e} + I_{\rm other} \tag{5.15}$$

where $I_{\rm other}$ (unit: A) includes the current from other devices (like autopilot and camera) and other current loss, usually, $I_{\rm other} \approx 1$ A. From above, the functions are expressed as

$$\sigma = f_\sigma\left(\Theta_{\rm e}, U_{\rm m}, I_{\rm m}, U_{\rm b}\right) \triangleq \frac{U_{\rm m} + I_{\rm m} R_{\rm e}}{U_{\rm b}} \tag{5.16}$$

$$I_{\rm e} = f_{I_{\rm e}}\left(\sigma, I_{\rm m}\right) \triangleq \sigma I_{\rm m} \tag{5.17}$$

$$U_{\rm e} = f_{U_{\rm e}}\left(\Theta_{\rm b}, I_{\rm b}\right) \triangleq U_{\rm b} - I_{\rm b} R_{\rm b}. \tag{5.18}$$

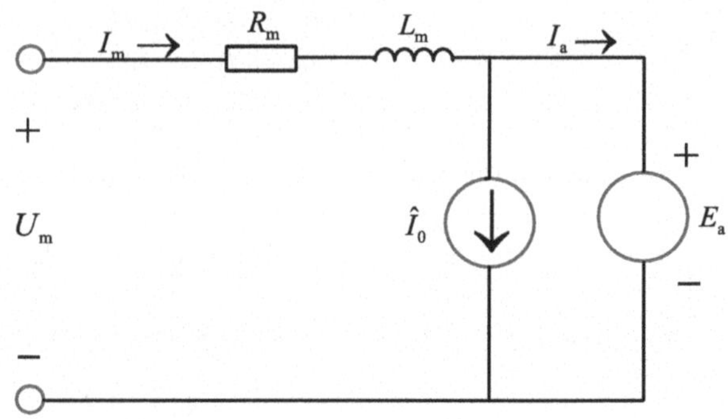

Fig. 5.14 Equivalent motor model

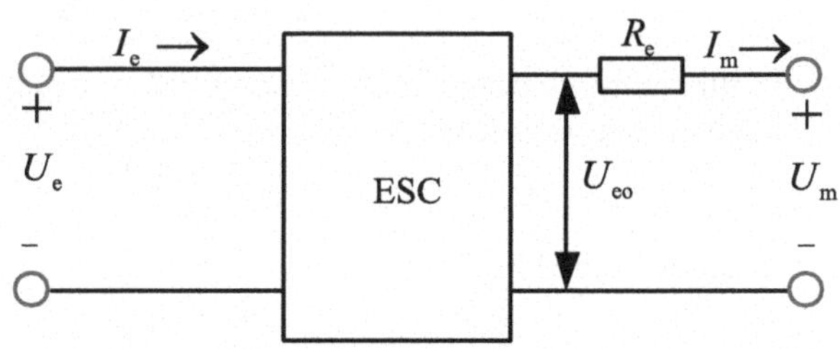

Fig. 5.15 Equivalent ESC model

(4) Battery Modeling

The battery model mainly aims to obtain the endurance time T_b (unit: min) by using I_b and Θ_b, where I_b is expressed in Eq. (5.15) and Θ_b is shown in Table 5.1. For clarity, the model is expressed in an abstract function as

$$T_b = f_{T_b}(\Theta_b, I_b). \tag{5.19}$$

The battery discharge process is simplified so that the battery voltage remains constant and the battery capacity decreases linearly. Thus, the model f_{T_b} is as follows

$$T_b = f_{T_b}(\Theta_b, I_b) \triangleq \frac{C_b - C_{min}}{I_b} \cdot \frac{60}{1000} \tag{5.20}$$

where C_{min} (unit: mAh) denotes the minimum battery capacity, set according to the safety margin. Generally, it can be chosen from 0.15 C_b–0.2 C_b. Finally, the battery current I_b cannot exceed the maximum current that the battery can withstand, which is expressed by the maximum discharge rate K_b (no unit, from Θ_b), so

$$I_b \le \frac{C_b K_b}{1000}. \tag{5.21}$$

5.2 Basic Experiment

5.2.1 Experimental Objective

(1) Things to prepare
The multicopter performance evaluation website: https://flyeval.com/paper/.
(2) Objectives
A multicopter (e.g., a tricopter, coaxial hexacopter, quadcopter, hexacopter, coaxial octocopter, or octocopter) should be configured with hover endurance longer than 10 min by using the evaluation website, where the flight environment parameters are "Altitude": 0 m, and "Air Temperature": 25 °C. In addition, all configuration parameters and basic performance parameters of the designed multicopters should be recorded from the multicopter performance evaluation website https://flyeval.com/paper/.

5.2.2 Experimental Procedure

(1) **Step1: Configure a tricopter**
As shown in Fig. 5.16, the first step is to configure a tricopter in which the "Total Weight" is set to "1.0 kg", the "Frame Size" is set to "450 mm", the "Altitude" is set to "0 m", the "Air Temperature" is set to "25 °C", and the "Aero Design" is set to "medium". In the second row for the autopilot or Flight Controller Unit (FCU) configuration, the "Min. Battery Capacity" is set to "15%" for the remaining capacity to prevent excessive discharge; to ensure enough thrust for the basic altitude control for the multicopter, the "Max. Takeoff Throttle" is set to "85%", meaning that we should keep at least 15% thrust margin for basic attitude control; the "FCU Max. Tilt Limit" is set to "No Limit", meaning that the tilted angle of the autopilot or FCU is unrestricted. The "FCU & Attaches Current" is set to "0.5 A" to define the current consumption by other electronic components on the multicopter (such as FCU) other than the propulsion system.

(2) **Step2: Select the brands and type specification of the motor, propeller, ESC, and battery to comprise a feasible propulsion system**
As an example, Fig. 5.17 presents a propulsion system in which the "Motor Brand" is set to "DJI 2212 KV920"; the "Propeller Brand" is set to "DJI CFP 10 × 3.8" (Diameter $D_p = 10$ inches, Pitch $H_p = 3.8$ inches, Blade Number $B_p = 2$); the "ESC Brand" is set to "Common max 30 A" (Maximum Current

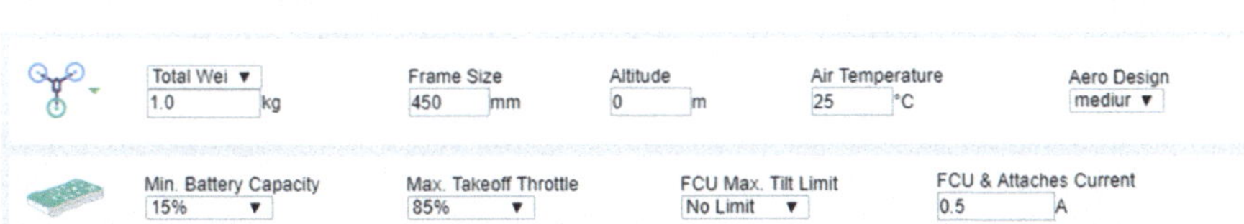

Fig. 5.16 Tricopter basic parameters

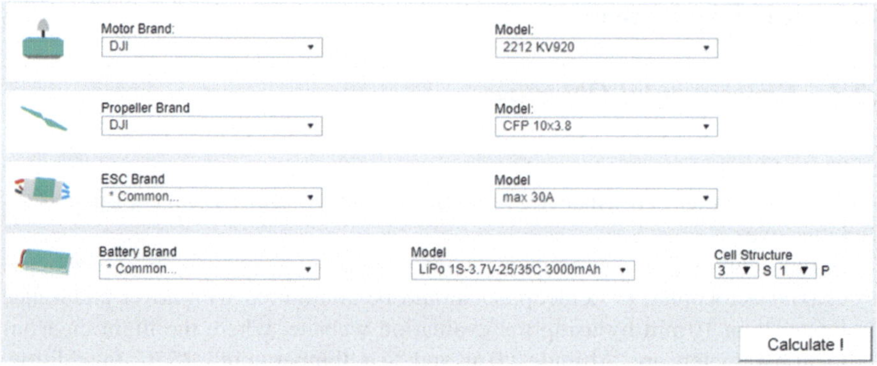

Fig. 5.17 Tricopter propulsion system parameters

$I_{\mathrm{eMax}} = 30\,\mathrm{A}$), commonly-used ESCs with average performance among products on the market; and the "Battery Brand" is set to "Common 1S-3.7V-25/35C-3000 mAh 3S1P". The detailed parameters (see Table 5.1) for the selected battery are listed as follows: the total voltage U_{b} is $1\,\mathrm{S} \times 3 = 3\,\mathrm{S} = 3 \times 3.7\,\mathrm{V} = 11.1\,\mathrm{V}$; the capacity C_{b} is 3000 mAh; and the maximum discharge rate K_{b} is 25 C, implying the battery has a maximum discharge current $K_{\mathrm{b}} \cdot C_{\mathrm{b}} = 75\,\mathrm{A}$. Other parameters, including the weight and the resistance of each component, are estimated by statistical models behind the website www.flyeval.com/paper/.

(3) **Step3: Calculate the parameters and performance of the multicopters**
The parameters and performance of the multicopters are obtained by clicking the "Calculate!" button on the website. As shown in Fig. 5.18, in the "Hovering Performance" tab, the obtained "Hovering Time" is 14.7 min; in the "Max.Throttle.Performance" tab, the obtained "Throttle Percentage" is 54.2%; the obtained "Total Lift" is 26 N; and the obtained "Motor Power" listed in the "Max.Throttle Performance" tab is 134.4 W.

(4) **Step4: Configuration examples for other types of multicopters**

1) A coaxial hexacopter is configured as shown in the red box in Fig. 5.19, in which the airframe configuration in the red box in Fig. 5.19 is set to be a coaxial hexacopter, the "Total Weight" is set to "1.5 kg", the "Motor Brand" is set to "DJI 2312E KV800", and the "Battery Brand" is set to "Common 1S-3.7V-20/30C-3700 mAh 3S1P". Other parameters, including the weight and the resistance of each component, are the same as in Figs. 5.16 and 5.17. As shown in Fig. 5.20, the obtained "Hovering Time" is 13.4 min, the obtained "Throttle Percentage" is 60%, the obtained "Total Lift" is 32.6 N, and the obtained "Motor Power" listed in the "Max.Throttle Performance" tab is 93.2 W.
2) A quadcopter is configured as shown in the red box in Fig. 5.21 in which the airframe configuration in the red box in Fig. 5.21 is set to be a quadcopter. Other parameters, including the weight and the resistance of each

Fig. 5.18 Tricopter performance

component, are the same as in Fig. 5.19. As shown in Fig. 5.22, the obtained "Hovering Time" is 12.3 min, the obtained "Throttle Percentage" is 66.4%, the obtained "Total Lift" is 27.7 N, and the obtained "Motor Power" listed in the "Max.Throttle Performance" tab is 95.6 W.

3) A hexacopter is configured according to Fig. 5.23, in which the airframe configuration in the red box in Fig. 5.23 is set to be a hexacopter and the "Frame Size" is set to "550 mm". Other parameters, including the weight and the resistance of each component, are the same as in Fig. 5.19. As shown in Fig. 5.24, the obtained "Hovering Time" is 14.8 min, the obtained "Throttle Percentage" is 53.2%, the obtained "Total Life" is 39.7 N, and the obtained "Motor Power" listed in the "Max.Throttle Performance" tab is 89.5 W.
4) A coaxial octocopter is configured according to Fig. 5.25, in which the airframe configuration in the red box in Fig. 5.25 is set to be a coaxial octocopter and the "Frame Size" is set to "550 mm". Other parameters, including the weight and the resistance of each component, are the same as in Fig. 5.19. As shown in Fig. 5.26, the obtained "Hovering Time" is 15.2 min, the obtained "Throttle Percentage" is 51.3%, the obtained "Total Lift" is 42 N, and the obtained "Motor Power" listed in the "Max.Throttle Performance" tab is 88.3 W.
5) An octocopter is configured according to Fig. 5.27, in which the airframe configuration in the red box in Fig. 5.25 is set to be an octocopter, the "Frame Size" is set to "650 mm", the "Propeller Brand" is set to "Quanum 8 × 6", and the "Battery Brand" is set to "Common 1S-3.7V-65/100C-5000 mAh

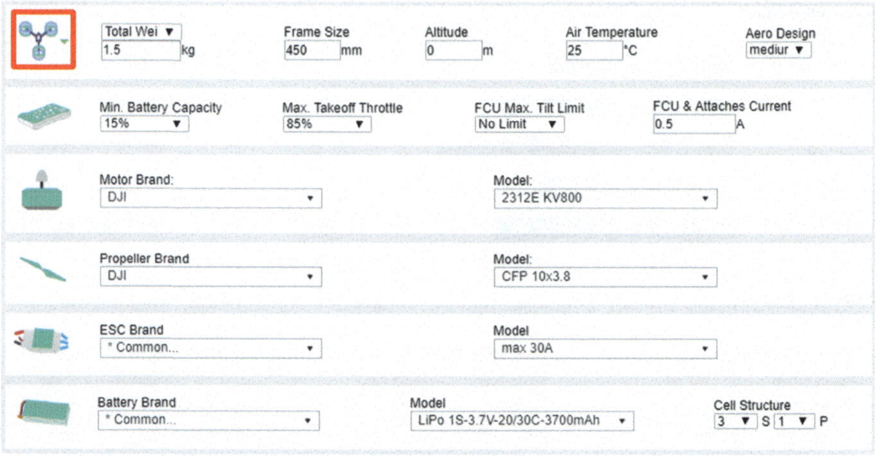

Fig. 5.19 Coaxial hexacopter configuration

Fig. 5.20 Coaxial hexacopter performance

3S1P". Other parameters, including the weight and the resistance of each component, are the same as in Fig. 5.19. As shown in Fig. 5.28, the obtained "Hovering Time" is 20.4 min, the obtained "Throttle Percentage" is 50%, the obtained "Total Lift" is 49.8 N, and the obtained "Motor Power" listed in the "Max.Throttle Performance" tab is 90.8 W.

Fig. 5.21 Quadcopter configuration

Fig. 5.22 Quadcopter performance

Fig. 5.23 Hexacopter configuration

Fig. 5.24 Hexacopter performance

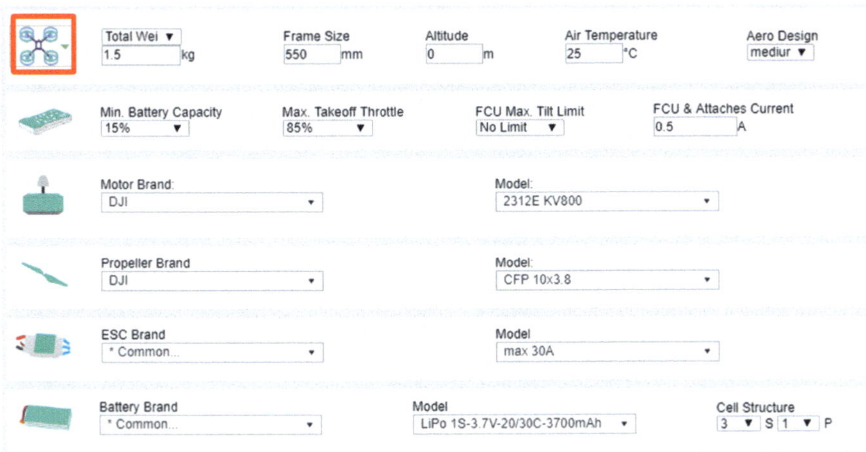

Fig. 5.25 Coaxial octocopter configuration

Fig. 5.26 Coaxial octocopter performance

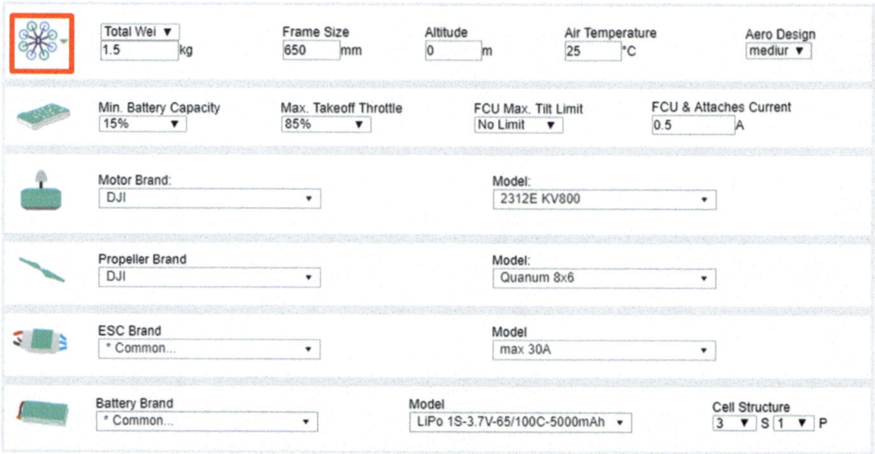

Fig. 5.27 Octocopter configuration

Fig. 5.28 Octocopter performance

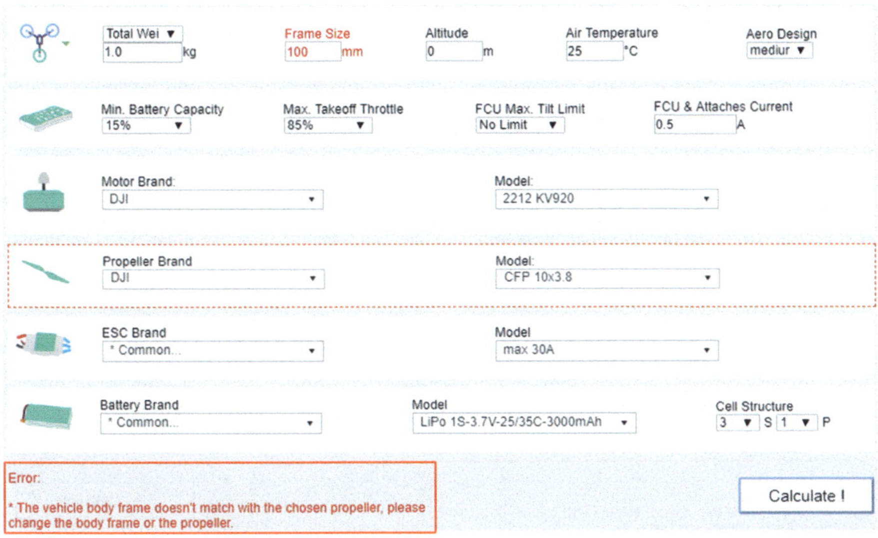

Fig. 5.29 "Diagonal size is too small" error

5.2.3 Remarks

(1) Diagonal size is too small
When the "Propeller Brand" option in the website is chosen as "CFP 10×3.8" (i.e, the diameter is 10 inches ≈ 25.4 mm), the minimum diagonal size for the tricopter is obtained from Eqs. (5.1) and (5.2) as

$$\text{diagnal size} = 10 \times 25.4/\sin(180°/3) \times 1.2 = 352\,\text{mm}.$$

As shown in Fig. 5.29, if the diagonal size is selected too small such as 100 mm, the website will return an error message stating "The vehicle body frame does not match with the chosen propeller, please change the body frame of the propeller".

(2) Current is too large
When a battery is selected with too large of a voltage, the ESCs and motors may be burnt out from exceeding their voltage and current limits. Figure 5.30 presents the error message for an incorrectly-configured tricopter whose battery voltage is selected as 5 S (1S = 3.7 V), whereas the maximum voltage for the selected motor is only 3 S.

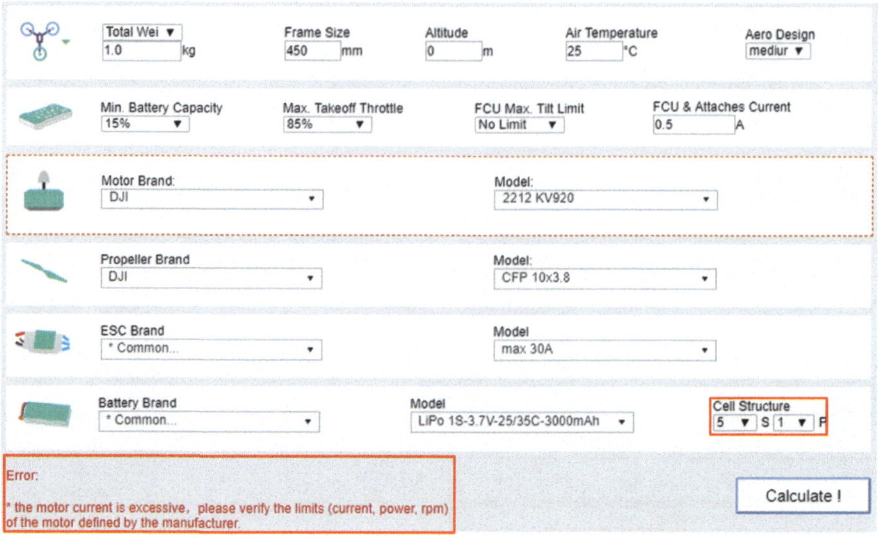

Fig. 5.30 "Current is too large" error

5.3 Analysis Experiment

5.3.1 Experimental Objective

(1) Things to prepare

1) The experiment requires a quadcopter whose total weight is 1.5 kg $\times$ 9.8 m/s^2 = 14.7 N, and a scenario where the flight altitude is 50 m and the local temperature is 25 °C.
2) The parameters of the propulsion system are listed in Table 5.2.

(2) Objectives

1) Show a detailed process for calculating the hover endurance and compare the obtained result with the result from the multicopter performance evaluation website https://flyeval.com/paper/;
2) Calculate the hover endurance values under different temperatures (e.g., 0, 10, 20, 30, 40 °C) at different locations such as Beijing, Shanghai, Lhasa, and Changsha. Then, based on the obtained results, analyze the hover endurance with respect to altitude and temperature;
3) Analyze the hover endurance with respect to the size and number of propellers.

Table 5.2 Propulsion system parameters

Component	Parameters
Propeller	APC1045 ($D_p = 10$ inches, $H_p = 4.5$ inches, $B_p = 2$), $C_T = 0.0984$, $C_M = 0.0068$
Motor	Sunnysky A2814-900 ($K_{V0} = 900$ RPM/V, $R_m = 0.08\ \Omega$, $W_{mMax} = 335$ W, $I_{eMax} = 0.6$ A, $U_{m0} = 10$ V)
ESC	$I_{eMax} = 30$ A, $R_e = 0.008\ \Omega$
Battery	ACE ($C_b = 4000$ mAh, $U_b = 12$ V), $R_b = 0.0084\ \Omega$, $K_b = 65$ C

5.3.2 Calculation and Analysis Procedure

5.3.2.1 Calculation Procedure for First Objective

(1) **Step1: Calculate the thrust**
The thrust generated by a single propeller is calculated based on the total weight of the quadcopter as follows

$$T = \frac{G}{n_r} = \frac{1.5 \times 9.8}{4} = 3.675\,\text{N}. \tag{5.22}$$

(2) **Step2: The motor speed and the propeller torque are calculated based on the propeller model**
First of all, by using the flight altitude and the temperature, based on Eq. (5.6), the atmospheric pressure P_a is obtained as

$$\begin{aligned} P_a &= 101325\left(1 - 0.0065\frac{h}{273 + T_t}\right)^{5.2561} \\ &= 101325\left(1 - 0.0065\frac{50}{273 + 25}\right)^{5.2561} \\ &= 100745.52\,\text{Pa}. \end{aligned} \tag{5.23}$$

Then, by using the obtained atmospheric pressure in Eq. (5.23), the air density ρ is obtained based on Eq. (5.5) as

$$\begin{aligned} \rho &= \frac{273 P_a}{101325(273 + T_t)}\rho_0 \\ &= \frac{273 \times 100745.52}{101325 \times (273 + 25)} \times 1.293 \\ &= 1.178\,\text{kg/m}^3. \end{aligned} \tag{5.24}$$

Next, by using the air density and the parameters of the propeller, the motor speed N is obtained based on Eq. (5.3) as

$$
\begin{aligned}
N &= 60\sqrt{\frac{T}{\rho {D_{\mathrm{p}}}^{4} C_{\mathrm{T}}}} \\
&= 60\sqrt{\frac{3.675}{1.178 \times (10 \times 25.4/1000)^{4} \times 0.0984}} \\
&= 5236.51\,\mathrm{RPM}.
\end{aligned}
\tag{5.25}
$$

Finally, based on Eq. (5.4), the torque M of the propeller is obtained as

$$
\begin{aligned}
M &= C_{\mathrm{M}}\rho\left(\frac{N}{60}\right)^{2} D_{\mathrm{p}}^{5} \\
&= 0.0068 \times 1.178 \times \left(\frac{5236.51}{60}\right)^{2} \times (10 \times 25.4/1000)^{5} \\
&= 0.0645\,\mathrm{N}\cdot\mathrm{m}.
\end{aligned}
\tag{5.26}
$$

(3) **Step3: The equivalent motor current I_{m} and the equivalent motor voltage U_{m} are calculated based on the motor model**
Based on Eq. (5.8), I_{m} and U_{m} are obtained as

$$
\begin{aligned}
I_{\mathrm{m}} &= \frac{M K_{\mathrm{V0}} U_{\mathrm{m0}}}{9.55(U_{\mathrm{m0}} - I_{\mathrm{m0}} R_{\mathrm{m}})} + I_{\mathrm{m0}}. \\
&= \frac{0.0645 \times 900 \times 10}{9.55(10 - 0.6 \times 0.08)} + 0.6 \\
&= 6.708\,\mathrm{A}
\end{aligned}
\tag{5.27}
$$

and

$$
\begin{aligned}
U_{\mathrm{m}} &= \left(\frac{M K_{\mathrm{V0}} U_{\mathrm{m0}}}{9.55(U_{\mathrm{m0}} - I_{\mathrm{m0}} R_{\mathrm{m}})} + I_{\mathrm{m0}}\right) R_{\mathrm{m}} + \frac{U_{\mathrm{m0}} - I_{\mathrm{m0}} R_{\mathrm{m}}}{K_{\mathrm{V0}} U_{\mathrm{m0}}} N \\
&= \left(\frac{0.0645 \times 900 \times 10}{9.55(10 - 0.6 \times 0.08)} + 0.6\right) \times 0.08 + \frac{10 - 0.6 \times 0.08}{900 \times 10} \times 5236.51 \\
&= 6.327\,\mathrm{V}.
\end{aligned}
\tag{5.28}
$$

(4) **Step4: The ESC input throttle σ, the ESC input current I_{e} and the ESC input voltage U_{e} are calculated based on the ESC model**
Based on Eq. (5.16), the ESC input throttle command is obtained as

$$
\sigma = \frac{U_{\mathrm{m}} + I_{\mathrm{m}} R_{\mathrm{e}}}{U_{\mathrm{b}}} = \frac{6.327 + 6.708 \times 0.008}{12} = 0.532. \tag{5.29}
$$

Subsequently, based on Eq. (5.12), the ESC input current is obtained as

$$I_e = \sigma I_m = 0.532 \times 6.708 = 3.567\,\text{A}. \tag{5.30}$$

Finally, based on Eq. (5.14), the ESC input voltage is obtained as

$$U_e = U_b - I_b R_b = 12 - 14.768 \times 0.0084 = 11.876\,\text{V}. \tag{5.31}$$

(5) **Step5: The hover endurance T_b is calculated based on the battery model**
Based on Eq. (5.15), the battery current is obtained as

$$I_b = n_r I_e + I_{other} = 4 \times 3.567 + 0.5 = 14.768\,\text{A}. \tag{5.32}$$

Finally, the minimum battery capacity is taken as 15% of the total capacity, based on Eq. (5.20), the hover endurance is obtained as

$$T_b = \frac{C_b - C_{min}}{I_b} \cdot \frac{60}{1000} = \frac{4000 - 4000 \times 0.15}{14.768} \times \frac{60}{1000} = 13.8\ \text{min}. \tag{5.33}$$

(6) **Step6: The result calculated above is compared by using the multicopter performance evaluation website** https://flyeval.com/paper/
A quadcopter is configured according to Table 5.2, and the configuration of the multicopter performance evaluation website is shown in Fig. 5.31, where the "Total Weight" is set to "1.5 kg", the "Frame Size" is set to "450 mm", the "Altitude" is set to "50 m", the "Air Temperature" is set to "25 °C", and the "Aero Design" is set to "medium". The "Motor Brand" is set to "SunnySky Angle A2814-KV900", and the "Propeller Brand" is set to "APC 10 × 4.5 MR" (Diameter $D_p = 10$ inches, Pitch $H_p = 4.5$ inches, Blade Number $B_p = 2$). The "ESC Brand" is set to "Customized...", where "Constant Discharge Current" is set to "30 A", "Max.Lipo Cells" is set to "3 S", and "Resistance" is set to "8 mΩ". The "Battery Brand" is also set to "Customized...", where "Capacity" is set to "4000 mAh", "Max.Const.C" is set to "65 C", and "Resistance" is set to "8.4 mΩ". As shown in Fig. 5.32, the obtained "Hovering Time" is 13.8 min, the obtained "Throttle Percentage" is 53.7%, the obtained "Total Lift" is 41.4 N, and the obtained "Motor Power" as listed in the "Max. Throttle Performance" tab is 166.6 W. It can be observed that the calculated hover endurance is consistent with the result on the multicopter performance evaluation website https://flyeval.com/paper/.

5.3.2.2 Analysis Procedure for Second Objective

(1) **Hover endurance with respect to altitude**
The basic configuration parameters of our chosen testing multicopter are shown in Fig. 5.33. The "Total Weight" is set to"1.5 kg", the "Frame Size" is set to

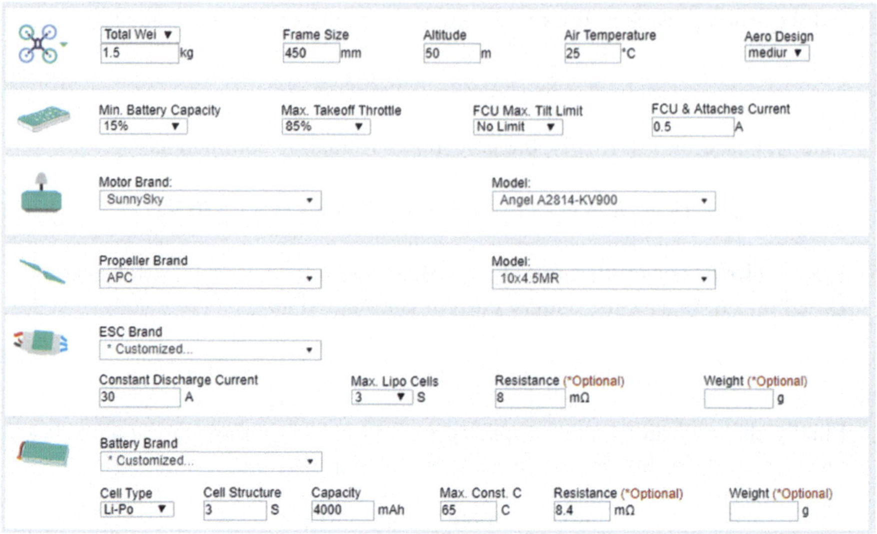

Fig. 5.31 Quadcopter configuration for comparison

Detail Information

Hovering Performance :		Max. Throttle Performance :		Integral Performance :	
Hovering Time	: 13.8 min.	Flying Time	: 2.9 min.	Normal Operation	: 10.3 min.
Throttle Percentage	: 53.7 %	Total Lift	: 41.4 N	Total Weight	: 1.5 kg
Motor Current	: 3.56 A	Motor Current	: 17.7 A	Remaining Load	: 1.76 kg
Motor Speed	: 5235.8 rpm	Motor Speed	: 8788.1 rpm	Max. Takeoff Altitude	: 6.14 km
Motor Power	: 35.2 W	Motor Power	: 166.6 W	Max. Tilt Angle	: 62.6 °
Battery Voltage	: 11.9 V	Battery Voltage	: 11.4 V	Max. Forward Speed	: 14.7 m/s
Battery Current	: 14.7 A	Battery Current	: 71 A	Max. Flying Range	: 4.43 km
Power Efficiency	: 79.6 %	Power Efficiency	: 78.3 %	Wind Resistance	: 5 Degree

Fig. 5.32 Results of the multicopter performance evaluation website

"450 mm", and the "Altitude" is set to "4 m". The "Motor Brand" is set to "DJI 2212 KV920"; the "Propeller Brand" is set to "DJI Turnigy slow fly 9.4 × 5" (Diameter $D_p = 9.4$ inches, Pitch $H_p = 5$ inches, Blade Number $B_p = 2$); the "ESC Brand" is set to "Common max 30 A" (Maximum Current $I_{eMax} = 30$ A) with average performance among products on the market; and the "Battery Brand" is set to "Common 1S-3.7V-20/30C-5000 mAh 3S1P". Other parameters, including the weight and resistance of each component, are estimated by using statistical models behind the multicopter performance evaluation website www.flyeval.com/paper/.

With the other configurations unchanged, Table 5.3 presents the changing trend of hover endurance in different cities. From Table 5.3, a conclusion can be drawn that the higher the altitude is, the shorter the hover endurance will be.

Fig. 5.33 Multicopter configuration for study on hover endurance with respect to altitude

Table 5.3 Hover endurance with respect to altitude

Site	Altitude (m)	Hover endurance (min)
Shanghai	4	16.5
Beijing	43.5	16.5
Changsha	500	16.1
Lhasa	3658	13.5

This is because the air density is decreased as the altitude is increased. When the propeller thrust is the same, the smaller the air density is, the greater the motor speed will be. When the propeller torque is constant, the higher the motor speed is, the larger the equivalent motor voltage will be. Furthermore, the ESC input current will be increased, and then the battery discharge current will also be increased. Therefore, when the battery capacity is constant, the higher the battery current is, the shorter the hover endurance is. These results, as given by the website, are consistent with the theoretical analysis.

(2) **Hover endurance with respect to temperature**
The basic configuration parameters of our chosen testing multicopter are shown in Fig. 5.34. The "Total Weight" is set to "1.5 kg", the "Frame Size" is set to "450 mm", and the "Altitude" is set to "43.5 m". The "Motor Brand" is set to "DJI 2212 KV920"; the "Propeller Brand" is set to "DJI Turnigy slow fly 9.4x5" (Diameter $D_p = 9.4$ inches, Pitch $H_p = 5$ inches, Blade Number $B_p = 2$); the "ESC Brand" is set to "Common max 30 A" (Maximum Current $I_{eMax} = 30$ A); and the "Battery Brand" is "Common 1S-3.7V-20/30C-5000 mAh 3S1P". Other parameters, including the weight and resistance of each component, are estimated by statistical models behind the multicopter performance evaluation website www.flyeval.com/paper/.

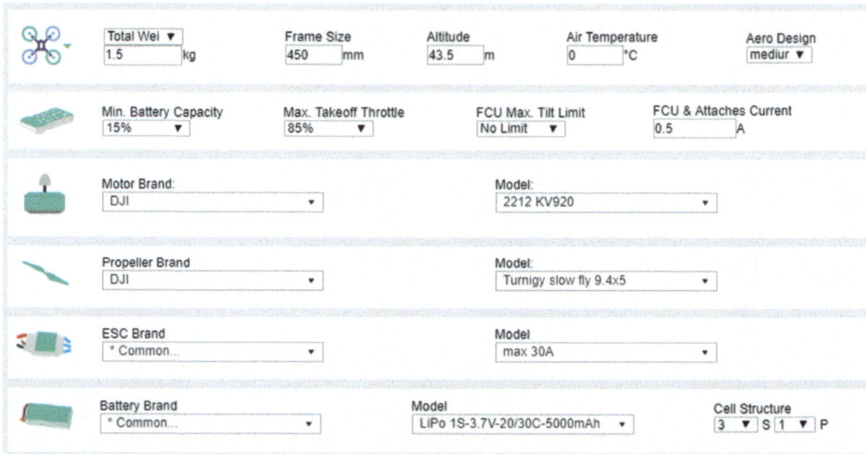

Fig. 5.34 Multicopter configuration for study on hover endurance with respect to temperature

Table 5.4 Hover endurance with respect to temperature

Temperature (°C)	Hover endurance (min)
0	17.1
10	16.8
20	16.6
30	16.3
40	16.1

With the other configurations unchanged, Table 5.4 presents the changing trend of hover endurance with respect to different temperatures. From Table 5.4, one can conclude that the higher the temperature is, the shorter the hover endurance is, as the air density is decreased with increased temperature. By recalling the conclusion of the altitude test, one can further conclude that the lower the air density is, the shorter the hover endurance is.

5.3.2.3 Analysis Procedure for Third Objective

(1) **Hover endurance with respect to propeller size**
The basic configuration parameters of the chosen testing multicopter are shown in Fig. 5.35, where the "Air Temperature" is set to 0 °C, and other configuration parameters are the same as in Fig. 5.34. As the "Frame Size" is set to 450 mm, according to Eq. (5.2), $r_{\max} = 1.1r_{\mathrm{p}}$, and the radius r_{p} is selected from choices up to 11.4 inches. The hover endurance with respect to propeller size is obtained as shown in Table 5.5. From Table 5.5, one can conclude that the larger the propeller is, the longer the hover endurance is.

Fig. 5.35 Multicopter configuration for study on hover endurance with respect to propeller size

Table 5.5 Hover endurance with respect to propeller size

Propeller size (inch)	Hover endurance (min)
10	17
9.4	16.5
9	15.9
8	14.5

(2) **Hover endurance with respect to the number of propellers**
The basic configuration parameters of the chosen testing multicopter are shown in Fig. 5.36. The "Total Weight" is set to "1.5 kg", the "Frame Size" is set to "550 mm", and the "Altitude" is set to "43.5 m". The "Motor Brand" is set to "DJI 2212 KV920"; the "Propeller Brand" is set to "DJI Turnigy slow fly 8 × 4.5" (Diameter $D_p = 8$ inches, Pitch $H_p = 4.5$ inches, Blade Number $B_p = 2$); the "ESC Brand" is set to "Common max 30 A" (Maximum Current $I_{eMax} = 30$ A); and the "Battery Brand" is set to "Common 1S-3.7V-20/30C-5000 mAh 3S1P". Other parameters, including the weight and resistance of each component, are estimated by statistical models behind the multicopter performance evaluation website https://flyeval.com/paper/. Table 5.6 concerns the hover endurance, and is summarized by modifying the "Frame Shape" parameter in Fig. 5.36. From the table, one can conclude that when the number of propellers is equal, the hover endurance is shorter for a coaxial multicopter. For example, a coaxial octocopter has a shorter hover endurance than a corresponding octocopter. However, it should be noted that the coaxial multicopter has a smaller size than a corresponding multicopter with the same propellers. In general, when the total weight is the same, the more the number of propellers is, the longer the hover endurance is.

Fig. 5.36 Multicopter configuration for study on hover endurance with respect to the number of propellers

Table 5.6 Hover endurance with respect to the number of propellers

Type	Hover endurance (min)
Octocopter	18.4
Coaxial octocopter	17.2
Hexacopter	16.8
Quadcopter	14.5
Coaxial hexacopter	15.5
Tricopter	Too heavy to take off

5.4 Design Experiment

5.4.1 Experimental Objective

(1) Things to prepare
The multicopter performance evaluation website https://flyeval.com/paper/.
(2) Objectives

1) Design a multicopter. The altitude is 0 m, the local temperature is 25 °C, the load weight is $1.0\,\mathrm{kg} \times 9.8\,\mathrm{m/s^2} = 9.8$ N, the weight of airframe, autopilot, and other accessories is also $1.0\,\mathrm{kg} \times 9.8\,\mathrm{m/s^2} = 9.8$ N, the total weight is lighter than $5\,\mathrm{kg} \times 9.8\,\mathrm{m/s^2} = 49$ N, the circumferential circle radius is smaller than 39.37 inches (approximately 1 m), the hover endurance is longer than 15 min, and the hover throttle is less than 65% of the full-throttle (Note: the throttle here refers to the ratio of the hover thrust force to the maximum thrust force, which is different from the actual throttle definitions of RC and ESC);
2) The configuration parameters and basic flight performance parameters of the multicopters should be listed and compared with the results from multicopter the performance evaluation website https://flyeval.com/paper/.

5.4.2 Design Procedure

(1) **Step1: Choose to design a quadcopter**
A quadcopter is designed to meet the objectives in the design experiment, i.e., the altitude is 0 m, the local temperature is 25 °C, the load weight is $1.0\,\text{kg} \times 9.8\,\text{m/s}^2 = 9.8\,\text{N}$, the weight of airframe, autopilot and other accessories are also $1.0\,\text{kg} \times 9.8\,\text{m/s}^2 = 9.8\,\text{N}$, the total weight is lighter than $5\,\text{kg} \times 9.8\,\text{m/s}^2 = 49\,\text{N}$, the circumferential circle radius is smaller than 39.37 inches (approximately 1 m), the hover endurance is longer than 15 min, and the hover throttle is less than 65% of the full-throttle.

(2) **Step2: Determine the thrust of a single propeller**
When the quadcopter is hovering, the thrust of a single propeller for the total weight of $5\,\text{kg} \times 9.8\,\text{m/s}^2 = 49\,\text{N}$ is

$$T_{\text{h,up}} = \frac{5}{4} \times 9.8 = 12.25\,\text{N}. \tag{5.34}$$

To leave a safety control margin, a hover thrust ratio (propeller hover thrust/maximum thrust) should be determined according to the maneuverability requirement. By taking the hover thrust ratio to be 0.65, the maximum thrust of a single propeller is obtained as $T_{\text{max,up}} = T_{\text{h,up}}/0.65 = 18.85\,\text{N}$. When the propulsion system weight is not considered, the propeller should provide at least the total weight of the payload and the airframe, autopilot, and accessories, etc., which is $2\,\text{kg} \times 9.8\,\text{m/s}^2 = 19.6\,\text{N}$. Thus, the thrust of the single propeller is

$$T_{\text{h,down}} = \frac{2}{4} \times 9.8 = 4.9\,\text{N}. \tag{5.35}$$

To leave a safety margin, the maximum thrust of a single propeller is $T_{\text{max,down}} = T_{\text{h,down}}/0.65 = 7.54\,\text{N}$. In other words, when selecting a motor, only motors with thrust between $T_{\text{max,down}}$ and $T_{\text{max,up}}$ are considered. This greatly reduces the selection range.

(3) **Step3: Calculate the maximum size of the propeller**
Based on the maximum size limit of the quadcopter and Eqs. (5.1) and (5.2), the following relationship is obtained as

$$r_{\max} + R < 1\,\text{m} \tag{5.36}$$

namely,

$$r_{\max} + \frac{r_{\max}}{\sin(180/4)} < 1\,\text{m}. \tag{5.37}$$

Then $r_{\max} < 0.414\,\text{m}$. In order to leave a safety margin, the maximum size of propeller has to satisfy as

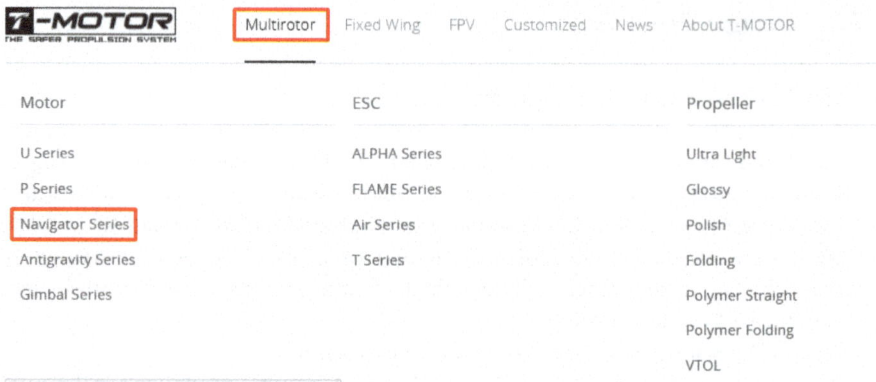

Fig. 5.37 T-MOTOR motor selection

$$r_p = r_{max}/(1.05 \sim 1.2) = 345 \sim 394\,\text{mm}. \tag{5.38}$$

The size of the propeller is often expressed in diameter and the maximum size of the propeller is between 27.2 inches and 31 inches.

(4) **Step4: Select a motor**
Appropriate motors are selected by browsing the manufacturer's official website, such as T-MOTOR motors. Readers can go to the T-MOTOR official website: http://uav-en.tmotor.com/ to select multicopters such as the one shown in Fig. 5.37. According to the maximum thrust range calculated above, i.e. from 1.0 kg to 2.5 kg, the MN4014 motor in the MN series is selected. Then, click the "Parameter" button, and the detailed information of the motor will be obtained as shown in Fig. 5.38. Furthermore, readers can observe parameters they are interested in, such as the motor weight and maximum current, as shown in Fig. 5.39. In "Load Testing Data", readers can observe the thrust of the motor under different throttle values, as also shown in Fig. 5.39. When the voltage is 22.2 V and the propeller is T-MOTOR 15 × 5 CF, the max thrust is 1.92 kg (18.82 N).

(5) **Step5: Select an ESC**
If "T-MOTOR ESC" is selected and the maximum current of the selected motor is 25 A, then the "AIR 40 A ESC" with a continuous current of 40 A (as shown in Fig. 5.40) is selected.

(6) **Step6: Select a propeller**
Here, the T-MOTOR propeller is selected. Moreover, according to the matches being offered, the P15 × 5 propeller is selected, as shown in Fig. 5.41. From the basic parameters, the weight of a single propeller is 26.5 g, as shown in Fig. 5.42.

(7) **Step7: Select a battery**
Here, if GENS ACE batteries are selected, readers can go to search the desired products on E-commerce sites (e.g., www.ebay.com or www.taobao.com), and then select the high voltage version, as shown in Fig. 5.43.

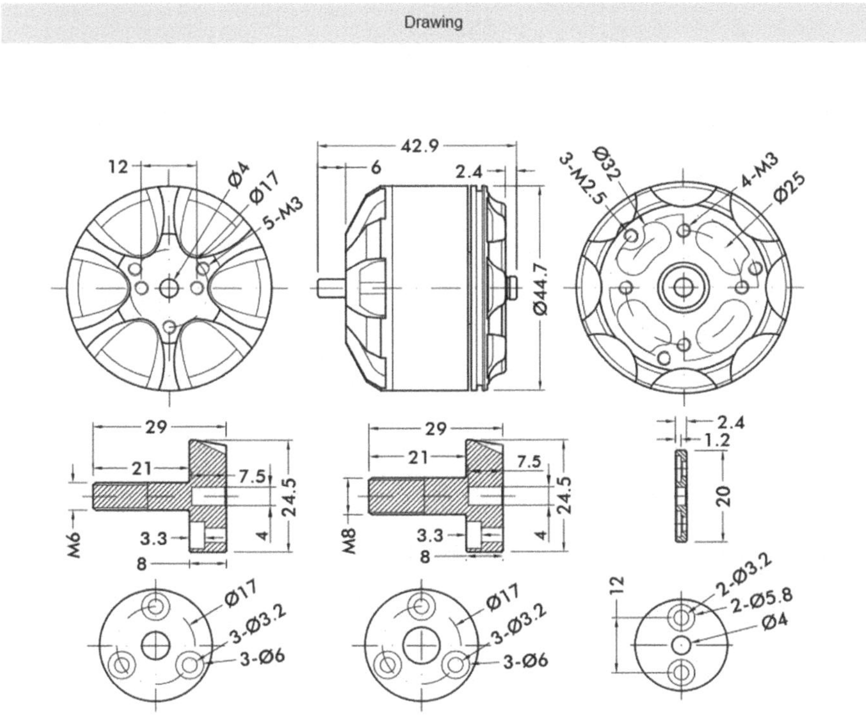

Fig. 5.38 Motor geometry parameters

After the battery is selected, readers can determine the total weight of the multicopter, and then calculate whether the hover throttle and hover endurance meet the requirements in this experiment. The total weight, hover throttle and hover endurance are calculated for the three batteries, If a 12000 mAh battery is selected, according to the motor, ESC and propeller selected above, the total weight is as follows

$$G = (2 + (0.171 + 0.026 + 0.0265) \times 4 + 1.46) \times 9.8 = 42.6692\,\text{N}.$$

Then, the thrust provided by a single propeller is as follows

$$T_{\mathrm{h}} = \frac{4.354}{4} \times 9.8 = 10.6673\,\text{N}. \tag{5.39}$$

As shown in Fig. 5.39, the throttle at this time is less than 65% of the full-throttle. Then, it can be determined that the thrust is 10.662 N and the motor current is 5.78 A, by using simple linear interpolation between 50% and 65% of the throttle. The hover endurance is calculated as

Load Testing Data									
Ambient Temperature		/			Voltage			DC Power Supplier	
Item No.	Voltage (V)	Prop	Throttle	Current (A)	Power (W)	Thrust (G)	RPM	Efficiency (G/W)	Operating Temperature (°C)
MN4014 KV330	22.2	T-MOTOR 15*5CF	50%	3.6	79.92	830	3900	10.39	45
			65%	5.9	130.98	1150	4600	8.78	
			75%	7.8	173.16	1430	5100	8.26	
			85%	10.1	224.22	1690	5600	7.54	
			100%	11.9	264.18	1920	6000	7.27	
		T-MOTOR 16*5.4CF	50%	4.3	95.46	950	3700	9.95	50
			65%	7	155.40	1420	4400	9.14	
			75%	9.6	213.12	1750	4900	8.21	
			85%	12.5	277.50	2060	5400	7.42	
			100%	14.7	326.34	2390	5600	7.32	
		T-MOTOR 17*5.8CF	50%	4.7	104.34	1050	3400	10.06	55
			65%	8	177.60	1580	4100	8.90	
			75%	10.7	237.54	1970	4600	8.29	
			85%	14.4	319.68	2300	5100	7.19	
			100%	17	377.40	2600	5400	6.89	

Fig. 5.39 Motor specifications

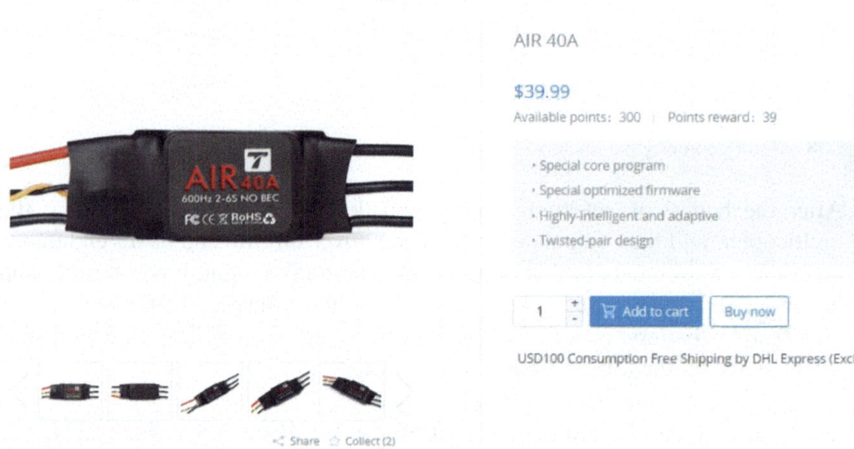

SPECIFICATIONS

Model	Con,Current	Peak Current (10S)	BEC	Lipo	programmable Item	Weight	Size(L*W*H)
AIR 40A	40A	60A	NO	2-6S	Timing (Intermediate/High)	26g	55.6mm*25.2mm*11.3mm

Fig. 5.40 T-MOTOR ESC specifications

Fig. 5.41 T-MOTOR propellers

Specifications			
Diameter/Pitch	15" *5 (381mm*127mm)	Working Temp	-40℃ ~ 65℃
Weight (single propeller)	26.5g	Storage Temp	-10℃ ~ 50℃
Material	CF+Epoxy	Storage Humidity	<85%
Surface Treatment	Polished	Optimum RPM	5200-7000 RPM/min
Propeller type	2blades-integrated	Thrust Limitation	6kg

Fig. 5.42 A T-MOTOR propeller's specification

$$T_{\mathrm{b}} = \frac{C_{\mathrm{b}} - C_{\mathrm{min}}}{I_{\mathrm{b}}} \cdot \frac{60}{1000} = \frac{0.85 \times 12000}{5.78 \times 4 + 0.5} \times 0.06 = 25.9 \text{ min}. \tag{5.40}$$

The calculated results for the other two batteries are shown in Table 5.7.

(8) **Step8: Recalculate the diagonal size**

The diagonal size is recalculated based on Eq. (5.3) that is

$$2R = \frac{2 \times r_{\mathrm{p}}}{\sin(\frac{180}{n_{\mathrm{r}}})} = \frac{15 \times 25.4}{\sin(180/4)} = 539\,\text{mm}. \tag{5.41}$$

To leave a safety margin, a diagonal size of 600 mm is adopted.

(9) **Step9: Compare with the flight performance calculated by the performance evaluation website**

The basic configuration set on the performance evaluation website https://flyeval.com/paper/ is presented in Fig. 5.44. The "Frame+Load Weight" is set to "2.0 kg", the "Frame Size" is set to "600 mm", and the "Altitude" is set to "0 m". The "Motor Brand" is set to "T-MOTOR MN4014 KV330", the "Propeller Brand" is set to "T-MOTOR 15 × 5 CF" (Diameter $D_{\mathrm{p}} = 15$ inches, Pitch

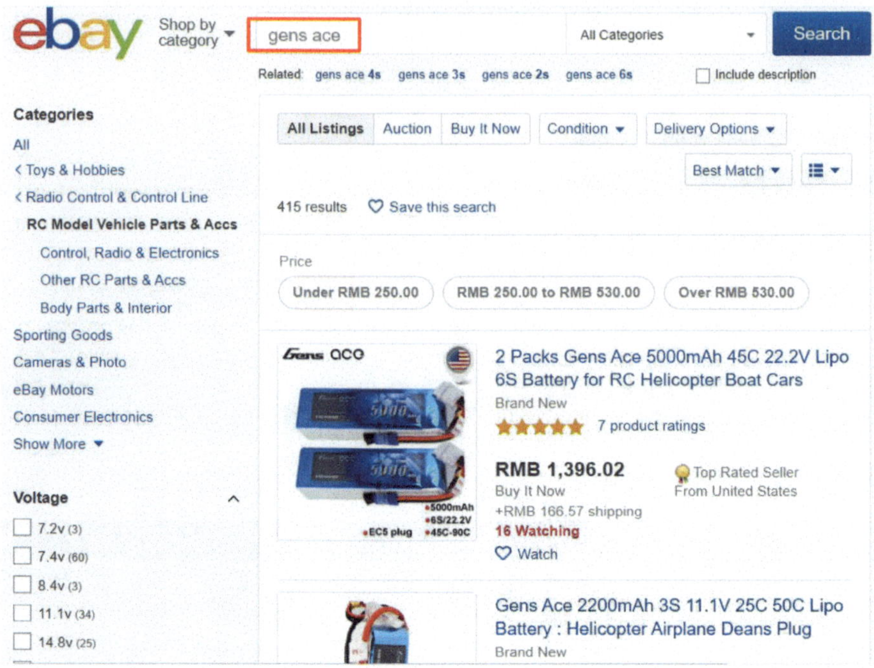

Fig. 5.43 GENS ACE batteries

Table 5.7 Hover throttle and hover endurance

	Total weight (kg)	Hover throttle (%)	Hover endurance (min)
Tattu 12000 mAh	4.354	<65	25.9
Tattu 16000 mAh	4.774	≈ 65	33.8
Tattu 22000 mAh	5.254	–	–

$H_p = 5$ inches, Blade Number $B_p = 2$), the "ESC Brand" is set to "Common max 40 A" (Maximum Current $I_{eMax} = 40$ A), and the "Battery Brand" is set to "ACE Lipo TATTU 6S-22.2V-15C-12000 mAh 1S1P". Other parameters, including the weight and resistance of each component, are estimated by using the statistical models behind the performance evaluation website https://flyeval.com/paper/. Additionally, as shown in Fig. 5.45, the obtained "Hovering Time" is 22.5 min, the obtained "Throttle Percentage" is 63.6%, the obtained "Total Lift" is 94.3 N, and the obtained "Motor Power" listed in the "Max.Throttle Performance" tab is 417.8 W. These meet the requirements from the design experiment. In addition, the website also lists the parameters for the control model of the designed multicopter as shown in Fig. 5.46. These control modeling parameters will be used in Chap. 6.

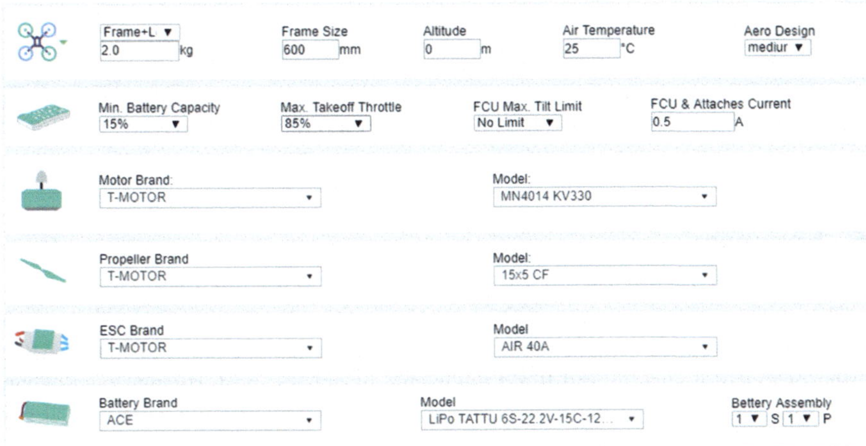

Fig. 5.44 Multicopter configuration for design experiment

Detail Information

Hovering Performance :		Max. Throttle Performance :		Integral Performance :	
Hovering Time	: 22.5 min.	Flying Time	: 7 min.	Normal Operation	: 17.8 min.
Throttle Percentage	: 63.6 %	Total Lift	: 94.3 N	Total Weight	: 4.56 kg
Motor Current	: 6.69 A	Motor Current	: 21.8 A	Remaining Load	: 2.8 kg
Motor Speed	: 4623.5 rpm	Motor Speed	: 6716.3 rpm	Max. Takeoff Altitude	: 3.85 km
Motor Power	: 132.2 W	Motor Power	: 417.8 W	Max. Tilt Angle	: 51.7 °
Battery Voltage	: 23.7 V	Battery Voltage	: 22.9 V	Max. Forward Speed	: 12.4 m/s
Battery Current	: 27.2 A	Battery Current	: 87.3 A	Max. Flying Range	: 8.5 km
Power Efficiency	: 80.9 %	Power Efficiency	: 79.8 %	Wind Resistance	: 4 Degree

Fig. 5.45 Flight performance of designed multicopter

Mathematical Model (Doc)

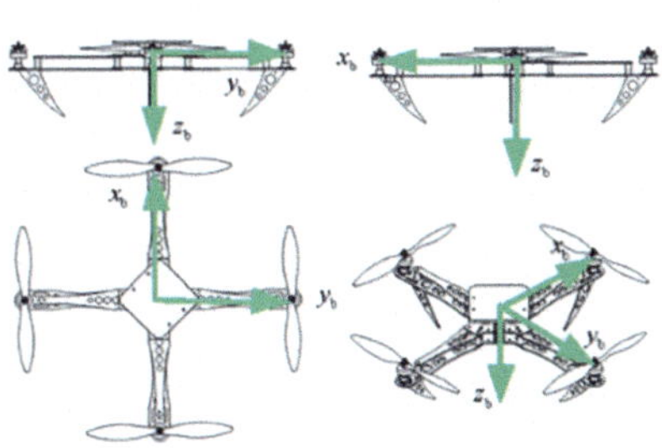

MultiCopter Mass	: m = 4.56 kg
Acceleration of Gravity	: g = 9.8 m/s²
MultiCopter Inertia Matrix J=diag(J_{xx}, J_{yy}, J_{zz})	: J_{xx} = 6.615e-2 kg.m²
	J_{yy} = 6.615e-2 kg.m²
	J_{zz} = 1.179e-1 kg.m²
Distance of Motor to Center	: d = 0.3 m
Propeller Thrust Coef. (T_p/ω^2)	: C_T = 4.764e-5 N/(rad/s)²
Propeller Moment Coef. (M_p/ω^2)	: C_M = 1.165e-6 N.m/(rad/s)²
Throttle σ to Motor steady speed ω_{ss}	: C_R = 586.11 rad/s
($\omega_{ss}=C_R*\sigma+\omega_b$)	: ω_b = 111.3 rad/s
Motor-Propeller Inertia	: J_m = 1.32e-3 kg.m²
Motor Response Time Constant	: T_m = 0.0646 s
MultiCopter Air-Drag Coef. (D/v²)	: C_d = 0.407 N/(m/s)²

Fig. 5.46 Model parameters

5.4.3 Remarks

Through the above design and calculation procedure, a multicopter can be built according to the actual payload, size limit, hover endurance and budget.

5.5 Summary

(1) The performance evaluation of a multicopter can be easily obtained through the multicopter performance evaluation website https://flyeval.com/paper/. After the propulsion system and flight environment are set, performance results can be obtained, such as the hover endurance, available payload, one-way flight distance, and maximum forward flight speed.
(2) Based on the propeller, motor, ESC, and battery model we have established, the hover endurance of a multicopter can be estimated with the given propeller parameters, motor parameters, ESC parameters, and battery parameters. Under the same condition, according to the above model analysis, it can be inferred that the higher the altitude or the higher the temperature is, the shorter the hover endurance is; the larger the radius of the propeller or the more the number of propellers is, the longer the hover endurance is.
(3) Given the flight environment of a multicopter, load capacity, maximum weight, maximum size, and minimum hover endurance, readers can select the propulsion system that meets their design requirements based on the product data provided by the manufacturers of the motor, ESC, propeller, and battery.
(4) The following modeling experiments (in Chap. 6) are based on the parameters generated by this experiment.

If you have any question, please go to https://rflysim.com for your information.

Chapter 6
Dynamic Modeling Experiment

In this chapter, multicopter dynamic modeling is studied and realized by MATLAB/Simulink. A deep understanding of the multicopter dynamic model aids in understanding its motion and further designing filters and controllers. The aim of the chapter is to introduce the multicoper dynamic modeling via three step-by-step experiments from shallow to deep, namely a basic experiment, an analysis experiment, and a design experiment. In the basic experiment, readers can intuitively understand multicopters' parameters on dynamic response during flight. In the analysis experiment, readers can analyze the dynamic response with respect to multicopter parameters during flight in theory. Finally, in the design experiment, readers can build a multicopter dynamic model in MATLAB and add a multicopter 3D model to FlightGear for visualization purposes.

6.1 Preliminary

In order to make this chapter self-contained, the preliminary is from Chaps. 5 and 6 of *Introduction to Multicopter Design and Control* [8].

6.1.1 Coordinate Frame

6.1.1.1 Right-Hand Rule

The right-hand rule should be introduced prior to defining the coordinate frames. As shown in Fig. 6.1a, the thumb of the right hand points to the positive direction of the ox axis, the first finger points to the positive direction of the oy axis, and the middle finger points to the positive direction of the oz axis. Furthermore, as shown

© Publishing House of Electronics Industry 2020, corrected publication 2021

Q. Quan et al., *Multicopter Design and Control Practice*,
https://doi.org/10.1007/978-981-15-3138-5_6

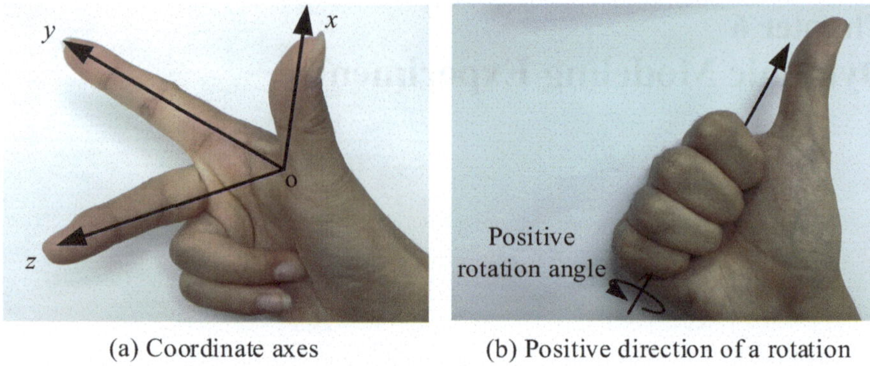

Fig. 6.1 Coordinate axes and the positive direction of a rotation using the right-hand rule

in Fig. 6.1b, in order to determine the positive direction of rotation, the thumb of the right hand points the positive direction of the rotation axis and the direction of the bent fingers denotes the positive direction of rotation. All the coordinate frames used in the chapter and the positive direction of angles defined later obey the right-hand rule.

6.1.1.2 Earth-Fixed Coordinate Frame and Aircraft-Body Coordinate Frame

The Earth-Fixed Coordinate Frame (EFCF) $o_e x_e y_e z_e$ is used to study multicopter's dynamic states with respect to the Earth's surface and to determine its three-dimensional (3D) position. The Earth's curvature is ignored, namely the Earth's surface is assumed as flat. The initial position of the multicopter or the center of the Earth is often set as the coordinate origin o_e, the $o_e x_e$ axis points to a certain direction in the horizontal plane, and the $o_e z_e$ axis points perpendicularly to the ground. Subsequently, the $o_e y_e$ axis is determined based on the right-hand rule.

The Aircraft-Body Coordinate Frame (ABCF) $o_b x_b y_b z_b$ is fixed to the multicopter. The Center of Gravity (CoG) of the multicopter is selected as the origin o_b of $o_b x_b y_b z_b$. The $o_b x_b$ axis points to the nose direction in the symmetric plane of the multicopter. The $o_b z_b$ axis is in the symmetric plane of the multicopter, pointing downward, and perpendicular to the $o_b x_b$ axis. The $o_b y_b$ axis is determined based on the right-hand rule. The relationship between the ABCF and the EFCF is shown in Fig. 6.2.

Define the following unit vectors

$$\mathbf{e}_1 \triangleq \begin{bmatrix} 1 \\ 0 \\ 0 \end{bmatrix}, \mathbf{e}_2 \triangleq \begin{bmatrix} 0 \\ 1 \\ 0 \end{bmatrix}, \mathbf{e}_3 \triangleq \begin{bmatrix} 0 \\ 0 \\ 1 \end{bmatrix}.$$

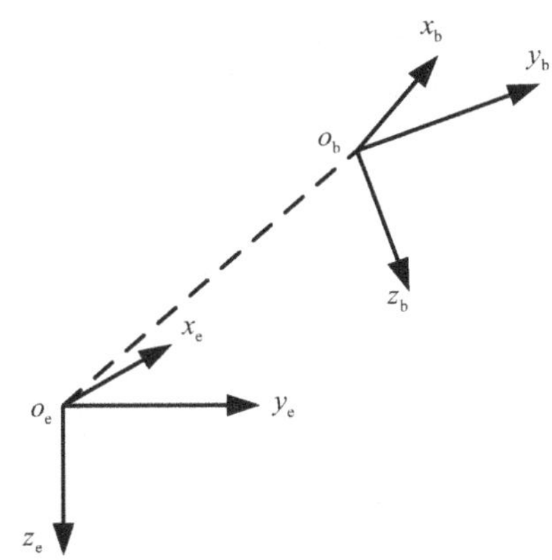

Fig. 6.2 Relationship between the ABCF and the EFCF

In the EFCF, the unit vectors along the $o_e x_e$ axis, $o_e y_e$ axis, and $o_e z_e$ axis are expressed as $\mathbf{e}_1$, $\mathbf{e}_2$, and $\mathbf{e}_3$ respectively. In the ABCF, the unit vectors along the $o_b x_b$ axis, $o_b y_b$ axis, and $o_b z_b$ axis satisfy the following relationship

$$^{b}\mathbf{b}_1 = \mathbf{e}_1, {}^{b}\mathbf{b}_2 = \mathbf{e}_2, {}^{b}\mathbf{b}_3 = \mathbf{e}_3.$$

In the EFCF, the unit vectors along the $o_b x_b$ axis, $o_b y_b$ axis, and $o_b z_b$ axis are expressed as ${}^{e}\mathbf{b}_1$, ${}^{e}\mathbf{b}_2$, and ${}^{e}\mathbf{b}_3$, respectively.

6.1.2 *Attitude Representations*

This section introduces the three representations of Euler angles, rotation matrices, and quaternions.

6.1.2.1 Euler Angles

(1) Definition of Euler Angles
The Euler angles are an intuitive way to represent the attitude. Their physical meanings are quite clear. Additionally, they are widely-used in attitude control. Based on Euler's theorem, the rotation of a rigid body around a fixed point can be considered as the composition of several finite rotations around the fixed point. The ABCF can be achieved by three elemental rotations of the EFCF around one fixed point. During the elemental rotations, each rotation axis is one of the coordinate axes of the rotating coordinate frame and each rotation angle corresponds to one of the Euler angles. Thus, the attitude matrix is closely related to the sequence of three elemental rotations, and it can be represented by the product of three elemental rotation matrices. Intuitively, we assumed that

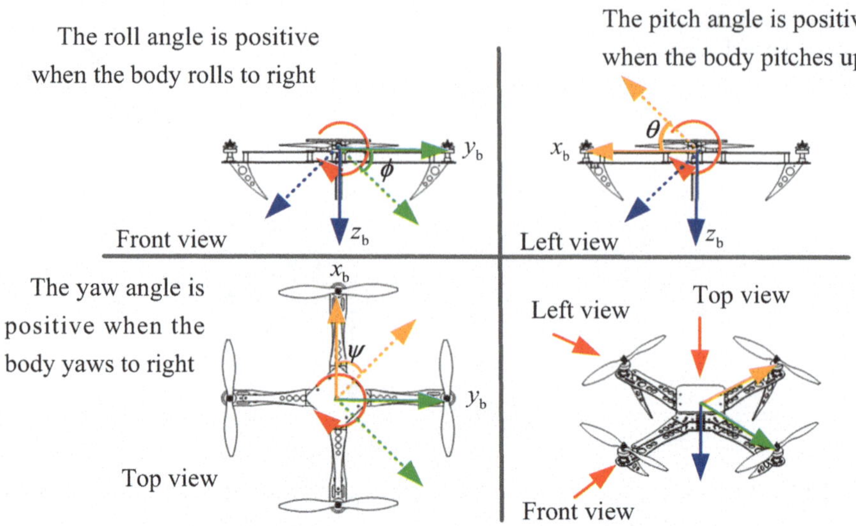

Fig. 6.3 Intuitive representation of Euler angles

the EFCF is aligned with the ABCF. Subsequently, the yaw angle ψ, pitch angle θ, and roll angle ϕ are shown in Fig. 6.3 with their directions determined by the right-hand rule.

The rotation is composed of three elemental rotations about $\mathbf{e}_3$ axis, $\mathbf{k}_2$ axis, and $\mathbf{n}_1$ axis by ψ, θ, ϕ separately. More specifically in Fig. 6.4;

(1) As shown in Fig. 6.4a, the rotation angle around $\mathbf{e}_3$ axis denotes the yaw angle ψ (provided that ψ is positive when the aircraft body turns to right and its value range is $[-\pi, \pi]$);
(2) As shown in Fig. 6.4b, the rotation angle around $\mathbf{k}_2$ axis denotes the pitch angle θ (provided that θ is positive when the aircraft nose pitches up and its value range is $[-\pi/2, \pi/2]$);
(3) As shown in Fig. 6.4c, the rotation angle around $\mathbf{n}_1$ axis denotes the roll angle ϕ (provided that ϕ is positive when the aircraft body rolls to right and its value range is $[-\pi, \pi]$).

(2) Relationship Between the Attitude Rate and the Aircraft Body's Angular Velocity
As shown in Fig. 6.4, if the angular velocity of the aircraft body is ${}^{\mathrm{b}}\boldsymbol{\omega} = [\omega_{x_b}\ \omega_{y_b}\ \omega_{z_b}]^{\mathrm{T}}$, and then the relationship between the attitude rate and the angular velocity of the aircraft body is expressed as

$$\begin{bmatrix} \omega_{x_b} \\ \omega_{y_b} \\ \omega_{z_b} \end{bmatrix} = \begin{bmatrix} 1 & 0 & -\sin\theta \\ 0 & \cos\phi & \cos\theta\sin\phi \\ 0 & -\sin\phi & \cos\theta\cos\phi \end{bmatrix} \begin{bmatrix} \dot{\phi} \\ \dot{\theta} \\ \dot{\psi} \end{bmatrix}. \tag{6.1}$$

Furthermore

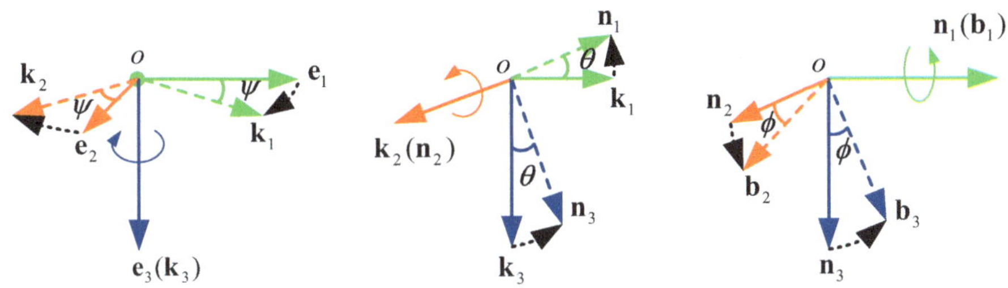

Fig. 6.4 Euler angles and frame transformation

$$\dot{\boldsymbol{\Theta}} = \mathbf{W} \cdot {}^{\mathrm{b}}\boldsymbol{\omega} \tag{6.2}$$

where

$$\boldsymbol{\Theta} \triangleq \begin{bmatrix} \phi \\ \theta \\ \psi \end{bmatrix}, \mathbf{W} \triangleq \begin{bmatrix} 1 & \tan\theta\sin\phi & \tan\theta\cos\phi \\ 0 & \cos\phi & -\sin\phi \\ 0 & \sin\phi/\cos\theta & \cos\phi/\cos\theta \end{bmatrix}. \tag{6.3}$$

Observed from Eq. (6.3), the denominators of some elements of matrix $\mathbf{W}$ are $\cos\theta$. In this case, the singularity problem, which should be avoided, may arise when $\cos\theta = 0$.

6.1.2.2 Rotation Matrix

(1) Definition of Rotation Matrix

The rotation matrix satisfies

$$\begin{aligned} {}^{\mathrm{e}}\mathbf{b}_1 &= \mathbf{R}_{\mathrm{b}}^{\mathrm{e}} \cdot {}^{\mathrm{b}}\mathbf{b}_1 = \mathbf{R}_{\mathrm{b}}^{\mathrm{e}} \cdot \mathbf{e}_1 \\ {}^{\mathrm{e}}\mathbf{b}_2 &= \mathbf{R}_{\mathrm{b}}^{\mathrm{e}} \cdot {}^{\mathrm{b}}\mathbf{b}_2 = \mathbf{R}_{\mathrm{b}}^{\mathrm{e}} \cdot \mathbf{e}_2 \\ {}^{\mathrm{e}}\mathbf{b}_3 &= \mathbf{R}_{\mathrm{b}}^{\mathrm{e}} \cdot {}^{\mathrm{b}}\mathbf{b}_3 = \mathbf{R}_{\mathrm{b}}^{\mathrm{e}} \cdot \mathbf{e}_3. \end{aligned} \tag{6.4}$$

Therefore, the rotation matrix is defined as follows

$$\mathbf{R}_{\mathrm{b}}^{\mathrm{e}} \triangleq \left[{}^{\mathrm{e}}\mathbf{b}_1 \; {}^{\mathrm{e}}\mathbf{b}_2 \; {}^{\mathrm{e}}\mathbf{b}_3 \right] \tag{6.5}$$

where $\mathbf{R}_{\mathrm{b}}^{\mathrm{e}} \in \mathrm{SO}(3)$, $\mathrm{SO}(3) \triangleq \{\mathbf{R} | \mathbf{R}^{\mathrm{T}}\mathbf{R} = \mathbf{I}_3, \det(\mathbf{R}) = 1, \mathbf{R} \in \mathbb{R}^{3\times3}\}$. Based on the definition of Euler angles, the rotation matrix $\mathbf{R}_{\mathrm{b}}^{\mathrm{e}}$, which represents the rotation from the ABCF to the EFCF, is expressed as

$$\mathbf{R}_{\mathrm{b}}^{\mathrm{e}} = \begin{bmatrix} \cos\theta\cos\psi & \cos\psi\sin\theta\sin\phi - \sin\psi\cos\phi & \cos\psi\sin\theta\cos\phi + \sin\psi\sin\phi \\ \cos\theta\sin\psi & \sin\psi\sin\theta\sin\phi + \cos\psi\cos\phi & \sin\psi\sin\theta\cos\phi - \cos\psi\sin\phi \\ -\sin\theta & \sin\phi\cos\theta & \cos\phi\cos\theta \end{bmatrix}. \tag{6.6}$$

Conversely, Euler angles can be solved according to the rotation matrix. First, the rotation matrix $\mathbf{R}_{\mathrm{b}}^{\mathrm{e}}$ is defined as follows

$$\mathbf{R}_{\mathrm{b}}^{\mathrm{e}} \triangleq \begin{bmatrix} r_{11} & r_{12} & r_{13} \\ r_{21} & r_{22} & r_{23} \\ r_{31} & r_{32} & r_{33} \end{bmatrix}.$$

According to Eq. (6.6), one has

$$\begin{aligned} \tan\psi &= \frac{r_{21}}{r_{11}} \\ \sin\theta &= -r_{31} \\ \tan\phi &= \frac{r_{32}}{r_{33}}. \end{aligned} \tag{6.7}$$

By considering Euler angles' value ranges $\psi \in [-\pi, \pi]$, $\theta \in [-\pi/2, \pi/2]$, $\phi \in [-\pi, \pi]$, the solutions to Eq. (6.7) are

$$\begin{aligned} \psi &= \arctan 2\frac{r_{21}}{r_{11}} \\ \theta &= \arcsin(-r_{31}) \\ \phi &= \arctan 2\frac{r_{32}}{r_{33}} \end{aligned} \tag{6.8}$$

where the function arctan2 is defined as

$$\arctan 2(y, x) \triangleq \begin{cases} \arctan(y/x) & x > 0 \\ \arctan(y/x) + \pi & y \geq 0, x < 0 \\ \arctan(y/x) - \pi & y < 0, x < 0 \\ +\pi/2 & y > 0, x = 0 \\ -\pi/2 & y < 0, x = 0 \\ \text{undefined} & y = 0, x = 0. \end{cases} \tag{6.9}$$

Here, the value ranges for $\arctan(\cdot)$ and $\arcsin(\cdot)$ are $[-\pi/2, \pi/2]$.

(2) Relationship Between the Derivative of the Rotation Matrix and the Aircraft Body's Angular Velocity
If the rigid body's rotation (without translation) is only considered, then the derivative of a vector ${}^{\mathrm{e}}\mathbf{r} \in \mathbb{R}^3$ satisfies the following expression

$$\frac{\mathrm{d}{}^{\mathrm{e}}\mathbf{r}}{\mathrm{d}t} = {}^{\mathrm{e}}\boldsymbol{\omega} \times {}^{\mathrm{e}}\mathbf{r} \tag{6.10}$$

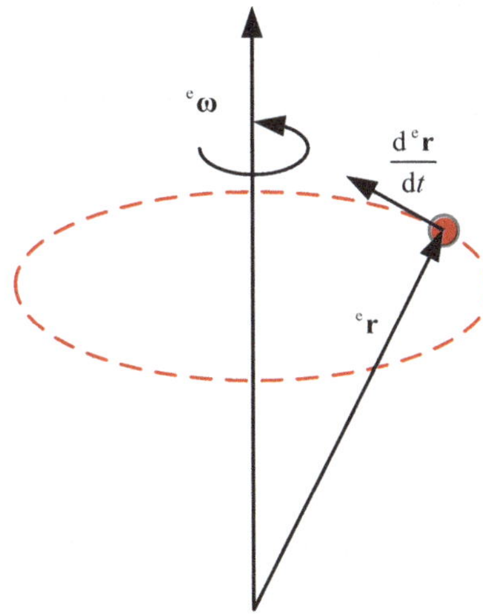

Fig. 6.5 Derivative of a vector presented by a circular motion

where the symbol $\times$ represents the vector cross product. The circular motion in Fig. 6.5 can illustrate Eq. (6.10) intuitively.

The cross product of two vectors $\mathbf{a} \triangleq [a_x\ a_y\ a_z]^{\mathrm{T}}$ and $\mathbf{b} \triangleq [b_x\ b_y\ b_z]^{\mathrm{T}}$ is defined as follows

$$\mathbf{a} \times \mathbf{b} = [\mathbf{a}]_{\times}\mathbf{b} \tag{6.11}$$

where

$$[\mathbf{a}]_{\times} \triangleq \begin{bmatrix} 0 & -a_z & a_y \\ a_z & 0 & -a_x \\ -a_y & a_x & 0 \end{bmatrix} \tag{6.12}$$

is a skew symmetric matrix.

Based on Eq. (6.10), one has

$$\frac{\mathrm{d}\left[{}^{e}\mathbf{b}_1\ {}^{e}\mathbf{b}_2\ {}^{e}\mathbf{b}_3\right]}{\mathrm{d}t} = \left[{}^{e}\boldsymbol{\omega} \times {}^{e}\mathbf{b}_1\quad {}^{e}\boldsymbol{\omega} \times {}^{e}\mathbf{b}_2\quad {}^{e}\boldsymbol{\omega} \times {}^{e}\mathbf{b}_3\right]. \tag{6.13}$$

Since ${}^{e}\boldsymbol{\omega} = \mathbf{R}_{b}^{e} \cdot {}^{b}\boldsymbol{\omega}$, by using the properties of the cross product, Eq. (6.13) is further written as

$$\frac{\mathrm{d}\mathbf{R}_{b}^{e}}{\mathrm{d}t} = \mathbf{R}_{b}^{e}\left[{}^{b}\boldsymbol{\omega}\right]_{\times} \tag{6.14}$$

where $\left[{}^{b}\boldsymbol{\omega}\right]_{\times}$ is the skew symmetric form of ${}^{b}\boldsymbol{\omega}$.

The use of the rotation matrix avoids the singularity problem. However, given that $\mathbf{R}_{b}^{e}$ has nine unknown variables, the computational burden of solving Eq. (6.14) is heavy. In the following section, another method that is known as quaternion representation is described.

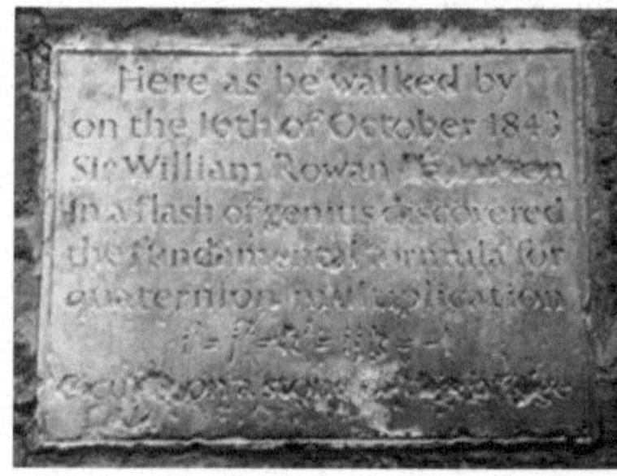

Fig. 6.6 Quaternion plaque on Brougham (Broom) Bridge, Dublin. Photo from https://en.wikipedia.org

6.1.2.3 Quaternions

So far, quaternion representation is one of the most widely-used methods of attitude representation. Quaternion algebra was introduced by William Rowan Hamilton in 1843, and some related theorems were established, but neither of these theorems could be applied in real systems due to the limited computational resources at that time. In memory of Hamilton, the formula for the quaternions was carved into the stone of Brougham Bridge,[1] as shown in Fig. 6.6. Recently, increasing research attention is focused on quaternions given the wide application of high-performance computers and the rapid development of the aircraft's attitude control technologies.

(1) Definition of Quaternions
Quaternions are normally expressed as follows

$$\mathbf{q} \triangleq \begin{bmatrix} q_0 \\ \mathbf{q}_\mathrm{v} \end{bmatrix} \tag{6.15}$$

where $q_0 \in \mathbb{R}$ denotes the scalar part of $\mathbf{q} \in \mathbb{R}^4$ and $\mathbf{q}_\mathrm{v} = [\, q_1 \; q_2 \; q_3 \,]^\mathrm{T} \in \mathbb{R}^3$ is the vector part. For a real number $s \in \mathbb{R}$, the corresponding quaternion is defined as $\mathbf{q} = [\, s \; \mathbf{0}_{1\times 3} \,]^\mathrm{T}$. For a vector $\mathbf{v} \in \mathbb{R}^3$, the corresponding quaternion is $\mathbf{q} = [\, 0 \; \mathbf{v}^\mathrm{T} \,]^\mathrm{T}$.

(2) Quaternions' Basic Operation Rules

1) Addition and subtraction

$$\mathbf{p} \pm \mathbf{q} = \begin{bmatrix} p_0 \\ \mathbf{p}_\mathrm{v} \end{bmatrix} \pm \begin{bmatrix} q_0 \\ \mathbf{q}_\mathrm{v} \end{bmatrix} = \begin{bmatrix} p_0 \pm q_0 \\ \mathbf{p}_\mathrm{v} \pm \mathbf{q}_\mathrm{v} \end{bmatrix}.$$

[1] It reads: "Here as he walked by on the 16th of October 1843 Sir William Rowan Hamilton in a flash of genius discovered the fundamental formula for quaternion multiplication $i^2 = j^2 = k^2 = ijk = -1$ and cut it on a stone of this bridge.".

2) Multiplication

$$\mathbf{p} \otimes \mathbf{q} = \begin{bmatrix} p_0 \\ \mathbf{p}_\mathrm{v} \end{bmatrix} \otimes \begin{bmatrix} q_0 \\ \mathbf{q}_\mathrm{v} \end{bmatrix} = \begin{bmatrix} p_0 q_0 - \mathbf{q}_\mathrm{v}^\mathrm{T} \mathbf{p}_\mathrm{v} \\ \mathbf{p}_\mathrm{v} \times \mathbf{q}_\mathrm{v} + p_0 \mathbf{q}_\mathrm{v} + q_0 \mathbf{p}_\mathrm{v} \end{bmatrix}.$$

The product of two quaternions is expressed as two equivalent matrix products, namely

$$\mathbf{p} \otimes \mathbf{q} = \mathbf{p}^+ \mathbf{q} = \mathbf{q}^- \mathbf{p}$$

where

$$\mathbf{p}^+ = p_0 \mathbf{I}_4 + \begin{bmatrix} 0 & -\mathbf{p}_\mathrm{v}^\mathrm{T} \\ \mathbf{p}_\mathrm{v} & [\mathbf{p}_\mathrm{v}]_\times \end{bmatrix}, \mathbf{q}^- = q_0 \mathbf{I}_4 + \begin{bmatrix} 0 & -\mathbf{q}_\mathrm{v}^\mathrm{T} \\ \mathbf{q}_\mathrm{v} & -[\mathbf{q}_\mathrm{v}]_\times \end{bmatrix}.$$

3) Conjugate
The conjugate of a quaternion is defined by

$$\mathbf{q}^* = \begin{bmatrix} q_0 \\ -\mathbf{q}_\mathrm{v} \end{bmatrix} \tag{6.16}$$

with $\mathbf{q} = [\, q_0 \; \mathbf{q}_\mathrm{v}^\mathrm{T} \,]^\mathrm{T}$.

4) Norm
The norm of a quaternion is defined as follows

$$\begin{aligned} \|\mathbf{q}\|^2 &= \|\mathbf{q} \otimes \mathbf{q}^*\| \\ &= \|\mathbf{q}^* \otimes \mathbf{q}\| \\ &= q_0^2 + \mathbf{q}_\mathrm{v}^\mathrm{T} \mathbf{q}_\mathrm{v} \\ &= q_0^2 + q_1^2 + q_2^2 + q_3^2. \end{aligned}$$

5) Inverse
The inverse quaternion $\mathbf{q}^{-1}$ satisfies

$$\mathbf{q} \otimes \mathbf{q}^{-1} = \begin{bmatrix} 1 \\ \mathbf{0}_{3\times1} \end{bmatrix}$$

which can be computed by

$$\mathbf{q}^{-1} = \frac{\mathbf{q}^*}{\|\mathbf{q}\|}.$$

6) Unit or normalized quaternion
For a unit quaternion, it satisfies $\|\mathbf{q}\| = 1$. Let $\|\mathbf{p}\| = \|\mathbf{q}\| = 1$. Subsequently, we obtain the following expression

$$\begin{aligned} \|\mathbf{p} \otimes \mathbf{q}\| &= 1 \\ \mathbf{q}^{-1} &= \mathbf{q}^*. \end{aligned}$$

(3) Quaternions as Rotations

As shown in Fig. 6.7, a vector $\mathbf{v}_0 \in \mathbb{R}^3$ is fixed in the coordinate frame $oxyz$. The coordinate of $\mathbf{v}_0$ in $oxyz$ is $\mathbf{v}_1 = \left[x\ y\ z\right]^{\mathrm{T}} \in \mathbb{R}^3$. The rotation of $oxyz$ about the axis $\mathbf{v} \in \mathbb{R}^3$ by the angle θ results in a new coordinate frame $ox'y'z'$. The coordinate of $\mathbf{v}_0$ in $ox'y'z'$ is $\mathbf{v}_1' = \left[x'\ y'\ z'\right]^{\mathrm{T}} \in \mathbb{R}^3$. Subsequently, the transformation of coordinates of the vector $\mathbf{v}_0$ between two frames is expressed as follows

$$\begin{bmatrix} 0 \\ \mathbf{v}_1' \end{bmatrix} = \mathbf{q}^{-1} \otimes \begin{bmatrix} 0 \\ \mathbf{v}_1 \end{bmatrix} \otimes \mathbf{q} \tag{6.17}$$

where $\mathbf{q} = [\cos\theta/2\ \mathbf{v}^{\mathrm{T}} \sin\theta/2]^{\mathrm{T}}$.

(4) Quaternions and Rotation Matrix

It is assumed that the rotation from the EFCF to the ABCF is represented by the quaternion $\mathbf{q}_{\mathrm{e}}^{\mathrm{b}} = \left[q_0\ q_1\ q_2\ q_3\right]^{\mathrm{T}}$. Based on Eq. (6.17), one has

$$\begin{aligned} \begin{bmatrix} 0 \\ {}^{\mathrm{e}}\mathbf{r} \end{bmatrix} &= \left(\mathbf{q}_{\mathrm{b}}^{\mathrm{e}}\right)^{-1} \otimes \begin{bmatrix} 0 \\ {}^{\mathrm{b}}\mathbf{r} \end{bmatrix} \otimes \mathbf{q}_{\mathrm{b}}^{\mathrm{e}} \\ &= \mathbf{q}_{\mathrm{e}}^{\mathrm{b}} \otimes \begin{bmatrix} 0 \\ {}^{\mathrm{b}}\mathbf{r} \end{bmatrix} \otimes \left(\mathbf{q}_{\mathrm{e}}^{\mathrm{b}}\right)^{-1} \end{aligned} \tag{6.18}$$

where ${}^{\mathrm{e}}\mathbf{r}, {}^{\mathrm{b}}\mathbf{r} \in \mathbb{R}^3$ denote the representations of the vector $\mathbf{r} \in \mathbb{R}^3$ in the two frames. The quaternions and the rotation matrix are related to each other. Based on Quaternions' Basic Operation Rules, Eq. (6.18) is expressed as follows

$$\begin{bmatrix} 0 \\ {}^{\mathrm{e}}\mathbf{r} \end{bmatrix} = \begin{bmatrix} q_0 & q_1 & q_2 & q_3 \\ -q_1 & q_0 & -q_3 & q_2 \\ -q_2 & q_3 & q_0 & -q_1 \\ -q_3 & -q_2 & q_1 & q_0 \end{bmatrix} \begin{bmatrix} q_0 & -q_1 & -q_2 & -q_3 \\ q_1 & q_0 & -q_3 & q_2 \\ q_2 & q_3 & q_0 & -q_1 \\ q_3 & -q_2 & q_1 & q_0 \end{bmatrix} \begin{bmatrix} 0 \\ {}^{\mathrm{b}}\mathbf{r} \end{bmatrix}.$$

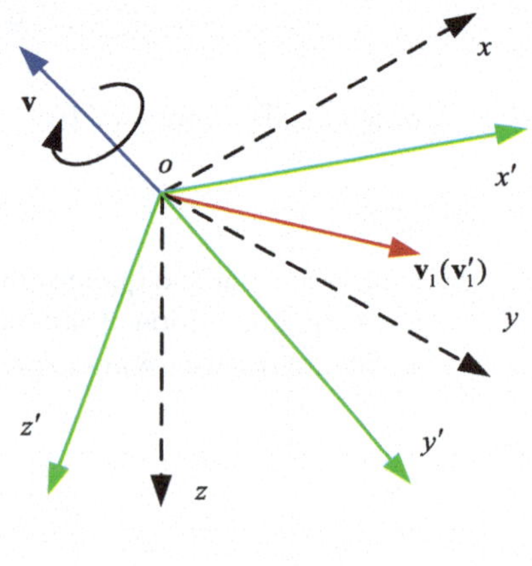

Fig. 6.7 Coordinate frame rotation by quaternions

Then,

$$ {}^{\mathrm{e}}\mathbf{r} = \mathbf{C}(\mathbf{q}_{\mathrm{e}}^{\mathrm{b}}) \cdot {}^{\mathrm{b}}\mathbf{r} \tag{6.19} $$

where

$$ \mathbf{C}(\mathbf{q}_{\mathrm{e}}^{\mathrm{b}}) \triangleq \begin{bmatrix} {q_0}^2 + {q_1}^2 - {q_2}^2 - {q_3}^2 & 2(q_1 q_2 - q_0 q_3) & 2(q_1 q_3 + q_0 q_2) \\ 2(q_1 q_2 + q_0 q_3) & {q_0}^2 - {q_1}^2 + {q_2}^2 - {q_3}^2 & 2(q_2 q_3 - q_0 q_1) \\ 2(q_1 q_3 - q_0 q_2) & 2(q_2 q_3 + q_0 q_1) & {q_0}^2 - {q_1}^2 - {q_2}^2 + {q_3}^2 \end{bmatrix}. $$

Since $\mathbf{r}$ is an arbitrary vector, one has

$$ \mathbf{R}_{\mathrm{b}}^{\mathrm{e}} = \mathbf{C}(\mathbf{q}_{\mathrm{e}}^{\mathrm{b}}). \tag{6.20} $$

Furthermore, if $\mathbf{R}_{\mathrm{b}}^{\mathrm{e}}$ is assumed to be known as

$$ \mathbf{R}_{\mathrm{b}}^{\mathrm{e}} \triangleq \begin{bmatrix} r_{11} & r_{12} & r_{13} \\ r_{21} & r_{22} & r_{23} \\ r_{31} & r_{32} & r_{33} \end{bmatrix}. $$

Then the corresponding quaternion is expressed as

$$ \begin{aligned} q_0 &= \operatorname{sign}(q_0)\,\frac{1}{2}\sqrt{1 + r_{11} + r_{22} + r_{33}} \\ q_1 &= \operatorname{sign}(q_1)\,\frac{1}{2}\sqrt{1 + r_{11} - r_{22} - r_{33}} \\ q_2 &= \operatorname{sign}(q_2)\,\frac{1}{2}\sqrt{1 - r_{11} + r_{22} - r_{33}} \\ q_3 &= \operatorname{sign}(q_3)\,\frac{1}{2}\sqrt{1 - r_{11} - r_{22} + r_{33}} \end{aligned} \tag{6.21} $$

where $\operatorname{sign}(q_0)$ is 1 or -1. And

$$ \begin{aligned} \operatorname{sign}(q_1) &= \operatorname{sign}(q_0)\operatorname{sign}(r_{32} - r_{23}) \\ \operatorname{sign}(q_2) &= \operatorname{sign}(q_0)\operatorname{sign}(r_{13} - r_{31}) \\ \operatorname{sign}(q_3) &= \operatorname{sign}(q_0)\operatorname{sign}(r_{21} - r_{12}). \end{aligned} \tag{6.22} $$

It should be noted that one rotation matrix corresponds to two quaternions, namely $\mathbf{q}_{\mathrm{e}}^{\mathrm{b}}$ and $-\mathbf{q}_{\mathrm{e}}^{\mathrm{b}}$.

(5) Quaternions and Euler Angles

The rotation of a coordinate frame about $\mathbf{r} = \begin{bmatrix} a & b & c \end{bmatrix}^{\mathrm{T}}$ $(a^2 + b^2 + c^2 = 1)$ by an angle α, is expressed as a quaternion

$$ \mathbf{q}(\alpha, \mathbf{r}) = \begin{bmatrix} \cos\frac{\alpha}{2} & a\sin\frac{\alpha}{2} & b\sin\frac{\alpha}{2} & c\sin\frac{\alpha}{2} \end{bmatrix}^{\mathrm{T}}. \tag{6.23} $$

Based on the rotation order in Sect. 6.1.2.1 in this chapter, one has

$$\mathbf{q}_{\mathrm{e}}^{\mathrm{b}} = \mathbf{q}_z(\psi) \otimes \mathbf{q}_y(\theta) \otimes \mathbf{q}_x(\phi) \tag{6.24}$$

where

$$\begin{aligned}
\mathbf{q}_x(\phi) &= \begin{bmatrix} \cos\frac{\phi}{2} & \sin\frac{\phi}{2} & 0 & 0 \end{bmatrix}^{\mathrm{T}} \\
\mathbf{q}_y(\theta) &= \begin{bmatrix} \cos\frac{\theta}{2} & 0 & \sin\frac{\theta}{2} & 0 \end{bmatrix}^{\mathrm{T}} \\
\mathbf{q}_z(\psi) &= \begin{bmatrix} \cos\frac{\psi}{2} & 0 & 0 & \sin\frac{\psi}{2} \end{bmatrix}^{\mathrm{T}}.
\end{aligned}$$

Furthermore, Eq. (6.24) is expressed as

$$\mathbf{q}_{\mathrm{e}}^{\mathrm{b}} = \begin{bmatrix}
\cos\frac{\phi}{2}\cos\frac{\theta}{2}\cos\frac{\psi}{2} + \sin\frac{\phi}{2}\sin\frac{\theta}{2}\sin\frac{\psi}{2} \\
\sin\frac{\phi}{2}\cos\frac{\theta}{2}\cos\frac{\psi}{2} - \cos\frac{\phi}{2}\sin\frac{\theta}{2}\sin\frac{\psi}{2} \\
\cos\frac{\phi}{2}\sin\frac{\theta}{2}\cos\frac{\psi}{2} + \sin\frac{\phi}{2}\cos\frac{\theta}{2}\sin\frac{\psi}{2} \\
\cos\frac{\phi}{2}\cos\frac{\theta}{2}\sin\frac{\psi}{2} - \sin\frac{\phi}{2}\sin\frac{\theta}{2}\cos\frac{\psi}{2}
\end{bmatrix}. \tag{6.25}$$

Based on Eq. (6.17), the rotation from ${}^{\mathrm{e}}\mathbf{r} \in \mathbb{R}^3$ to ${}^{\mathrm{b}}\mathbf{r}$ is written as

$$\begin{aligned}
\begin{bmatrix} 0 \\ {}^{\mathrm{b}}\mathbf{r} \end{bmatrix} &= \left(\mathbf{q}_{\mathrm{e}}^{\mathrm{b}}\right)^{-1} \otimes \begin{bmatrix} 0 \\ {}^{\mathrm{e}}\mathbf{r} \end{bmatrix} \otimes \mathbf{q}_{\mathrm{e}}^{\mathrm{b}} \\
&= \left(\mathbf{q}_z(\psi) \otimes \mathbf{q}_y(\theta) \otimes \mathbf{q}_x(\phi)\right)^{-1} \otimes \begin{bmatrix} 0 \\ {}^{\mathrm{e}}\mathbf{r} \end{bmatrix} \otimes \left(\mathbf{q}_z(\psi) \otimes \mathbf{q}_y(\theta) \otimes \mathbf{q}_x(\phi)\right) \\
&= \left(\mathbf{q}_x(\phi)\right)^{-1} \otimes \left(\left(\mathbf{q}_y(\theta)\right)^{-1} \otimes \left(\left(\mathbf{q}_z(\psi)\right)^{-1} \otimes \begin{bmatrix} 0 \\ {}^{\mathrm{e}}\mathbf{r} \end{bmatrix} \otimes \mathbf{q}_z(\psi)\right) \otimes \mathbf{q}_y(\theta)\right) \otimes \mathbf{q}_x(\phi).
\end{aligned}$$

It should be noted that the rotation order is $\psi \to \theta \to \phi$. This is consistent with the definition in Sect. 6.1.2.1 in this chapter (see Fig. 6.4).

In turn, Euler angles can be derived by using quaternions. Based on Eq. (6.25), one has

$$\begin{aligned}
\tan\phi &= \frac{2(q_0q_1 + q_2q_3)}{1 - 2\left(q_1^2 + q_2^2\right)} \\
\sin\theta &= 2\left(q_0q_2 - q_1q_3\right) \\
\tan\psi &= \frac{2\left(q_0q_3 + q_1q_2\right)}{1 - 2\left(q_2^2 + q_3^2\right)}.
\end{aligned} \tag{6.26}$$

By considering that Euler angles' value ranges are $\psi \in [-\pi, \pi]$, $\theta \in [-\pi/2, \pi/2]$, $\phi \in [-\pi, \pi]$, the solutions to Eq. (6.26) are

$$\begin{aligned}\phi &= \arctan 2\frac{2(q_0q_1 + q_2q_3)}{1 - 2\left(q_1^2 + q_2^2\right)} \\ \theta &= \arcsin\left(2\left(q_0q_2 - q_1q_3\right)\right) \\ \psi &= \arctan 2\left(\frac{2\left(q_0q_3 + q_1q_2\right)}{1 - 2\left(q_2^2 + q_3^2\right)}\right).\end{aligned} \tag{6.27}$$

(6) Relationship Between the Derivative of the Quaternions and the Aircraft Body's Angular Velocity
First, what should be known is the relationship between $\mathbf{q}_{\mathrm{e}}^{\mathrm{b}}(t + \Delta t)$ and $\mathbf{q}_{\mathrm{e}}^{\mathrm{b}}(t)$. According to Eq. (6.17), under the action of $\mathbf{q}_{\mathrm{e}}^{\mathrm{b}}(t)$, the representation of ${}^{\mathrm{b}}\mathbf{v}(t)$ with respect to ${}^{\mathrm{e}}\mathbf{v}(t)$ and ${}^{\mathrm{b}}\mathbf{v}(t)$ is

$$\begin{bmatrix} 0 \\ {}^{\mathrm{b}}\mathbf{v}(t) \end{bmatrix} = \mathbf{q}_{\mathrm{e}}^{\mathrm{b}}(t)^{-1} \otimes \begin{bmatrix} 0 \\ {}^{\mathrm{e}}\mathbf{v}(t) \end{bmatrix} \otimes \mathbf{q}_{\mathrm{e}}^{\mathrm{b}}(t). \tag{6.28}$$

During the rotation from the EFCF to the ABCF, it is assumed that the ABCF has a tiny perturbation. Here, $\Delta\mathbf{q}$ is used to represent the perturbation from t to $t + \Delta t$. According to Eq. (6.23), when Δt is small enough, $\Delta\mathbf{q}$ is expressed as

$$\Delta\mathbf{q} = \begin{bmatrix} 1 & \frac{1}{2}{}^{\mathrm{b}}\boldsymbol{\omega}^{\mathrm{T}}\Delta t \end{bmatrix}^{\mathrm{T}} \tag{6.29}$$

which can be written in the form of Eq. (6.23) with $\alpha = \left\|{}^{\mathrm{b}}\boldsymbol{\omega}\right\| \Delta t$, $\mathbf{r} = {}^{\mathrm{b}}\boldsymbol{\omega} / \left\|{}^{\mathrm{b}}\boldsymbol{\omega}\right\|$. Then, the derivative of $\mathbf{q}_{\mathrm{e}}^{\mathrm{b}}(t)$ is obtained as

$$\dot{\mathbf{q}}_{\mathrm{e}}^{\mathrm{b}} = \frac{1}{2}\begin{bmatrix} 0 & -{}^{\mathrm{b}}\boldsymbol{\omega}^{\mathrm{T}} \\ {}^{\mathrm{b}}\boldsymbol{\omega} & -\left[{}^{\mathrm{b}}\boldsymbol{\omega}\right]_{\times} \end{bmatrix} \mathbf{q}_{\mathrm{e}}^{\mathrm{b}}. \tag{6.30}$$

In practice, ${}^{\mathrm{b}}\boldsymbol{\omega}$ can be measured approximately by a three-axis gyroscope and can be considered as a known value. In this sense, Eq. (6.30) is linear. With $\mathbf{q}_{\mathrm{e}}^{\mathrm{b}} = [q_0 \; \mathbf{q}_{\mathrm{v}}^{\mathrm{T}}]^{\mathrm{T}}$, Eq. (6.30) is further rewritten as

$$\begin{aligned}\dot{q}_0 &= -\frac{1}{2}\mathbf{q}_{\mathrm{v}}^{\mathrm{T}} \cdot {}^{\mathrm{b}}\boldsymbol{\omega} \\ \dot{\mathbf{q}}_{\mathrm{v}} &= \frac{1}{2}\left(q_0\mathbf{I}_3 + \left[\mathbf{q}_{\mathrm{v}}\right]_{\times}\right){}^{\mathrm{b}}\boldsymbol{\omega}\end{aligned} \tag{6.31}$$

where $-\left[{}^{\mathrm{b}}\boldsymbol{\omega}\right]_{\times}\mathbf{q}_{\mathrm{v}} = \left[\mathbf{q}_{\mathrm{v}}\right]_{\times}{}^{\mathrm{b}}\boldsymbol{\omega}$ is used.
The attitude representation using Euler angles has a singularity problem in the case of large angles (e.g., $\theta = \pm\pi/2$), whereas quaternions can help to keep the

linearity of the equation and avoid the singularity problem. Compared with Euler angles, quaternions require only simple calculation and work for all attitudes. Moreover, Eq. (6.30) which represents the relationship between quaternions' derivative and the aircraft body's angular velocity has only four unknown variables, whereas the differential Eq. (6.14) expressed by using the rotation matrix has nine unknown variables. Therefore, compared with the rotation matrix, quaternions have better numerical stability as well as higher efficiency.

6.1.3 Multicopter Flight Control Rigid-Body Model

6.1.3.1 Assumptions

For the purpose of convenience, when modeling a multicopter, the following assumptions are introduced.

Assumption 6.1 The multicopter is a rigid body.

Assumption 6.2 The mass and the moment of inertia are constant.

Assumption 6.3 The geometric center and the CoG of the multicopter are the same.

Assumption 6.4 The multicopter is only forced by the gravity, propeller thrust force, and aerodynamic drag. Furthermore, the gravity points along the positive direction of the $o_e z_e$ axis and the propeller thrust points along the negative direction of the $o_b z_b$ axis.

Assumption 6.5 Propellers with odd indices rotate counterclockwise and propellers with even indices rotate clockwise.

The main difference between the model and other rigid-body dynamic models is that the thrust produced by the propellers is always perpendicular to the fuselage plane. Thus, the thrust direction is always consistent with the negative direction of the $o_b z_b$ axis.

6.1.3.2 Rigid-Body Kinematic Model

Let the vector of the multicopter's CoG be ${}^{e}\mathbf{p} \in \mathbb{R}^3$. Subsequently, we obtain the following expression

$$ {}^{e}\dot{\mathbf{p}} = {}^{e}\mathbf{v} \tag{6.32} $$

where ${}^{e}\mathbf{v} \in \mathbb{R}^3$ represents the velocity of the multicopter. The attitude kinematic model is divided into the following three types: Euler angle model, rotation matrix model, and quaternions model. Therefore, the following three rigid-body kinematic models are presented.

(1) Euler angle model
Based on **Assumption** 6.1, combining Eq. (6.2) with Eq. (6.32), we obtain the following expression

$$\begin{aligned} {}^{\mathrm{e}}\dot{\mathbf{p}} &= {}^{\mathrm{e}}\mathbf{v} \\ \dot{\boldsymbol{\Theta}} &= \mathbf{W} \cdot {}^{\mathrm{b}}\boldsymbol{\omega}. \end{aligned} \tag{6.33}$$

(2) Rotation matrix model
Based on **Assumption** 6.1, combining Eq. (6.14) with Eq. (6.32), we obtain the following expression

$$\begin{aligned} {}^{\mathrm{e}}\dot{\mathbf{p}} &= {}^{\mathrm{e}}\mathbf{v} \\ \dot{\mathbf{R}} &= \mathbf{R}\left[{}^{\mathrm{b}}\boldsymbol{\omega}\right]_{\times} \end{aligned} \tag{6.34}$$

where $\mathbf{R} \triangleq \mathbf{R}_{\mathrm{b}}^{\mathrm{e}}$ in order to simplify the descriptions.

(3) Quaternions model
Based on **Assumption** 6.1, combining Eq. (6.31) and Eq. (6.32), we obtain the following expression

$$\begin{aligned} {}^{\mathrm{e}}\dot{\mathbf{p}} &= {}^{\mathrm{e}}\mathbf{v} \\ \dot{q}_0 &= -\frac{1}{2}\mathbf{q}_{\mathrm{v}}^{\mathrm{T}} \cdot {}^{\mathrm{b}}\boldsymbol{\omega} \\ \dot{\mathbf{q}}_{\mathrm{v}} &= \frac{1}{2}\left(q_0\mathbf{I}_3 + \left[\mathbf{q}_{\mathrm{v}}\right]_{\times}\right){}^{\mathrm{b}}\boldsymbol{\omega}. \end{aligned} \tag{6.35}$$

6.1.3.3 Rigid-Body Dynamic Model

(1) Position dynamic model
Assuming all forces acting on the vehicle body are presented by a resultant force ${}^{\mathrm{e}}\mathbf{F} \in \mathbb{R}^3$, by analyzing the forces on a multicopter based on **Assumption** 6.4, we obtain the following expression

$$ {}^{\mathrm{e}}\dot{\mathbf{v}} = {}^{\mathrm{e}}\mathbf{F}/m \tag{6.36}$$

where $m > 0$ represents the multicopter mass. The total force ${}^{\mathrm{e}}\mathbf{F}$ is composed of gravity $\mathbf{G}$ defined in EFCF, propeller control force ${}^{\mathrm{b}}\mathbf{T}$ defined in the ABCF, and aerodynamic force ${}^{\mathrm{b}}\mathbf{F}_{\mathrm{d}}$ defined in the ABCF, which are expressed as follows

$$\begin{aligned} {}^{\mathrm{e}}\mathbf{F} &= m\mathbf{G} + \mathbf{R}\left({}^{\mathrm{b}}\mathbf{T} + {}^{\mathrm{b}}\mathbf{F}_{\mathrm{d}}\right) \\ \mathbf{G} &= \left[0\ 0\ g\right]^{\mathrm{T}} = g\mathbf{e}_3 \\ {}^{\mathrm{b}}\mathbf{T} &= \left[0\ 0\ -f\right]^{\mathrm{T}} = -f\mathbf{b}_3 \end{aligned} \tag{6.37}$$

where $f \geq 0$ represents the magnitude of the total propeller thrust and the thrust is unidirectional (the situation of negative thrust caused by variable-pitch propellers is not considered here) and $g > 0$ represents the acceleration of gravity. Furthermore, given that

$$^{\mathrm{e}}\mathbf{v} = \mathbf{R} \cdot {}^{\mathrm{b}}\mathbf{v} \tag{6.38}$$

combining Eq. (6.36) with Eq. (6.38), we obtain the following expression

$$\begin{aligned} {}^{\mathrm{b}}\dot{\mathbf{v}} &= -\left[{}^{\mathrm{b}}\boldsymbol{\omega}\right]_{\times} {}^{\mathrm{b}}\mathbf{v} + {}^{\mathrm{b}}\mathbf{F}/m \\ {}^{\mathrm{b}}\mathbf{F} &= \mathbf{R}^{-1} \cdot {}^{\mathrm{e}}\mathbf{F} = m\mathbf{R}^{-1}\mathbf{G} + {}^{\mathrm{b}}\mathbf{T} + {}^{\mathrm{b}}\mathbf{F}_{\mathrm{d}}. \end{aligned} \tag{6.39}$$

(2) Attitude dynamic model
Based on **Assumptions** 6.1–6.3, the attitude dynamic equation in the ABCF is established as follows

$$\begin{aligned} \mathbf{J} \cdot {}^{\mathrm{b}}\dot{\boldsymbol{\omega}} &= -{}^{\mathrm{b}}\boldsymbol{\omega} \times \left(\mathbf{J} \cdot {}^{\mathrm{b}}\boldsymbol{\omega}\right) + {}^{\mathrm{b}}\mathbf{M} \\ {}^{\mathrm{b}}\mathbf{M} &= \mathbf{G}_{\mathrm{a}} + \boldsymbol{\tau} + {}^{\mathrm{b}}\mathbf{M}_{\mathrm{d}} \end{aligned} \tag{6.40}$$

where $\boldsymbol{\tau} \triangleq [\,\tau_x \;\tau_y \;\tau_z\,]^{\mathrm{T}} \in \mathbb{R}^3$ denotes the moments generated by the propellers in the body axes; $\mathbf{J} \in \mathbb{R}^{3\times3}$ denotes the multicopter moment of inertia; and ${}^{\mathrm{b}}\mathbf{M}_{\mathrm{d}}$ denotes the aerodynamic moment acting on the body. For a multicopter with n_{r} propellers, $\mathbf{G}_{\mathrm{a}} \triangleq [G_{\mathrm{a},\phi} \; G_{\mathrm{a},\theta} \; G_{\mathrm{a},\psi}]^{\mathrm{T}} \in \mathbb{R}^3$ denotes the gyroscopic torques. Their signs are related to the rotation direction of propellers. Based on **Assumption** 6.5 and the definition of coordinate frames, a single propeller's angular velocity vector is $(-1)^k \varpi_k {}^{\mathrm{b}}\mathbf{b}_3$, $k = 1, \cdots, n_{\mathrm{r}}$, where $\varpi_k > 0$ denotes the angular speed (rad/s) of the kth propeller. Therefore, the gyroscopic torques caused by the rotation of a single propeller are expressed as follows

$$\begin{aligned} \mathbf{G}_{\mathrm{a},k} &= {}^{\mathrm{b}}\mathbf{L}_k \times {}^{\mathrm{b}}\boldsymbol{\omega} \\ &= J_{\mathrm{RP}}(-1)^k \varpi_k {}^{\mathrm{b}}\mathbf{b}_3 \times {}^{\mathrm{b}}\boldsymbol{\omega} \\ &= J_{\mathrm{RP}}\left({}^{\mathrm{b}}\boldsymbol{\omega} \times \mathbf{e}_3\right)(-1)^{k+1}\varpi_k \end{aligned} \tag{6.41}$$

where $J_{\mathrm{RP}} > 0$ (N·m·s^2) denotes the total moments of inertia of the entire rotor and the propeller about the axis of rotation. Specifically, equations ${}^{\mathrm{b}}\mathbf{b}_3 \times {}^{\mathrm{b}}\boldsymbol{\omega} = -{}^{\mathrm{b}}\boldsymbol{\omega} \times {}^{\mathrm{b}}\mathbf{b}_3$, ${}^{\mathrm{b}}\mathbf{b}_3 = \mathbf{e}_3$ are applied to derivation. For a multicopter with n_{r} propellers, we obtain the following expression

$$\mathbf{G}_{\mathrm{a}} = \sum_{k=1}^{n_{\mathrm{r}}} J_{\mathrm{RP}}({}^{\mathrm{b}}\boldsymbol{\omega} \times \mathbf{e}_3)(-1)^{k+1}\varpi_k. \tag{6.42}$$

Furthermore, since

$$^{\mathrm{b}}\boldsymbol{\omega} \times \mathbf{e}_3 = \begin{bmatrix} \omega_{y_{\mathrm{b}}} \\ -\omega_{x_{\mathrm{b}}} \\ 0 \end{bmatrix}$$

one has

$$\begin{aligned} G_{\mathrm{a},\phi} &= \sum_{k=1}^{n_{\mathrm{r}}} J_{\mathrm{RP}}\omega_{y_{\mathrm{b}}}(-1)^{k+1}\varpi_k \\ G_{\mathrm{a},\theta} &= \sum_{k=1}^{n_{\mathrm{r}}} J_{\mathrm{RP}}\omega_{x_{\mathrm{b}}}(-1)^{k}\varpi_k \\ G_{\mathrm{a},\psi} &= 0. \end{aligned} \tag{6.43}$$

As shown above, the gyroscopic torque does not exist in the yaw channel.

6.1.3.4 Multicopter Flight Control Rigid-Body Model

By combining Eqs. (6.33), (6.34), (6.35) and (6.40), the multicopter flight control rigid-body model is expressed as follows

$$\begin{aligned} {}^{\mathrm{e}}\dot{\mathbf{p}} &= {}^{\mathrm{e}}\mathbf{v} = \mathbf{R}\cdot{}^{\mathrm{b}}\mathbf{v} \\ {}^{\mathrm{b}}\dot{\mathbf{v}} &= -\left[{}^{\mathrm{b}}\boldsymbol{\omega}\right]_{\times}\cdot{}^{\mathrm{b}}\mathbf{v}+{}^{\mathrm{b}}\mathbf{F}/m \\ \dot{\boldsymbol{\Theta}} &= \mathbf{W}\cdot{}^{\mathrm{b}}\boldsymbol{\omega} \\ \mathbf{J}\cdot{}^{\mathrm{b}}\dot{\boldsymbol{\omega}} &= -{}^{\mathrm{b}}\boldsymbol{\omega}\times\left(\mathbf{J}\cdot{}^{\mathrm{b}}\boldsymbol{\omega}\right)+{}^{\mathrm{b}}\mathbf{M} \end{aligned} \tag{6.44}$$

or

$$\begin{aligned} {}^{\mathrm{e}}\dot{\mathbf{p}} &= {}^{\mathrm{e}}\mathbf{v} = \mathbf{R}\cdot{}^{\mathrm{b}}\mathbf{v} \\ {}^{\mathrm{b}}\dot{\mathbf{v}} &= -\left[{}^{\mathrm{b}}\boldsymbol{\omega}\right]_{\times}\cdot{}^{\mathrm{b}}\mathbf{v}+{}^{\mathrm{b}}\mathbf{F}/m \\ \dot{\mathbf{R}} &= \mathbf{R}\left[{}^{\mathrm{b}}\boldsymbol{\omega}\right]_{\times} \\ \mathbf{J}\cdot{}^{\mathrm{b}}\dot{\boldsymbol{\omega}} &= -{}^{\mathrm{b}}\boldsymbol{\omega}\times\left(\mathbf{J}\cdot{}^{\mathrm{b}}\boldsymbol{\omega}\right)+{}^{\mathrm{b}}\mathbf{M} \end{aligned} \tag{6.45}$$

or

$$\begin{aligned} {}^{\mathrm{e}}\dot{\mathbf{p}} &= {}^{\mathrm{e}}\mathbf{v} = \mathbf{R}\cdot{}^{\mathrm{b}}\mathbf{v} \\ {}^{\mathrm{b}}\dot{\mathbf{v}} &= -\left[{}^{\mathrm{b}}\boldsymbol{\omega}\right]_{\times}\cdot{}^{\mathrm{b}}\mathbf{v}+{}^{\mathrm{b}}\mathbf{F}/m \\ \dot{q}_0 &= -\frac{1}{2}\mathbf{q}_{\mathrm{v}}^{\mathrm{T}}\cdot{}^{\mathrm{b}}\boldsymbol{\omega} \\ \dot{\mathbf{q}}_{\mathrm{v}} &= \frac{1}{2}\left(q_0\mathbf{I}_3+\left[\mathbf{q}_{\mathrm{v}}\right]_{\times}\right){}^{\mathrm{b}}\boldsymbol{\omega} \\ \mathbf{J}\cdot{}^{\mathrm{b}}\dot{\boldsymbol{\omega}} &= -{}^{\mathrm{b}}\boldsymbol{\omega}\times\left(\mathbf{J}\cdot{}^{\mathrm{b}}\boldsymbol{\omega}\right)+{}^{\mathrm{b}}\mathbf{M}. \end{aligned} \tag{6.46}$$

These models contain both ABCF and EFCF. The position and the velocity of the multicopter are expected to be described in the EFCF. This can help remote pilots to determine the flight position and flight velocity. Additionally, the presentation is consistent with GPS measurement. Conversely, in the ABCF, the presentations of thrust and moments are intuitive, and the measurements by sensors are always also expressed in the ABCF. The salient feature of the multicopter flight control rigid-body model is embedded in the thrust force vector ${}^{\mathrm{b}}\mathbf{T}$ of Eq. (6.37). This implies that the thrust direction is always consistent with the negative direction of the $o_{\mathrm{b}}z_{\mathrm{b}}$ axis. In many existing studies, controllers were designed based on Eqs. (6.44) and (6.46) directly. In the following, the control effectiveness model is considered to distinguish the quadcopter from the hexacopter.

6.1.4 *Control Effectiveness Model*

6.1.4.1 Single Propeller Thrust and Reaction Torque Model

When a multicopter hovers without wind, the propeller thrust is expressed as follows

$$T_i = c_{\mathrm{T}} \varpi_i^2 \tag{6.47}$$

where $c_{\mathrm{T}} = 1/(4\pi^2) \cdot \rho D_{\mathrm{p}}^4 C_{\mathrm{T}}$ is modeled as a constant that is easily determined by experiments. The reaction torque is expressed as follows

$$M_i = c_{\mathrm{M}} \varpi_i^2 \tag{6.48}$$

where M_i denotes the reaction torque of the ith propeller acting on the fuselage, $c_{\mathrm{M}} = 1/(4\pi^2) \cdot \rho D_{\mathrm{p}}^5 C_{\mathrm{M}}$ is also determined by experiments. Equation (6.48) is a static model about the reaction torque. Its dynamic model is

$$J_{\mathrm{RP}} \dot{\varpi}_i = -c_{\mathrm{M}} \varpi_i^2 + \tau_i \tag{6.49}$$

where $\tau_i \in \mathbb{R}$ denotes the torque acting on the ith propeller. Based on Newton's third law, the reaction torque is as large as the torque acting on the ith propeller. Subsequently, this is expressed as follows

$$M_i = c_{\mathrm{M}} \varpi_i^2 + J_{\mathrm{RP}} \dot{\varpi}_i . \tag{6.50}$$

6.1.4.2 Thrust and Moments Model

The flight of the multicopter is driven by multiple propellers. The propeller angular speeds $\varpi_i, i = 1, 2, \cdots n_{\mathrm{r}}$ will determine the total thrust f and moments $\boldsymbol{\tau}$. For ease of understanding, this section starts with quadcopters.

(1) Quadcopters
As shown in Fig. 6.8a, the total thrust that acts on the quadcopter is as follows

$$f = \sum_{i=1}^{4} T_i = c_{\mathrm{T}} (\varpi_1^2 + \varpi_2^2 + \varpi_3^2 + \varpi_4^2). \tag{6.51}$$

For a plus-configuration quadcopter, the moments produced by propellers are as follows

$$\begin{aligned} \tau_x &= d c_{\mathrm{T}} \left(-\varpi_2^2 + \varpi_4^2\right) \\ \tau_y &= d c_{\mathrm{T}} \left(\varpi_1^2 - \varpi_3^2\right) \\ \tau_z &= c_{\mathrm{M}} \left(\varpi_1^2 - \varpi_2^2 + \varpi_3^2 - \varpi_4^2\right) \end{aligned} \tag{6.52}$$

where $d > 0$ represents the distance between the body center and any motor. Based on Eqs. (6.51) and (6.52), we obtain the following matrix form

$$\begin{bmatrix} f \\ \tau_x \\ \tau_y \\ \tau_z \end{bmatrix} = \begin{bmatrix} c_{\mathrm{T}} & c_{\mathrm{T}} & c_{\mathrm{T}} & c_{\mathrm{T}} \\ 0 & -dc_{\mathrm{T}} & 0 & dc_{\mathrm{T}} \\ dc_{\mathrm{T}} & 0 & -dc_{\mathrm{T}} & 0 \\ c_{\mathrm{M}} & -c_{\mathrm{M}} & c_{\mathrm{M}} & -c_{\mathrm{M}} \end{bmatrix} \begin{bmatrix} \varpi_1^2 \\ \varpi_2^2 \\ \varpi_3^2 \\ \varpi_4^2 \end{bmatrix}. \tag{6.53}$$

As shown in Fig. 6.8b, for an X-configuration quadcopter, the total thrust produced by propellers is still

$$f = c_{\mathrm{T}} \left(\varpi_1^2 + \varpi_2^2 + \varpi_3^2 + \varpi_4^2 \right). \tag{6.54}$$

But the moments are different, as shown in the following

$$\begin{cases} \tau_x = dc_{\mathrm{T}} \left(-\frac{\sqrt{2}}{2}\varpi_1^2 + \frac{\sqrt{2}}{2}\varpi_2^2 + \frac{\sqrt{2}}{2}\varpi_3^2 - \frac{\sqrt{2}}{2}\varpi_4^2 \right) \\ \tau_y = dc_{\mathrm{T}} \left(\frac{\sqrt{2}}{2}\varpi_1^2 - \frac{\sqrt{2}}{2}\varpi_2^2 + \frac{\sqrt{2}}{2}\varpi_3^2 - \frac{\sqrt{2}}{2}\varpi_4^2 \right) \\ \tau_z = c_{\mathrm{M}} \left(\varpi_1^2 + \varpi_2^2 - \varpi_3^2 - \varpi_4^2 \right) \end{cases} \tag{6.55}$$

$$\Rightarrow \begin{bmatrix} f \\ \tau_x \\ \tau_y \\ \tau_z \end{bmatrix} = \begin{bmatrix} c_{\mathrm{T}} & c_{\mathrm{T}} & c_{\mathrm{T}} & c_{\mathrm{T}} \\ -\frac{\sqrt{2}}{2}dc_{\mathrm{T}} & \frac{\sqrt{2}}{2}dc_{\mathrm{T}} & \frac{\sqrt{2}}{2}dc_{\mathrm{T}} & -\frac{\sqrt{2}}{2}dc_{\mathrm{T}} \\ \frac{\sqrt{2}}{2}dc_{\mathrm{T}} & -\frac{\sqrt{2}}{2}dc_{\mathrm{T}} & \frac{\sqrt{2}}{2}dc_{\mathrm{T}} & -\frac{\sqrt{2}}{2}dc_{\mathrm{T}} \\ c_{\mathrm{M}} & c_{\mathrm{M}} & -c_{\mathrm{M}} & -c_{\mathrm{M}} \end{bmatrix} \begin{bmatrix} \varpi_1^2 \\ \varpi_2^2 \\ \varpi_3^2 \\ \varpi_4^2 \end{bmatrix}$$

(2) Multicopters
For multicopters, in order to implement the control allocation, the positions of all motors in the ABCF need to be determined. For a multicopter with n_{r} propellers, the propellers are marked in clockwise fashion from $i = 1$ to $i = n_{\mathrm{r}} \geq 5$, as shown in Fig. 6.9. The angle between the $o_{\mathrm{b}}x_{\mathrm{b}}$ axis and the supported arm of each motor is denoted by $\varphi_i \geq 0$. The distance between the body center and the ith motor is denoted by $d_i \geq 0, i = 1, 2, \cdots, n_{\mathrm{r}}$. Then, the thrust and moments produced by propellers are expressed as

$$\begin{bmatrix} f \\ \tau_x \\ \tau_y \\ \tau_z \end{bmatrix} = \underbrace{\begin{bmatrix} c_{\mathrm{T}} & c_{\mathrm{T}} & \cdots & c_{\mathrm{T}} \\ -d_1 c_{\mathrm{T}} \sin\varphi_1 & -d_2 c_{\mathrm{T}} \sin\varphi_2 & \cdots & -d_{n_r} c_{\mathrm{T}} \sin\varphi_{n_{\mathrm{r}}} \\ d_1 c_{\mathrm{T}} \cos\varphi_1 & d_2 c_{\mathrm{T}} \cos\varphi_2 & \cdots & d_{n_r} c_{\mathrm{T}} \cos\varphi_{n_{\mathrm{r}}} \\ c_{\mathrm{M}}\delta_1 & c_{\mathrm{M}}\delta_2 & \cdots & c_{\mathrm{M}}\delta_{n_r} \end{bmatrix}}_{\mathbf{M}_{n_{\mathrm{r}}}} \begin{bmatrix} \varpi_1^2 \\ \varpi_2^2 \\ \vdots \\ \varpi_{n_{\mathrm{r}}}^2 \end{bmatrix} \tag{6.56}$$

where $\mathbf{M}_{n_{\mathrm{r}}} \in \mathbb{R}^{4 \times n_r}$ represents the control effectiveness matrix and $\delta_i = (-1)^{i+1}, i = 1, \cdots, n_{\mathrm{r}}$.

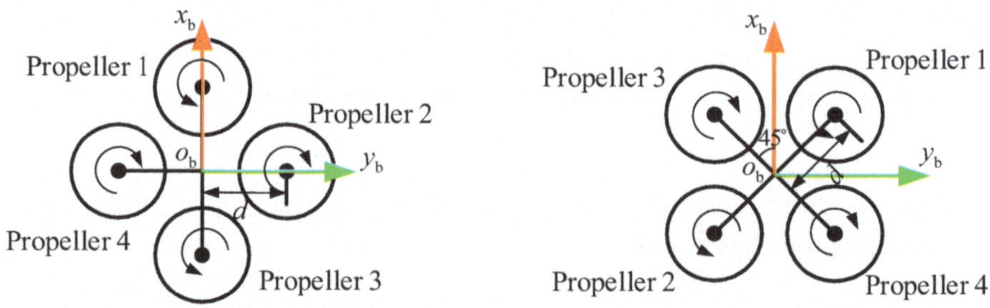

(a) Plus-configuration (b) X-configuration

Fig. 6.8 Two configurations of quadcopters

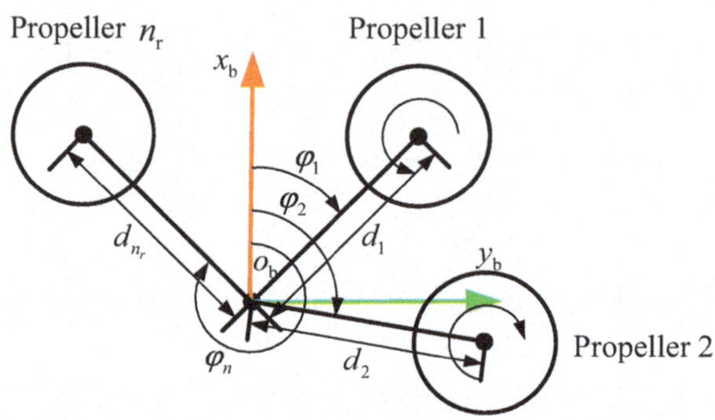

Fig. 6.9 Airframe geometry parameters of a multicopter

6.1.5 Propulsor Model

As shown in Fig. 6.10, the propulsor model includes not only a brushless Direct Current (DC) motor but also an Electronic Speed Controller (ESC) and a propeller. Throttle command σ is an input signal between 0 and 1, while it is not easy to control the battery output voltage U_b.

The ESC generates an equivalent average voltage $U_m = \sigma U_b$ after receiving the throttle command σ the battery output voltage U_b. First, when given a voltage signal,

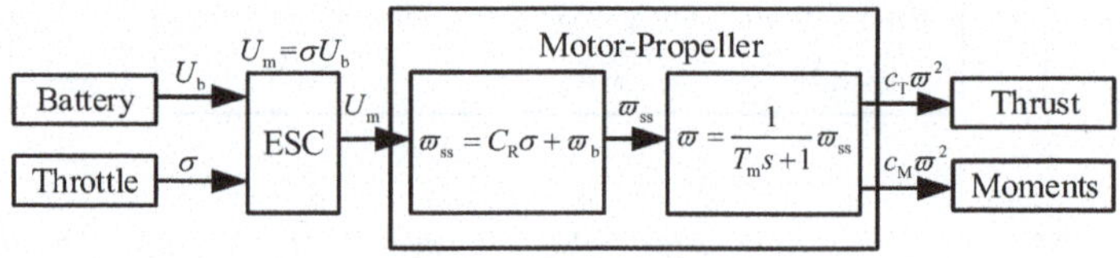

Fig. 6.10 Signal flow of propulsor model

the motor achieves a steady-state speed ϖ_{ss}. The relation is normally linear, which is expressed as follows

$$\varpi_{\text{ss}} = C_{\text{b}} U_{\text{m}} + \varpi_{\text{b}} = C_{\text{R}} \sigma + \varpi_{\text{b}} \tag{6.57}$$

where $C_{\text{R}} = C_{\text{b}} U_{\text{b}}$, C_{b}, and ϖ_{b} denotes constant parameters. Secondly, when given a throttle command, the motor needs some time to achieve the steady-state speed ϖ_{ss}. This time constant denoted by T_{m} determines the dynamic response. Generally, the dynamics of a brushless DC motor can be simplified as a first-order low-pass filter. Its transfer function is expressed as follows

$$\varpi = \frac{1}{T_{\text{m}} s + 1} \varpi_{\text{ss}}. \tag{6.58}$$

Thus, when given a desired steady-state speed ϖ_{ss}, the motor speed cannot achieve ϖ_{ss} immediately. The process requires some time to adjust. Combine Eq. (6.57) with Eq. (6.58) to obtain a complete propulsor model as follows

$$\varpi = \frac{1}{T_{\text{m}} s + 1} (C_{\text{R}} \sigma + \varpi_{\text{b}}) \tag{6.59}$$

where the input is the throttle command σ and the output is motor speed ϖ.

Currently, the relationship between the throttle command perturbation and the reaction torque perturbation is studied further. When a multicopter is hovering, for the ith propeller, it is assumed that the propeller angular speed at equilibrium points ϖ_i^*, the throttle command in equilibrium is σ_i^*, and the reaction torque at equilibrium points is M_i^*. Subsequently, the expression is as follows

$$\begin{aligned} \varpi_i &= \varpi_i^* + \Delta \varpi_i \\ \sigma_i &= \sigma_i^* + \Delta \sigma_i \\ M_i &= M_i^* + \Delta M_i \end{aligned} \tag{6.60}$$

where $\Delta \varpi_i$, $\Delta \sigma_i$, and ΔM_i denote the perturbations of propeller angular speed, throttle command, and reaction torque, respectively. Linearization of Eqs. (6.50) and (6.59) at equilibrium points is as follows

$$\begin{aligned} \Delta M_i &= 2 c_{\text{M}} \varpi_i^* \Delta \varpi_i + J_{\text{RP}} \Delta \dot{\varpi}_i \\ \Delta \varpi_i &= \frac{1}{T_{\text{m}} s + 1} C_{\text{R}} \Delta \sigma_i. \end{aligned}$$

Subsequently, the transfer function from $\Delta \sigma_i$ to ΔM_i is further obtained as follows

$$\Delta M_i = \frac{C_{\text{R}} \left(2 c_{\text{M}} \varpi_i^* + J_{\text{RP}} s \right)}{T_{\text{m}} s + 1} \Delta \sigma_i. \tag{6.61}$$

It is suggested that the dynamic model (6.61) should be considered in the performance optimization for multicopters.

6.1.6 Aerodynamic Model

The airflow speed ${}^{\mathrm{b}}\mathbf{v}_{\mathrm{a}}$ with respect to the body is expressed as follows

$$\begin{aligned} {}^{\mathrm{b}}\mathbf{v}_{\mathrm{a}} &= -{}^{\mathrm{b}}\mathbf{v} + \mathbf{R}_{\mathrm{e}}^{\mathrm{b}} \cdot {}^{\mathrm{e}}\mathbf{w} \\ &= -{}^{\mathrm{b}}\mathbf{v} + \mathbf{R}_{\mathrm{b}}^{\mathrm{e}-1} \cdot {}^{\mathrm{e}}\mathbf{w} \end{aligned}$$

where ${}^{\mathrm{e}}\mathbf{w}$ denotes the wind speed defined in the earth frame, and ${}^{\mathrm{b}}\mathbf{v}_{\mathrm{a}}$ is defined as the vehicle speed minus the local wind speed. The wind speed is superposed by many factors as follows

$${}^{\mathrm{e}}\mathbf{w} = {}^{\mathrm{e}}\mathbf{w}_{\mathrm{tur}} + {}^{\mathrm{e}}\mathbf{w}_{\mathrm{gust}} + {}^{\mathrm{e}}\mathbf{w}_{\mathrm{sheer}} + {}^{\mathrm{e}}\mathbf{w}_{\mathrm{const}} + \cdots$$

where ${}^{\mathrm{e}}\mathbf{w}_{\mathrm{tur}}$ denotes the atmospheric turbulence, ${}^{\mathrm{e}}\mathbf{w}_{\mathrm{gust}}$ denotes the wind gust, ${}^{\mathrm{e}}\mathbf{w}_{\mathrm{sheer}}$ denotes the sheer wind, and ${}^{\mathrm{e}}\mathbf{w}_{\mathrm{const}}$ denotes a constant wind. The relative airflow speed ${}^{\mathrm{b}}\mathbf{v}_{\mathrm{a}}$ is expressed into the following component form as follows

$${}^{\mathrm{b}}\mathbf{v}_{\mathrm{a}} \triangleq \begin{bmatrix} u & v & w \end{bmatrix}^{\mathrm{T}}.$$

Thus, its airspeed V_{a}, angle of attack α, angle of sideslip β are given as follows

$$\begin{aligned} V_{\mathrm{a}} &\triangleq \left\| {}^{\mathrm{b}}\mathbf{v}_{\mathrm{a}} \right\| = \sqrt{u^2 + v^2 + w^2} \\ \alpha &\triangleq \arcsin\left(-\tfrac{w}{V_{\mathrm{a}}}\right) = \arctan\left(\tfrac{w}{u}\right) \in \left(-\tfrac{\pi}{2}, \tfrac{\pi}{2}\right) \\ \beta &\triangleq \arcsin\left(\tfrac{v}{V_{\mathrm{a}}}\right) \in \left(-\tfrac{\pi}{2}, \tfrac{\pi}{2}\right). \end{aligned} \tag{6.62}$$

By using the above values, we can obtain the desired aerodynamic parameters, such as the drag coefficient C_{d}, and the drag moment coefficient C_{dm}. Given that both aerodynamic force and moment are proportional to the squares of the speed ${}^{\mathrm{b}}\mathbf{v}_{\mathrm{a}}$ and angular speed ${}^{\mathrm{b}}\boldsymbol{\omega}_{\mathrm{a}}$ respectively, by assuming that the relative air rotation speed is ${}^{\mathrm{b}}\boldsymbol{\omega}_{\mathrm{a}} \approx -{}^{\mathrm{b}}\boldsymbol{\omega} \triangleq \begin{bmatrix} \omega_x & \omega_y & \omega_z \end{bmatrix}^{\mathrm{T}}$, the aerodynamic force and moment in Eqs. (6.39) and (6.40) can be approximately obtained by

$$\begin{aligned} {}^{\mathrm{b}}\mathbf{F}_{\mathrm{d}} &= C_{\mathrm{d}} \begin{bmatrix} u|u| \\ v|v| \\ w|w| \end{bmatrix} \\ {}^{\mathrm{b}}\mathbf{M}_{\mathrm{d}} &= C_{\mathrm{dm}} \begin{bmatrix} \omega_x|\omega_x| \\ \omega_y|\omega_y| \\ \omega_z|\omega_z| \end{bmatrix}. \end{aligned} \tag{6.63}$$

6.2 Basic Experiment

6.2.1 Experimental Objective

(1) Things to prepare
Software: MATLAB R2017b or above, Instruction Package "e2.1" (https://rflysim.com/course).
(2) Objectives
Analyze the flight performance with respect to the total mass, moment of inertia matrix, and propeller parameters of a multicopter.

6.2.2 Experimental Procedure

(1) **Step1**: **Flight state with respect to total mass changing**

1) Open the file "e2\e2.1\e2_1.slx", as shown in Fig. 6.11. Subsequently, open the file "Init_control.m" and click "Run" button in MATLAB (as shown in the box "c" in Fig. 6.12c) to initialize the parameters in the file.
2) Altitude response with respect to the hover throttle. Select "z" signal line on "PosE" output by clicking the left mouse, as shown between "c" and "b" in Fig. 6.11. Next, move the mouse to "three blue dots", which is next to "b" in Fig. 6.11 and select the second "Enable Data Logging" (the icon is shown in the box "c" in Fig. 6.11). Subsequently, in order to run Simulink, click "Run" button in Simulink (as shown in the box "b" in Fig. 6.12a).

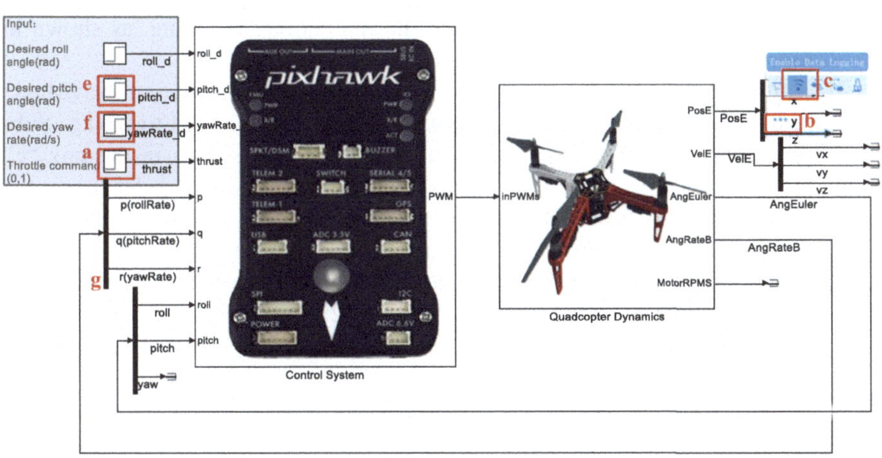

Fig. 6.11 SIL test for flight state with respect to total mass changing, Simulink model "e2_1.slx"

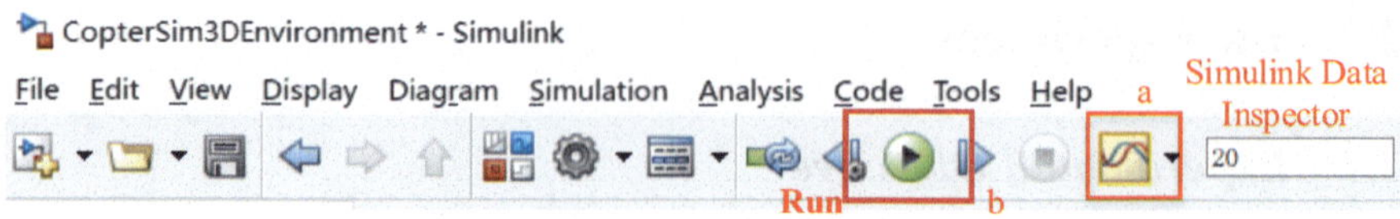

(a) Simulink "Run" button on MATLAB 2017b - 2019a

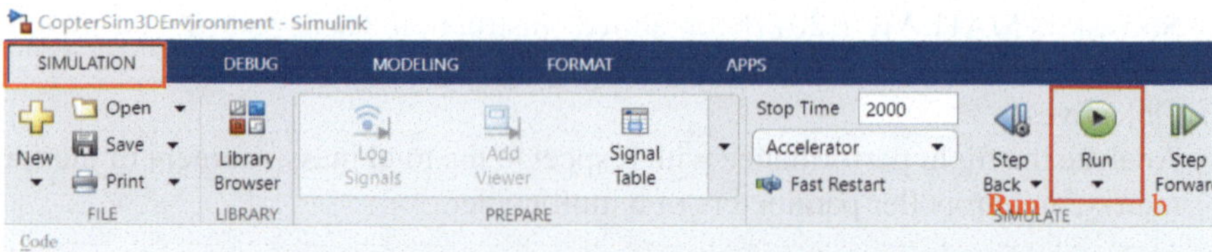

(b) Simulink "Run" button on MATLAB 2019b and above

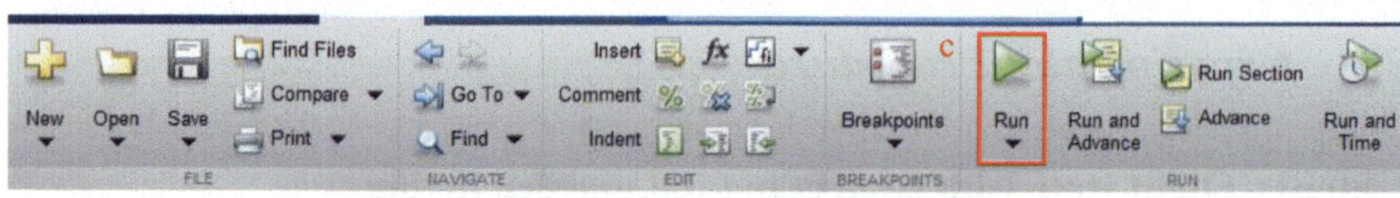

(c) MATLAB "Run" button on MATLAB 2017b and above

Fig. 6.12 "Run" button in MATLAB, "Run" button and "Simulink Data Inspector" button in Simulink

Finally, open the corresponding signal output in the "Simulink Data Inspector" as shown in the box "a" in Fig. 6.12a. Modify "Throttle command" to change the throttle value, where "Throttle command" location is shown in the box "a" in Fig. 6.11. When the mass is 1.4 kg ("ModelParam_uavMass" in file "Init_control.m" is set to 1.4, which represents the mass of the multicopter) and "Throttle command" is 0.6085 (60.85% throttle percentage) in Simulink, the multicopter can keep the multicopter hovering, as shown in Fig. 6.13.

3) Altitude response with respect to mass. Change "ModelParam_uavMass" in file "Init_control.m" to 2.0, namely the mass is changed to 2.0 kg. Subsequently, click the "Run" button in MATLAB (as shown in the box "c" in Fig. 6.12c) to update the parameter. Run the Simulink model to get the result shown in Fig. 6.14. The conclusion is that the altitude of the multicopter decreases when "Throttle command" is constant. Given the increased weight, the same input throttle is unable to provide sufficient thrust to keep the multicopter hovering. When the mass is 2 kg, the "Throttle command" is set to 0.7032 (70.32 % throttle percentage) to keep the multicopter hovering.

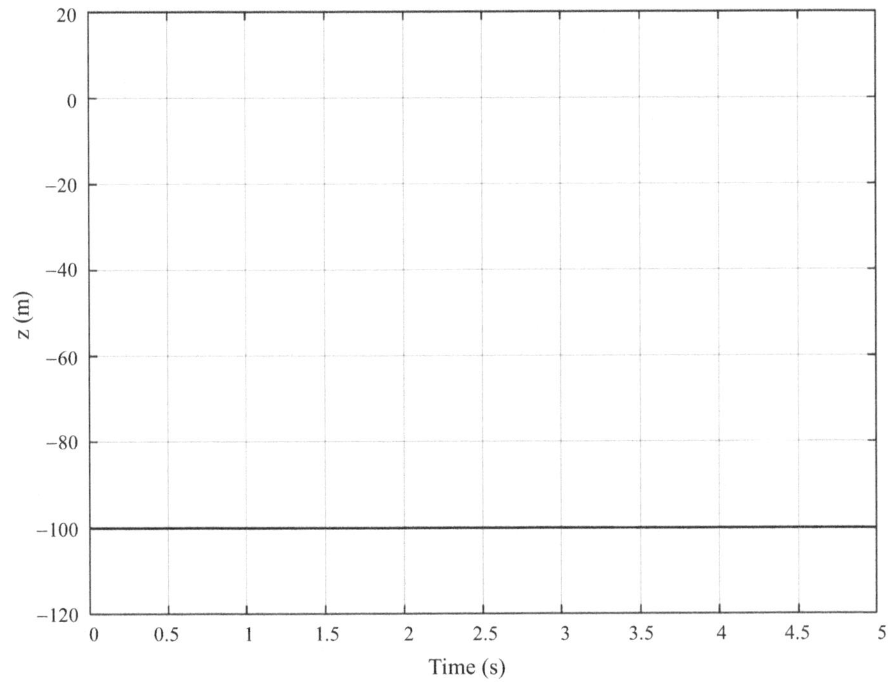

Fig. 6.13 Altitude response when mass is 1.4 kg

4) Attitude control performance with respect to mass, where two cases are considered. In file "Init_control.m", the mass is set to 1.4 kg or 2.0 kg and click "Run" button in MATLAB (as shown in the box "c" in Fig. 6.12c) to update the parameter. Correspondingly, the "Throttle command" is set to 0.6085 or 0.7032. Besides that, as shown in Fig. 6.15, "pitch_d" in the file "e2_1.slx" is set to 0.2 rad and the output pitch angle is observed with "Scope" in Simulink. Run the Simulink model to obtain the results shown in Fig. 6.16. A conclusion is drawn that the attitude response is almost independent of the mass when the remaining parameters such as the moment of inertia are constant.

(2) **Step2**: **Yaw rate with respect to moment of inertia**
Modify "ModelParam_uavJzz" in file "Init_control.m" to double the moment of inertia about the $o_b z_b$ axis ("ModelParam_uavJzz = 0.0732") and click "Run" button in MATLAB (as shown in the box "c" in Fig. 6.12c) to update the parameter. "yawRate_d" (shown in "f" in Fig. 6.11) in Simulink is set to 0.2 rad and the output yaw rate (shown in "g" in Fig. 6.11) is observed with "Scope" in Simulink. Run the Simulink model to obtain the results shown in Fig. 6.17. The conclusion is drawn that the system yaw rate response becomes slower when the moment of inertia about the $o_b z_b$ axis increases.

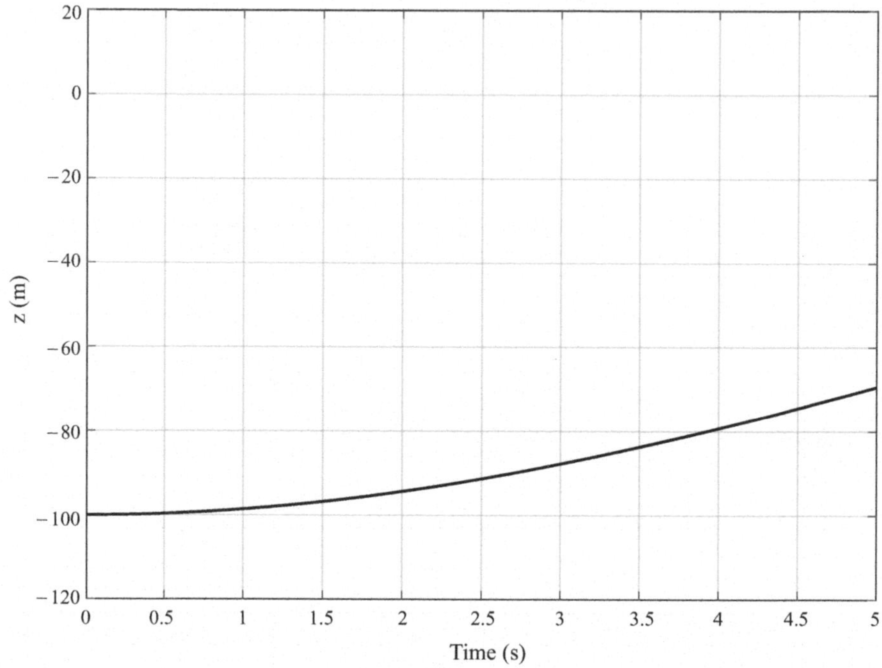

Fig. 6.14 Altitude response when mass is 2 kg

(3) **Step3**: **Altitude with respect to propeller thrust coefficient**
The propeller thrust coefficient "ModelParam_rotorCt" is doubled ("ModelParam_rotorCt = 2.21e-05") and other parameters correspond to the initial value. Subsequently, click "Run" button in MATLAB (as shown in the box "c" in Fig. 6.12c) to update these parameters. Evidently, with the propeller thrust parameter increased, the thrust provided by propellers is increased under the same throttle command. The altitude response (shown in "b" in Fig. 6.11) is shown in Fig. 6.18. In order to maintain the multicopter hovering, the "Throttle command" is set to 0.4663 (46.63% throttle percentage).

(4) **Step4**: **Yaw rate with respect to propeller torque**
The torque coefficient parameter "ModelParam.rotorCm" is doubled ("ModelParam.rotorCm = 3.558e-07") and "yawRate_d" (shown in "f" in Fig. 6.11) in Simulink is set to 0.2 rad. Subsequently, click the "Run" button in MATLAB (as shown in the box "c" in Fig. 6.12c) to update the parameter. The result of yaw rate output (shown in "g" in Fig. 6.11) is shown in Fig. 6.19. It is concluded that the larger torque coefficient is, the faster yaw rate response is.

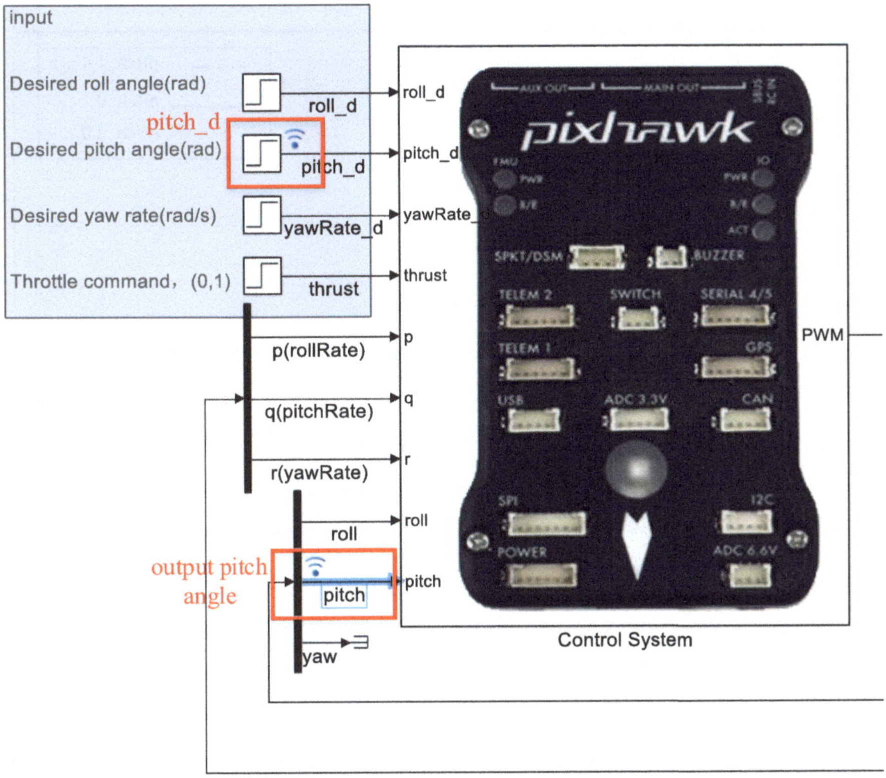

Fig. 6.15 Setting input and observation signal in Simulink model "e2_1.slx"

6.3 Analysis Experiment

6.3.1 Experimental Objective

When an X-configuration quadcopter is hovering, calculate the equilibrium point of the dynamic system in the following

$$\begin{aligned} {}^{\mathrm{e}}\dot{\mathbf{p}} &= {}^{\mathrm{e}}\mathbf{v} \\ {}^{\mathrm{e}}\dot{\mathbf{v}} &= {}^{\mathrm{e}}\mathbf{F}/m \\ \dot{\boldsymbol{\Theta}} &= \mathbf{W}\cdot{}^{\mathrm{b}}\boldsymbol{\omega} \\ \mathbf{J}\cdot{}^{\mathrm{b}}\dot{\boldsymbol{\omega}} &= -{}^{\mathrm{b}}\boldsymbol{\omega}\times\left(\mathbf{J}\cdot{}^{\mathrm{b}}\boldsymbol{\omega}\right)+{}^{\mathrm{b}}\mathbf{M} \end{aligned} \tag{6.64}$$

and the control effectiveness matrix is Eq. (6.55). We express the linearization model at the equilibrium point by considering the motor dynamics. Subsequently, compare and analyze the conclusions from the basic experiment and the analysis experiment.

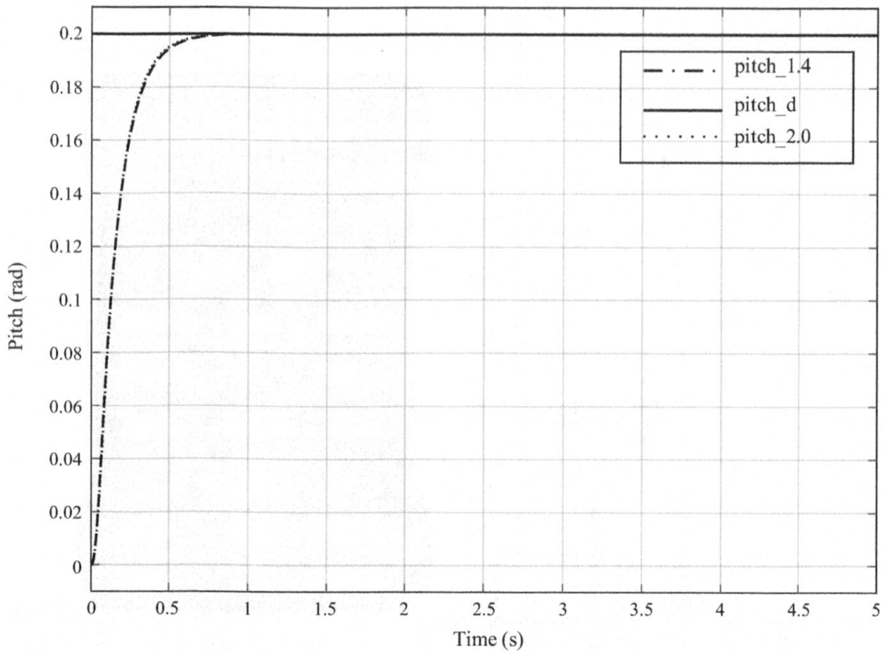

Fig. 6.16 Pitch angle response with respect to mass

6.3.2 *Calculation Procedure*

(1) **Step1**: **Calculate the hover throttle command**
When a multicopter is hovering, based on Eq. (6.51), the weight satisfies the following expression

$$mg = T_1 + T_2 + T_3 + T_4. \tag{6.65}$$

Based on Eq. (6.47), it leads to the following expression

$$mg = c_T(\varpi_1^2 + \varpi_2^2 + \varpi_3^2 + \varpi_4^2). \tag{6.66}$$

Based on **Assumption** 6.3, the multicopter is symmetrical, it includes $\varpi_1 = \varpi_2 = \varpi_3 = \varpi_4$. So obtain the following expression

$$\varpi_1 = \sqrt{\frac{mg}{4c_T}}. \tag{6.67}$$

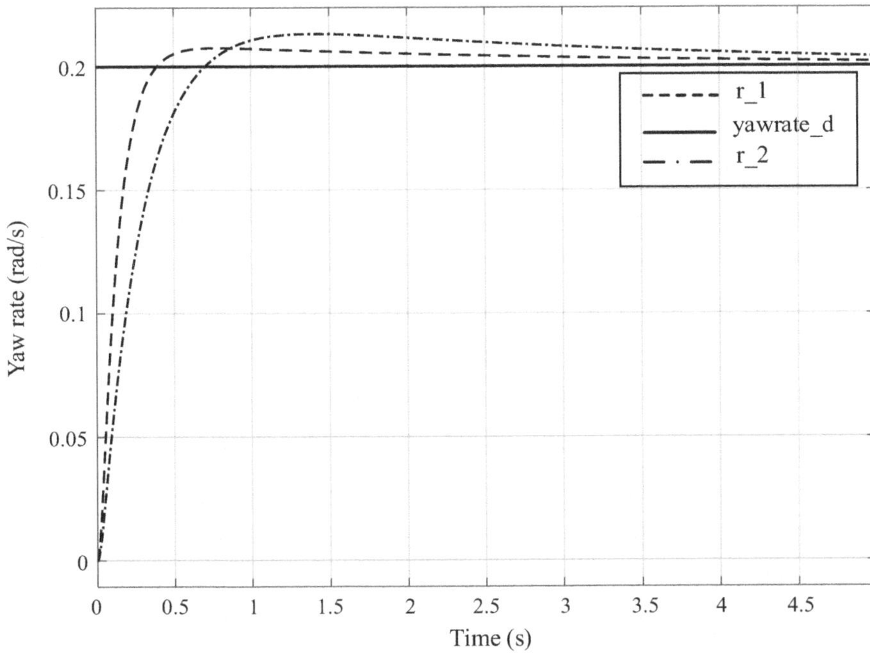

Fig. 6.17 Yaw rate response with respect to the moment of inertia about the $o_b z_b$ axis. Parameter "yawrate_d" denotes the desired yaw rate, "r_1" denotes the yaw rate when "ModelParam_uavJzz" is the initial value ("ModelParam_uavJzz=0.0366"), and "r_2" denotes the yaw rate when "ModelParam_uavJzz" is doubled ("ModelParam_uavJzz=0.0732")

Based on Eq. (6.57), it further obtains the following expression

$$\sigma = \frac{\varpi_1 - \varpi_b}{c_R} = \frac{\sqrt{\frac{mg}{4c_T}} - \varpi_b}{c_R}. \tag{6.68}$$

Finally, substitute parameter $m = 1.4\,\text{kg}$, $g = 9.8\,\text{m/s}^2$, $c_T = 1.105 \times 10^{-7}$ N/RPM^2, $c_R = 1148\,\text{RPM}$, $\varpi_b = -141.4\,\text{RPM}$ into Eq. (6.68), thereby resulting in $\sigma = 0.6085$ (60.85 % throttle percentage).

(2) **Step2**: **Calculate the linearization model of the equilibrium point**

1) Linear model simplification. Based on Eqs. (6.42) and (6.63), assuming ${}^e\mathbf{v} \approx \mathbf{0}$, ${}^b\boldsymbol{\omega} \approx \mathbf{0}$, and ${}^e\mathbf{w} \approx \mathbf{0}$ with the small perturbation hypothesis [9] around the hovering mode, one has

$$\begin{aligned} -{}^b\boldsymbol{\omega} \times \left(\mathbf{J}\,{}^b\boldsymbol{\omega}\right) &\approx \mathbf{0} \\ \mathbf{G}_a &\approx \mathbf{0} \\ {}^b\mathbf{M}_d &\approx \mathbf{0} \\ {}^b\mathbf{F}_d &\approx \mathbf{0}. \end{aligned}$$

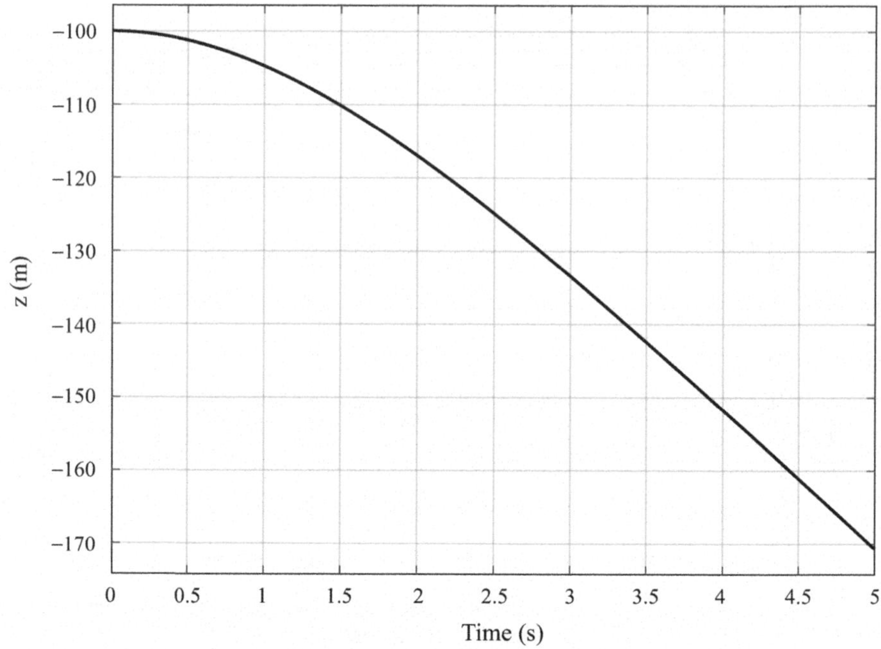

Fig. 6.18 Altitude response with respect to propeller thrust coefficient

The simplified model of Eq. (6.64) is given as follows

$$\begin{aligned} {}^{\mathrm{e}}\dot{\mathbf{p}} &= {}^{\mathrm{e}}\mathbf{v} \\ {}^{\mathrm{e}}\dot{\mathbf{v}} &= g\mathbf{e}_3 - \frac{f}{m}\mathbf{R}\mathbf{e}_3 \\ \dot{\boldsymbol{\Theta}} &= \mathbf{W}\cdot{}^{\mathrm{b}}\boldsymbol{\omega} \\ \mathbf{J}\cdot{}^{\mathrm{b}}\dot{\boldsymbol{\omega}} &= \boldsymbol{\tau}. \end{aligned} \tag{6.69}$$

The assumptions are further expressed as $\sin\phi \approx \phi, \cos\phi \approx 1,\ \sin\theta \approx \theta, \cos\theta \approx 1,\ f \approx mg, \boldsymbol{\tau} \approx \mathbf{0}$. Thus, $\mathbf{R}\mathbf{e}_3$ is simplified as follows

$$\mathbf{R}\mathbf{e}_3 \approx \begin{bmatrix} \theta\cos\psi + \phi\sin\psi \\ \theta\sin\psi - \phi\cos\psi \\ 1 \end{bmatrix}.$$

Based on Eq. (6.69), the original model is decoupled into three linear models, namely horizontal position channel model, altitude channel model, and attitude model. They are introduced as follows.

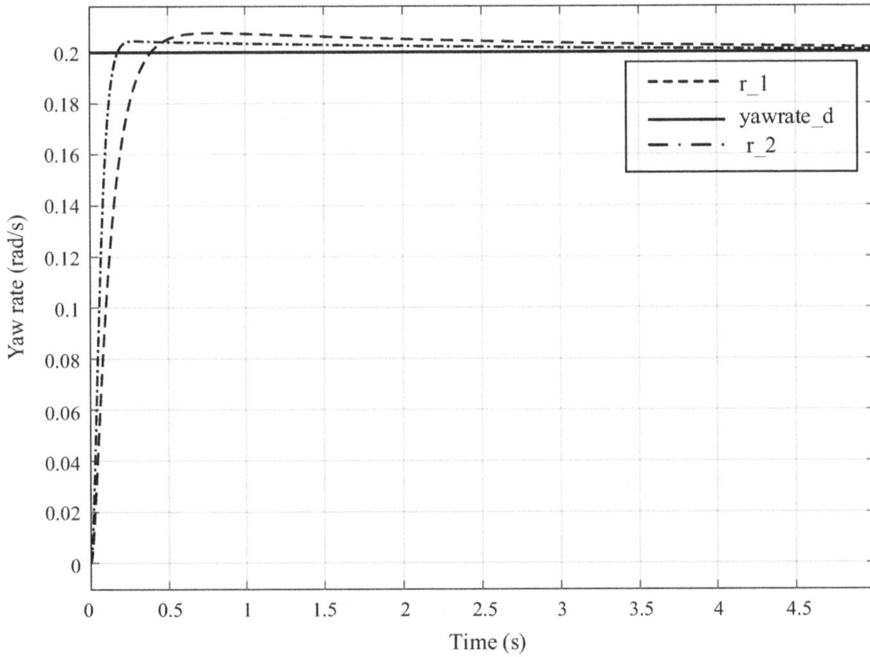

Fig. 6.19 Yaw rate response with respect to the torque coefficient. Parameter "yawrate_d" denotes the desired yaw rate, "r_1" denotes the yaw rate when "ModelParam.rotorCm" is the initial value ("ModelParam.rotorCm = 1.779e-07"), and "r_2" denotes the yaw rate when "ModelParam.rotorCm" is doubled ("ModelParam.rotorCm = 3.558e-07")

(a) Horizontal position channel model

$$\begin{aligned}\dot{\mathbf{p}}_{\mathrm{h}} &= \mathbf{v}_{\mathrm{h}} \\ \dot{\mathbf{v}}_{\mathrm{h}} &= -g\mathbf{A}_{\psi}\boldsymbol{\Theta}_{\mathrm{h}}\end{aligned} \tag{6.70}$$

where $\mathbf{p}_{\mathrm{h}} = \begin{bmatrix} p_x \\ p_y \end{bmatrix}$, $\mathbf{A}_{\psi} = \begin{bmatrix} \sin\psi & \cos\psi \\ -\cos\psi & \sin\psi \end{bmatrix}$, $\boldsymbol{\Theta}_{\mathrm{h}} = \begin{bmatrix} \phi \\ \theta \end{bmatrix}$. In the horizontal position channel model, $\boldsymbol{\Theta}_{\mathrm{h}}$ is viewed as the input. Furthermore, $-g\mathbf{A}_{\psi}$ can be obtained. So, $-g\mathbf{A}_{\psi}\boldsymbol{\Theta}_{\mathrm{h}}$ can be viewed as the input and $\mathbf{p}_{\mathrm{h}}$ is viewed as the output. Thus, the horizontal position channel model is, in fact, a linear model.

(b) Altitude channel model

$$\begin{aligned}\dot{p}_{\mathrm{z}} &= v_{\mathrm{z}} \\ \dot{v}_{\mathrm{z}} &= -g - \frac{f}{m}.\end{aligned} \tag{6.71}$$

In Eq. (6.71), in contrast to the simplification of the horizontal position channel model, the term $g - f/m$ is not a higher-order infinite small term. Thus, it cannot be ignored. Otherwise, there is no input in the altitude channel. Evidently, the altitude channel model (6.71) is also a linear model.

(c) Attitude model

The attitude model in Eq. (6.69) is simplified as follows

$$\begin{aligned}\dot{\boldsymbol{\Theta}} &= {}^{\mathrm{b}}\boldsymbol{\omega}\\ \mathbf{J}\cdot{}^{\mathrm{b}}\dot{\boldsymbol{\omega}} &= \boldsymbol{\tau}.\end{aligned}\tag{6.72}$$

Thus, the attitude model (6.72) is also a linear model.

2) Hover state linearization. When a multicopter is hovering, the angular speed of the ith propeller at the equilibrium point is $\varpi_i^* = \varpi_0^*$, throttle command at the equilibrium point is $\sigma_i^* = \sigma_0^*$, reaction torque at the equilibrium point is $M_i^* = M_0^*$ and thrust at the equilibrium point is $T_i^* = T_0^*$. Furthermore, the attitude angle and speed at the equilibrium point are zero, and the position at equilibrium points is $\mathbf{p}_{\mathrm{d}}$. Based on Eq. (6.60), any variable can be decomposed into a sum of the value at the equilibrium point and the perturbation in the form as follows

$$\begin{aligned}\boldsymbol{\Theta} &= \mathbf{0} + \Delta\boldsymbol{\Theta}\\ \boldsymbol{\omega} &= \mathbf{0} + \Delta\boldsymbol{\omega}\\ \varpi_i &= \varpi_0^* + \Delta\varpi_i\\ \sigma_i &= \sigma_0^* + \Delta\sigma_i\\ M_i &= M_0^* + \Delta M_i\\ T_i &= T_0^* + \Delta T_i.\end{aligned}\tag{6.73}$$

Based on Eqs. (6.47), (6.48), (6.61) and (6.73), the perturbation ΔT_i of thrust and the perturbation of reaction torque ΔM_i are

$$\begin{aligned}\Delta T_i &= \frac{C_{\mathrm{R}} 2 c_{\mathrm{T}} \varpi_0^*}{T_{\mathrm{m}} s + 1}\Delta\sigma_i\\ \Delta M_i &= \frac{C_{\mathrm{R}}\left(2 c_{\mathrm{M}} \varpi_0{}^* + J_{\mathrm{RP}} s\right)}{T_{\mathrm{m}} s + 1}\Delta\sigma_i.\end{aligned}\tag{6.74}$$

Based on Eqs. (6.55), (6.72) and (6.73), the perturbations on $\boldsymbol{\Theta}$ and $\boldsymbol{\omega}$ are

$$\begin{aligned}\Delta\dot{\boldsymbol{\Theta}} &= \Delta\boldsymbol{\omega}\\ \mathbf{J}\Delta\dot{\boldsymbol{\omega}} &= \Delta\boldsymbol{\tau}\end{aligned}\tag{6.75}$$

where

$$\Delta\boldsymbol{\tau} \approx \begin{bmatrix} \sqrt{2}d\frac{C_{\mathrm{R}}c_{\mathrm{T}}\varpi_0^*}{T_{\mathrm{m}}s+1} & 0 & 0 \\ 0 & \sqrt{2}d\frac{C_{\mathrm{R}}c_{\mathrm{T}}\varpi_0^*}{T_{\mathrm{m}}s+1} & 0 \\ 0 & 0 & \frac{C_{\mathrm{R}}(2c_{\mathrm{M}}\varpi_0{}^*+J_{\mathrm{RP}}s)}{T_{\mathrm{m}}s+1} \end{bmatrix} \begin{bmatrix} \Delta\bar{\tau}_x \\ \Delta\bar{\tau}_y \\ \Delta\bar{\tau}_z \end{bmatrix}$$

$$\begin{bmatrix} \Delta\bar{\tau}_x \\ \Delta\bar{\tau}_y \\ \Delta\bar{\tau}_z \end{bmatrix} \approx \begin{bmatrix} -1 & 1 & 1 & -1 \\ 1 & -1 & 1 & -1 \\ 1 & 1 & -1 & -1 \end{bmatrix} \begin{bmatrix} \Delta\sigma_1 \\ \Delta\sigma_2 \\ \Delta\sigma_3 \\ \Delta\sigma_4 \end{bmatrix}.$$

Furthermore,

$$\begin{aligned} \Delta\phi(s) &= \sqrt{2}\frac{dC_{\mathrm{R}}c_{\mathrm{T}}\varpi_0^*}{J_x}\frac{1}{s^2}\frac{1}{T_{\mathrm{m}}s+1}\Delta\bar{\tau}_x \\ \Delta\theta(s) &= \sqrt{2}\frac{dC_{\mathrm{R}}c_{\mathrm{T}}\varpi_0^*}{J_y}\frac{1}{s^2}\frac{1}{T_{\mathrm{m}}s+1}\Delta\bar{\tau}_y \\ \Delta\dot{\psi}(s) &= \frac{C_{\mathrm{R}}(2c_{\mathrm{M}}\varpi_0{}^*+J_{\mathrm{RP}}s)}{J_z}\frac{1}{s}\frac{1}{T_{\mathrm{m}}s+1}\Delta\bar{\tau}_z. \end{aligned} \tag{6.76}$$

As for the position model, it has

$$\begin{aligned} \mathbf{p}_{\mathrm{h}} &= \mathbf{p}_{\mathrm{hd}} + \Delta\mathbf{p}_{\mathrm{h}} \\ p_z &= p_{z\mathrm{d}} + \Delta p_z \\ \boldsymbol{v}_{\mathrm{h}} &= 0 + \Delta\boldsymbol{v}_{\mathrm{h}} \\ v_z &= 0 + \Delta v_z. \end{aligned} \tag{6.77}$$

Eqs. (6.70) and (6.71) are linearized at the equilibrium point that

$$\begin{bmatrix} \Delta p_x \\ \Delta p_y \end{bmatrix} = \begin{bmatrix} -\frac{g}{s^2}\Delta\theta \\ \frac{g}{s^2}\Delta\phi \end{bmatrix} \tag{6.78}$$

$$\Delta p_z = -\frac{\Sigma\Delta T_i}{ms^2}. \tag{6.79}$$

Substituting Eq. (6.76) into Eq. (6.78) and substituting Eq. (6.74) into Eq. (6.79) result in

$$\begin{aligned} \Delta p_x &= -\sqrt{2}g\frac{dC_{\mathrm{R}}c_{\mathrm{T}}\varpi_0^*}{J_y}\frac{1}{s^4}\frac{1}{T_{\mathrm{m}}s+1}\Delta\bar{\tau}_y \\ \Delta p_y &= \sqrt{2}g\frac{dC_{\mathrm{R}}c_{\mathrm{T}}\varpi_0^*}{J_x}\frac{1}{s^4}\frac{1}{T_{\mathrm{m}}s+1}\Delta\bar{\tau}_x \\ \Delta p_z &= -\frac{2C_{\mathrm{R}}c_{\mathrm{T}}\varpi_0^*}{ms^2(T_{\mathrm{m}}s+1)}\Delta\bar{\tau} \end{aligned} \tag{6.80}$$

where $\Delta\bar{\tau} = \Delta\sigma_1 + \Delta\sigma_2 + \Delta\sigma_3 + \Delta\sigma_4$.

(3) **Step3**: **Compare theoretical analysis with conclusion from the basic experiment**

1) Analyze the yaw rate response
Based on Eq. (6.76), the transfer function of yaw rate is as follow

$$\Delta\dot{\psi}\,(s) = \frac{C_{\mathrm{R}}\,(2c_{\mathrm{M}}\varpi_0{}^{*} + J_{\mathrm{RP}}s)}{J_z}\frac{1}{s}\frac{1}{T_{\mathrm{m}}s+1}\Delta\bar{\tau}_z.$$

It is concluded that, with the slope of the linear relationship from the throttle command to the motor speed C_{R}, the torque coefficient c_{M} and torque speed ϖ_0^* increased, the yaw angle rate response gets faster. If the motor response time constant T_{m} and the moment of inertia J_z about the $o_{\mathrm{b}}z_{\mathrm{b}}$ axis increase, the yaw angle rate response decreases. These are consistent with the conclusions from Step2 and Step4 in the basic experiment.

2) Analyze the altitude response
Based on Eq. (6.80), the transfer function of the altitude channel model is as follows

$$\Delta p_z = -\frac{2C_{\mathrm{R}}c_{\mathrm{T}}\varpi_0^*}{ms^2(T_{\mathrm{m}}s+1)}\Delta\bar{\tau}.$$

It is concluded that with the propeller thrust coefficient c_{T} and the slope of the linear relationship from the throttle command to the motor speed C_{R} increased, the altitude gets higher. If the mass m is increased, the altitude response gets slower. The conclusions can be verified by readers yourselves with corresponding simulation experiments.

6.4 Design Experiment

6.4.1 Experimental Objective

(1) Things to prepare
Software: MATLAB R2017b or above, a designed multicopter model in Chap. 5 (the hardware configuration of the quadcopter used in Sect. 5.4 in Chap. 5 is shown in Fig. 5.46), the model parameters provided on the multicopter performance evaluation website https://flyeval.com/paper/ (The model parameters are shown in Table 6.1 and all parameters are in file "e2\e2.3\Init.m").

Table 6.1 Model parameters of a given quadcopter

	Notation	Parameter name in code	Value
MultiCopter mass	m	ModelParam _uavMass	1.4 (kg)
Acceleration of gravity	g	ModelParam _envGravityAcc	9.8 (m/s^2)
Multicopter inertia Matrix	$\mathbf{J}$	ModelParam_uavJ	$\begin{bmatrix} 0.0211 & 0 & 0 \\ 0 & 0.0219 & 0 \\ 0 & 0 & 0.0366 \end{bmatrix}$ (m/s^2)
Radius of airframe	$d/2$	ModelParam_uavR	0.225 (m)
Lumped thrust coefficient	c_{T}	ModelParam_rotorCt	1.105e-05 (N/rad/s^2)
Lumped torque coefficient	c_{M}	ModelParam_rotorCm	1.779e-07 (N.m/(rad/s)^2)
Slope of the linear relationship from the throttle command to the motor speed	C_{R}	ModelParam_motorCr	1148
Constant of linear relationship from the throttle command to the motor speed	ϖ_{b}	ModelParam_motorWb	−141.4 (rad/s)
Motor propeller moment of inertia	J_{RP}	ModelParam_motorJm	0.0001287 (kg/m^2)
Motor response time constant	T_{m}	ModelParam_motorT	0.02 (s)
Time constant	C_{d}	ModelParam_uavCd	0.073 (N/(m/s)^2)
Damping moment coefficient	C_{dm}	ModelParam_uavCCm	$\begin{bmatrix} 0.01 & 0.01 & 0.0055 \end{bmatrix}$ (N/(rad/s)^2)

(2) Objectives

1) Establish a control model for the designed quadcopter with its attitude control model based on quaternions, rotation matrix or Euler angles;
2) Add a quadcopter 3D model to FlightGear flight simulator.[2]

[2]It is a very popular open-source flight simulator software that can receive flight status from Simulink via UDP and easily observe the flight status of the aerial vehicle during Simulink simulation. Reference: http://home.flightgear.org.

6.4.2 General Description of the Multicopter Dynamic Model

As shown in Fig. 6.20, the multicopter modeling mainly includes four parts.

(1) **Rigid-body kinematic model.** Kinematics are independent of the mass and force. It only examines variables such as position, velocity, attitude and angular velocity. For the multicopter kinematic model, the inputs include velocity and angular velocity, and the outputs include position and attitude.
(2) **Rigid-body dynamic model**. Dynamics involve both the movement and the force. They are related to the object's mass and moments of inertia. Equations such as Newton's second law, law of kinetic energy, and law of momentum, are typically used to investigate the mutual effect among different objects. For the multicopter dynamic model, the inputs include thrust and moments (pitching moment, rolling moment, and yawing moment), and the outputs include velocity and angular velocity. The rigid-body kinematic model and dynamic model constitute the general flight control rigid-body model of multicopters.
(3) **Control effectiveness model.** The inputs include propeller angular speeds, and the outputs include thrust and moments. For either a quadcopter or a hexacopter, the thrust and moments are all generated by propellers. Given the propeller angular speeds, the thrust and moments can be calculated by using control effectiveness model. The inversion of the control effectiveness model is termed as the control allocation model. When the thrust and moments are obtained by controller design, the propeller angular speeds are calculated by using the control allocation model.
(4) **Propulsor model.** The propulsor model is a whole power mechanism that includes a brushless DC motor, an ESC, and a propeller. The input is a throttle command between 0 and 1 and the outputs are propeller angular speeds. In practice, the model with throttle command as input and propeller thrust as output can also be established.

6.4.3 Modeling Procedure

6.4.3.1 Establish a Complete Multicopter Dynamic Model in MATLAB/Simulink

(1) **Step1**: **Propulsor model**
In order to facilitate to simulate the case when the initial state is nonzero, Eq. (6.58) is expressed in the form of a state equation. Let the state variable $x = T_{\mathrm{m}}\varpi$, output $y = \varpi$ and input $u = \varpi_{ss}$. Then,

$$\begin{cases} \dot{x} = -\frac{1}{T_{\mathrm{m}}}x + u \\ y = \frac{1}{T_{\mathrm{m}}}x. \end{cases}$$

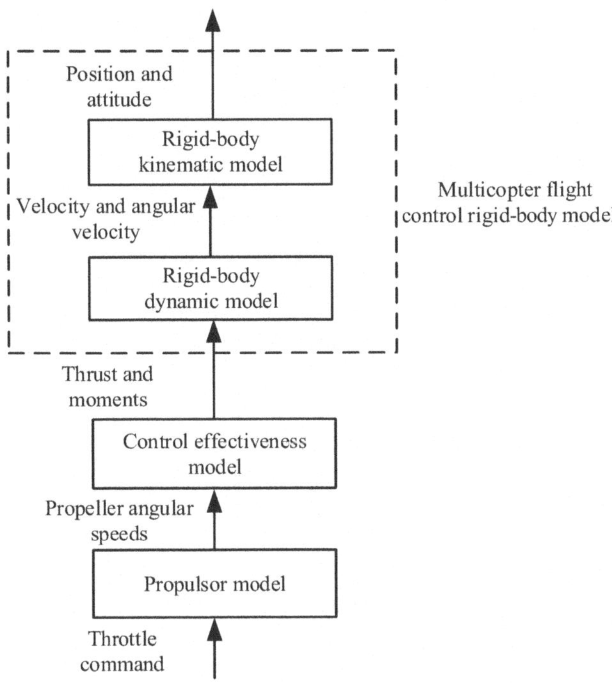

Fig. 6.20 Architecture of multicopter dynamic model

Design the propulsor model by combining Eq. (6.57) with Eq. (6.58) as shown in Fig. 6.21. Additionally, the module "Throttle to RPM Gain" is shown in Fig. 6.22.

(2) **Step2**: **Control effectiveness model**
Based on Eq. (6.55), the force and moment of the propeller acting on the body are obtained; aerodynamic force and moment are obtained based on Eq. (6.63); and the code is shown in Table 6.2. The gravity force is also added here to obtain the external force "Fb" and the external torque "Mb" acting on the aircraft in ABCF. The control effectiveness model is shown in Fig. 6.23.

(3) **Step3**: **Rigid-body dynamic model**
The position dynamic model and attitude dynamic model as shown in Fig. 6.24, is established based on Eqs. (6.36) and (6.40), respectively.

(4) **Step4**: **Rigid-body kinematic model**
Specifically, the kinematic model shown in Fig. 6.25 is established based on Eq. (6.35) which is a attitude control model based on quaternions. The relationships from quaternions to rotation matrix and Euler angles are obtained by Eqs. (6.19) and (6.26), respectively.

(5) **Step5**: **Model connection**
The connection between kinematics model and dynamic model is shown as Fig. 6.26, which consists of the multicopter flight control rigid-body model corresponding to Fig. 6.20. The connection relationship among rigid-body model,

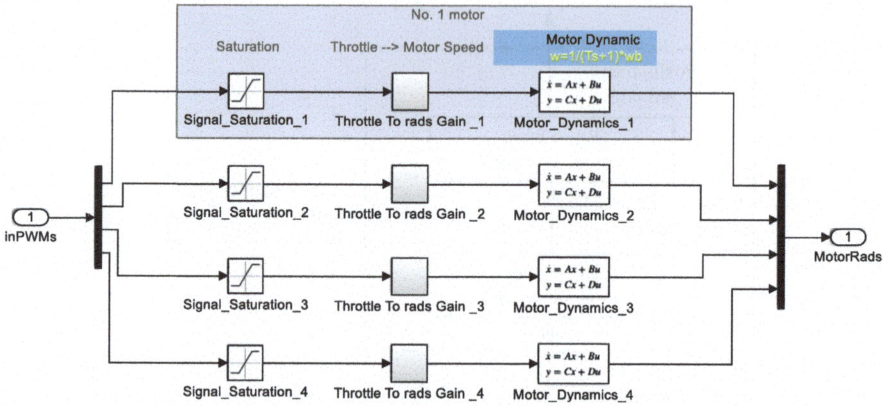

Fig. 6.21 Propulsor model, Simulink model "dynamics.slx"

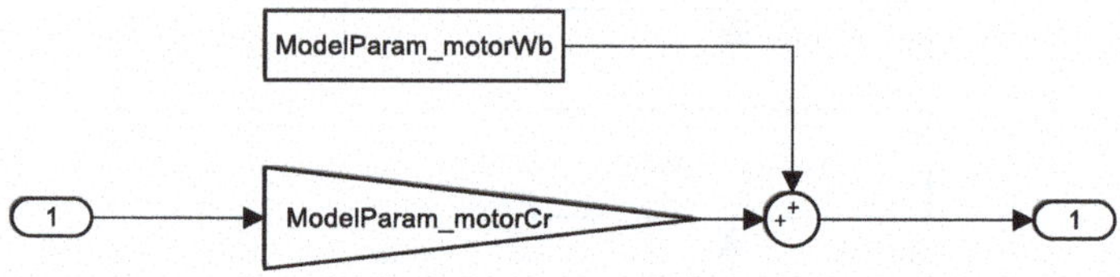

Fig. 6.22 "Throttle to RPM Gain" module, Simulink model "dynamics.slx"

propulsor model, and control effectiveness model is shown as Fig. 6.27. Additionally, in the actual flight system, the final outputs of the autopilot to the ESC are PWM signals with an amplitude ranging from 1000 to 2000. Since the range of the throttle command in the propulsor model is from 0 to 1, a PWM signal ranging from1000 to 2000 should be normalized. The outputs include the state of the multicopter, such as position, velocity, attitude angle, angular rate, and propeller speed, where the propeller speed is mainly for driving the propeller of the corresponding quadcopter 3D model in FlightGear to rotate.

6.4.3.2 Add a Quadcopter 3D Model in FlightGear

The sample steps to add a quadcopter 3D model in FlightGear are as follows. For details, refer to websites: http://www.inivis.com/tutorials.html and http://wiki.flightgear.org/Howto:3D_Aircraft_Models.

(1) **Step1**: **Prepare a quadcopter 3D model**
A quadcopter 3D model drawn by AC3D software is in the "FlightGear 2016.1.2\data\Aircraft\F450\Models\F450.ac" at the installation folder ("C:\PX4PSP" by default) for readers, as shown in Fig. 6.28. Generally, a simpler 3D model is also available, as shown in Fig. 6.29 for example.
The parameters of the four rotors of the quadcopter model are shown in Table 6.3.

Table 6.2 Propeller model

```
function [Fb,Fd,Mb] = fcn(w,R,Cm,Ct,Vb,Cd)
%%The X-configuration quadcopter
%3  1
%2  4
M_rctcm = [-sin(pi/4)*R*Ct,sin(pi/4)*R*Ct,sin(pi/4)*R*Ct,-sin(pi/4)*R*Ct;
           sin(pi/4)*R*Ct,-sin(pi/4)*R*Ct,sin(pi/4)*R*Ct,-sin(pi/4)*R*Ct;
           Cm,Cm,-Cm,-Cm];
Mb = M_rctcm*(w.^2);%Torque under the body coordinate system

Fb = [0; 0; -sum(Ct*(w.^2))]; %Thrust
Fd = -Cd*Vb.*abs(Vb)*0.5;%Aerodynamic force
end
```

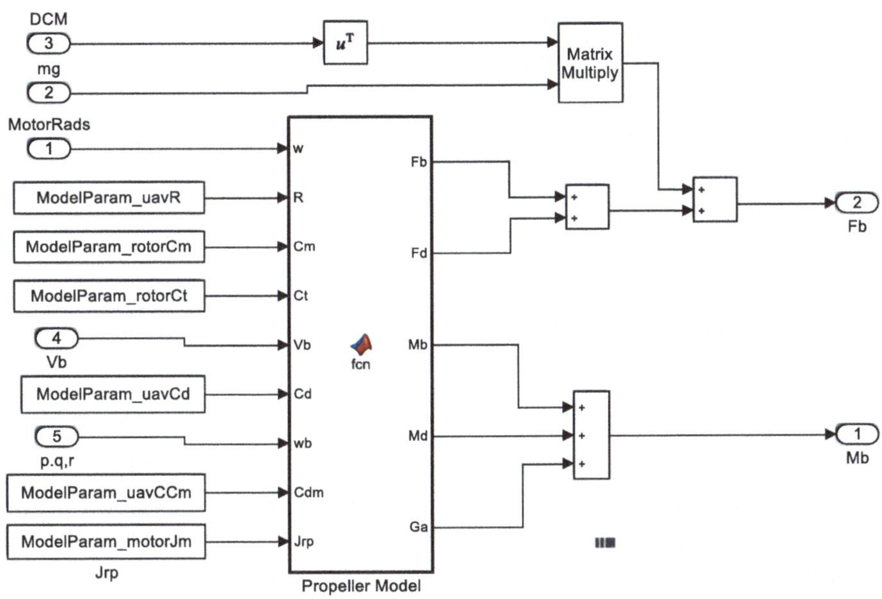

Fig. 6.23 Control effectiveness model, Simulink model "dynamics.slx"

(2) **Step2**: **Parameter configuration**

The type of the main configuration file in FlightGear is XML, such as "myownUAV-set.xml" in the design experiment. The configuration file mainly describes the corresponding parameters in the multicopter 3D model, as shown in Table 6.4.

In the system, file "myownUAV.ac" describes the multicopter model, which is located in the directory "model". File "myownUAV.xml" is also located in the directory "model" and is configured to realize that the multicopter model. In order to make flight realistic, FlightGear allows controlling any part of the quadcopter.

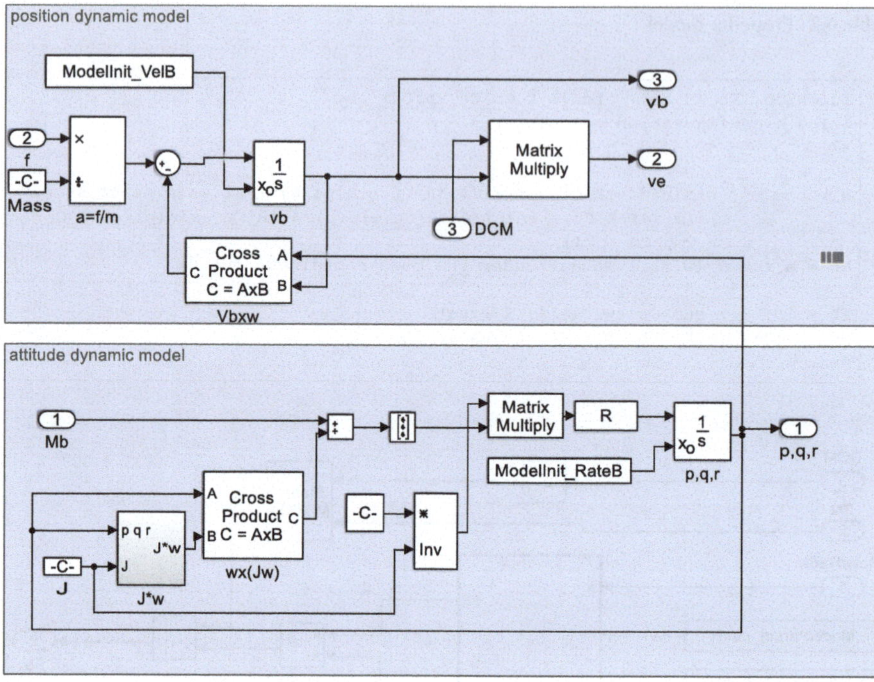

Fig. 6.24 Position dynamic and attitude dynamic model, Simulink model "dynamics.slx"

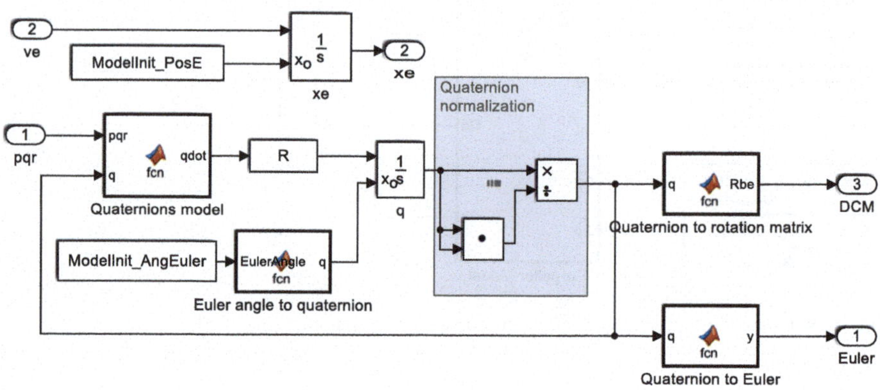

Fig. 6.25 Attitude kinematic model, Simulink model "dynamics.slx"

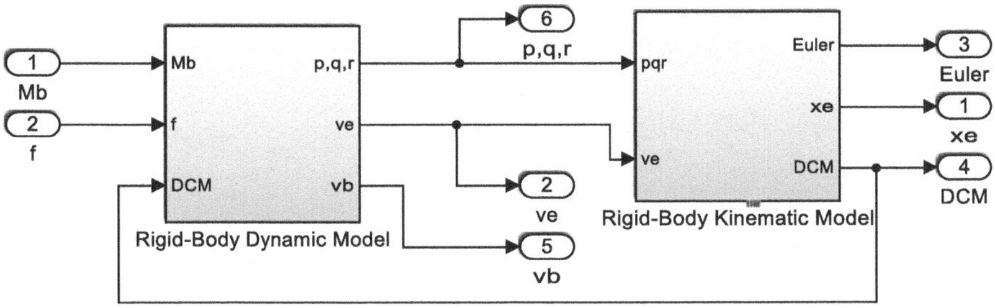

Fig. 6.26 Multicopter flight control rigid-body model, Simulink model "dynamics.slx"

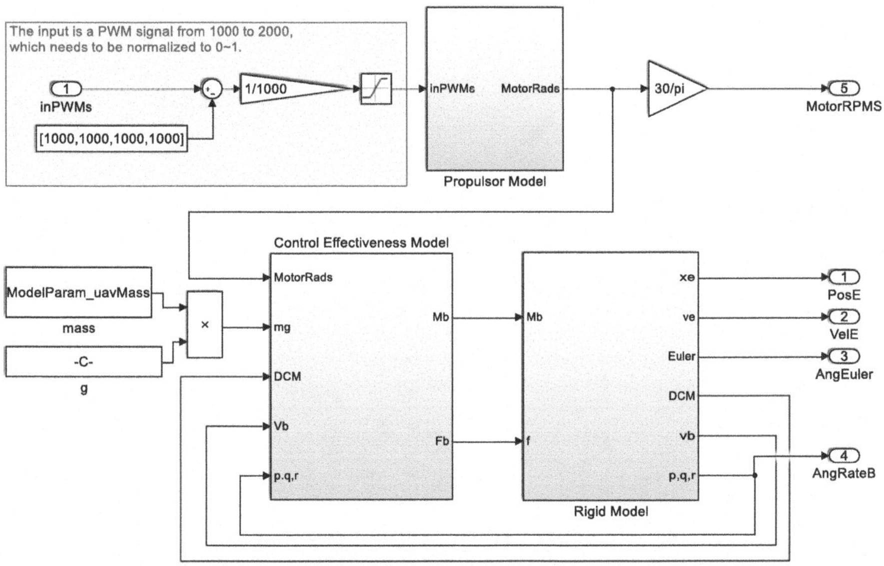

Fig. 6.27 Quadcopter dynamics model, Simulink model "dynamics.slx"

The only requirement is that the corresponding "object name" is set in the 3D model. The simulator now supports the following types of actions, such as none, billboard, rotate, scale, blend, select, spin, timed, translate, texrotate, textmultiple, material, range, alpha-text and noshadow. One of the most common actions is "spin". In order to introduce the meaning and usage of this type of action with code, take one of the rotors as an example, as shown in Table 6.5. The other rotors have similar meaning and usage. The rotation center in the file is expressed by the ABCF. The actual coordinates of each propeller are shown in Table 6.3.

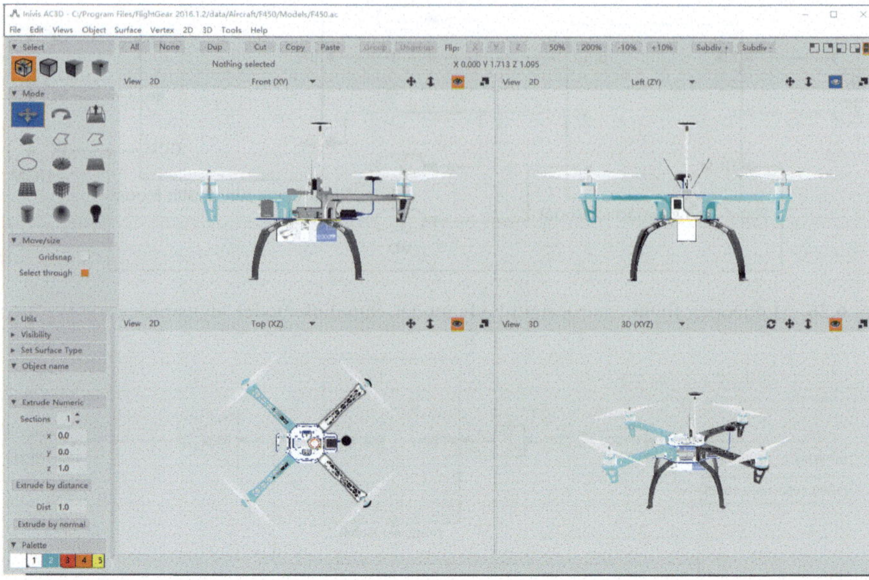

Fig. 6.28 A quadcopter 3D model

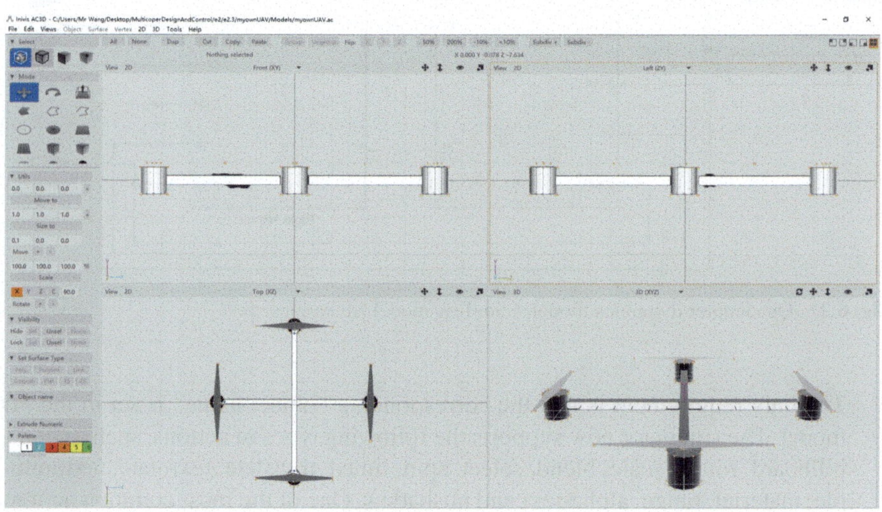

Fig. 6.29 A simple quadcopter 3D model

Table 6.3 Actual coordinates of each propeller

Type	Name	Location attribute		
		X	Y	Z
Propeller	Propeller 1	0	7.5	1
	Propeller 2	7.5	0	1
	Propeller 3	0	−7.5	1
	Propeller 4	−7.5	0	1

(3) **Step3**: **Configuration files**
New a directory "myownUAV\models". Subsequently, put file "Models\myownUAV.ac" and file "myownUAV.xml" in the directory "myownUAV\models\". Additionally, put file "myownUAV-set.xml" in the directory "myownUAV\". Next, copy the file "myownUAV" to the folder "data\Aircraft" in the FlightGear installation path. The default installation path of FlightGear is "C:\PX4PSP\FlightGear 2016.1.2" (the folder "C:\PX4PSP" denotes the installation directory by readers).

(4) **Step4**: **Drive FlightGear by MATLAB**
The process of driving FlightGear by MATLAB is shown in Fig. 6.30. The rigid-body dynamics model built in Sect. 6.3 provides the quadcopter's information, such as position, attitude and rotor speed of the aircraft. As shown in Fig. 6.31, the top six items of data interface block are longitude, dimension, height, roll angle, pitch angle, and rotation angle. It should be noted that the position is measured in meters in the rigid-body dynamic model. In order to calculate correctly, the position unit needs to be converted into latitude and longitude for FlightGear. As shown in Fig. 6.31, FlightGear supports only four engines at most, which are sufficient for the quadcopter. But, a hexacopter requires two more ports. For such a purpose, "left_aileron" and "right_aileron" ports can be used instead.
Prior to running FlightGear, double-click the module "Generate Run Script" and, then open and set the script, which includes name, FlightGear position, model name, port, flight airport background and others. Subsequently, click "Generate Script" to generate a script in the current workspace of MATLAB. Open the script with a text editor and make the following changes.

1) The "time" following "-start-date-lat" in the script is changed to 2004:06:01:01:00:00
2) Find "freeze" and then change "enable" to "disable". Modify and save the script. In order to drive FlightGear by MATLAB, run the script and then click "Run" button in the Simulink to run the Simulink model.

Table 6.4 Main configuration file, where "spin" is a type of rotary motion, which denotes the most important type of motion of the multicopters model and controls the rotation center of the propeller and the rotation axis; "property" denotes the speed; "factor" denotes the rotation direction; "center" denotes the rotation center and "axis" denotes the rotation axis

```
<?xml version="1.0"?>
<PropertyList>
 <sim>
  <description>myownUAV</description>
  <flight-model>network</flight-model>
  <model>
    <path>Aircraft/myownUAV/Models/myownUAV.xml</path>
  </model>
<chase-distance-m type="double"> -40</chase-distance-m>
       <current-view>
    <view-number type="int">2</view-number>
       </current-view>
 </sim>
</PropertyList>
```

Table 6.5 XML Configuration file of a quadcopter

```
<?xml version="1.0"?>
<PropertyList>
 <path>myownUAV.ac</path>
<animation>
  <type>spin</type>
  <object-name>propeller1</object-name>
  <property>/engines/engine[0]/rpm</property>
  <factor>-1</factor>
  <center>
    <x-m>0</x-m>
    <y-m>7.5</y-m>
    <z-m>1</z-m>
  </center>
  <axis>
    <x>0.0</x>
    <y>0.0</y>
    <z>1.0</z>
  </axis>
</animation>
...
</PropertyList>
```

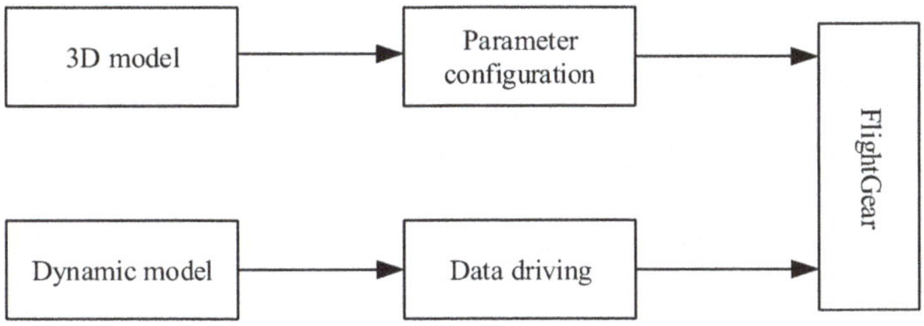

Fig. 6.30 Flow chart of driving FlightGear flight simulator by MATLAB

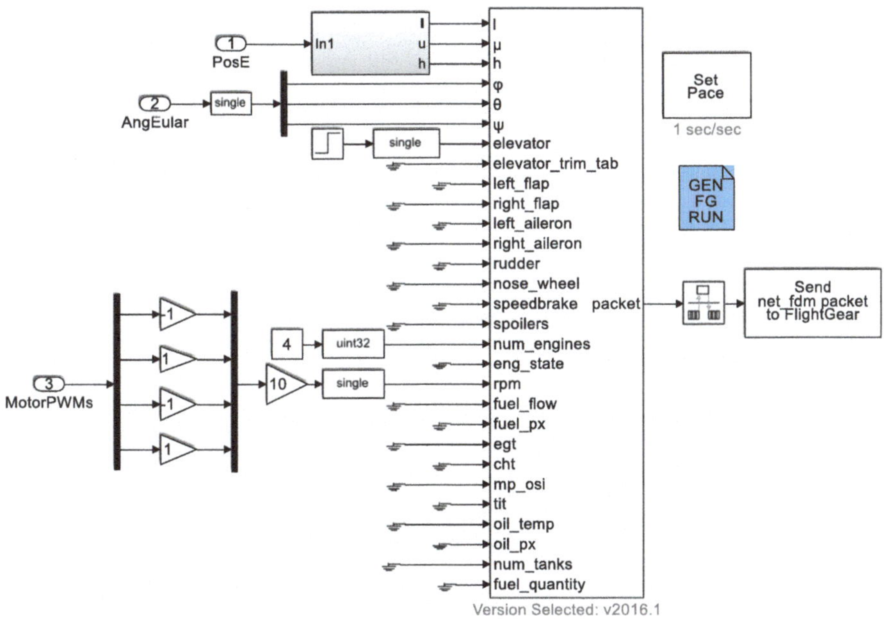

Fig. 6.31 Data interface, Simulink model "dynamics.slx"

6.4.4 Remark

If the background in FlightGear is dim or the view position needs to be adjusted, then readers should set up the background in FlightGear.

(1) Set environment time As shown in Fig. 6.32, select "Environment"—"Time Setting"—"noon".
(2) Set view position Select "View"—"Adjust View Position", as shown in Fig. 6.33. Adjust the angle and distance of the observation by adjusting the three markers, namely "Left/Right", "Down/Up", and "Fwd/Back".

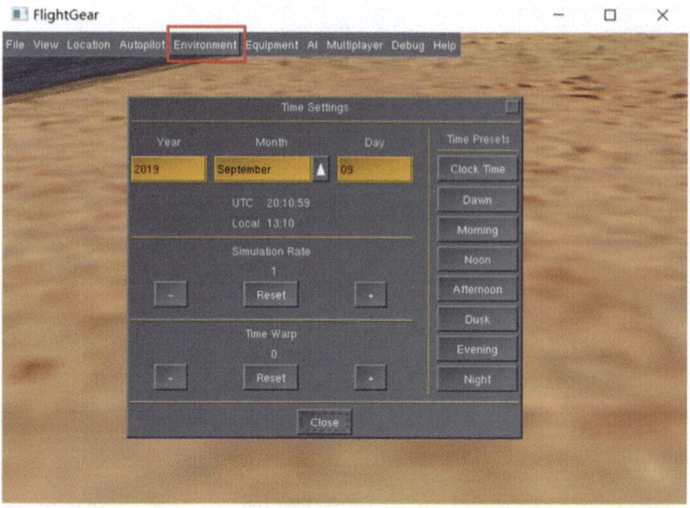

Fig. 6.32 Setting environment time in FlightGear flight simulator

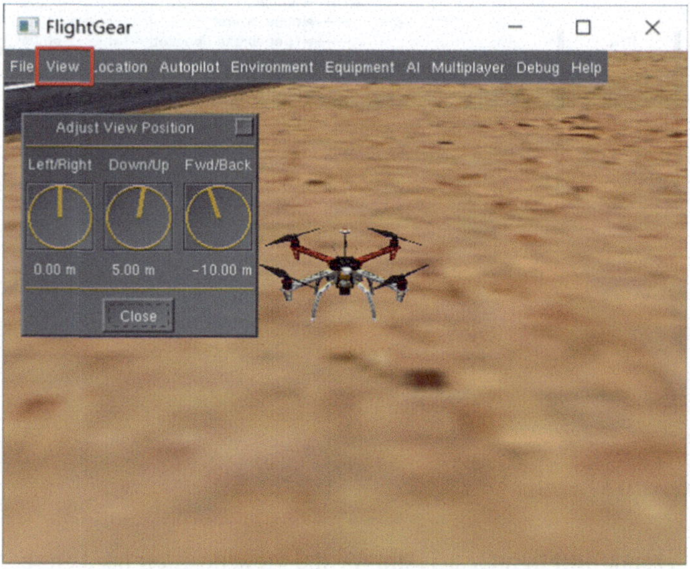

Fig. 6.33 Setting view position in FlightGear flight simulator

6.5 Summary

(1) The flight control rigid-body model of multicopters includes a rigid-body kinematic model, rigid-body dynamic model, control effectiveness model, and propulsor model.
(2) Kinematics is not directly related to mass and force, in which the inputs are only speed and angular velocity and the outputs include position and attitude. Dynamics modeling involves both force in ABCF and motions in EFCF based on Newton's second law and Euler's equations.
(3) At the equilibrium point where a multicopter is hovering, pitch and roll angles are often confined within a small range. In order to simplify the equations in the model, the whole nonlinear model is linearized. In the analysis experiment, yaw angle rate and altitude with respect to different parameters are considered wherein the conclusions are consistent with those from the basic experiment.
(4) The experiments in the following chapters, such as Chaps. 9–12, are based on the model built in this chapter.

If you have any question, please go to https://rflysim.com for your information.

Chapter 7
Sensor Calibration Experiment

There are many sensors mounted in one multicopter, such as the three-axis accelerometer, three-axis gyroscope, three-axis magnetometer, ultrasonic range finder, 2D laser range finder, Global Positioning System (GPS) receiver, and camera. A multicopter can obtain its position and attitude using these sensors. However, these sensors, based on Micro-Electro-Mechanical System (MEMS) [10], are inaccurate. For example, the electric propulsion system can affect the measurement of the magnetometer. Thus, it is impossible to obtain an accurate orientation, much less reliable flight control. This chapter will introduce the calibration principles and methods for some sensors in detail, with the hope that readers gradually master this part of the knowledge through three step-by-step experiments from shallow to deep, namely a basic experiment, an analysis experiment, and a design experiment. In the basic experiment, readers can repeat the calibration of a three-axis accelerometer. In the analysis experiment, readers can analyze if the calibration result is independent of different accelerations of gravity. In the design experiment, readers can design a calibration method for a three-axis magnetometer and perform the calibration.

7.1 Preliminary

In order to make this chapter self-contained, the preliminary is from Chap. 7 of *Introduction to Multicopter Design and Control* [8].

7.1.1 Three-Axis Accelerometer

7.1.1.1 Fundamental Principle

The three-axis accelerometer is a type of inertial sensor which measures the specific force, namely, the non-gravitational force per unit mass or the total acceleration

© Publishing House of Electronics Industry 2020, corrected publication 2021

Q. Quan et al., *Multicopter Design and Control Practice*,
https://doi.org/10.1007/978-981-15-3138-5_7

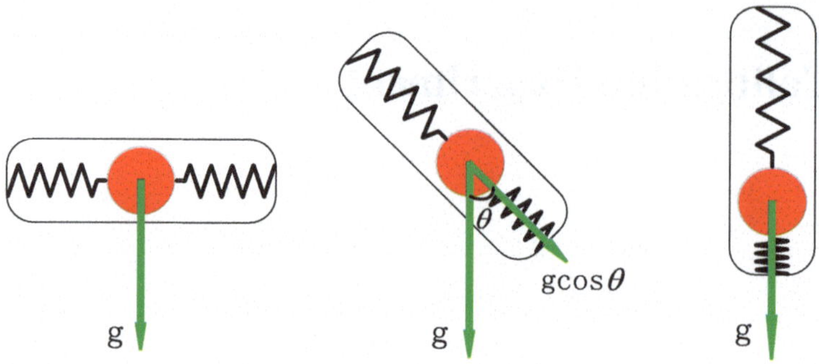

(a) Indicating value is 0 (b) Indicating value is $g\cos\theta$ (c) Indicating value is g

Fig. 7.1 Measuring principle of MEMS accelerometers

while eliminating gravity. When an accelerometer remains still, the acceleration of gravity can still be sensed by the accelerometer, but the total acceleration is zero. In free-fall, the total acceleration will be the acceleration owing to gravity, but a three-axis accelerometer will output zero. The principle of three-axis accelerometers can be used to measure angles. Intuitively, as shown in Fig. 7.1, the amount of spring compression is decided by the angle between the accelerometer and the ground. The specific force can be measured by the length of the spring compressed. Accordingly, in the absence of external forces, accurate pitch and roll angles can be measured without accumulative error.

The three-axis accelerometer is fixed to the multicopter and is aligned with the aircraft-body coordinate frame. Therefore, observation of low-frequency pitch and roll angles can be acquired by accelerometer measurements, illustrated as

$$\begin{aligned} \theta_{\mathrm{m}} &= \arcsin\left(\frac{a_{x_{\mathrm{b}}\mathrm{m}}}{g}\right) \\ \phi_{\mathrm{m}} &= -\arcsin\left(\frac{a_{y_{\mathrm{b}}\mathrm{m}}}{g\cos\theta_{\mathrm{m}}}\right) \end{aligned} \tag{7.1}$$

where ${}^{\mathrm{b}}\mathbf{a}_{\mathrm{m}} = [\, a_{x_{\mathrm{b}}\mathrm{m}} \; a_{y_{\mathrm{b}}\mathrm{m}} \; a_{z_{\mathrm{b}}\mathrm{m}} \,]^{\mathrm{T}}$ denotes the measurement from the accelerometer.

The accelerometers used in multicopters are often based on MEMS technology so that the size of the accelerometers can be as small as a fingernail. Moreover, the energy consumption is low. Nowadays, MEMS technology is widely applied in producing accelerometers for smartphones. The MEMS accelerometers are based on the piezoresistive effect, piezoelectric effect, or capacitive principle; the specific force of those effects is proportional to the resistance, voltage, or capacitance, respectively. These values can be collected through an amplification circuit and filter circuit. The output error induced by vibration is large, which is the shortcoming of this type of sensor.

7.1.1.2 Calibration

The method provided here is only one of several calibration methods.

(1) Error Model
Let ${}^{\mathrm{b}}\mathbf{a}_{\mathrm{m}} \in \mathbb{R}^3$ be the calibrated specific force and ${}^{\mathrm{b}}\mathbf{a}'_{\mathrm{m}} \in \mathbb{R}^3$ be the specific force without calibration. They have the following relationship

$$ {}^{\mathrm{b}}\mathbf{a}_{\mathrm{m}} = \mathbf{T}_{\mathrm{a}}\mathbf{K}_{\mathrm{a}}\left({}^{\mathrm{b}}\mathbf{a}'_{\mathrm{m}} + \mathbf{b}'_{\mathrm{a}}\right). \tag{7.2} $$

Here,

$$ \mathbf{T}_{\mathrm{a}} = \begin{bmatrix} 1 & \Delta\psi_{\mathrm{a}} & -\Delta\theta_{\mathrm{a}} \\ -\Delta\psi_{\mathrm{a}} & 1 & \Delta\phi_{\mathrm{a}} \\ \Delta\theta_{\mathrm{a}} & -\Delta\phi_{\mathrm{a}} & 1 \end{bmatrix}, \mathbf{K}_{\mathrm{a}} = \begin{bmatrix} s_{\mathrm{ax}} & 0 & 0 \\ 0 & s_{\mathrm{ay}} & 0 \\ 0 & 0 & s_{\mathrm{az}} \end{bmatrix}, \mathbf{b}'_{\mathrm{a}} = \begin{bmatrix} b'_{\mathrm{ax}} \\ b'_{\mathrm{ay}} \\ b'_{\mathrm{az}} \end{bmatrix} \tag{7.3} $$

where $\mathbf{T}_{\mathrm{a}} \in \mathbb{R}^{3\times3}$ denotes the tiny tilt in the process of mounting sensors, $\mathbf{K}_{\mathrm{a}} \in \mathbb{R}^{3\times3}$ denotes the scale factor, and $\mathbf{b}'_{\mathrm{a}} \in \mathbb{R}^3$ denotes the bias.

(2) Calibration Principle
To calibrate an accelerometer, the following unknown parameters

$$ \boldsymbol{\Theta}_{\mathrm{a}} \triangleq \left[\, \Delta\psi_{\mathrm{a}}\ \Delta\theta_{\mathrm{a}}\ \Delta\phi_{\mathrm{a}}\ s_{\mathrm{ax}}\ s_{\mathrm{ay}}\ s_{\mathrm{az}}\ b'_{\mathrm{ax}}\ b'_{\mathrm{ay}}\ b'_{\mathrm{az}} \,\right]^{\mathrm{T}} \tag{7.4} $$

need to be estimated. Then, the right side of the model (7.2) can be written as a function

$$ \mathbf{h}_{\mathrm{a}}\left(\boldsymbol{\Theta}_{\mathrm{a}}, {}^{\mathrm{b}}\mathbf{a}'_{\mathrm{m}}\right) \triangleq \mathbf{T}_{\mathrm{a}}\mathbf{K}_{\mathrm{a}}\left({}^{\mathrm{b}}\mathbf{a}'_{\mathrm{m}} + \mathbf{b}'_{\mathrm{a}}\right). \tag{7.5} $$

Here, the measurement noise is eliminated because the specific force is measured many times in a single attitude. Concretely, the accelerometer is rotated at different angles for $M \in \mathbb{Z}_+$ times. At each angle, the accelerometer remains still for a short time. The average specific force in the kth interval is denoted as ${}^{\mathrm{b}}\mathbf{a}'_{\mathrm{m},k} \in \mathbb{R}^3, k = 1, 2, \ldots, M$. The calibration principle is that the magnitude of specific force keeps constant with different attitude of accelerometers, i.e. the local gravity, denoted as g. According to this principle, the following optimization is performed to estimate the unknown parameters

$$ \boldsymbol{\Theta}^*_{\mathrm{a}} = \arg\min_{\boldsymbol{\Theta}_{\mathrm{a}}} \sum_{k=1}^{M} \left(\left\| \mathbf{h}_{\mathrm{a}}\left(\boldsymbol{\Theta}_{\mathrm{a}}, {}^{\mathrm{b}}\mathbf{a}'_{\mathrm{m},k}\right)\right\| - g\right)^2. \tag{7.6} $$

The Levenberg–Marquardt (LM) algorithm [11] can be used to find the optimal solution $\boldsymbol{\Theta}^*_{\mathrm{a}}$.

7.1.2 Three-Axis Magnetometer

7.1.2.1 Fundamental Principle

Magnetometers are important components for providing navigation and location-based services. Magnetometers use anisotropic magnetoresistance or the Hall effect to detect the strength of the magnetic induction. In addition, Lorentz-force magnetometers have also been studied and developed. Based on the Lorenz principle, the electric and magnetic fields generate a force to change the capacitance in the circuit.

7.1.2.2 Calibration

The method given here is only one of calibration methods.

(1) Error Model
Similar to the case with accelerometers, there exist some errors in magnetometers. Assume that ${}^{\mathrm{b}}\mathbf{m}_{\mathrm{m}} \in \mathbb{R}^3$ denotes the magnetic induction value after calibration and ${}^{\mathrm{b}}\mathbf{m}'_{\mathrm{m}} \in \mathbb{R}^3$ denotes the magnetic induction value without calibration. They have the following relationship

$$ {}^{\mathrm{b}}\mathbf{m}_{\mathrm{m}} = \mathbf{T}_{\mathrm{m}}\mathbf{K}_{\mathrm{m}}\left({}^{\mathrm{b}}\mathbf{m}'_{\mathrm{m}} + \mathbf{b}'_{\mathrm{m}}\right). \tag{7.7} $$

Here,

$$ \mathbf{T}_{\mathrm{m}} = \begin{bmatrix} 1 & \Delta\psi_{\mathrm{m}} & -\Delta\theta_{\mathrm{m}} \\ -\Delta\psi_{\mathrm{m}} & 1 & \Delta\phi_{\mathrm{m}} \\ \Delta\theta_{\mathrm{m}} & -\Delta\phi_{\mathrm{m}} & 1 \end{bmatrix}, \mathbf{K}_{\mathrm{m}} = \begin{bmatrix} s_{\mathrm{mx}} & 0 & 0 \\ 0 & s_{\mathrm{my}} & 0 \\ 0 & 0 & s_{\mathrm{mz}} \end{bmatrix}, \mathbf{b}'_{\mathrm{m}} = \begin{bmatrix} b'_{\mathrm{mx}} \\ b'_{\mathrm{my}} \\ b'_{\mathrm{mz}} \end{bmatrix} $$

where $\mathbf{T}_{\mathrm{m}} \in \mathbb{R}^{3\times 3}$ denotes the tiny tilt in the process of mounting sensors, $\mathbf{K}_{\mathrm{m}} \in \mathbb{R}^{3\times 3}$ denotes the scale factor, and $\mathbf{b}'_{\mathrm{m}} \in \mathbb{R}^3$ denotes the bias.

(2) Calibration Principle
To calibrate a magnetometer, the following unknown parameters

$$ \mathbf{\Theta}_{\mathrm{m}} \triangleq \left[\, \Delta\psi_{\mathrm{m}}\ \Delta\theta_{\mathrm{m}}\ \Delta\phi_{\mathrm{m}}\ s_{\mathrm{mx}}\ s_{\mathrm{my}}\ s_{\mathrm{mz}}\ b'_{\mathrm{mx}}\ b'_{\mathrm{my}}\ b'_{\mathrm{mz}} \,\right]^{\mathrm{T}} \tag{7.8} $$

need to be estimated. Then, the right side of the model (7.7) can be written as a function

$$ \mathbf{h}_{\mathrm{m}}\left(\mathbf{\Theta}_{\mathrm{m}}, {}^{\mathrm{b}}\mathbf{m}'_{\mathrm{m}}\right) \triangleq \mathbf{T}_{\mathrm{m}}\mathbf{K}_{\mathrm{m}}\left({}^{\mathrm{b}}\mathbf{m}'_{\mathrm{m}} + \mathbf{b}'_{\mathrm{m}}\right). \tag{7.9} $$

Here, the measurement noise is eliminated, as the magnetic induction value is measured many times at one attitude. To be specific, the magnetometer is rotated at different angles for $M \in \mathbb{Z}_+$ times. At each angle, the magnetic induction remains still for a short time. Generally, the magnetic induction keeps constant with different attitude of magnetometer at the same place. For simplicity, let $\left\|{}^{\mathrm{b}}\mathbf{m}_{\mathrm{m},k}\right\|^2 = 1, k = 1, 2, ..., M$. Here, the magnetic induction value is normalized. So, ${}^{\mathrm{b}}\mathbf{m}_{\mathrm{m},k}(k = 1, 2, ..., M)$ are all on the unit circle. The calibration principle is that the value of $\mathbf{\Theta}_{\mathrm{m}}$ should cause the value of $\left\|\mathbf{T}_{\mathrm{m}}\mathbf{K}_{\mathrm{m}}\left({}^{\mathrm{b}}\mathbf{m}'_{\mathrm{m}} + \mathbf{b}'_{\mathrm{m}}\right)\right\|$ to be as close to 1 as possible. According to this principle, the following optimization is performed for calibration

$$\mathbf{\Theta}^*_{\mathrm{m}} = \arg\min_{\mathbf{\Theta}_{\mathrm{m}}} \sum_{k=1}^{M} \left(\left\|\mathbf{h}_{\mathrm{m}}\left(\mathbf{\Theta}_{\mathrm{m}}, {}^{\mathrm{b}}\mathbf{m}'_{\mathrm{m},k}\right)\right\| - 1\right)^2. \tag{7.10}$$

7.2 Basic Experiment

7.2.1 Experimental Objective

(1) Things to prepare

1) Hardware: Pixhawk autopilot system;
2) Software: MATLAB R2017b or above, Pixhawk Support Package (PSP) toolbox, QGroundControl (QGC), Instructional Package "e3.1" (https://rflysim.com/course);
3) Data for calibration are prepared in Instructional Package "e3.1" for readers without hardware to collect data.

(2) Objectives
Repeat the given calibration steps to calibrate an accelerometer in the given Pixhawk autopilot system. Subsequently, make a comparison between the calibrated results and uncalibrated results.

7.2.2 Experimental Procedure

(1) Step1: Obtain accelerometer data via the Pixhawk autopilot system

The experiment is divided into the following five steps.

1) Hardware connection. Referring to the procedure shown in Sect. 2.3 in Chap. 2, the connection between the RC receiver and the Pixhawk autopilot is shown in Fig. 7.2. Open the QGC, as shown in Fig. 7.3, and Click "Vehicle Setup"—"Airfram"—"Quadcopter x"—"DJI Flame Wheel F450" in turn to select the DJI Flame Wheel F450 quadrotor frame used in the flight experiment of this book. Then, click "Apply and Restart" in the upper-right corner. After that, the system will ask for confirmation again. Finally, click the "Apply" button and the autopilot will automatically reboot to make the adopted configuration available. If the

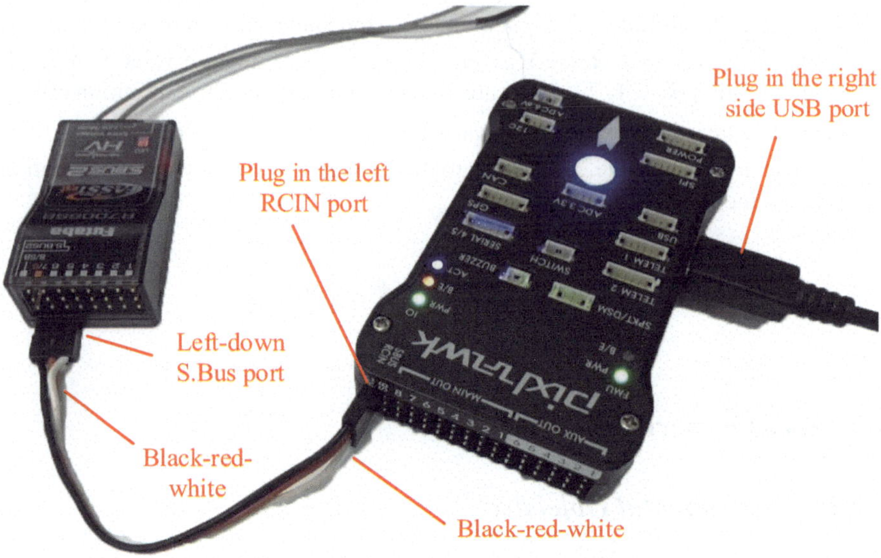

Fig. 7.2 Pixhawk and RC transmitter connection

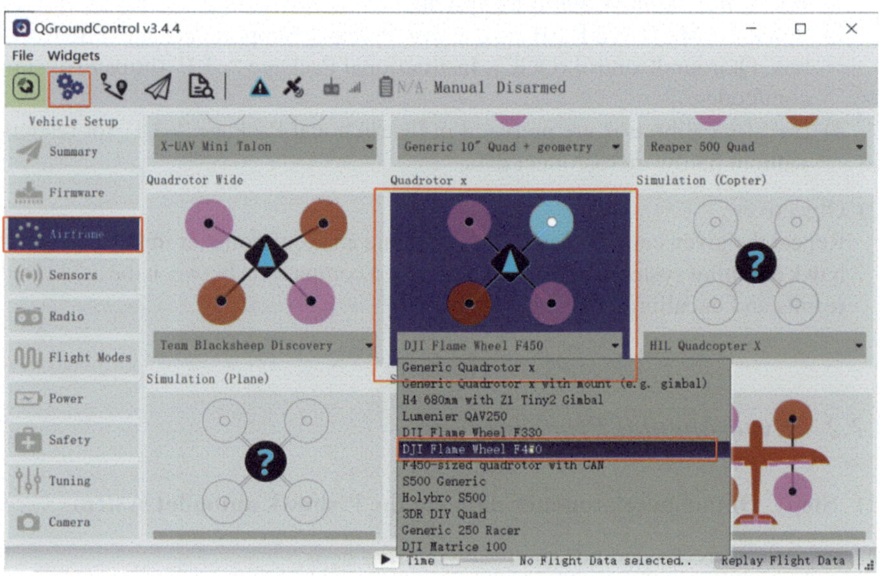

Fig. 7.3 Airframe configuration on QGC

adopted configuration is modified, the system will require another re-calibration of the sensor data, as shown in Fig. 7.4. Finally, sensor calibration is performed according to the prompt given in QGC.

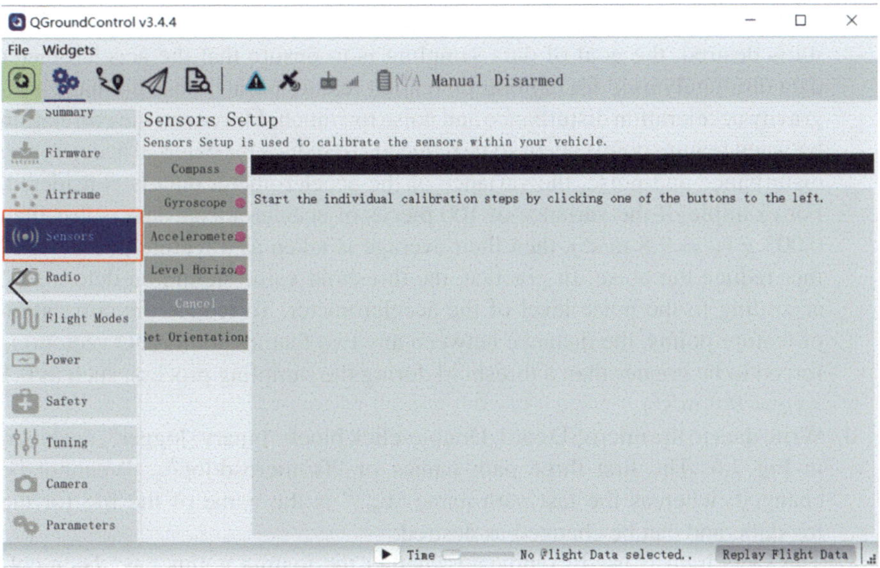

Fig. 7.4 Sensor calibration on QGC

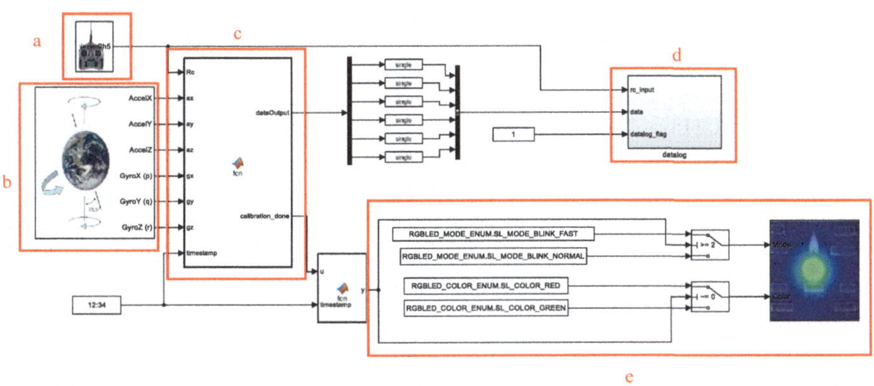

Fig. 7.5 Accelerometer data logging, Simulink model "acquire_data_ag.slx"

2) Open the data collection model. To record the sensor and Radio Control (RC) data in the Pixhawk microSD card, a file "acquire_data_ag.slx" is created, as shown in Fig. 7.5. The execution process of the file is composed of five parts, corresponding to the boxes "a", "b", "c","d" and "e" in Fig. 7.5.

   a. Receive CH5 (refer to the Sect. 2.3 for the definition of CH5) data of the RC transmitter, to control the start and end of the data collection process.
   b. Receive data from the accelerometer and gyroscope.

c. Implementation of data sampling logic. As only the acceleration of gravity is desired, the goal of data sampling is to ensure that the accelerometer data ultimately used for calibration (called feature points) does not have non-gravity acceleration disturbance and noise too much. Moreover, the correlation between feature points needs to be reduced. To reduce the external acceleration disturbance and noise, the variance of the accelerometer data is considered. For example, if the variance of 100 pieces of accelerometer data is less than 0.003 g ($g = 9.8\ \text{m/s}^2$), then their average is taken as a feature point to further reduce the noise. In practice, the threshold value should be determined according to the noise level of the accelerometer. To reduce the correlation of feature points, the distance between any two feature points (3D vectors) is forced to be greater than a threshold during the sampling process, such as 0.1 g ($g = 9.8\ \text{m/s}^2$).

d. Write data to the microSD card. Double-click block "binary_logger", as shown in Fig. 7.6. The first three path names of "fs/microsd/log/e_{acc}" cannot be changed, whereas the last path name "e_{acc}" is the name of the file for the log data, and can be changed as desired.

e. The LED light is used to indicate whether the feature points have been collected.

3) Compile the file "acquire_data_ag.slx" and upload it to the Pixhawk autopilot. The process of compiling and uploading is shown in Fig. 7.7. For details, see Sect. 4.2.3 in Chap. 4.

4) Rotate the Pixhawk autopilot to log data. Pull back the upper-left switch corresponding to CH5 > 1500 (see the definition in Sect. 2.3.1), to start writing data to the microSD card. Place the Pixhawk autopilot as guided by Fig. 7.8 and hold the Pixhawk autopilot still with each orientation for a period of time. Meanwhile, the Pixhawk autopilot logs data to a file called "e3_A.bin" on the microSD card. Once a feature point is collected, the Pixhawk LED status light will slowly blink in red. By recalling the feature point collection method, one feature point corresponds to one orientation that the Pixhawk is placed at. Repeat the logging process for all orientations. Once ten feature points corresponding to ten orientations are collected, the Pixhawk LED status light will begin quickly blinking in red. Then, pull forward the upper-left switch (CH5 < 1500) to stop writing data to the microSD card.

5) Read data. Take out the microSD card, read the data by a card reader, copy the file "e3_A.bin" to the folder "e3\e3.1". Use the function

```
[datapoints, numpoints] = px4_read_binary_file('e3_A.bin')
```

to decode the data. The data are saved in "datapoints" and the number of the data is saved in "numpoints". The x-axis accelerometer sampling data and feature data are shown in Fig. 7.9, where "*" represents the feature points.

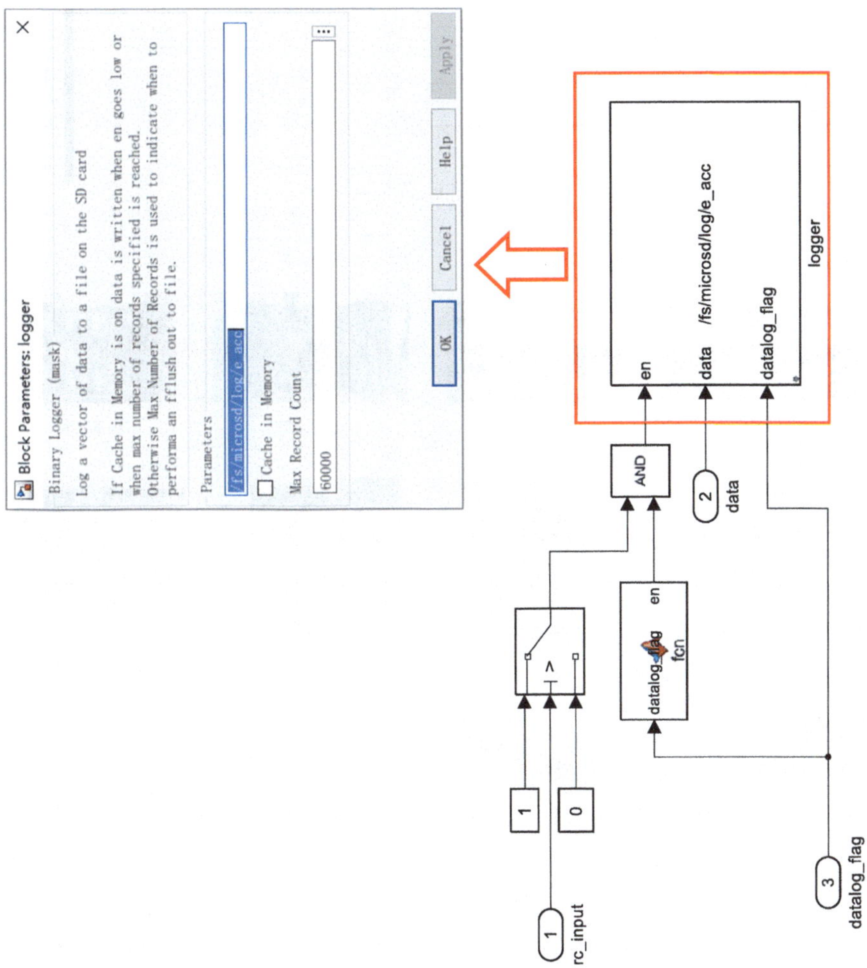

Fig. 7.6 Logging block in Simulink model "acquire_data_ag.slx"

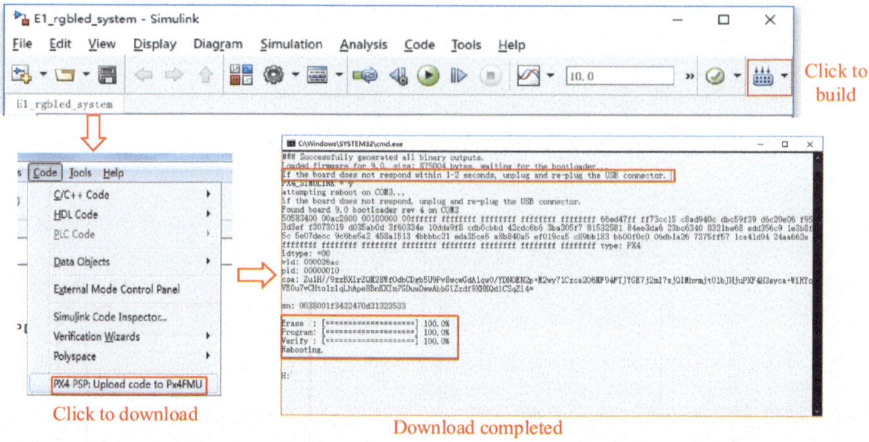

Fig. 7.7 Compiling and uploading process

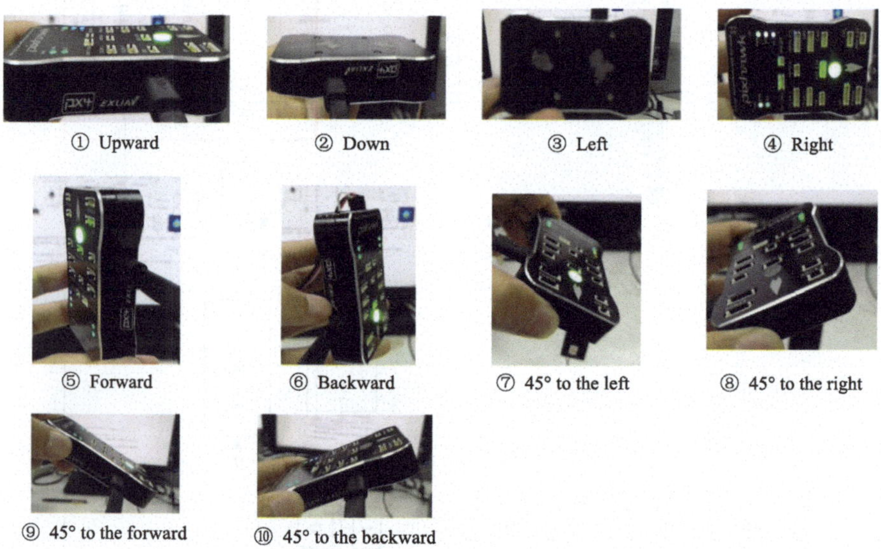

Fig. 7.8 Ten different orientations of Pixhawk autopilot

(2) Step2: Parameter calibration

It should be noted that the obtained data have been calibrated by the Pixhawk autopilot. In order to obtain uncalibrated original data, the accelerometer calibration parameters in the Pixhawk autopilot are first read by QGC. In QGC, select "Parameters"—"Sensor Calibration" to obtain the accelerometer calibration parameters, as shown in Fig. 7.10. Then, the data are restored to uncalibrated data, based on the obtained accelerometer calibration parameters from QGC. With these restored data, readers now can use the calibration codes from the basic experiment. Before this, a key function "LM" should be introduced.

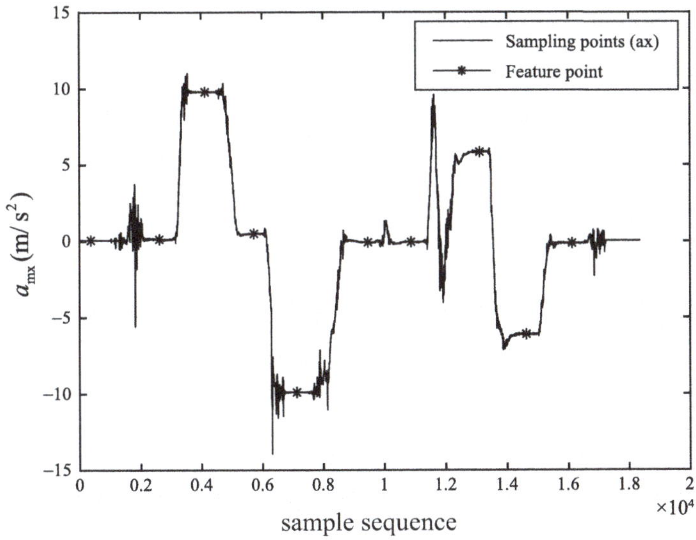

Fig. 7.9 x-axis accelerometer sampling data and feature data

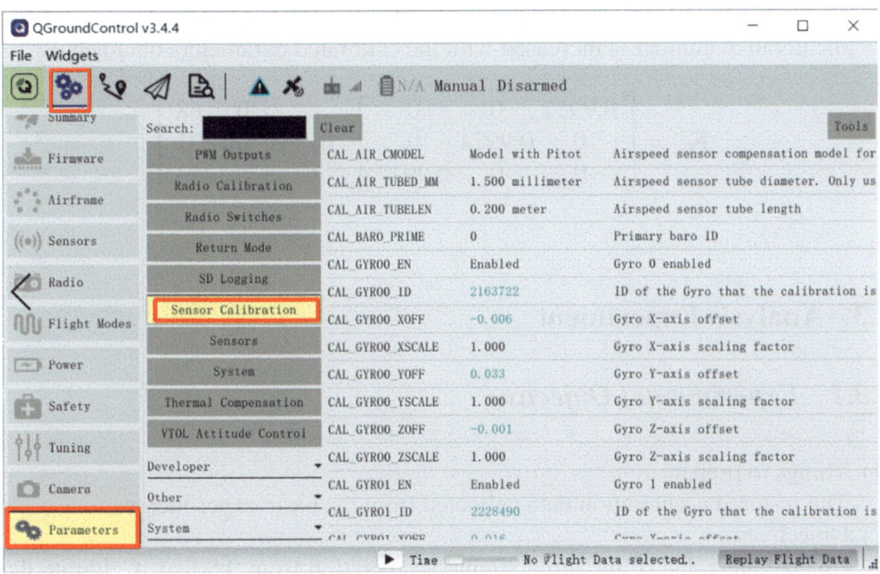

Fig. 7.10 Calibration parameter in QGC

(1) The usage of function "p = lm(func,p,x,y_dat,dp,p_min,p_max)"

1. Input parameters
   - func function name: y_hat = func(x,p); the functional relationship is $\left\| \mathbf{T}_{\mathrm{a}}\mathbf{K}_{\mathrm{a}}\left({}^{\mathrm{b}}\mathbf{a}_{\mathrm{m}}' + \mathbf{b}_{\mathrm{a}}'\right) \right\|$, where $\mathbf{T}_{\mathrm{a}} = \mathbf{I}_3$.

- p: the initial value of the parameter to be estimated, which denotes the initial value in optimization (7.6);
- x: feature points;
- dp: related to the Jacobian matrix;
- p_min: the minimum norm of the unknown parameter, the default is -$100 \times$ abs(p);
- p_max: the maximum norm of the unknown parameter, the default is $100 \times$ abs(p);

2. Output
   - p: The estimated parameter value by the algorithm iteration, i.e. Θ_a^*.

(2) Calibrate the accelerometer by the LM algorithm. The main code is shown in Table 7.1. For simplicity, only six are parameters are calibrated in this experiment, namely $\mathbf{K}_a$, $\mathbf{b}_a$, leaving $\mathbf{T}_a = \mathbf{I}_3$ [12]. This is because the value of $\Delta\psi_a$, $\Delta\theta_a$, $\Delta\phi_a$ are often very small for a Pixhawk autopilot. The optimization objective $f_a = \sum_{k=1}^{M} \left(\left\|\mathbf{h}_a\left(\mathbf{\Theta}_a, {}^{\mathrm{b}}\mathbf{a}'_{\mathrm{m},k}\right)\right\| - g\right)^2$ is used, where $M = 10$ corresponds to 10 feature points. As shown in Fig. 7.11, the optimization objective gets smaller as compared to that with uncalibrated parameters. Moreover, as shown in Fig. 7.12, the optimization objective converges to zero very quickly as the iterative number is increased with the calibrated parameters obtained as

$$\mathbf{K}_a^* = \begin{bmatrix} 0.9912 & 0 & 0 \\ 0 & 0.9974 & 0 \\ 0 & 0 & 0.9947 \end{bmatrix}, \mathbf{b}_a^* = \begin{bmatrix} 0.0168 \\ 0.2691 \\ 0.1253 \end{bmatrix}.$$

7.3 Analysis Experiment

7.3.1 *Experimental Objective*

(1) Things to prepare
The restored acceleration data collected from the basic experiments.

(2) Objectives
Change the value of the gravitational acceleration from 9.8 to 1. Calibrate the accelerometer parameters again; with the calibrated data, calculate the pitch angle. Compare and analyze the calibration parameters and pitch angles computed based on different calibrated parameters when the gravitational acceleration is set from 9.8 to 1.

Table 7.1 calLM.m

```
%File Description:
% According to the accelerometer error model, the accelerometer error model parameters are
      calculated using the lm optimization algorithm.
close all
clc
clear

load AccRaw %Load uncalibrated accelerometer data
g = 9.8;
m = length(AccRaw);

y_dat = g*ones(m, 1); %Expected gravitational acceleration data
p0 = [1 1 1 0 0 0]';
p_init = [1.0 1.0 1.0 0.1 0.1 0.1]'; %Accelerometer error model parameter initial data

y_raw = calFunc(AccRaw, p0); %2−norm of uncalibrated accelerometer data
y_raw = y_raw(:);

%The difference between the uncalibrated gravitational acceleration measured by the
      accelerometer and the standard gravitational acceleration
r_raw = y_dat − y_raw;
p_fit = lm('calFunc', p_init, AccRaw, y_dat);
y_lm = calFunc(AccRaw, p_fit); %2−norm of calibrated accelerometer value
y_lm = y_lm(:);
r_lm = y_dat − y_lm;

kx = p_fit(1);
ky = p_fit(2);
kz = p_fit(3);
bx = p_fit(4);
by = p_fit(5);
bz = p_fit(6);

Ka9_8 = [kx 0 0; 0 ky 0; 0 0 kz]
ba9_8 = [bx by bz]'
save('calP9_8', 'Ka9_8', 'ba9_8')
```

7.3.2 *Experimental Analysis*

The optimization of $\boldsymbol{\Theta}_{\mathrm{a}}^{*}$ is the desired objective. Then

$$\left.\frac{\partial \sum\limits_{k=1}^{M}\left(\left\|\mathbf{T}_{\mathrm{a}}\mathbf{K}_{\mathrm{a}}\left({}^{\mathrm{b}}\mathbf{a}_{\mathrm{m}}^{\prime}+\mathbf{b}_{\mathrm{a}}^{\prime}\right)\right\|-g\right)^{2}}{\partial \boldsymbol{\Theta}_{\mathrm{a}}}\right|_{\boldsymbol{\Theta}_{\mathrm{a}}=\boldsymbol{\Theta}_{\mathrm{a}}^{*}}=\mathbf{0} \tag{7.11}$$

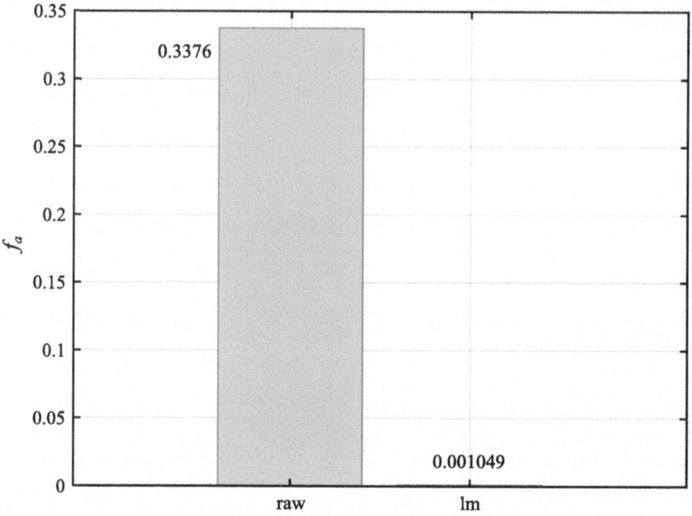

Fig. 7.11 Values of optimization objective f_a using calibrated and uncalibrated accelerometers

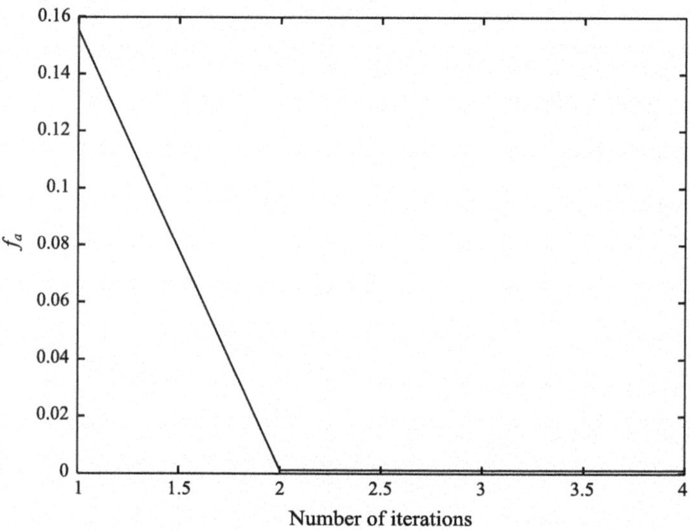

Fig. 7.12 Value of optimization objective f_a changes with number of iterations

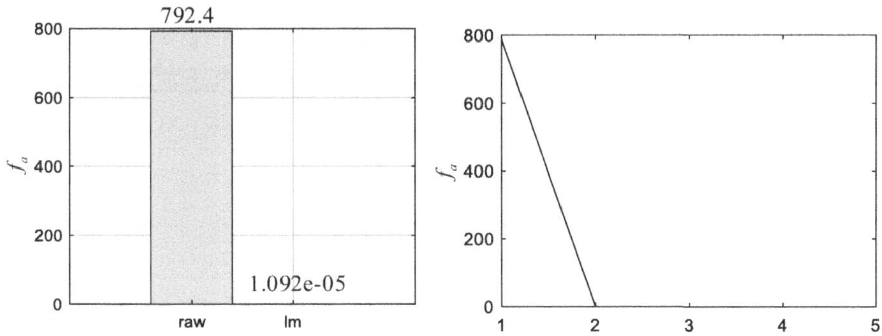

(a) Values of optimazation objective f_a using calibrated and uncalibrated accelerameters

(b) f_a changes with number of iterations

Fig. 7.13 Calibration results when "g" is 1

where $\mathbf{T}_\text{a} = \mathbf{I}_3$. In that regard, the gravitational acceleration changes. For example, g is changed to αg, where $\alpha > 0$. Multiplying the left and right sides of Eq. (7.11) by α yields

$$\left.\frac{\partial \sum\limits_{k=1}^{M}\left(\left\|\mathbf{T}_\text{a}\alpha\mathbf{K}_\text{a}\left({}^\text{b}\mathbf{a}_\text{m}' + \boldsymbol{b}_\text{a}'\right)\right\| - \alpha g\right)^2}{\partial \boldsymbol{\Theta}_\text{a}}\right|_{\boldsymbol{\Theta}_\text{a}=\boldsymbol{\Theta}_\text{a}^*} = \mathbf{0}.$$

This implies that, after the gravitational acceleration is changed, the calibration parameters are $\mathbf{K}_\text{a}^{*\prime} = \alpha\mathbf{K}_\text{a}^*$ and $\mathbf{b}_\text{a}^{*\prime} = \mathbf{b}_\text{a}^*$ based on the optimization (7.6).

7.3.3 *Experimental Procedure*

(1) Step1: Open the file "calLM.m" and change the value of gravitational acceleration "g" from 9.8 to 1.
(2) Step2: Copy the sensor data obtained in the basic experiment for file "calLM.m".
(3) Step3: Run the file "calLM.m" to obtain the calibration results and curves as shown in Fig. 7.13.

The calibration parameters are

$$\mathbf{K}_\text{a}^{*\prime} = \begin{bmatrix} 0.1012 & 0 & 0 \\ 0 & 0.1017 & 0 \\ 0 & 0 & 0.1014 \end{bmatrix}, \mathbf{b}_\text{a}^{*\prime} = \begin{bmatrix} 0.0168 \\ 0.2691 \\ 0.1253 \end{bmatrix}.$$

One can conclude that when "g" is 1 and 9.8, the calibration parameters $\mathbf{b}_\text{a}'$ and $\mathbf{b}_\text{a}$ correspondingly remain the same, but $\mathbf{K}_\text{a}^{*\prime}$ is smaller. Concretely, $\mathbf{K}_\text{a}^{*\prime}$ is reduced to approximately 1/9.8 of $\mathbf{K}_\text{a}^*$, which is consistent with the theoretical analysis. The three-axis accelerometer is fixed to the multicopter and aligned with the aircraft-body

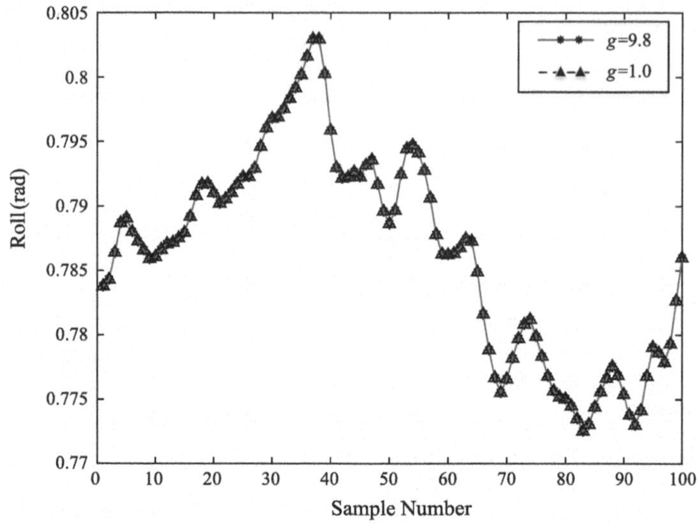

Fig. 7.14 Pitch angle with respect two different "g" values

coordinate frame. To obtain the roll angle when "g" is 9.8 and 1, calculate the roll angle by using the calibration acceleration value and Eq. (7.1). To better illustrate the roll angle in a continuous process, a set of accelerometer data is logged again with the Pixhawk autopilot being slowly turned, as shown in Fig. 7.14. One can conclude that the two calibration solutions result in the same angle. This implies that the pitch angle measurement is independent of the acceleration of gravity.

7.4 Design Experiment

7.4.1 Experimental Objective

(1) Things to prepare

1) Hardware: Pixhawk Autopilot System;
2) Software: MATLAB R2017b or above, Pixhawk Support Package (PSP) Toolbox, QGroundControl (QGC) and Instructional Package "e3.3" (https://rflysim.com/course);
3) Data for calibration are prepared in Instructional Package "e3.3" for readers without hardware to collect data.

(2) Objectives

Design the magnetometer data logging block, following the procedure in the basic experiment. With the obtained data, calibrate the magnetometer, and compare the calibrated and uncalibrated results.

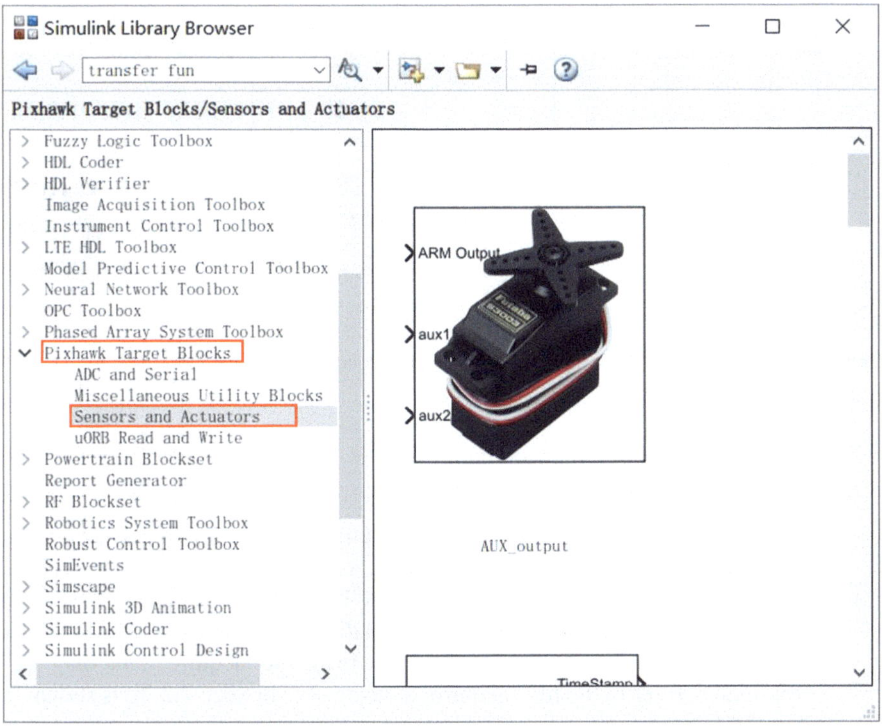

Fig. 7.15 PSP toolbox in Simulink library browser

7.4.2 Experimental Procedure

The design experiment has two steps: (1) Obtain the data of the magnetometer, and (2) calibrate the parameters.

(1) **Step1: Obtain magnetometer data from Pixhawk Autopilot**
The method for logging the magnetometer data in Sect. 7.1.2 is similar to the method for logging the accelerometer feature sampling in the basic experiment. Only slight changes are made. Specifically, the accelerometer data output in the "b" block in Fig. 7.5 is replaced with the magnetometer data output; and the method for obtaining the data variance and the minimum distance in the "c" block in Fig. 7.5 is modified. In this experiment, a different method for eliminating noise is used. When the data of the magnetometer is logged, the average data is not used to eliminate data noise; rather, a significant amount of raw data is used to eliminate noise. It should be very careful to rotate the Pixhawk autopilot to avoid non-gravitational acceleration. So, only ten feature points are collected to save time. However, the magnetic field is independent of the movement of the Pixhawk autopilot. Based on this, more data can be collected by rotating the Pixhawk autopilot freely.

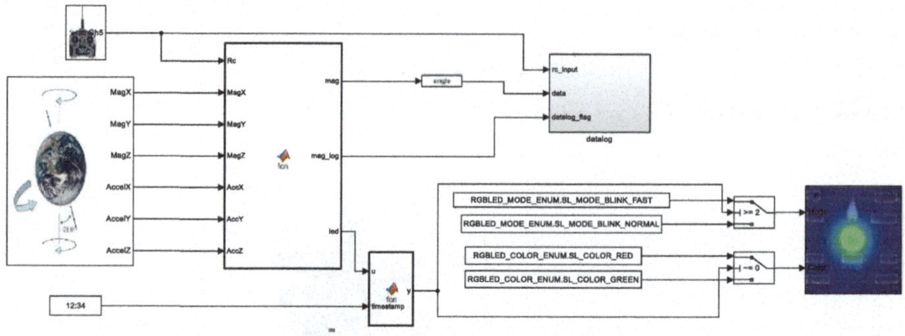

Fig. 7.16 Magnetometer data logging, Simulink model “acquire_data_mag.slx”

The process of getting magnetometer data is divided into the following five steps.

1) Data logging block. Create a new Simulink file and drag out the corresponding modules from the Simulink PSP Toolbox, as shown in Fig. 7.15. To log data, use the corresponding blocks in “Pixhawk Target Blocks” that log data from the inertial sensor and RC transmitter. That data can be saved into the Pixhawk microSD card. An appropriate model given in the file “acquire_data_mag.slx” (for details, refer to the file “acquire_data_ag.slx” in Sect. 7.2.2.) is shown in Fig. 7.16.
2) Hardware connection. By referring to the procedure from Sect. 2.3.1 in Chap. 2, the connection between the RC receiver and the Pixhawk autopilot can be determined, as is shown in Fig. 7.2.
3) Compile the file “acquire_data_mag.slx” and upload it to the Pixhawk autopilot. The process of compiling and uploading is shown in Fig. 7.7. For details, see Sect. 4.2.3 in Chap. 4.
4) Log the data. Pull back the upper-left switch corresponding to CH5 > 1500, (see the definition in Sect. 2.3.1) to start writing data to the microSD card. Place the Pixhawk autopilot as guided by Fig. 7.17. Starting at each orientation, rotate the Pixhawk autopilot a circle clockwise or counterclockwise around its principal axes of the moment of inertia, where 40 sampling data are logged. Meanwhile, the Pixhawk autopilot logs data to a file called “e3_m_A.bin” on the microSD card. Once the process of logging is completed for the current orientation, the Pixhawk LED status light will be slowly blinking in red. Then, repeat the logging process for all orientations. Once all data corresponding to all six orientation is collected, the Pixhawk LED status light will be quickly blinking in red and a total 240 sampling data are logged. Then, pull forward the upper-left switch (CH5 < 1500) to stop writing data to the microSD card.

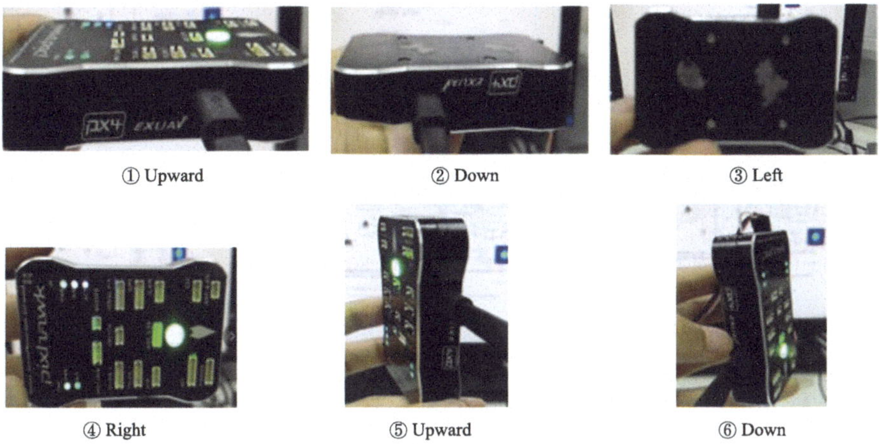

Fig. 7.17 Pixhawk autopilot placement facing six different directions

Fig. 7.18 Magnetometer calibration sampling points

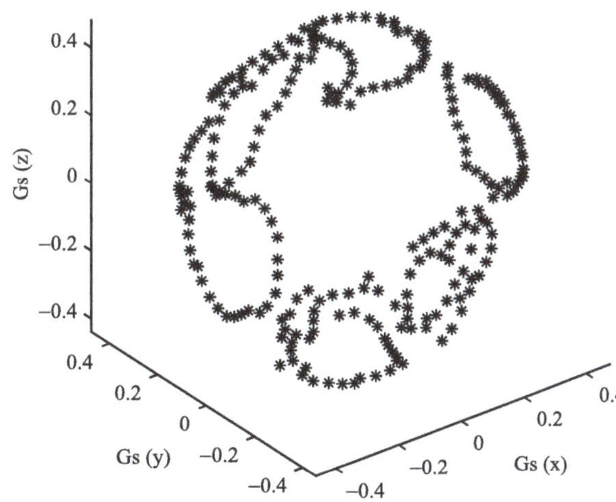

5) Read the data. First, remove the microSD card from Pixhawk autopilot. Read the data using a card reader. Copy the file "e3_m_A.bin" to the folder "e3\e3.3". Use the function

```
[datapoints, numpoints] = px4_read_binary_file('e3_m_A.bin')
```

to decode the data. The data are saved in "datapoints" and the number of the data is saved in "numpoints". The logged data is shown in Fig. 7.18.

(2) **Step2: Parameter calibration**
Calibrate the magnetometer using the LM algorithm. For details, see the Step 2 in the basic experiment. The main code is shown in Table 7.2. For simplicity, only six parameters are calibrated in this experiment, namely $\mathbf{K}_{\mathrm{m}}, \mathbf{b}_{\mathrm{m}}$, leaving $\mathbf{T}_{\mathrm{m}} = \mathbf{I}_3$.

Table 7.2 calLM.m

```
close all
clc
clear
load MagRaw.mat
CAL_MAG_SCALE = [1, 1, 1]’; %Calibration value in flight controller
CAL_MAG_OFF = [0.064, 0.014, -0.053]’;
MagRaw = (mag + CAL_MAG_OFF)./CAL_MAG_SCALE; %Original magnetometer data

m = length(MagRaw);
MagSum = 0;
for k = 1 : m
    MagSum = MagSum + norm(MagRaw(:, k));
end
MagAver = MagSum/m; %Estimated magnetic field strength
Vdata = MagRaw/MagAver; %Normalization

y_dat = ones(m, 1);
p0 = [1 1 1 0 0 0]’;
p_init = [1 1 1 0.01 0.01 0.01]’; %Initial value of the parameter to be estimated

y_raw = calFunc(Vdata, p0); %2-norm of uncalibrated magnetometer value
y_raw = y_raw(:);
r_raw = y_dat - y_raw;
p_fit = lm(’calFunc’, p_init, Vdata, y_dat, 0.001);
y_lm = calFunc(Vdata, p_fit); %2-norm of calibrated magnetometer value
y_lm = y_lm(:);
r_lm = y_dat - y_lm;
y_px4 = calFunc(mag/MagAver, p0); %2-norm of PX4 Calibrated magnetometer value
y_px4 = y_px4(:);
r_px4 = y_dat - y_px4;

kx = p_fit(1);
ky = p_fit(2);
kz = p_fit(3);
bx = p_fit(4);
by = p_fit(5);
bz = p_fit(6);

Km = [kx 0 0;0 ky 0;0 0 kz]
bm = [bx by bz]’
```

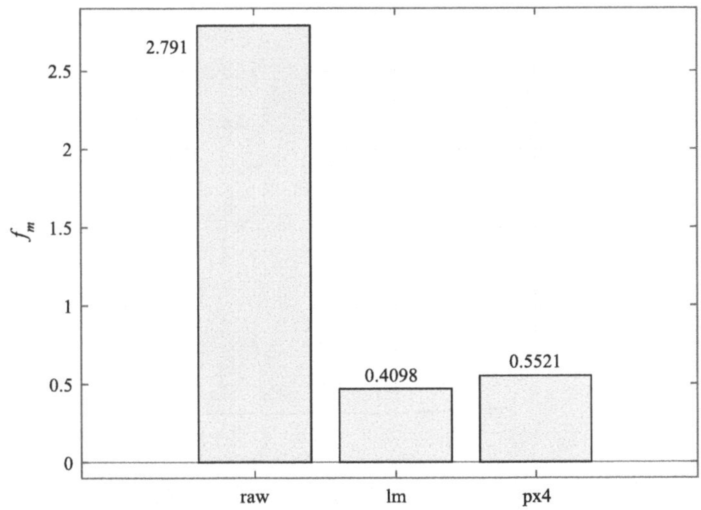

Fig. 7.19 Values of optimization objective f_{m} using calibrated, uncalibrated magnetometers and PX4

This is because the value of $\Delta\psi_{\mathrm{m}}, \Delta\theta_{\mathrm{m}}, \Delta\phi_{\mathrm{m}}$ are often very small for a Pixhawk autopilot. The optimization objective $f_{\mathrm{m}} = \sum_{k=1}^{M} \left(\left\| \mathbf{h}_{\mathrm{m}} \left(\mathbf{\Theta}_{\mathrm{m}}, {}^{\mathrm{b}}\mathbf{m}'_{\mathrm{m},k} \right) \right\| - 1 \right)^2$ is used. As shown in Fig. 7.19, the optimization objective becomes smaller than that by using uncalibrated parameters, and as shown in Fig. 7.20, the optimization objective converges to 0.5 very quickly as the iterative number is increased. Finally, the calibrated parameters are obtained as

$$\mathbf{K}_{\mathrm{m}}^{*} = \begin{bmatrix} 0.9853 & 0 & 0 \\ 0 & 1.0202 & 0 \\ 0 & 0 & 1.0004 \end{bmatrix}, \mathbf{b}_{\mathrm{m}}^{*} = \begin{bmatrix} -0.1448 \\ -0.0334 \\ -0.0898 \end{bmatrix}.$$

Notably, when the flight environment changes greatly (i.e., the magnetic field environment where the Pixhawk autopilot is located may change greatly), the magnetometer must be recalibrated.

7.5 Summary

(1) An accelerometer calibration model and a magnetometer calibration model are illustrated using the multicopter sensor calibration experiment, where the PSP Toolbox is used for data logging.

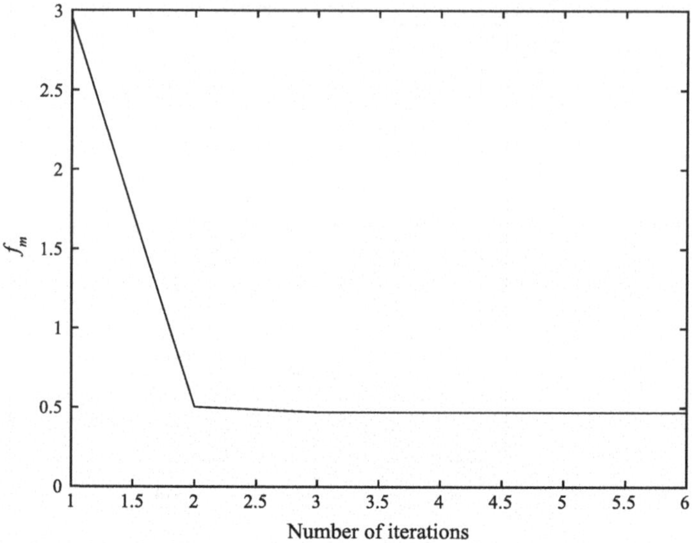

Fig. 7.20 Value of optimization objective f_{m} changes with number of iterations

(2) When recording accelerometer data, to avoid additional non-gravity acceleration, it is necessary to hold the Pixhawk still. To reduce the external acceleration noise, some extracted feature points are used for calibration optimization, rather than all sample points. The calibration results are satisfactory.

(3) After the gravity acceleration "g" is changed from 9.8 to 1, the same code from the basic experiment is executed again. The results show that the scale factor $\mathbf{K}_{\mathrm{a}}$ narrows 1/9.8, but the attitude angle is consistent with that when "g" is 9.8. This implies that the pitch angle measurement is independent of the acceleration of gravity.

(4) As for recording the magnetometer data, the Pixhawk autopilot is rotated starting from six different orientations. The calibration results are satisfactory.

If you have any question, please go to https://rflysim.com for your information.

Chapter 8
State Estimation and Filter Design Experiment

The state estimation [13] is fundamental because it is the base for control and decision-making. This chapter introduces the principle and method of attitude estimation in detail via three step-by-step experiments from shallow to deep, namely a basic experiment, an analysis experiment, and a design experiment. In the basic experiment, readers can repeat the simulation of the complementary filter to estimate Euler angles so that readers can understand how the basic filter works. In the analysis experiment, readers can adjust a key parameter of the complementary filter to observe the estimation results and analyze the reason. In the design experiment, for a similar problem, readers can design a Kalman filter based on a new model with the estimate compared with those by the complementary filter and self-contained filter in PX4 software of Pixhawk autopilot.

8.1 Preliminary

In order to make this chapter self-contained, the preliminary is from Chap. 9 of *Introduction to Multicopter Design and Control* [8].

8.1.1 Measurement Principle

The three-axis accelerometer is fixed to the multicopter, aligned with the aircraft-body frame. Therefore, the observation of low-frequency pitch and roll angle acquired by accelerometer measurement illustrated as

© Publishing House of Electronics Industry 2020, corrected publication 2021

Q. Quan et al., *Multicopter Design and Control Practice*,
https://doi.org/10.1007/978-981-15-3138-5_8

$$\begin{aligned} \theta_{\mathrm{m}} &= \arcsin\left(\frac{a_{x_{\mathrm{b}}\mathrm{m}}}{g}\right) \\ \phi_{\mathrm{m}} &= -\arcsin\left(\frac{a_{y_{\mathrm{b}}\mathrm{m}}}{g\cos\theta_{\mathrm{m}}}\right) \end{aligned} \tag{8.1}$$

where ${}^{\mathrm{b}}\mathbf{a}_{\mathrm{m}} = [a_{x_{\mathrm{b}}\mathrm{m}}\ a_{y_{\mathrm{b}}\mathrm{m}}\ a_{z_{\mathrm{b}}\mathrm{m}}]^{\mathrm{T}}$ denotes the measurement from the accelerometer. Several further considerations are as follows

(1) It is better to eliminate the slow time-varying drift of the accelerometer to obtain a more accurate angle.
(2) If the amplitude of the vibration is large, $a_{x_{\mathrm{b}}\mathrm{m}}, a_{y_{\mathrm{b}}\mathrm{m}}$ are polluted by noise severely and further affect the estimation of $\theta_{\mathrm{m}}, \phi_{\mathrm{m}}$. Thus, the vibration damping is very important. Additionally, the attitude rates $\dot{\theta}, \dot{\phi}, \dot{\psi}$ and angular velocity ${}^{\mathrm{b}}\boldsymbol{\omega}$ exhibit the following relationship

$$\begin{bmatrix} \dot{\phi} \\ \dot{\theta} \\ \dot{\psi} \end{bmatrix} = \begin{bmatrix} 1 & \tan\theta\sin\phi & \tan\theta\cos\phi \\ 0 & \cos\phi & -\sin\phi \\ 0 & \sin\phi/\cos\theta & \cos\phi/\cos\theta \end{bmatrix} \begin{bmatrix} \omega_{x_{\mathrm{b}}} \\ \omega_{y_{\mathrm{b}}} \\ \omega_{z_{\mathrm{b}}} \end{bmatrix}. \tag{8.2}$$

Multicopters typically work under condition that θ, ϕ are small, and thus the above equation is approximated as follows

$$\begin{bmatrix} \dot{\phi} \\ \dot{\theta} \\ \dot{\psi} \end{bmatrix} \approx \begin{bmatrix} \omega_{x_{\mathrm{b}}} \\ \omega_{y_{\mathrm{b}}} \\ \omega_{z_{\mathrm{b}}} \end{bmatrix}. \tag{8.3}$$

According to the working principle, the attitude is estimated by the accelerometers and magnetometers with large noise but small drifts.

8.1.2 *Linear Complementary Filter*

The basic idea of complementary filtering involves using accelerometer and gyroscope complementary characteristics to obtain more accurate attitude estimation.

8.1.2.1 Pitch Angle

The Laplace transform of pitch angle $\theta \in \mathbb{R}$ is expressed as follows

$$\theta(s) = \frac{1}{\tau s + 1}\theta(s) + \frac{\tau s}{\tau s + 1}\theta(s) \tag{8.4}$$

where $1/(\tau s + 1)$ denotes the transfer function of a low-pass filter, $\tau > 0$ denotes a time constant, and $\tau s/(\tau s + 1) = 1 - 1/(\tau s + 1)$ denotes the transfer function of a high-pass filter. Given that the pitch angle obtained by an accelerometer has high noise but a low drift; for simplicity, it is modeled as follows

$$\theta_{\mathrm{m}} = \theta + n_{\theta} \tag{8.5}$$

where $n_{\theta} \in \mathbb{R}$ denotes high-frequency noise, and θ denotes the true pitch angle. Given that the pitch angle estimated by integrating angular velocity exhibits a little noise but a large drift, and the integration is modeled as follows

$$\frac{\omega_{y_{\mathrm{b}}\mathrm{m}}(s)}{s} = \theta(s) + c\frac{1}{s} \tag{8.6}$$

where $\omega_{y_{\mathrm{b}}\mathrm{m}}(s)/s$ denotes the Laplace transform of the integration of angular velocity $\omega_{y_{\mathrm{b}}\mathrm{m}}$, c/s represents the Laplace transform of the constant drift, and $\omega_{y_{\mathrm{b}}\mathrm{m}}$ denotes the gyroscope measurement. Therefore, for the pitch angle, the standard form of a linear complementary filter is expressed as follows

$$\hat{\theta}(s) = \frac{1}{\tau s + 1}\theta_{\mathrm{m}}(s) + \frac{\tau s}{\tau s + 1}\left(\frac{1}{s}\omega_{y_{\mathrm{b}}\mathrm{m}}(s)\right). \tag{8.7}$$

Next, the explanation is provided as to why the linear complementary filter can obtain a more accurate attitude estimation. By using Eqs. (8.5), (8.6) and (8.7), $\hat{\theta}(s)$ is as follows

$$\hat{\theta}(s) = \theta(s) + \left(\frac{1}{\tau s + 1}n_{\theta}(s) + \frac{\tau s}{\tau s + 1}c\frac{1}{s}\right) \tag{8.8}$$

where the high-frequency noise n_{θ} can be attenuated near zero after passing the low-pass filter $1/(1+\tau s)$, and the low-frequency signal c/s can be eliminated after passing the high-pass filter $\tau s/(1+\tau s)$. Thus, $\hat{\theta}(s) \approx \theta(s)$. During the process, the low-frequency filter exhibits the advantage that θ_{m} has a small drift; while the high-pass filter maintains the advantage that $\omega_{y_{\mathrm{b}}\mathrm{m}}(s)/s$ has a little noise. The process is shown in Fig. 8.1.

In order to realize the filter (8.7) with digital computers, the filter should be transformed into a discrete-time differential form. For example, through the first-order backward difference, s is expressed as $s = \left(1 - z^{-1}\right)/T_{\mathrm{s}}$, where $T_{\mathrm{s}} > 0$ denotes the sampling period. Subsequently, Eq. (8.7) is as follows

$$\hat{\theta}(z) = \frac{1}{\tau\frac{1-z^{-1}}{T_{\mathrm{s}}} + 1}\theta_{\mathrm{m}}(z) + \frac{\tau}{\tau\frac{1-z^{-1}}{T_{\mathrm{s}}} + 1}\omega_{y_{\mathrm{b}}\mathrm{m}}(z). \tag{8.9}$$

The above equation is further transformed into a discrete-time difference form as follows

$$\hat{\theta}(k) = \frac{\tau}{\tau + T_{\mathrm{s}}}\left(\hat{\theta}(k-1) + T_{\mathrm{s}}\omega_{y_{\mathrm{b}}\mathrm{m}}(k)\right) + \frac{T_{\mathrm{s}}}{\tau + T_{\mathrm{s}}}\theta_{\mathrm{m}}(k). \tag{8.10}$$

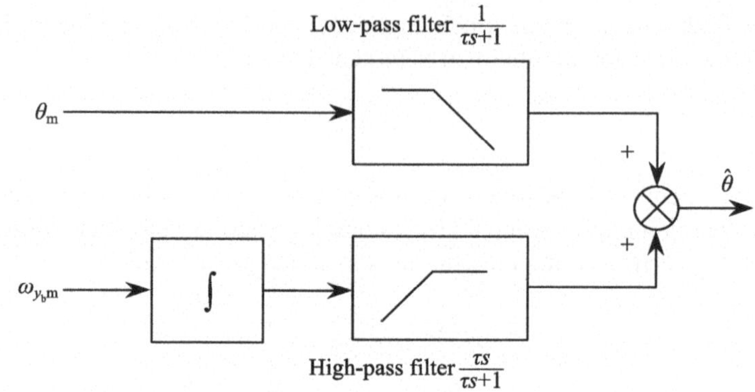

Fig. 8.1 Structure of complementary filter

If $\tau/(\tau+T_{\mathrm{s}})=0.95$, then $T_{\mathrm{s}}/(\tau+T_{\mathrm{s}})=0.05$. Thus, the complementary filter for the pitch angle becomes

$$\hat{\theta}(k)=0.95\left(\hat{\theta}(k-1)+T_{\mathrm{s}}\omega_{y_{\mathrm{b}}\mathrm{m}}(k)\right)+0.05\theta_{\mathrm{m}}(k). \tag{8.11}$$

8.1.2.2 Roll Angle

Similarly, the complementary filter of the roll angle is as follows

$$\hat{\phi}(k)=\frac{\tau}{\tau+T_{\mathrm{s}}}\left(\hat{\phi}(k-1)+T_{\mathrm{s}}\omega_{x_{\mathrm{b}}\mathrm{m}}(k)\right)+\frac{T_{\mathrm{s}}}{\tau+T_{\mathrm{s}}}\phi_{\mathrm{m}}(k). \tag{8.12}$$

Let $\tau/(\tau+T_{\mathrm{s}})=0.95$, and $T_{\mathrm{s}}/(\tau+T_{\mathrm{s}})=0.05$. Subsequently, Eq. (8.12) becomes

$$\hat{\phi}(k)=0.95\left(\hat{\phi}(k-1)+T_{\mathrm{s}}\omega_{x_{\mathrm{b}}\mathrm{m}}(k)\right)+0.05\phi_{\mathrm{m}}(k). \tag{8.13}$$

8.1.3 Kalman Filter

The "truth" model for discrete-time cases is given as follows

$$\begin{aligned}\mathbf{x}_k&=\mathbf{\Phi}_{k-1}\mathbf{x}_{k-1}+\mathbf{u}_{k-1}+\mathbf{\Gamma}_{k-1}\mathbf{w}_{k-1}\\ \mathbf{z}_k&=\mathbf{H}_k\mathbf{x}_k+\mathbf{v}_k\end{aligned} \tag{8.14}$$

where $\mathbf{x}_k\in\mathbb{R}^n$ denotes the system state vector at the discrete-time k; $\mathbf{z}_k\in\mathbb{R}^m$ denotes the measurement vector at the time k; $\mathbf{u}_{k-1}\in\mathbb{R}^n$ denotes the control vector at the time $k-1$; $\mathbf{\Phi}_{k-1}\in\mathbb{R}^{n\times n}$ denotes the transition matrix taking the state $\mathbf{x}_k$ from time

$k-1$ to time k, and it remains constant for system (8.14) when the corresponding system is time-invariant; $\mathbf{w}_{k-1} \in \mathbb{R}^n$ denotes the process noise at the time $k-1$; $\boldsymbol{\Gamma}_{k-1} \in \mathbb{R}^{n\times n}$ denotes the system noise matrix to reflect the contribution of each noise signal made to the system state; $\mathbf{H}_k \in \mathbb{R}^{m\times n}$ denotes the measurement matrix at the time k; $\mathbf{v}_k \in \mathbb{R}^m$ denotes the measurement noise at the time k. Additionally, $\mathbf{w}_k$ and $\mathbf{v}_k$ are assumed as zero-mean Gaussian white noise processes. This means that the errors are uncorrelated forward or backward in time such that

$$\begin{aligned}
\mathrm{E}\left(\mathbf{w}_{k-1}\right) &= \mathbf{0}_{n\times 1}, \mathrm{E}\left(\mathbf{v}_k\right) = \mathbf{0}_{m\times 1}, \mathbf{R}_{\mathbf{wv}}\left(k,j\right) = \mathbf{0}_{n\times m} \\
\mathbf{R}_{\mathbf{ww}}\left(k,j\right) &= \mathrm{E}\left(\mathbf{w}_k\mathbf{w}_j^{\mathrm{T}}\right) = \mathbf{Q}_k\delta_{kj} = \begin{cases}\mathbf{Q}_k, k=j \\ \mathbf{0}_{n\times n}, k\neq j\end{cases} \\
\mathbf{R}_{\mathbf{vv}}\left(k,j\right) &= \mathrm{E}\left(\mathbf{v}_k\mathbf{v}_j^{\mathrm{T}}\right) = \mathbf{R}_k\delta_{kj} = \begin{cases}\mathbf{R}_k, \ k=j \\ \mathbf{0}_{m\times m}, k\neq j\end{cases}
\end{aligned} \tag{8.15}$$

where $\mathbf{Q}_k \in \mathbb{R}^{n\times n}$ denotes the system noise covariance matrix and $\mathbf{R}_k \in \mathbb{R}^{m\times m}$ denotes the measurement noise covariance matrix. In a Kalman filter, they are positive semi-definite matrices at least, represented as follows

$$\mathbf{Q}_k \geqslant \mathbf{0}_{n\times n}, \mathbf{R}_k > \mathbf{0}_{m\times m} \tag{8.16}$$

where $\mathbf{Q}_k \geqslant \mathbf{0}_{n\times n}$ indicates that there may be no noise in some system states, while $\mathbf{R}_k > \mathbf{0}_{m\times m}$ is necessary, thereby implying that each measurement must contain noise; δ_{kj} denotes the Kronecker δ function, and thus the following expression is obtained

$$\delta_{kj} = \begin{cases} 1, k=j \\ 0, k\neq j. \end{cases} \tag{8.17}$$

Suppose that the initial condition of state $\mathbf{x}_0$ satisfies following expression

$$\mathrm{E}\left(\mathbf{x}_0\right) = \hat{\mathbf{x}}_0, \mathrm{Cov}\left(\mathbf{x}_0\right) = \mathbf{P}_0 \tag{8.18}$$

where $\hat{\mathbf{x}}_0$ and $\mathbf{P}_0$ are known as prior knowledge. Besides, $\mathbf{x}_0$, $\mathbf{u}_k$ and $\mathbf{w}_{k-1}$, $\mathbf{v}_k$, $k \geqslant 1$ are such uncorrelated that the following expression is obtained

$$\begin{aligned}
\mathbf{R}_{\mathbf{xw}}\left(0,k\right) &= \mathrm{E}\left(\mathbf{x}_0\mathbf{w}_k^{\mathrm{T}}\right) = \mathbf{0}_{n\times n} \\
\mathbf{R}_{\mathbf{xv}}\left(0,k\right) &= \mathrm{E}\left(\mathbf{x}_0\mathbf{v}_k^{\mathrm{T}}\right) = \mathbf{0}_{n\times m} \\
\mathbf{R}_{\mathbf{uw}}\left(k,j\right) &= \mathrm{E}\left(\mathbf{u}_k\mathbf{w}_j^{\mathrm{T}}\right) = \mathbf{0}_{n\times n}.
\end{aligned} \tag{8.19}$$

Suppose that the measurements $\mathbf{z}_1, \mathbf{z}_2, \ldots, \mathbf{z}_{k-1}$ are given, and the optimal estimate $\mathbf{x}_{k-1|k-1}$ of the state $\mathbf{x}_{k-1}$ denotes the Minimum-variance unbiased estimate, thereby implying the following expression

$$\mathrm{E}\left(\mathbf{x}_{k-1} - \mathbf{x}_{k-1|k-1}\right) = \mathbf{0}_{n\times 1}. \tag{8.20}$$

8.1.3.1 Summary of the Kalman Filter

(1) Step1: Process model

$$\mathbf{x}_k = \boldsymbol{\Phi}_{k-1}\mathbf{x}_{k-1} + \mathbf{u}_{k-1} + \boldsymbol{\Gamma}_{k-1}\mathbf{w}_{k-1},\ \mathbf{w}_k \sim \mathcal{N}\left(\mathbf{0}_{n\times 1}, \mathbf{Q}_k\right). \tag{8.21}$$

Measurement model:

$$\mathbf{z}_k = \mathbf{H}_k\mathbf{x}_k + \mathbf{v}_k,\ \mathbf{v}_k \sim \mathcal{N}\left(\mathbf{0}_{m\times 1}, \mathbf{R}_k\right) \tag{8.22}$$

where $\mathbf{w}_k$ and $\mathbf{v}_k$ are independent, zero-mean, Gaussian noise processes with covariance matrices being $\mathbf{Q}_k$ and $\mathbf{R}_k$, respectively.

(2) Step2: Initial state

$$\begin{aligned} \hat{\mathbf{x}}_0 &= \mathrm{E}\left(\mathbf{x}_0\right) \\ \mathbf{P}_0 &= \mathrm{E}\left[\left(\mathbf{x}_0 - \mathrm{E}\left(\mathbf{x}_0\right)\right)\left(\mathbf{x}_0 - \mathrm{E}\left(\mathbf{x}_0\right)\right)^{\mathrm{T}}\right]. \end{aligned} \tag{8.23}$$

(3) Step3: For $k = 0$, set $\mathbf{P}_{0|0} = \mathbf{P}_0$, $\mathbf{x}_{0|0} = \mathbf{x}_0$.

(4) Step4: $k = k + 1$.

(5) Step5: State estimate propagation

$$\mathbf{x}_{k|k-1} = \boldsymbol{\Phi}_{k-1}\mathbf{x}_{k-1|k-1} + \mathbf{u}_{k-1}. \tag{8.24}$$

(6) Step6: Error covariance propagation

$$\mathbf{P}_{k|k-1} = \boldsymbol{\Phi}_{k-1}\mathbf{P}_{k-1|k-1}\boldsymbol{\Phi}_{k-1}^{\mathrm{T}} + \boldsymbol{\Gamma}_{k-1}\mathbf{Q}_{k-1}\boldsymbol{\Gamma}_{k-1}^{\mathrm{T}}. \tag{8.25}$$

(7) Step7: Kalman gain matrix

$$\mathbf{K}_k = \mathbf{P}_{k|k-1}\mathbf{H}_k^{\mathrm{T}}\left(\mathbf{H}_k\mathbf{P}_{k|k-1}\mathbf{H}_k^{\mathrm{T}} + \mathbf{R}_k\right)^{-1}. \tag{8.26}$$

(8) Step8: State estimate update

$$\hat{\mathbf{x}}_{k|k} = \hat{\mathbf{x}}_{k|k-1} + \mathbf{K}_k\left(\mathbf{z}_k - \hat{\mathbf{z}}_{k|k-1}\right) \tag{8.27}$$

where $\hat{\mathbf{z}}_{k|k-1} = \mathbf{H}_k\hat{\mathbf{x}}_{k|k-1}$.

(9) Step9: Error covariance update

$$\mathbf{P}_{k|k} = \left(\mathbf{I}_n - \mathbf{K}_k\mathbf{H}_k\right)\mathbf{P}_{k|k-1}. \tag{8.28}$$

(10) Step10: Go back to Step4.

The Kalman filter is a type of recursive algorithm in time domain. It can estimate the current state based on the previous estimate and the current measurement. Furthermore, it does not necessitate the storage a large amount of prior data, and it is

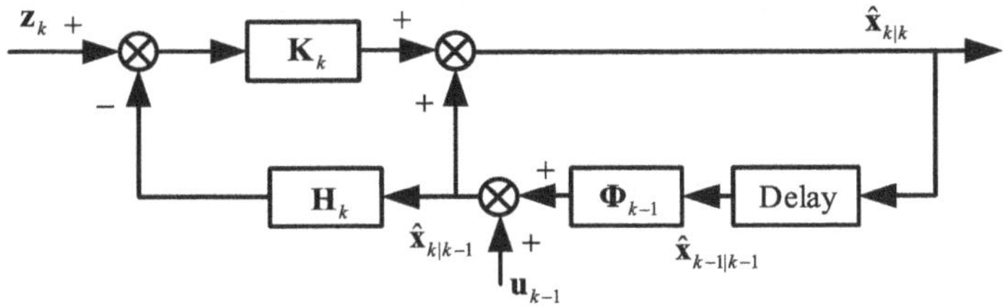

Fig. 8.2 Structure of Kalman filter algorithm

easy to realize with a computer. The essence of the estimation algorithm involves minimizing the trace of state estimate error matrix $\mathbf{P}_{k|k}$. The structure of the Kalman filter algorithm is plotted in Fig. 8.2. In practice, it is necessary to consider how to determine the initial state $\hat{\mathbf{x}}_0$ and the initial variance matrix $\mathbf{P}_0$. The value of the initial state $\hat{\mathbf{x}}_0$ is often obtained by experience, but $\mathbf{P}_0$ cannot. Instead, it should be obtained via statistical methods with several initial measurements. Fortunately, the stable estimate result does not depend on the choice of the initial value of $\hat{\mathbf{x}}_0$ and $\mathbf{P}_0$ if the filtering algorithm is stable.

In order to better understand the Kalman filter, a few remarks are given as follows

(1) It is observed that the error covariance matrix $\mathbf{P}_{k|k}$ can be obtained using the filter, which represents the estimation accuracy. Additionally, it can be used to evaluate the health of sensors.
(2) Generally speaking, if a reasonable sampling time is adopted and the continuous-time system is observable, then the corresponding discrete-time system is also observable. Conversely, the system can also lose controllability and observability when an improper sampling time is adopted. Thus, it is necessary to check the observability of the discrete system after sampling.
(3) The matrix $\mathbf{H}_k\mathbf{P}_{k|k-1}\mathbf{H}_k^{\mathrm{T}} + \mathbf{R}_k$ needs to be non-singular. Otherwise, the solution expressed by $\mathbf{K}_k = \mathbf{P}_{k|k-1}\mathbf{H}_k^{\mathrm{T}}\left(\mathbf{H}_k\mathbf{P}_{k|k-1}\mathbf{H}_k^{\mathrm{T}} + \mathbf{R}_k\right)^{-1}$ does not make sense. In general, $\mathbf{R}_k$ is required to be nonsingular, i.e., the measurements must always contain noise, and thus $\mathbf{H}_k\mathbf{P}_{k|k-1}\mathbf{H}_k^{\mathrm{T}} + \mathbf{R}_k$ is non-singular.
(4) If the system $(\mathbf{\Phi}_{k-1}, \mathbf{H}_k)$ is unobservable, the filter also works without causing numerical problems. Only, the unobservable mode will not be corrected. In an extreme case, the whole system is completely unobservable if $\mathbf{H}_k = \mathbf{0}_{m\times n}$. Subsequently, the filter gain $\mathbf{K}_k = \mathbf{0}_{n\times m}$. Thus, the Kalman filter degenerates as follows

$$\begin{aligned} \hat{\mathbf{x}}_{k|k} &= \mathbf{\Phi}_{k-1}\hat{\mathbf{x}}_{k-1|k-1} + \mathbf{u}_{k-1} \\ \mathbf{P}_{k|k} &= \mathbf{\Phi}_{k-1}\mathbf{P}_{k-1|k-1}\mathbf{\Phi}_{k-1}^{\mathrm{T}} + \mathbf{\Gamma}_{k-1}\mathbf{Q}_{k-1}\mathbf{\Gamma}_{k-1}^{\mathrm{T}}. \end{aligned} \tag{8.29}$$

Thus, only prediction is performed for the next state.

8.1.4 Extended Kalman Filter

Kalman filter is an efficient recursive filter that estimates the internal state of a linear dynamical system from a series of noisy measurements. However, the disadvantage of the Kalman filter is that the noise is assumed to be with Gaussian distribution and the algorithm is only applicable to linear systems. Specifically, a larger class of estimation problems are involved in nonlinear systems. Thus, the state estimation for nonlinear systems is significantly more difficult and admits a wide variety of solutions than the linear problem. There are many possible ways to produce a linearized version of the Kalman filter. In this section, the most common approach, namely the EKF, is considered. The EKF, although not precisely optimum, is successfully applied to many nonlinear systems over past many years.

8.1.4.1 Basic Principle

The main idea of EKF denotes the linearization of nonlinear functions, which ignores the higher-order terms. The nonlinear problem is transformed into a linear problem through Taylor expansion and first order linear truncation. The EKF is a suboptimal filter since the processing of linearization will cause an additional error.

8.1.4.2 Theoretical Derivation

The following general nonlinear system is first considered that is described as follows

$$\begin{aligned} \mathbf{x}_k &= \mathbf{f}\left(\mathbf{x}_{k-1}, \mathbf{u}_{k-1}, \mathbf{w}_{k-1}\right) \\ \mathbf{z}_k &= \mathbf{h}\left(\mathbf{x}_k, \mathbf{v}_k\right) \end{aligned} \tag{8.30}$$

where the functions are defined as $\mathbf{f} : \mathbb{R}^n \times \mathbb{R}^m \times \mathbb{R}^n \rightarrow \mathbb{R}^n$ and $\mathbf{h} : \mathbb{R}^n \times \mathbb{R}^m \rightarrow \mathbb{R}^m$; the random vector $\mathbf{w}_{k-1}$ captures uncertainties in the system model and $\mathbf{v}_k$ denotes the measurement noise, both of which are temporally uncorrelated (white noise), zero-mean random sequences with covariance matrices $\mathbf{Q}_k$ and $\mathbf{R}_k$, respectively. Their properties are found in Eq. (8.15) in Sect. 8.1.3. In the derivation of the EKF, $\mathbf{f}\left(\mathbf{x}_{k-1}, \mathbf{u}_{k-1}, \mathbf{w}_{k-1}\right)$ and $\mathbf{h}\left(\mathbf{x}_k, \mathbf{v}_k\right)$ are expanded via Taylor expansion. Assume that an optimal estimate $\hat{\mathbf{x}}_{k-1|k-1}$ with covariance matrix $\mathbf{P}_{k-1|k-1}$ at time $k-1$ is obtained. By ignoring the higher-order terms, $\mathbf{f}\left(\mathbf{x}_{k-1}, \mathbf{u}_{k-1}, \mathbf{w}_{k-1}\right)$ is expanded via Taylor series at the state $\hat{\mathbf{x}}_{k-1|k-1}$ as follows

$$\begin{aligned} \mathbf{f}\left(\mathbf{x}_{k-1}, \mathbf{u}_{k-1}, \mathbf{w}_{k-1}\right) = &\ \mathbf{f}(\hat{\mathbf{x}}_{k-1|k-1}, \mathbf{u}_{k-1}, \mathbf{0}_{n\times 1}) \\ &+ \left.\frac{\partial \mathbf{f}(\mathbf{x}, \mathbf{u}_{k-1}, \mathbf{w})}{\partial \mathbf{x}}\right|_{\mathbf{x}=\hat{\mathbf{x}}_{k-1|k-1}, \mathbf{w}=\mathbf{0}_{n\times 1}} \left(\mathbf{x}_{k-1} - \hat{\mathbf{x}}_{k-1|k-1}\right) \\ &+ \left.\frac{\partial \mathbf{f}(\mathbf{x}, \mathbf{u}_{k-1}, \mathbf{w})}{\partial \mathbf{w}}\right|_{\mathbf{x}=\hat{\mathbf{x}}_{k-1|k-1}, \mathbf{w}=\mathbf{0}_{n\times 1}} \mathbf{w}_{k-1}. \end{aligned} \tag{8.31}$$

Similarly,

$$\begin{aligned}\mathbf{h}(\mathbf{x}_k, \mathbf{v}_k) = &\ \mathbf{h}\left(\hat{\mathbf{x}}_{k|k-1}, \mathbf{0}_{m\times 1}\right) \\ &+ \left.\frac{\partial \mathbf{h}(\mathbf{x},\mathbf{v})}{\partial \mathbf{x}}\right|_{\mathbf{x}=\hat{\mathbf{x}}_{k|k-1},\mathbf{v}=\mathbf{0}_{m\times 1}} \left(\mathbf{x}_k - \hat{\mathbf{x}}_{k|k-1}\right) \\ &+ \left.\frac{\partial \mathbf{h}(\mathbf{x},\mathbf{v})}{\partial \mathbf{v}}\right|_{\mathbf{x}=\hat{\mathbf{x}}_{k|k-1},\mathbf{v}=\mathbf{0}_{m\times 1}} \mathbf{v}_k.\end{aligned} \tag{8.32}$$

It is noticed that $\mathbf{h}(\mathbf{x}_k, \mathbf{v}_k)$ is linearized at $\hat{\mathbf{x}}_{k|k-1}$ instead of $\hat{\mathbf{x}}_{k-1|k-1}$. In order to simplify the expression of the EKF, the following notation is defined

$$\begin{aligned}\mathbf{\Phi}_{k-1} &\triangleq \left.\frac{\partial \mathbf{f}(\mathbf{x},\mathbf{u}_{k-1},\mathbf{w})}{\partial \mathbf{x}}\right|_{\mathbf{x}=\hat{\mathbf{x}}_{k-1|k-1},\mathbf{w}=\mathbf{0}_{n\times 1}} \\ \mathbf{H}_k &\triangleq \left.\frac{\partial \mathbf{h}(\mathbf{x},\mathbf{v})}{\partial \mathbf{x}}\right|_{\mathbf{x}=\hat{\mathbf{x}}_{k|k-1},\mathbf{v}=\mathbf{0}_{m\times 1}} \\ \mathbf{\Gamma}_{k-1} &\triangleq \left.\frac{\partial \mathbf{f}(\mathbf{x},\mathbf{u}_{k-1},\mathbf{w})}{\partial \mathbf{w}}\right|_{\mathbf{x}=\hat{\mathbf{x}}_{k-1|k-1},\mathbf{w}=\mathbf{0}_{n\times 1}} \\ \mathbf{u}'_{k-1} &\triangleq \mathbf{f}(\hat{\mathbf{x}}_{k-1|k-1}, \mathbf{u}_{k-1}, \mathbf{0}_{n\times 1}) - \mathbf{\Phi}_{k-1}\hat{\mathbf{x}}_{k-1|k-1} \\ \mathbf{z}'_k &\triangleq \mathbf{z}_k - \mathbf{h}\left(\hat{\mathbf{x}}_{k|k-1}, \mathbf{0}_{m\times 1}\right) + \mathbf{H}_k\hat{\mathbf{x}}_{k|k-1} \\ \mathbf{v}'_k &\triangleq \left.\frac{\partial \mathbf{h}(\mathbf{x},\mathbf{v})}{\partial \mathbf{v}}\right|_{\mathbf{x}=\hat{\mathbf{x}}_{k|k-1},\mathbf{v}=\mathbf{0}_{m\times 1}} \mathbf{v}_k.\end{aligned} \tag{8.33}$$

Based on the definitions above, the simplified system model is as follows

$$\begin{aligned}\mathbf{x}_k &= \mathbf{\Phi}_{k-1}\mathbf{x}_{k-1} + \mathbf{u}'_{k-1} + \mathbf{\Gamma}_{k-1}\mathbf{w}_{k-1} \\ \mathbf{z}'_k &= \mathbf{H}_k\mathbf{x}_k + \mathbf{v}'_k.\end{aligned} \tag{8.34}$$

Here, the statistical property of $\mathbf{v}'_k$ is $\mathrm{E}\left(\mathbf{v}'_k\right) = \mathbf{0}_{m\times 1}$ and

$$\mathbf{R}_{\mathbf{v}'\mathbf{v}'}(k, j) \triangleq \mathrm{E}\left(\mathbf{v}'_k\mathbf{v}'^{\mathrm{T}}_j\right) = \begin{cases}\mathbf{R}'_k, k = j \\ \mathbf{0}_{m\times m}, k \neq j\end{cases}$$

where

$$\mathbf{R}'_k \triangleq \left.\frac{\partial \mathbf{h}(\mathbf{x},\mathbf{v})}{\partial \mathbf{v}}\right|_{\mathbf{x}=\hat{\mathbf{x}}_{k|k-1},\mathbf{v}=\mathbf{0}_{m\times 1}} \mathbf{R}_k \left(\left.\frac{\partial \mathbf{h}(\mathbf{x},\mathbf{v})}{\partial \mathbf{v}}\right|_{\mathbf{x}=\hat{\mathbf{x}}_{k|k-1},\mathbf{v}=\mathbf{0}_{m\times 1}}\right)^{\mathrm{T}}. \tag{8.35}$$

Then, the EKF algorithm is summarized as follows.

(1) Step1: Process model

$$\mathbf{x}_k = \mathbf{f}(\mathbf{x}_{k-1}, \mathbf{u}_{k-1}, \mathbf{w}_{k-1}), \ \mathbf{w}_k \sim \mathcal{N}(\mathbf{0}_{n\times 1}, \mathbf{Q}_k). \tag{8.36}$$

Measurement model

$$\mathbf{z}_k = \mathbf{h}\left(\mathbf{x}_k, \mathbf{v}_k\right), \ \mathbf{v}_k \sim \mathcal{N}\left(\mathbf{0}_{m\times 1}, \mathbf{R}_k\right) \tag{8.37}$$

where $\mathbf{w}_k$ and $\mathbf{v}_k$ are independent, zero-mean, Gaussian noise processes with covariance matrices being $\mathbf{Q}_k$ and $\mathbf{R}_k$, respectively. The definitions of $\mathbf{\Phi}_{k-1}, \mathbf{H}_k, \mathbf{\Gamma}_{k-1}, \mathbf{z}'_k$ are given as in Eq. (8.33).

(2) Step 2: Initial state

$$\begin{aligned} \hat{\mathbf{x}}_0 &= \mathrm{E}\left(\mathbf{x}_0\right) \\ \mathbf{P}_0 &= \mathrm{E}\left[\left(\mathbf{x}_0 - \mathrm{E}\left(\mathbf{x}_0\right)\right)\left(\mathbf{x}_0 - \mathrm{E}\left(\mathbf{x}_0\right)\right)^{\mathrm{T}}\right]. \end{aligned} \tag{8.38}$$

(3) Step3: For $k = 0$, set $\mathbf{P}_{0|0} = \mathbf{P}_0, \hat{\mathbf{x}}_{0|0} = \hat{\mathbf{x}}_0$.

(4) Step4: $k = k + 1$

(5) Step5: State estimate propagation

$$\hat{\mathbf{x}}_{k|k-1} = \mathbf{f}\left(\hat{\mathbf{x}}_{k-1|k-1}, \mathbf{u}_{k-1}, \mathbf{0}_{n\times 1}\right). \tag{8.39}$$

(6) Step6: Error covariance propagation

$$\mathbf{P}_{k|k-1} = \mathbf{\Phi}_{k-1}\mathbf{P}_{k-1|k-1}\mathbf{\Phi}_{k-1}^{\mathrm{T}} + \mathbf{\Gamma}_{k-1}\mathbf{Q}_{k-1}\mathbf{\Gamma}_{k-1}^{\mathrm{T}}. \tag{8.40}$$

(7) Step7: Kalman gain matrix

$$\mathbf{K}_k = \mathbf{P}_{k|k-1}\mathbf{H}_k^{\mathrm{T}}\left(\mathbf{H}_k\mathbf{P}_{k|k-1}\mathbf{H}_k^{\mathrm{T}} + \mathbf{R}'_k\right)^{-1} \tag{8.41}$$

where $\mathbf{R}'_k$ is defined in Eq. (8.35).

(8) Step8: State estimate update

$$\hat{\mathbf{x}}_{k|k} = \hat{\mathbf{x}}_{k|k-1} + \mathbf{K}_k\left(\mathbf{z}'_k - \hat{\mathbf{z}}_{k|k-1}\right) \tag{8.42}$$

where $\hat{\mathbf{z}}_{k|k-1} = \mathbf{H}_k\hat{\mathbf{x}}_{k|k-1}$.

(9) Step9: Error covariance update

$$\mathbf{P}_{k|k} = \left(\mathbf{I}_n - \mathbf{K}_k\mathbf{H}_k\right)\mathbf{P}_{k|k-1}. \tag{8.43}$$

(10) Step10: Go back to Step4.

8.2 Basic Experiment

8.2.1 Experimental Objective

(1) Things to prepare

1) Hardware: Pixhawk autopilot system;
2) Software: MATLAB R2017b or above, Pixhawk Support Package (PSP) Toolbox, Instructional Package "e4.1" (https://rflysim.com/course);
3) Data for experiment are prepared in Instructional Package "e4.1" for readers without hardware to collect data.

(2) Objectives
Repeat the given steps to log accelerometer and gyroscope data via the given Pixhawk autopilot system. Subsequently, run the offered code to compare the estimated attitude of complementary filter with that of raw data and the self-contained filter in PX4 software of the Pixhawk autopilot (the estimate by the PX4 software is taken as ground truth).

8.2.2 Experimental Procedure

(1) **Step1: Log data from accelerometers and gyroscopes**

1) Hardware and software connection. By referring to the procedure shown in Sect. 2.3.1 in Chap. 2, the connection between the Radio Control (RC) receiver and the Pixhawk autopilot is shown in Fig. 8.3. Additionally, by referring to Sect. 7.2.2 in Chap. 7 and, make the configuration "DJI Flame Wheel F450" available in the experiment.
2) Open file "log_data.slx" (for details, refer to the file "acquire_data_ag.slx" in Sect. 7.2.2), as shown in Fig. 8.4, which can obtain data including acceleration, angular velocity, time stamp, and attitude in the Pixhawk autopilot. The data logged is saved in the Pixhawk microSD card via placing the upper-left stick (CH5) at a corresponding position. For details, see Sect. 7.2.2 in Chap. 7. The function of block "datalog" is used to log data down to the microSD card as shown in a box "a" in Fig. 8.4.
3) Subsequently, compile the file "log_data.slx" and upload it to the Pixhawk autopilot. The process is shown in Fig. 8.5. For details, see 4.2.3 in Chap. 4.
4) Log data. The Pixhawk LED status light lighting red denotes that PX4 software does not work. Thus, after connecting the RC receiver to the Pixhawk autopilot, wait for a while until Pixhawk LED status light gets green (if Pixhawk LED status light does not get green, the Pixhawk is required to be re-plugged). Pull back the upper-left switch corresponding to CH5 > 1500 (see the definition in Sect. 2.3.1) to start writing data to the microSD card.

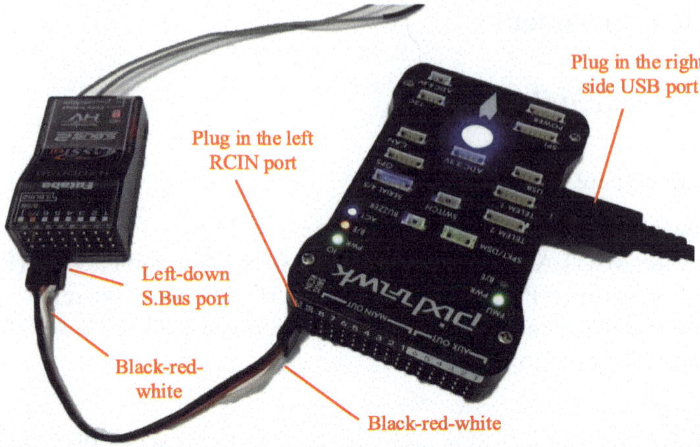

Fig. 8.3 Pixhawk and RC transmitter connection diagram

Subsequently, manually shake the Pixhawk autopilot. After data logging finished, pull forward the upper-left switch (CH5 < 1500) to stop writing data to the microSD card.

5) Read data. First, take the microSD card out from the Pixhawk autopilot. Read the data by a card reader. Copy the file "e4_A.bin" to folder "e4\e4.1". Use the function

```
[datapoints, numpoints] = px4_read_binary_file('e4_A.bin')
```

to decode the data. The data are saved in "datapoints" and the number of data is saved in "numpoints".

(2) **Step2**: **Design a complementary filter**
Compare the attitude estimated by a complementary filter with that by PX4 software in the Pixhawk autopilot. The procedure of a complementary filter in MATLAB is shown in Table 8.1. The 17th and 18th lines correspond to the pitch and roll angles calculated by the raw data from the accelerometer based on Eq. (8.1). The 21st and 22nd lines correspond the pitch and roll angles filtered by the complementary filter based on Eqs. (8.10) and (8.12), respectively.

(3) **Step3**: **Analyze filtering results**
Two sets of sensor data are given, where the file "e4_A.bin" stores the data directly from the Pixhawk autopilot rotating by hands, and the other file "logdata.mat" stores the data from a practical flight of a quadcopter. Run the file "Attitude_estimator0.m" with the results shown in Fig. 8.6. Several observations are obtained as follows: 1) cumulative errors of the pure integral of the angle velocity from the gyroscope are large and diverge; 2) based on the raw data from the accelerometer, the estimate does not diverge although the noise is larger with obvious peaks and, especially using the data from a practical flight; (3) by using the complementary filter, the estimate is smooth and exhibits less cumulative error.

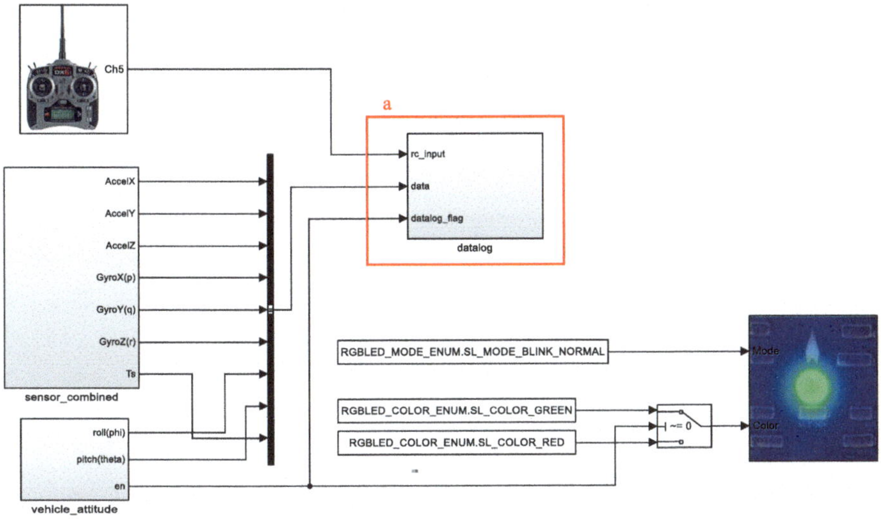

Fig. 8.4 Data logging, Simulink model "log_data.slx"

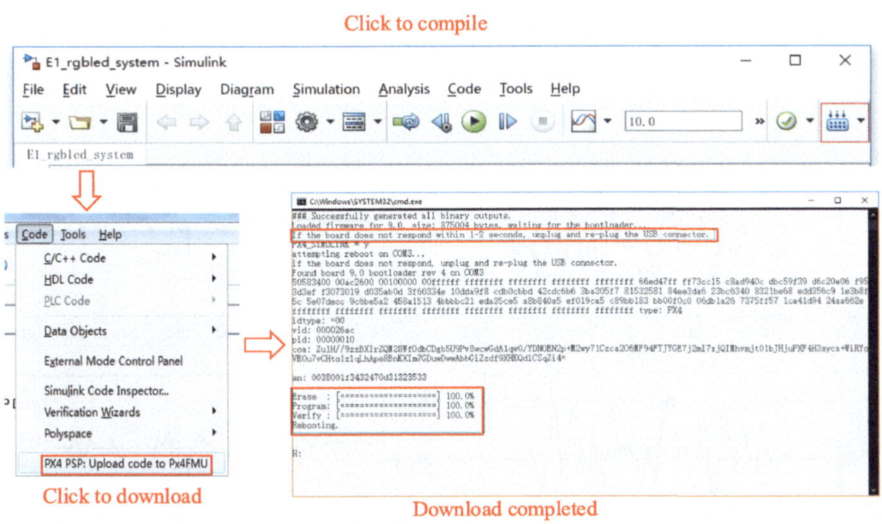

Fig. 8.5 Process of compiling and uploading

8.2.3 Remark

When collecting data using the Pixhawk autopilot, the time to start writing data and time to stop writing data should be appropriate to avoid logging a large amount of useless data.

Table 8.1 Attitude_cf.m

```
function [ phi_cf, theta_cf ] = Attitude_cf(dt, z, phi_cf_k, theta_cf_k, tao)
%Function description:
%  Complementary filter for attitude estimation.
%Input:
%  dt: sampling time, unit: s
%  z: three-axis angle gyroscope and three-axis accelerometer measurements, [gx,
      gy, gz, ax, ay, az]', unit: rad/s, m/s2
%  phi_cf_k, theta_cf_k: Angle value of the previous moment, unit: rad
%  tao: Filter parameter
%Output:
%  phi_cf, theta_cf: Attitude angle, unit: rad

    gx = z(1); gy = z(2);
    ax = z(4); ay = z(5); az = z(6);

    %Calculate the attitude angle using accelerometer measurement
    g = sqrt(ax*ax + ay*ay + az*az);
    theta_am = asin(ax/g);
    phi_am = -asin(ay/(g*cos(theta_am)));

    %Complementary filtering
    theta_cf = tao/(tao + dt)*(theta_cf_k + gy*dt) + dt/(tao + dt)*theta_am;
    phi_cf = tao/(tao + dt)*(phi_cf_k + gx*dt) + dt/(tao + dt)*phi_am;
end
```

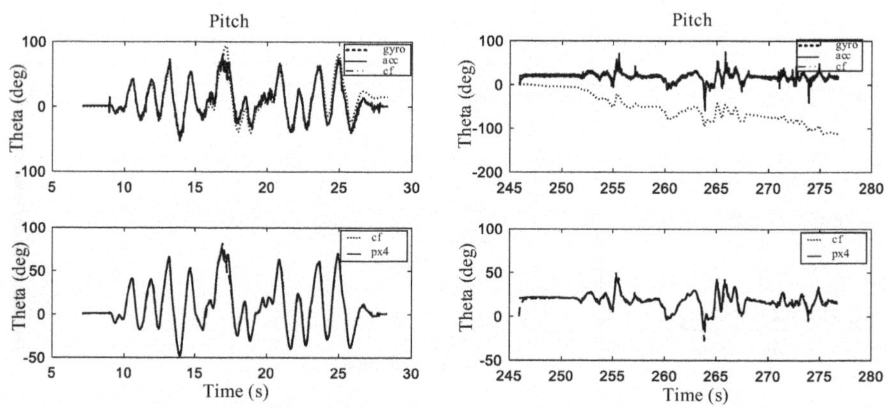

(a) Data from a rotating Pixhawk

(b) Data from an actual flight

Fig. 8.6 Estimate comparison of gyro (integral of the angle velocity from the gyroscope), acc (based on Eq. (8.1)), cf (based on Eqs. (8.10) and (8.12)) and PX4 (px4)

8.3 Analysis Experiment

8.3.1 Experimental Objective

(1) Things to prepare
Data logged in the basic experiment and Instructional Package "e4.2" (https://rflysim.com/course).
(2) Objectives
Based on the basic experiment, change the value of the parameter τ in the complementary filter

$$\hat{\theta}(k) = \frac{\tau}{\tau + T_s}(\hat{\theta}(k-1) + T_s\omega_{y_b\mathrm{m}}(k)) + \frac{T_s}{\tau + T_s}\theta_{\mathrm{m}}(k) \tag{8.44}$$

to observe the filtered result, and analyze the function of the parameter τ in the complementary filter.

8.3.2 Experimental Analysis

Write a program shown in Table 8.2, where the parameter τ in Eq. (8.44) corresponding to "tao" in Table 8.2 is modified. Subsequently, compare the estimate with respect to different values of the parameter τ. Obtain the estimate by the file "Attitude_cf_tao.m" with τ corresponding to 0.01, 0.1, and 1, respectively. The results are shown in Fig. 8.7.

Based on Eq. (8.7), the larger the parameter τ is, the more high-frequency noise is filtered. In particular, when $\tau \gg T_s$, namely

$$\frac{\tau}{\tau + T_s} \approx 1, \frac{T_s}{\tau + T_s} \approx 0.$$

Subsequently, Eqs. (8.10) and (8.12) are as follows

$$\begin{cases} \hat{\theta}(k) \approx \hat{\theta}(k-1) + T_s\omega_{y_b m}(k) \\ \hat{\phi}(k) \approx \hat{\phi}(k-1) + T_s\omega_{x_b m}(k). \end{cases}$$

This implies that the accelerometer no longer contributes to the estimate and only the pure integral of the angle velocity from the gyroscope is used. However, when $\tau \ll T_s$, namely

$$\frac{\tau}{\tau + T_s} \approx 0, \frac{T_s}{\tau + T_s} \approx 1.$$

Based on Eqs. (8.10) and (8.12), $\hat{\theta}(k)$ and $\hat{\phi}(k)$ become

Table 8.2 Attitude_cf_tao.m

```
%The influence of the parameter tao on the filtering performance
clear;
load logdata n = length(ax); %Number of data collected
Ts = zeros(1,n); %Sampling time
Ts(1) = 0.004;

for k =1:n-1
   Ts(k+1) = (timestamp(k + 1) - timestamp(k))*0.000001;
end

theta_cf = zeros(1, n); %Pitch angle obtained from complementary filtering
phi_cf = zeros(1, n); %Roll angle obtained from complementary filtering
tao = 0.001;
for i = 1 : 3
    tao = tao*10;
    for k = 2 : n

        g = sqrt(ax(k)*ax(k) + ay(k)*ay(k) + az(k)*az(k));
        theta_am = asin(ax(k)/g);
        phi_am = -asin(ay(k)/(g*cos(theta_am)));

        theta_cf(i, k) = tao/(tao + Ts(k))*(theta_cf(i, k - 1) + gy(k)*Ts(
             k)) + Ts(k)/(tao + Ts(k))*theta_am;
        phi_cf(i,k) = tao/(tao + Ts(k))*(phi_cf(i, k- 1) + gx(k)*Ts(k)) +
             Ts(k)/(tao + Ts(k))*phi_am;
    end
end
```

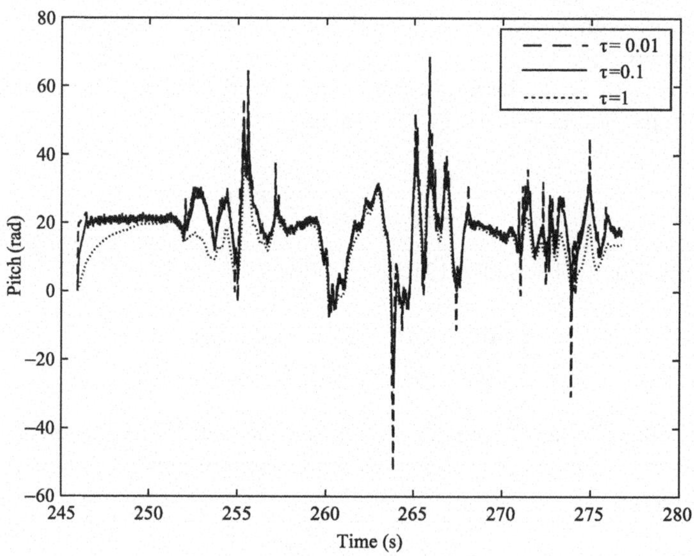

Fig. 8.7 Pitch estimate with respect to parameter τ

$$\begin{cases} \hat{\theta}(k) \approx \theta_{\mathrm{m}}(k) \\ \hat{\phi}(k) \approx \phi_{\mathrm{m}}(k). \end{cases}$$

This implies that the gyroscope no longer contributes to the estimate and only the calculation of the angle velocity from the accelerometer is used. Therefore, the parameter τ should be reasonably set to achieve a trade-off.

8.4 Design Experiment

8.4.1 Experimental Objective

(1) Things to prepare

1) Hardware: Pixhawk autopilot system;
2) Software: MATLAB R2017b or above, Pixhawk Support Package (PSP) toolbox, gyroscope and accelerometer data logged in Sect. 8.1 and Instructional Package "e4.3" (https://rflysim.com/course);
3) Data for experiment are prepared in the instructional package "e4.3" for readers without hardware to collect data.

(2) Objectives
With the obtained data, design a Kalman filter to estimate the pitch and roll angles, and compare the estimated attitude of Kalman filter with that of raw data and the self-contained filter in PX4 software of the Pixhawk autopilot (the estimate by the PX4 software is taken as ground truth).

8.4.2 Experimental Design

(1) **Step1**: **Kalman filter for attitude estimation**
A Kalman filter is adopted to estimate the attitude. The state vector $\mathbf{x} \in \mathbb{R}^3$ [14] is chosen as the third column of $\left(\mathbf{R}_{\mathrm{b}}^{\mathrm{e}}\right)^{\mathrm{T}}$, that is

$$\mathbf{x} = \begin{bmatrix} -\sin\theta \\ \sin\phi\cos\theta \\ \cos\phi\cos\theta \end{bmatrix}. \tag{8.45}$$

Based on the definition of the rotation matrix in Chap. 6, the third column of the rotation matrix $\left(\mathbf{R}_{\mathrm{b}}^{\mathrm{e}}\right)^{\mathrm{T}}$ represents the projection coordinates of the earth-fixed frame system $o_{\mathrm{e}}z_{\mathrm{e}}$ axis in the aircraft-body frame system, i.e, it includes the pitch and roll angles. As shown in Eq. (8.45), θ and ϕ can be solved by $\mathbf{x}$. Additionally, the process model required for Kalman filtering based on the properties of the rotation matrix is also obtained. Let $\mathbf{R} = \mathbf{R}_{\mathrm{b}}^{\mathrm{e}}$ for simplicity in the following. Given that $\mathbf{R}$ satisfies $\dot{\mathbf{R}} = \mathbf{R}\left[{}^{\mathrm{b}}\boldsymbol{\omega}\right]_{\times}$, one has $\dot{\mathbf{R}}^{\mathrm{T}} = -\left[{}^{\mathrm{b}}\boldsymbol{\omega}\right]_{\times}\mathbf{R}^{\mathrm{T}}$. Subsequently, the process model is established as follows

$$\dot{\mathbf{x}} = -\left[{}^{\mathrm{b}}\boldsymbol{\omega}\right]_{\times}\mathbf{x}. \tag{8.46}$$

When the multicopter is held still and horizontally, the measured value of the accelerometer is $[0\ 0\ -g]^{\mathrm{T}}$, and represents the information of the gravitational acceleration. Based on the coordinate system that is established, the direction of the gravitational acceleration coincides with the direction of $o_{\mathrm{e}}z_{\mathrm{e}}$ aixs. Subsequently $-{}^{\mathrm{b}}\mathbf{a}_{\mathrm{m}}/||{}^{\mathrm{b}}\mathbf{a}_{\mathrm{m}}||$ also represents the projected coordinates of the $o_{\mathrm{e}}z_{\mathrm{e}}$ axis of the earth-fixed frame in the aircraft-body frame, where ${}^{\mathrm{b}}\mathbf{a}_{\mathrm{m}}$ denotes the reading of a three-axis accelerometer. Namely, the assumption that the multicopter is stationary can be relaxed in the case that it is moving at a constant speed or exhibits a small acceleration and deceleration motion. The measurement model is built as follows

$${}^{\mathrm{b}}\mathbf{a}_{\mathrm{m}} = -g\mathbf{x} + \mathbf{n}_{\mathrm{a}} \tag{8.47}$$

where $\mathbf{n}_{\mathrm{a}} \in \mathbb{R}^3$ denotes a noise. Further considering the drift model of gyroscope, the process model of the Kalman filter is established as follows

$$\begin{aligned} \dot{\mathbf{x}} &= -\left[{}^{\mathrm{b}}\boldsymbol{\omega}_{\mathrm{m}} - \mathbf{b}_{\mathrm{g}} - \mathbf{n}_{\mathrm{g}}\right]_{\times}\mathbf{x} \\ \dot{\mathbf{b}}_{\mathrm{g}} &= \mathbf{n}_{\mathbf{b}_{\mathrm{g}}} \end{aligned} \tag{8.48}$$

where $\mathbf{n}_{\mathbf{b}_{\mathrm{g}}}$ denotes the noise of the drift model of gyroscope. A Kalman filter is adopted to obtain the estimation of $\mathbf{x}$. Furthermore, the estimated $\hat{\theta}$ and $\hat{\phi}$ are obtained by Eq. (8.45).

(2) **Step2: Design Kalman filter**

In order to run Kalman filtering on a computer, a discretization form should be used. So, first of all, Eqs. (8.47) and (8.48) are transformed into the discrete-time difference form through a first-order backward difference that

$$\begin{aligned} \begin{bmatrix} \mathbf{b}_{\mathrm{g},k} \\ \mathbf{x}_k \end{bmatrix} &= \begin{bmatrix} \mathbf{b}_{\mathrm{g},k-1} + \mathbf{n}_{\mathbf{b}_{\mathrm{g},k-1}} T_{\mathrm{s}} \\ (\mathbf{I}_3 - [{}^{\mathrm{b}}\boldsymbol{\omega}_{\mathrm{m},k} - \mathbf{b}_{\mathrm{g},k-1} - \mathbf{n}_{\mathrm{g},k-1}]_{\times} T_{\mathrm{s}})\mathbf{x}_{k-1} \end{bmatrix} \\ &= \begin{bmatrix} \mathbf{b}_{\mathrm{g},k-1} \\ (\mathbf{I}_3 - [{}^{\mathrm{b}}\boldsymbol{\omega}_{\mathrm{m},k} - \mathbf{b}_{\mathrm{g},k-1}]_{\times} T_{\mathrm{s}})\mathbf{x}_{k-1} \end{bmatrix} + \begin{bmatrix} \mathbf{n}_{\mathbf{b}_{\mathrm{g},k-1}} T_{\mathrm{s}} \\ [\mathbf{n}_{\mathrm{g},k-1}]_{\times} T_{\mathrm{s}})\mathbf{x}_{k-1} \end{bmatrix} \\ {}^{\mathrm{b}}\mathbf{a}_{\mathrm{m},k} &= \begin{bmatrix} \mathbf{0}_{3\times 3} & -g\mathbf{I}_3 \end{bmatrix} \begin{bmatrix} \mathbf{b}_{\mathrm{g},k} \\ \mathbf{x}_k \end{bmatrix} + \mathbf{n}_{\mathrm{a},k} \end{aligned} \tag{8.49}$$

where T_{s} denotes the sampling period. It can be seen that Eq. (8.49) is a nonlinear equation, the method in the Sect. 8.1.4 should be used. According to Eq. (8.33), the transition matrix is

$$\boldsymbol{\Phi}_{k-1} = \begin{bmatrix} \mathbf{I}_3 & \mathbf{0}_{3\times 3} \\ -[\mathbf{x}_{k-1}]_\times T_{\mathrm{s}} & (\mathbf{I}_3 - [{}^{\mathrm{b}}\boldsymbol{\omega}_{\mathrm{m},k} - \mathbf{b}_{\mathrm{g},k-1}]_\times T_{\mathrm{s}}) \end{bmatrix}.$$

The system noise matrix is

$$\boldsymbol{\Gamma}_{k-1} = \begin{bmatrix} T_{\mathrm{s}}\mathbf{I}_3 & \mathbf{0}_{3\times 3} \\ \mathbf{0}_{3\times 3} & -[\mathbf{x}_{k-1}]_\times T_{\mathrm{s}} \end{bmatrix}.$$

The measurement matrix is

$$\mathbf{H}_k = \begin{bmatrix} \mathbf{0}_{3\times 3} & -g\mathbf{I}_3 \end{bmatrix}.$$

(3) **Step3**: **Kalman filtering step**

1) State estimate propagation

$$\begin{bmatrix} \mathbf{b}_{\mathrm{g},k} \\ \mathbf{x}_k \end{bmatrix} = \begin{bmatrix} \mathbf{b}_{\mathrm{g},k-1} \\ (\mathbf{I}_3 - [{}^{\mathrm{b}}\boldsymbol{\omega}_{\mathrm{m},k} - \mathbf{b}_{\mathrm{g},k-1}]_\times T_{\mathrm{s}})\mathbf{x}_{k-1} \end{bmatrix}. \tag{8.50}$$

Calculate state transition matrix and system noise matrix

$$\boldsymbol{\Phi}_{k-1} = \begin{bmatrix} \mathbf{I}_3 & \mathbf{0}_{3\times 3} \\ -[\mathbf{x}_{k-1}]_\times T_{\mathrm{s}} & (\mathbf{I}_3 - [{}^{\mathrm{b}}\boldsymbol{\omega}_{\mathrm{m},k} - \mathbf{b}_{\mathrm{g},k-1}]_\times T_{\mathrm{s}}) \end{bmatrix} \tag{8.51}$$

$$\boldsymbol{\Gamma}_{k-1} = \begin{bmatrix} T_{\mathrm{s}}\mathbf{I}_3 & \mathbf{0}_{3\times 3} \\ \mathbf{0}_{3\times 3} & -[\mathbf{x}_{k-1}]_\times T_{\mathrm{s}} \end{bmatrix} \tag{8.52}$$

where ${}^{b}\boldsymbol{\omega}_{\mathrm{m},k}$ denotes the current measurement value of the gyroscope, and $\mathbf{x}_{k-1}$ denotes the former state estimate.

2) Error covariance propagation:

$$\mathbf{P}_{k|k-1} = \boldsymbol{\Phi}_{k-1}\mathbf{P}_{k-1|k-1}\boldsymbol{\Phi}_{k-1}^{\mathrm{T}} + \Gamma_{k-1}\mathbf{Q}_{k-1}\Gamma_{k-1}^{\mathrm{T}} \tag{8.53}$$

where $\mathbf{Q}_{k-1}$ denotes the variance of system noise.

3) Kalman gain matrix

$$\mathbf{K}_k = \mathbf{P}_{k|k-1}\mathbf{H}_k^{\mathrm{T}}\left(\mathbf{H}_k\mathbf{P}_{k|k-1}\mathbf{H}_k^{\mathrm{T}} + \mathbf{R}_k\right)^{-1} \tag{8.54}$$

where $\mathbf{R}_k$ denotes the variance of measurement noise.

4) State estimate update

$$\begin{bmatrix} \mathbf{b}_{g,k} \\ \mathbf{x}_k \end{bmatrix} = \begin{bmatrix} \mathbf{b}_{g,k-1} \\ \mathbf{x}_{k-1} \end{bmatrix} + \mathbf{K}_k \left(\mathbf{z}_k - \mathbf{H}_k \begin{bmatrix} \mathbf{b}_{g,k-1} \\ \mathbf{x}_{k-1} \end{bmatrix} \right) \tag{8.55}$$

where $\mathbf{z}_k$ denotes the accelerometer measurement.

5) Error covariance update

$$\mathbf{P}_{k|k} = (\mathbf{I} - \mathbf{K}_k\mathbf{H}_k)\mathbf{P}_{k|k-1}. \tag{8.56}$$

(4) **Step4**: **Kalman filter in procedure**
The procedure of Kalman filter are shown in Tables 8.3 and 8.4. The 30th and 31th lines in Table 8.3 represent the state estimate propagation, corresponding to Eq. (8.50); the 38th and 39th lines in Table 8.3 represent the state transition matrix, corresponding to Eq. (8.51); the 40th and 41th lines in Table 8.3 represent the system noise matrix, corresponding to Eq. (8.52); the 2nd line in Table 8.4 represents the error covariance propagation matrix, corresponding to Eq. (8.53); the 7th line in Table 8.4 represents the Kalman gain matrix, corresponding to Eq. (8.54); the 9th line in Table 8.4 represents the state estimate by using Kalman gain, the state propagation and the measurement, corresponding to Eq. (8.55); and the 11th line in Table 8.4 represents the error covariance of the state estimate error, corresponding to Eq. (8.56). Next, based on Eq. (8.45), the 15th and 16th lines in Table 8.4 represent the estimate of the roll and pitch angle, respectively.

8.4.3 Experimental Procedure

(1) **Step1**: **Algorithm simulation and verification**
Run simulation based on the data logged in file folder "e4.3". The main code is shown in Table 8.5. Run the file "Attitude_estimator.m" in file folder "e4.3" to obtain the estimate shown in Fig. 8.8. An observation is that the estimate by the Kalman filter algorithm is better than that by the complementary filtering when the data is from an actual flight.

(2) **Step2**: **Design the model for the HIL simulation**
Based on the linear complementary filter and Kalman filter designed above, design a Simulink model termed as "ekf_cf.slx" (For details, refer the file "acquire_data_ag.slx" in Sect. 7.2.2) as shown in Fig. 8.9, which includes the two filters and can save the data to the microSD card of the Pixhawk autopilot.

(3) **Step3**: **Hardware connection**
By referring to the procedure shown in Sect. 2.3.1 in Chap. 2, the connection between the RC receiver and the Pixhawk autopilot is shown in Fig. 8.3.

Table 8.3 Attitude_ekf.m

```
function [ x_aposteriori, P_aposteriori, roll, pitch] = Attitude_ekf( dt, z, q, r, x_aposteriori_k, P_aposteriori_k)
%Function description:
% Extended Kalman Filtering Method for State Estimation
%Input:
% dt: Sampling time
% z: Measured value
% q: System noise,r:Measuring noise
% x_aposteriori_k: State estimate at the last moment
% P_aposteriori_k: Estimate covariance at the last moment
%Output:
% x_aposteriori: State estimate of current time
% P_aposteriori: Estimated covariance at the current moment
% roll,pitch: Euler angle, unit: rad

w_m = z(1:3); %Angular velocity measurement
a_m = z(4:6); %Acceleration measurement
g = norm(a_m,2); %Gravitational acceleration
%  w_x_=[ 0,-(wz-bzg, wy-byg;
%        wz-bzg, 0 ,-(wx-bxg);
%        -(wy-byg), wx-bxg, 0];
w_x_ = [0, -(w_m(3) - x_aposteriori_k(3)), w_m(2) - x_aposteriori_k(2);
      w_m(3) - x_aposteriori_k(3), 0, -(w_m(1) - x_aposteriori_k(1));
      -(w_m(2) - x_aposteriori_k(2)), w_m(1) - x_aposteriori_k(1), 0];

bCn = eye(3, 3) - w_x_*dt;

% Predict
% The state estimate propagation
x_apriori = zeros(1, 6);
x_apriori(1: 3) = x_aposteriori_k(1 : 3); %The drift model of gyroscope
x_apriori(4 : 6) = bCn*x_aposteriori_k(4 : 6); %Acceleration normalized value

%[x]x
x_aposteriori_k_x = [0, -x_aposteriori_k(6), x_aposteriori_k(5);
                    x_aposteriori_k(6), 0, -x_aposteriori_k(4);
                   -x_aposteriori_k(5), x_aposteriori_k(4), 0];
% Update state transition matrix
PHI = [eye(3, 3), zeros(3, 3);
      -x_aposteriori_k_x*dt, bCn];
GAMMA = [eye(3, 3)*dt, zeros(3, 3); % System noise matrix
      zeros(3, 3), -x_aposteriori_k_x*dt];

Q = [eye(3, 3)*q(1), zeros(3, 3);
   zeros(3, 3), eye(3, 3)*q(2)];
```

Table 8.4 Attitude_ekf.m continued

```
%  Error covariance propagation matrix
 P_apriori = PHI*P_aposteriori_k*PHI' + GAMMA*Q*GAMMA';
% Update
R = eye(3, 3)*r(1);
H_k = [zeros(3, 3), -g*eye(3, 3)];
%Kalman gain
K_k = (P_apriori*H_k')/(H_k*P_apriori*H_k' + R);
% State estimation matrix
x_aposteriori = x_apriori' + K_k*(a_m - H_k*x_apriori');
% Estimation error covariance
P_aposteriori = (eye(6, 6) - K_k*H_k)*P_apriori;
% Calculate roll, pitch
k = x_aposteriori(4 : 6) /norm(x_aposteriori(4 : 6), 2);

roll = atan2(k(2), k(3)); % Roll angle
pitch = -asin(k(1));   %Pitch angle
end
```

(4) **Step4**: **Compile and upload the codes**
Compile the file "ekf_cf.slx" and upload it to the Pixhawk autopilot. The process of compiling and uploading is shown as Fig. 8.5. For details, see Sect. 4.2.3 in Chap. 4.

(5) **Step5**: **Data logging**
The Pixhawk LED status light being in red denotes that PX4 software does not work. Hence, after connecting the RC receiver and the Pixhawk autopilot, wait for a while until Pixhawk LED status light gets green (if Pixhawk LED status light does not get green, the Pixhawk requires to be re-plugged). Pull back the upper-left switch corresponding to CH5 > 1500 (see the definition in Sect. 2.3.1) to start writing data to the microSD card. Subsequently, manually rotate the Pixhawk autopilot. After data logging finished, pull forward the upper-left switch (CH5 < 1500) to stop writing data to the microSD card.

(6) **Step6**: **Read data**
Take out the microSD card, read the data by a card reader, copy the file "ekf1_A.bin" to the folder "e4\e4.3".

(7) **Step7**: **Draw data curves**
Run the file "plot_filter.m" to get the curves, as shown in Fig. 8.10. The observation is made that, during the first half time when the Pixhawk autopilot is rotated slowly, the estimate results by the complementary filter, Kalman filter and the self-contained filter in PX4 software of the Pixhawk autopilot are similar. During the second half time, the Pixhawk autopilot is rotated quickly. The estimate by the complementary filter is evidently bad although, the estimates by the designed Kalman filter and PX4 software in the Pixhawk autopilot are similar. This is reasonable because the Kalman filter fuses more information sufficiently.

Table 8.5 Attitude_estimator.m

```
clear;
load logdata n = length(ax); %Number of data collected
Ts = zeros(1,n); %Sampling time
Ts(1) =0.004;

for k = 1 : n-1
   Ts(k+1) = (timestamp(k + 1) - timestamp(k))*0.000001;
end

theta_am = zeros(1, n); %Pitch angle calculated from acceleration
phi_am = zeros(1, n); %Roll angle calculated from acceleration
theta_gm = zeros(1, n); %Pitch angle from the gyroscope
phi_gm = zeros(1, n); %Roll angle from the gyroscope
theta_cf = zeros(1, n); %Pitch angle obtained from complementary filtering
phi_cf = zeros(1, n); %Roll angle obtained from complementary filtering
phi_ekf = zeros(1, n);
theta_ekf = zeros(1, n);

tao = 0.3;
w = [0.08, 0.01]; %System noise
v = 50; %Measurement noise

P_aposteriori = zeros(6, 6, n);
P_aposteriori(:, :, 1)=eye(6, 6)*100; %P0
x_aposteriori = zeros(6, n);
x_aposteriori(:, 1) = [0, 0, 0, 0, 0, -1]; %X0

for k = 2 : n
  %Calculate Euler angles using accelerometer data
  g = sqrt(ax(k)*ax(k) + ay(k)*ay(k) + az(k)*az(k));
  theta_am(k) = asin(ax(k)/g);
  phi_am(k) = -asin(ay(k)/(g*cos(theta_am(k))));
  %Calculate Euler angles using gyroscope data
  theta_gm(k) = theta_gm(k - 1) + gy(k)*Ts(k);
  phi_gm(k) = phi_gm(k - 1) + gx(k)*Ts(k);
 %Complementary filtering and EKF
  z = [gx(k), gy(k), gz(k), ax(k), ay(k), az(k)];
 [ phi_cf(k), theta_cf(k) ] = Attitude_cf(Ts(k), z', phi_cf(k - 1), theta_
      cf(k - 1), tao);
  [x_aposteriori(1 : 6, k), P_aposteriori(1 : 6, 1 : 6, k), phi_ekf(k),
       theta_ekf(k)] = Attitude_ekf(Ts(k), z', w, v, x_aposteriori(1 : 6,
       k - 1), P_aposteriori(1 : 6, 1 : 6, k - 1));
end
t = timestamp*0.000001;
rad2deg = 180/pi;
```

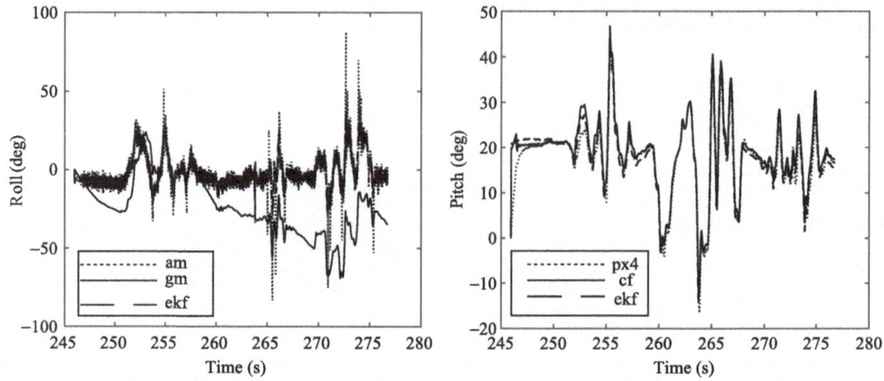

(a) Roll estimation comparison (b) Pitch estimation comparison

Fig. 8.8 Roll and pitch estimation comparison

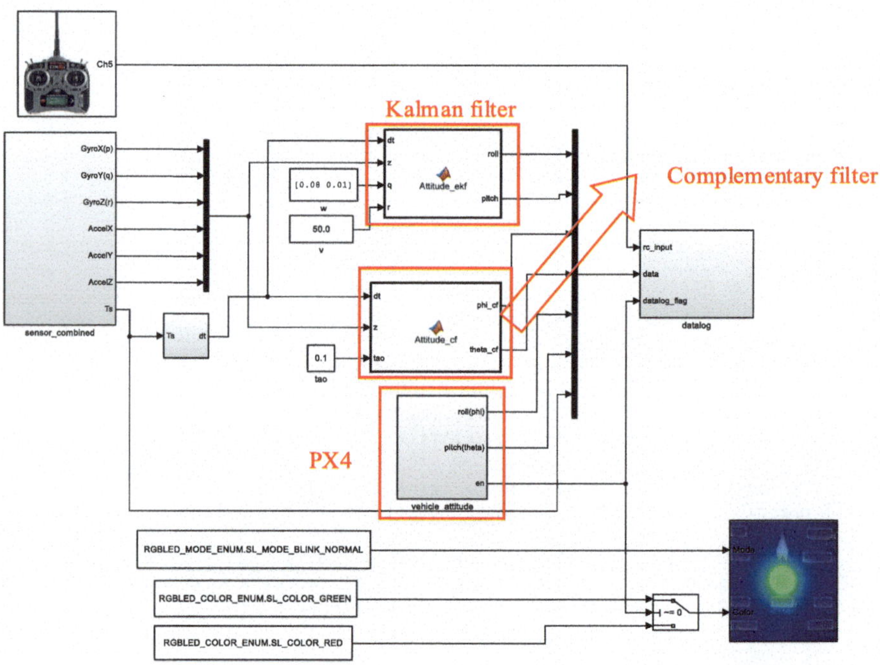

Fig. 8.9 Kalman filter and complementary filter comparison, Simulink model "ekf_cf.slx"

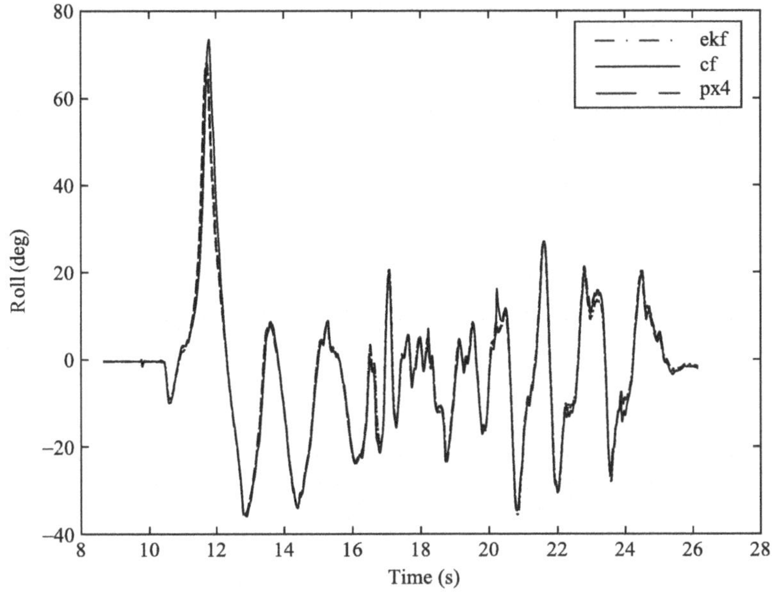

Fig. 8.10 Comparison among complementary filter (cf), Kalman filter (ekf) and filter in PX4 (px4)

8.5 Summary

(1) In order to obtain an accurate attitude angle, a complementary filter is designed to fuse the data from the gyroscope and accelerometer. This filter is equivalent to a high-pass filter for the gyroscope and a low-pass filter for the accelerometer, which effectively eliminates the measurement noise and improve accuracy.

(2) The contribution of the gyroscope and accelerometer in the complementary filter is controlled by parameter τ. Hence, the value of the parameter τ affects the complementary filter performance. When the parameter τ is high, the gyroscope plays a major role. In contrast, the accelerometer contributes more when the parameter τ is small.

(3) Design a Kalman filter including the process model and the measurement model. The experimental results indicate that the Kalman filter is better than the complementary filter and is similar to the self-contained filter in PX4 software of the Pixhawk autopilot.

If you have any question, please go to https://rflysim.com for your information.

Chapter 9
Attitude Controller Design Experiment

The objective of Chap. 8 is to estimate flight state accurately. In this chapter, it is assumed that the attitude of the multicopter has been estimated accurately. Feedback signals used in the controllers are the estimated values. However, for simplicity, true values are used instead of feedback according to the *separation principle*.[1] The purpose of this chapter is to enable the multicopter to maintain a specific attitude. The performance of the attitude controller directly determines whether or not the multicopter can fly smoothly; further it is also the basis of position control, which will be discussed in Chap. 10. Control performance is determined by the rise time, overshoot, settling time, steady-state error in the time domain, and the cut-off frequency and stability margin in the frequency domain. According to the criteria in both time and frequency domains, readers can design a controller based on the multicopter model established in Chap. 6. This chapter will describe in detail the design principles and method of the attitude controller of the multicopter in detail. It is hoped that readers can gradually master the subject via three step-by-step experiments, namely basic, analysis and design experiments. In the basic experiment, readers will repeat the simulation of the multicopter attitude controller. In the analysis experiment, readers will adjust the PID (Proportion Integration Differentiation) parameters of the attitude controller so that it exhibits better control performance in the time domain. Finally, in the design experiment, readers will design compensators based on well-known principles of automatic control so that multicopters can achieve better control in the frequency domain.

[1]This principle describes separating controller design into state estimation and specific feedback control. When applying this principle, system states are firstly estimated according to observation data; the estimated values are viewed as true values, and the controller is designed based on the given system.

© Publishing House of Electronics Industry 2020, corrected publication 2021
Q. Quan et al., *Multicopter Design and Control Practice*,
https://doi.org/10.1007/978-981-15-3138-5_9

9.1 Preliminary

In order to make this chapter self-contained, the preliminary is from Chap. 11 of *Introduction to Multicopter Design and Control* [8].

9.1.1 Attitude Control

The successive loop closure method is adopted in flight controller design for multicopters. The outer loop controller provides commands to the inner loop controller, i.e., outputs of the horizontal position channel controller provide reference values for the attitude control system. Here, $\mathbf{\Theta}_{\mathrm{hd}}$ or $\mathbf{R}_{\mathrm{d}}$ is viewed as the reference values. Therefore, the objective of attitude control is to accomplish $\lim_{t\to\infty}\left\|\mathbf{\Theta}_{\mathrm{h}}(t)-\mathbf{\Theta}_{\mathrm{hd}}(t)\right\|=0$ or $\lim_{t\to\infty}\left\|\mathbf{R}^{\mathrm{T}}\mathbf{R}_{\mathrm{d}}-\mathbf{I}_3\right\|=0$. It is required that the convergence rate of the attitude loop should be 4–10 times higher than that of the horizontal position channel dynamics. Then, it can be considered that the attitude control objective $\mathbf{\Theta}_{\mathrm{h}}(t)=\mathbf{\Theta}_{\mathrm{hd}}(t)$ or $\mathbf{R}(t)=\mathbf{R}_{\mathrm{d}}(t)$ has already been achieved from the control of the horizontal position channel. Thus, the remaining control objective is handed on to the attitude control. As long as the attitude control is achieved, the horizontal position control problem can be solved. Next, the attitude control design will be discussed.

9.1.1.1 Basic Concepts

Multicopter attitude control is the basis for position control. Currently, two common rigid-body attitude representation methods, namely Euler angles and rotation matrices, are used. The advantages and disadvantages of these two representations are summarized in Table 9.1.

Table 9.1 Comparison between two representation methods

Attitude representation	Advantages	Disadvantages
Euler angle	No redundancy parameters; Clear physical meaning	Singularity problem when the pitch angle is 90°; Plenty of transcendental function operation; Gimbal lock
Rotation matrix	No singularity problem; No transcendental function operation; Applicability for the continuous rotation; Global and unique; Easy to interpolate	Six redundant parameters

In this section, we design two attitude tracking controllers for the two attitude representations. First, for the Euler angle, a PID controller is designed, assuming all the angles are small. For the rotation matrix, an attitude controller based on the attitude error matrix is designed. In practice, depending on the requirements, it is necessary to select an appropriate attitude representation and the corresponding attitude controller according to specific requirements.

9.1.1.2 Euler-Angle-Based Attitude Control

The following controller design is based on the attitude model Eq. (6.72). The attitude control objective is as follows. Given desired attitude angle $\boldsymbol{\Theta}_{\mathrm{d}} = [\boldsymbol{\Theta}_{\mathrm{hd}}^{\mathrm{T}}\ \psi_{\mathrm{d}}]^{\mathrm{T}}$, design a controller $\boldsymbol{\tau} \in \mathbb{R}^3$ so that $\lim_{t\to\infty} \|\mathbf{e}_{\boldsymbol{\Theta}}(t)\| = 0$, where $\mathbf{e}_{\boldsymbol{\Theta}} \triangleq \boldsymbol{\Theta} - \boldsymbol{\Theta}_{\mathrm{d}}$. Here, $\boldsymbol{\Theta}_{\mathrm{hd}}$ is generated by the position controller and ψ_{d} is given by the mission planner. To achieve this objective, according to

$$\dot{\boldsymbol{\Theta}} = \boldsymbol{\omega} \tag{9.1}$$

the desired angle velocity $\boldsymbol{\omega}_{\mathrm{d}}$ is designed as

$$\boldsymbol{\omega}_{\mathrm{d}} = -\mathbf{K}_{\boldsymbol{\Theta}}\mathbf{e}_{\boldsymbol{\Theta}} \tag{9.2}$$

where $\mathbf{K}_{\Theta} \in \mathbb{R}^{3\times 3}$ is a diagonal positive definite matrix. Then, Eqs. (9.1) and (9.2) can be used to build the angle control loop. Under the assumption $\dot{\boldsymbol{\Theta}}_{\mathrm{d}} = \mathbf{0}_{3\times 1}$, if $\lim_{t\to\infty} \|\mathbf{e}_{\omega}(t)\| = 0$, then $\lim_{t\to\infty} \|\mathbf{e}_{\boldsymbol{\Theta}}(t)\| = 0$, where $\mathbf{e}_{\omega} \triangleq \boldsymbol{\omega} - \boldsymbol{\omega}_{\mathrm{d}}$. According to

$$\mathbf{J} \cdot {}^{\dot{\omega}} = \boldsymbol{\tau} \tag{9.3}$$

the remaining task is to design the desired moment $\boldsymbol{\tau}_{\mathrm{d}}$ to yield $\lim_{t\to\infty} \|\mathbf{e}_{\omega}(t)\| = 0$. The PID controller is then designed as

$$\boldsymbol{\tau}_{\mathrm{d}} = -\mathbf{K}_{\omega\mathrm{p}}\mathbf{e}_{\omega} - \mathbf{K}_{\omega\mathrm{i}} \int \mathbf{e}_{\omega} - \mathbf{K}_{\omega\mathrm{d}}\dot{\mathbf{e}}_{\omega} \tag{9.4}$$

where $\mathbf{K}_{\omega\mathrm{p}}$, $\mathbf{K}_{\omega\mathrm{i}}$, and $\mathbf{K}_{\omega\mathrm{d}} \in \mathbb{R}^{3\times 3}$. Equations (9.3) and (9.4) are used to build the angular velocity control loop. Under the assumption $\dot{\boldsymbol{\omega}}_{\mathrm{d}} = \mathbf{0}_{3\times 1}$, if $\lim_{t\to\infty} \|\boldsymbol{\tau}(t) - \boldsymbol{\tau}_{\mathrm{d}}(t)\| = 0$, then $\lim_{t\to\infty} \|\mathbf{e}_{\omega}(t)\| = 0$. To avoid noise caused by the differential, the differential term $\mathbf{K}_{\omega\mathrm{d}}\dot{\mathbf{e}}_{\omega}$ can be omitted. The Euler-angle-based attitude controller design is finished, which is combined with Eqs. (9.2) and (9.4). In practice, saturation also needs to be considered in attitude control. Similar to position control, the PID controller can be rewritten as

$$\begin{aligned} \mathbf{e}_{\omega} &= \mathrm{sat}_{\mathrm{gd}}\left(\boldsymbol{\omega} - \boldsymbol{\omega}_{\mathrm{d}}, a_{10}\right) \\ \boldsymbol{\tau}_{\mathrm{d}} &= \mathrm{sat}_{\mathrm{gd}}\left(-\mathbf{K}_{\omega\mathrm{p}}\mathbf{e}_{\omega} - \mathbf{K}_{\omega\mathrm{i}} \textstyle\int \mathbf{e}_{\omega} - \mathbf{K}_{\omega\mathrm{d}}\dot{\mathbf{e}}_{\omega}, a_{11}\right) \end{aligned} \tag{9.5}$$

where a_{10}, $a_{11} > 0$. The parameters a_{10} and a_{11}, related to saturation can be specified according to practical requirements. In particular, if $\tau_d \in [-\tau_{max}, \tau_{max}]$, then the saturated attitude controller (9.5) is defined as.

$$\boldsymbol{\tau}_{\mathrm{d}} = \mathrm{sat}_{\mathrm{gd}}\left(-\mathbf{K}_{\omega \mathrm{p}}\mathbf{e}_{\omega} - \mathbf{K}_{\omega \mathrm{i}}\int \mathbf{e}_{\omega} - \mathbf{K}_{\omega \mathrm{d}}\dot{\mathbf{e}}_{\omega}, \tau_{\max}\right). \quad (9.6)$$

9.1.1.3 Rotation Matrix-Based Attitude Control

Rotation matrix based attitude control is designed as follows. According to the rotation matrix $\mathbf{R}$ and desired rotation matrix $\mathbf{R}_{\mathrm{d}}$, the attitude error matrix is shown below

$$\tilde{\mathbf{R}} = \mathbf{R}^{\mathrm{T}}\mathbf{R}_{\mathrm{d}}.$$

From the above definition, if and only if $\mathbf{R} = \mathbf{R}_{\mathrm{d}}$, then $\tilde{\mathbf{R}} = \mathbf{I}_3$. Thus, the control objective of rotation matrix based attitude control is $\lim_{t\to\infty}\left\|\tilde{\mathbf{R}}(t) - \mathbf{I}_3\right\| = 0$. First, the attitude tracking error is defined as shown in

$$\mathbf{e}_{\mathbf{R}} \triangleq \frac{1}{2}\mathrm{vex}\left(\mathbf{R}_{\mathrm{d}}^{\mathrm{T}}\mathbf{R} - \mathbf{R}^{\mathrm{T}}\mathbf{R}_{\mathrm{d}}\right) \quad (9.7)$$

where $\mathrm{vex}([\mathbf{x}]_{\times}) \triangleq \mathbf{x}$. Furthermore, the error in angular velocity tracking error is defined as shown in

$$\mathbf{e}_{\omega} \triangleq \boldsymbol{\omega} - \mathbf{R}^{\mathrm{T}}\mathbf{R}_{\mathrm{d}}\boldsymbol{\omega}_{\mathrm{d}}. \quad (9.8)$$

Under the small-angle assumption, $\boldsymbol{\omega}_{\mathrm{d}} = \dot{\boldsymbol{\Theta}}_{\mathrm{d}}$ is the desired angular velocity, which can be ignored in general cases. Therefore, $\mathbf{e}_{\omega} = \boldsymbol{\omega}$. Based on the above definitions, the following PD controller is designed

$$\boldsymbol{\tau}_{\mathrm{d}} = -\mathbf{K}_{\mathbf{R}}\mathbf{e}_{\mathbf{R}} - \mathbf{K}_{\omega}\mathbf{e}_{\omega} \quad (9.9)$$

where $\mathbf{K}_{\mathbf{R}}$ and $\mathbf{K}_{\omega} \in \mathbb{R}^{3\times 3}$ are diagonal matrices. The PD controller can guarantee system stability only within a small neighborhood of the hover position. To obtain a wider range of stability, by bringing in the compensation error term, a nonlinear controller can be designed as follows

$$\boldsymbol{\tau}_{\mathrm{d}} = -\mathbf{K}_{\mathbf{R}}\mathbf{e}_{\mathbf{R}} - \mathbf{K}_{\omega}\mathbf{e}_{\omega} - \mathbf{J}\left([\boldsymbol{\omega}]_{\times}\mathbf{R}^{\mathrm{T}}\mathbf{R}_{\mathrm{d}}\boldsymbol{\omega}_{\mathrm{d}} - \mathbf{R}^{\mathrm{T}}\mathbf{R}_{\mathrm{d}}\dot{\boldsymbol{\omega}}_{\mathrm{d}}\right). \quad (9.10)$$

By this controller, the resulting attitude control system is guaranteed to be exponentially stable for almost all rotation actions. In practice, the last term in the controller (9.10) is often very small and may be ignored. However, for an aggressive maneuver flight, the last term is dominant and must be considered.

9.1.2 Implementation of Control Allocation in Autopilots

The control effectiveness models of a quadcopter with a plus-configuration and any multicopter are shown in Eqs. (6.53) and (6.56), respectively, in Chap. 6. For a quadcopter, because $\mathbf{M}_4 \in \mathbb{R}^{4\times4}$ is non-singular, the control allocation matrix can be directly obtained by matrix inversion,

$$\mathbf{P}_4 = \mathbf{M}_4^{-1}$$

where $\mathbf{P}_4 \in \mathbb{R}^{4\times4}$. The allocation is unique. Furthermore, if a multicopter has more than four propellers, the allocation may be non-unique. In open source autopilots, the control allocation matrix is often obtained by calculating the pseudo-inverse matrix, i.e.,

$$\mathbf{P}_{n_\mathrm{r}} = \mathbf{M}_{n_\mathrm{r}}^{\dagger} = \mathbf{M}_{n_\mathrm{r}}^{\mathrm{T}} \left(\mathbf{M}_{n_\mathrm{r}}\mathbf{M}_{n_\mathrm{r}}^{\mathrm{T}}\right)^{-1} \tag{9.11}$$

where $\mathbf{P}_{n_\mathrm{r}} \in \mathbb{R}^{n_\mathrm{r}\times4}$, $\mathbf{M}_{n_\mathrm{r}} \in \mathbb{R}^{4\times n_\mathrm{r}}$. After the desired thrust f_d and desired moment $\boldsymbol{\tau}_\mathrm{d}$ have been obtained, the desired propeller angular speeds can be obtained as follows:

$$\begin{bmatrix} \varpi_{\mathrm{d},1}^2 \\ \varpi_{\mathrm{d},2}^2 \\ \vdots \\ \varpi_{\mathrm{d},n_\mathrm{r}}^2 \end{bmatrix} = \mathbf{P}_{n_\mathrm{r}} \begin{bmatrix} f_\mathrm{d} \\ \boldsymbol{\tau}_\mathrm{d} \end{bmatrix}. \tag{9.12}$$

By a particular allocation, some propeller angular speeds may exceed the saturation value. Thus, a good allocation algorithm is often very important. Furthermore, some parameters in $\mathbf{M}_{n_\mathrm{r}}$, namely $c_\mathrm{T}, c_\mathrm{M}, d_i, i = 1, \ldots, n_\mathrm{r}$ (the concrete definitions can be referred to Sect. 6.1.3 in Chap. 6) are unknown. Under this condition, a question arises: how is control allocation conducted? To answer this question, the control effectiveness matrix of a standard multicopter is defined as a function matrix

$$\mathbf{M}_{n_\mathrm{r}}(c_\mathrm{T}, c_\mathrm{M}, d) = \begin{bmatrix} c_\mathrm{T} & c_\mathrm{T} & \cdots & c_\mathrm{T} \\ -dc_\mathrm{T}\sin\varphi_1 & -dc_\mathrm{T}\sin\varphi_2 & \cdots & -dc_\mathrm{T}\sin\varphi_{n_r} \\ dc_\mathrm{T}\cos\varphi_1 & dc_\mathrm{T}\cos\varphi_2 & \cdots & dc_\mathrm{T}\cos\varphi_{n_r} \\ c_\mathrm{M}\delta_1 & c_\mathrm{M}\delta_2 & \cdots & c_\mathrm{M}\delta_{n_r} \end{bmatrix}$$

which satisfies

$$\mathbf{M}_{n_\mathrm{r}}(c_\mathrm{T}, c_\mathrm{M}, d) = \mathbf{P}_\mathrm{a}\mathbf{M}_{n_\mathrm{r}}(1, 1, 1).$$

Here, $\mathbf{P}_\mathrm{a} = \mathrm{diag}(c_\mathrm{T},\ dc_\mathrm{T},\ dc_\mathrm{T},\ c_\mathrm{M})$. Thus, the following relation is satisfied

$$\mathbf{M}_{n_\mathrm{r}}^{\dagger}(c_\mathrm{T}, c_\mathrm{M}, d) = \mathbf{M}_{n_\mathrm{r}}^{\dagger}(1, 1, 1)\,\mathbf{P}_\mathrm{a}^{-1}.$$

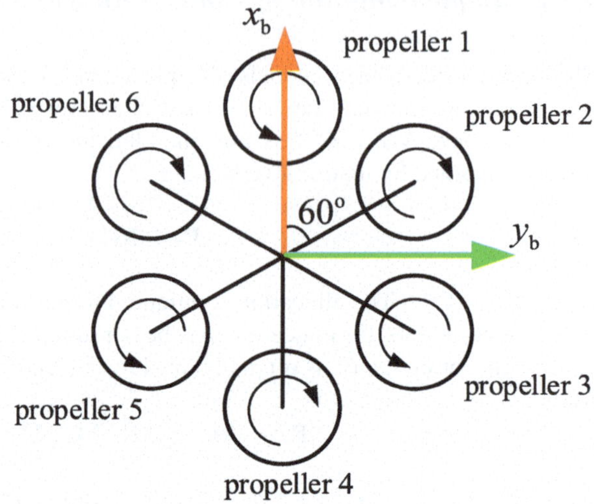

Fig. 9.1 Hexacopter with plus configuration

Taking the hexacopter in Fig. 9.1 as an example, $\mathbf{M}_6(c_T, c_M, d)$ is as follows

$$\mathbf{M}_6(c_T, c_M, d) = \begin{bmatrix} c_T & c_T & c_T & c_T & c_T & c_T \\ 0 & -\frac{\sqrt{3}dc_T}{2} & -\frac{\sqrt{3}dc_T}{2} & 0 & \frac{\sqrt{3}dc_T}{2} & \frac{\sqrt{3}dc_T}{2} \\ dc_T & \frac{dc_T}{2} & -\frac{dc_T}{2} & -dc_T & -\frac{dc_T}{2} & \frac{dc_T}{2} \\ c_M & -c_M & c_M & -c_M & c_M & -c_M \end{bmatrix}.$$

Then

$$\mathbf{M}_6^{\dagger}(1, 1, 1) = \frac{1}{6}\begin{bmatrix} 1 & 0 & 2 & 1 \\ 1 & -\sqrt{3} & 1 & -1 \\ 1 & -\sqrt{3} & -1 & 1 \\ 1 & 0 & -2 & -1 \\ 1 & \sqrt{3} & -1 & 1 \\ 1 & \sqrt{3} & 1 & -1 \end{bmatrix}.$$

Therefore

$$\begin{bmatrix} \varpi_{d,1}^2 \\ \varpi_{d,2}^2 \\ \vdots \\ \varpi_{d,6}^2 \end{bmatrix} = \mathbf{M}_6^{\dagger}(1, 1, 1)\,\mathbf{P}_a^{-1}\begin{bmatrix} f_d \\ \boldsymbol{\tau}_d \end{bmatrix} = \mathbf{M}_6^{\dagger}(1, 1, 1)\begin{bmatrix} f_d/c_T \\ \tau_{dx}/(dc_T) \\ \tau_{dy}/(dc_T) \\ \tau_{dz}/c_T \end{bmatrix}. \tag{9.13}$$

In most autopilots, controllers f_d and $\boldsymbol{\tau}_d$ are all PID controllers. Therefore, the unknown variables c_T, c_M, and d can be compensated for by adjusting the PID parameters.

9.2 Basic Experiment

9.2.1 Experimental Objective

(1) Things to prepare

1) Hardware: Multicopter System, Pixhawk Autopilot System;
2) Software: MATLAB R2017b or above, Simulink-based Controller Design and Simulation Platform, HIL(Hardware in the loop) Simulation Platform, Instructional Package "e5.1" (https://rflysim.com/course).

(2) Objectives

1) Repeat the Simulink simulation of a quadcopter to analyze the functions of the control allocation;
2) Record the step response of the attitude, obtain the open-loop Bode plot of the attitude control system and further analyze the stability margin of the closed-loop control system;
3) Perform the HIL simulation.

9.2.2 Experimental Procedure

(1) **Step1: SIL simulation—Control allocation**

1) Parameter Initialization
Run the file "e5\e5.1\Init_control.m" to initialize the parameters. Next, the Simulink file "AttitudeControl_Sim.slx" will open automatically as shown in Fig. 9.2. The model is similar to the one discussed in Sect. 4.3.1. The

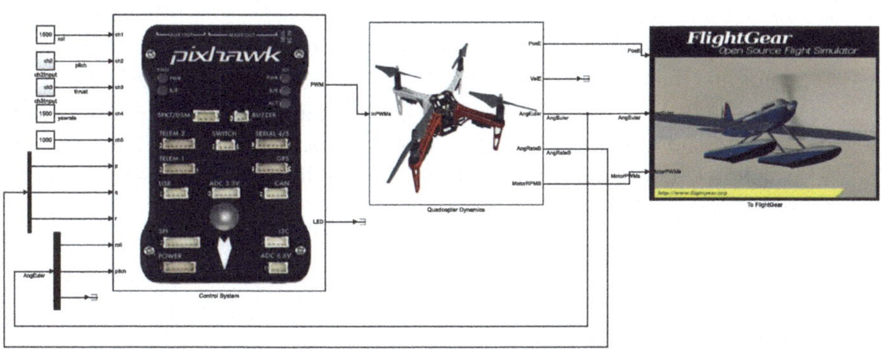

Fig. 9.2 SIL model for control allocation function testing, Simulink model "AttitudeControl_Sim.slx"

difference is that the quadcopter dynamic model used here is based on the model established in Sect. 6.4.3.1. Furthermore, the parameters of the controller used here exhibit better performance. The structure of the "Control System" is shown in Fig. 9.3. In addition to the basic "RC Signal Process", "Attitude Controller" and "Allocation" modules, two "Scope" modules are added to record the data in the experiment.

2) Run the simulation
Open the file "FlightGear-F450" and click on the Simulink "Run" button to run "AttitudeControl_Sim.slx". Subsequently, the motion of the quadcopter is observed in FlightGear, as shown in Fig. 9.4. The quadcopter in FlightGear climbs up for a short time, and then flies against the screen, corresponding to the $-o_b x_b$ axis according to the aircraft-body frame described in Sect. 6.1.1.2.

3) Simulation results
The inputs of "Control System", i.e., the values of "ch1–ch5"[2] shown in Fig. 9.2, are set differently. The PWM(Pulse Width Modulation) values of "ch1" (corresponding to the roll channel) and "ch4" (corresponding to the yaw channel) are equal to 1500. Recalling the experiment in Sect. 4.3, the control sticks are at the middle position. The settings of "ch2" (corresponding to the pitch channel) and "ch3" (corresponding to the thrust channel) are shown in Fig. 9.5. Both inputs pass "Delay" modules, where numbers represent the delay times. For example, if the value of "Delay" module for "ch3" is set to 500 and the sample time in the experiment is 0.001s, then the delay time is 0.5s. This implies that the output of "ch3" is 1600 in the first 0.5s, and then changes to be 1500. This is also the reason why the quadcopter performs at this time in FlightGear. The two Simulink "Scope" modules will appear automatically once the simulation is run(the action can be canceled through "Scope" module configuration; readers can click "Help" in MATLAB to get additional information). The output of "Scope" is shown in Fig. 9.6, where the dotted line represents the desired pitch angle and, the solid line represents the response.

4) Control allocation function
The control allocation module receives roll, pitch, yaw, and thrust control inputs and then assigns them to four motors, as shown in Fig. 9.3. When the pitch channel is given a value of 1600, the output of "Scope1" is shown in Fig. 9.7. The motor configuration of the quadcopter used here is the same as that in Chap. 6, as shown in Fig. 9.8. According to Eqs. (6.54) and (6.55), the following relationship is derived

[2]Corresponding to CH1–CH5 of the RC transmitter introduced in Sect. 2.3.1.1.

$$\mathbf{M}_4\left(c_\mathrm{T}, c_\mathrm{M}, d\right) = \begin{bmatrix} c_\mathrm{T} & c_\mathrm{T} & c_\mathrm{T} & c_\mathrm{T} \\ -\frac{\sqrt{2}}{2}dc_\mathrm{T} & \frac{\sqrt{2}}{2}dc_\mathrm{T} & \frac{\sqrt{2}}{2}dc_\mathrm{T} & -\frac{\sqrt{2}}{2}dc_\mathrm{T} \\ \frac{\sqrt{2}}{2}dc_\mathrm{T} & -\frac{\sqrt{2}}{2}dc_\mathrm{T} & \frac{\sqrt{2}}{2}dc_\mathrm{T} & -\frac{\sqrt{2}}{2}dc_\mathrm{T} \\ c_\mathrm{M} & c_\mathrm{M} & -c_\mathrm{M} & -c_\mathrm{M} \end{bmatrix}$$

i.e.,

$$\mathbf{M}_4\left(c_\mathrm{T}, c_\mathrm{M}, d\right) = \begin{bmatrix} c_\mathrm{T} & 0 & 0 & 0 \\ 0 & \frac{\sqrt{2}}{2}dc_\mathrm{T} & 0 & 0 \\ 0 & 0 & \frac{\sqrt{2}}{2}dc_\mathrm{T} & 0 \\ 0 & 0 & 0 & c_\mathrm{M} \end{bmatrix} \begin{bmatrix} 1 & 1 & 1 & 1 \\ -1 & 1 & 1 & -1 \\ 1 & -1 & 1 & -1 \\ 1 & 1 & -1 & -1 \end{bmatrix}.$$

Then, according to Eq. (9.13),

$$\begin{bmatrix} \varpi_{\mathrm{d},1}^2 \\ \varpi_{\mathrm{d},2}^2 \\ \varpi_{\mathrm{d},3}^2 \\ \varpi_{\mathrm{d},4}^2 \end{bmatrix} = \mathbf{M}_4^\dagger\left(1,1,1\right)\mathbf{P}_\mathrm{a}^{-1}\begin{bmatrix} f_\mathrm{d} \\ \boldsymbol{\tau}_\mathrm{d} \end{bmatrix} = \mathbf{M}_4^\dagger\left(1,1,1\right)\begin{bmatrix} f_\mathrm{d}/c_\mathrm{T} \\ \tau_{\mathrm{d}x}/\left(\frac{\sqrt{2}}{2}dc_\mathrm{T}\right) \\ \tau_{\mathrm{d}y}/\left(\frac{\sqrt{2}}{2}dc_\mathrm{T}\right) \\ \tau_{\mathrm{d}z}/c_\mathrm{T} \end{bmatrix}$$

Because d and c_T are compensated for by the PID controller,

$$\begin{aligned} \varpi_{\mathrm{d},1}^2 &\propto \left(f_\mathrm{d} - \tau_{\mathrm{d}x} + \tau_{\mathrm{d}y} + \tau_{\mathrm{d}z}\right) \\ \varpi_{\mathrm{d},2}^2 &\propto \left(f_\mathrm{d} + \tau_{\mathrm{d}x} - \tau_{\mathrm{d}y} + \tau_{\mathrm{d}z}\right) \\ \varpi_{\mathrm{d},3}^2 &\propto \left(f_\mathrm{d} + \tau_{\mathrm{d}x} + \tau_{\mathrm{d}y} - \tau_{\mathrm{d}z}\right) \\ \varpi_{\mathrm{d},4}^2 &\propto \left(f_\mathrm{d} - \tau_{\mathrm{d}x} - \tau_{\mathrm{d}y} - \tau_{\mathrm{d}z}\right) \end{aligned}.$$

In this case, if "ch2" >1500, i.e., $\tau_{\mathrm{d}x} = 0$, $\tau_{\mathrm{d}y} > 0$, $\tau_{\mathrm{d}z} = 0$, and $f_\mathrm{d} =$ constant, then changes in the motors provide a moment around the $o_\mathrm{b}y_\mathrm{b}$ axis.[3] Referring to Fig. 9.8, the control signals for motors #1 and #2 are increased, and the control signals for motors #3 and #4 are decreased. These observations are consistent with the expected control allocation results.

(2) **Step2: SIL simulation—Stability margin**[4]

1) Open the file "e5\e5.1\tune\AttitudeControl_tune.slx" as shown in Fig. 9.9.

---

[3] For details, see Sect. 6.1.1.2.

[4] "Linear Analysis Tool" in MATLAB is used here.

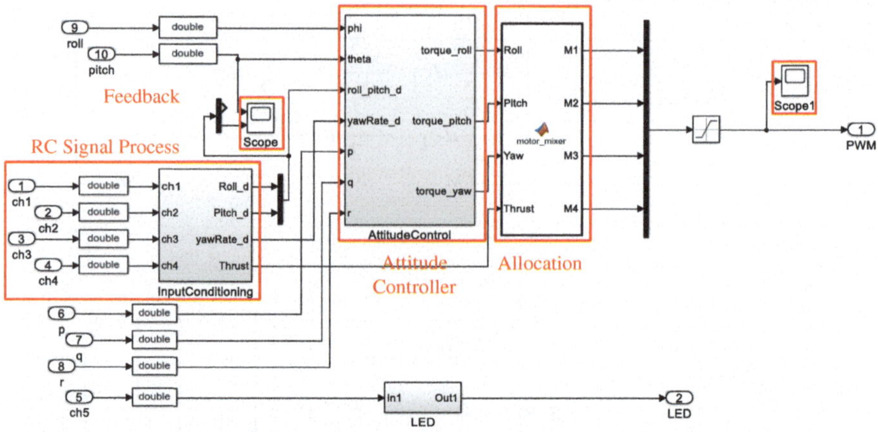

Fig. 9.3 Structure of "Control System" module, Simulink model "AttitudeControl_Sim.slx"

2) Specify signals as input and output for sweeping
The model of the attitude control system shown in Fig. 9.10 can be found in the submodule "Control System"—"Attitude Control". There are four channels in the attitude control system, namely roll, pitch, yaw, and thrust channels. Here, the thrust channel is set to 0.6085 (thrust percentage is 60.85%), which is the thrust value for just hovering.[5] Choose one of the other three channels such as the pitch channel for sweeping.
   - Specify a signal as the input using the following procedure. In the "Simulink Editor", right-click on the signal and then select "Linear Analysis Points"—"Open-loop Input" from the context menu.[6]
   - Specify a signal as the output. Using the procedure followed for setting: it is the same as the input signal, but select "Open-loop Output" as shown in Fig. 9.11.[7]

   The specified signals are shown in Fig. 9.12.
3) Get the Bode plot
Select "Analysis"—"Control Design"—"Linear Analysis", as shown in Fig. 9.13. From the context menu, select "Linear Analysis" and click "Bode" to get the Bode plot, as shown in Fig. 9.14. Right-click on the curve and select "Characteristics"—"All Stability Margins" to observe the blocked frequency, phase margin, and gain margin.

---

[5]The details can be found in Sect. 6.3.2.

[6]In this step, the signal selected is cut off by "Linear Analysis Tool" and replaced with a sweeping signal with a frequency of 0.1–10000 rad as a new input.

[7]In this step, the signal selected is blocked and becomes the new output. In fact, the closed-loop system becomes an open-loop system.

Fig. 9.4 Quadcopter in FlightGear

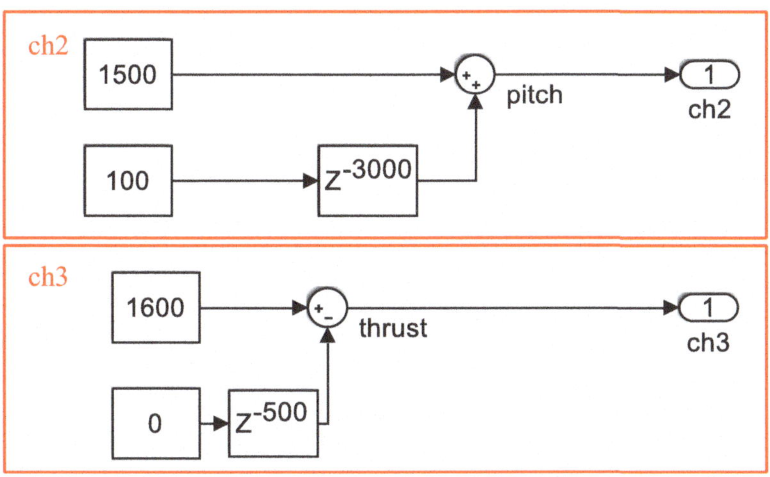

Fig. 9.5 Setting for "ch2" and "ch3" channels

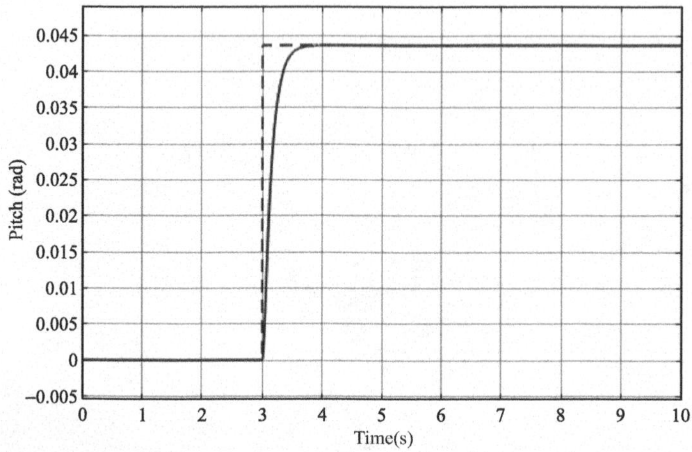

Fig. 9.6 Step response of pitch angle

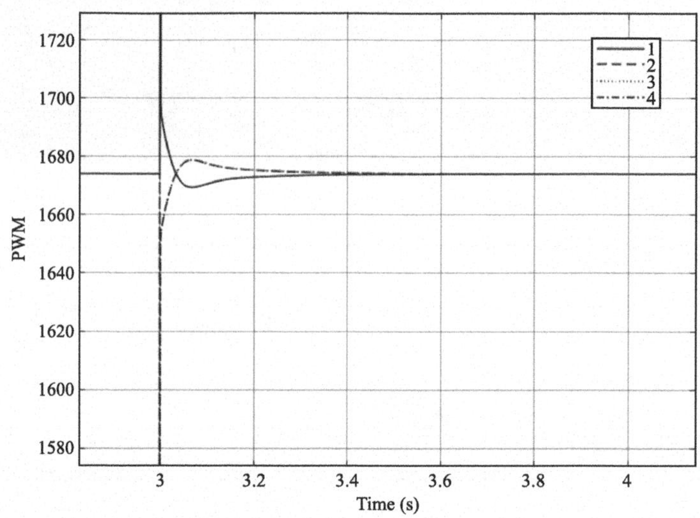

Fig. 9.7 PWM values after control allocation

(3) **Step3: SIL simulation—Step response**

1) Specify signals as input and output to obtain a step response curve of the pitch channel, as shown in Fig. 9.15. Specify a signal as the input using the following procedure. In "Simulink Editor", right-click the signal, then select "Linear Analysis Points"—"Open-loop Input" from the context menu. In the same way, specify a signal as the output expect for selecting "Output Measurement".

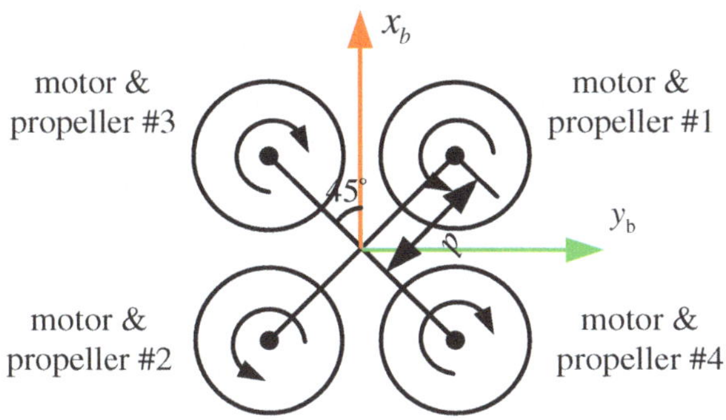

Fig. 9.8 Quadcopter configuration

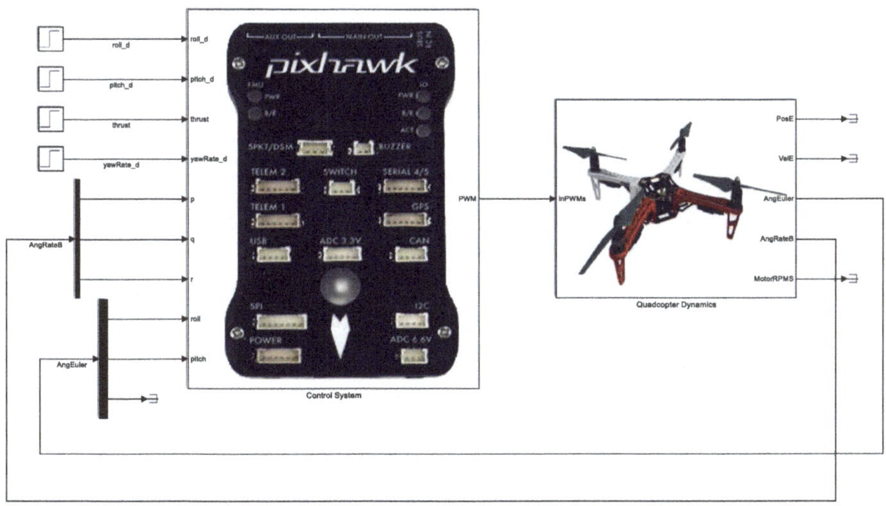

Fig. 9.9 SIL model for stability margin, Simulink model "AttitudeControl_tune.slx"

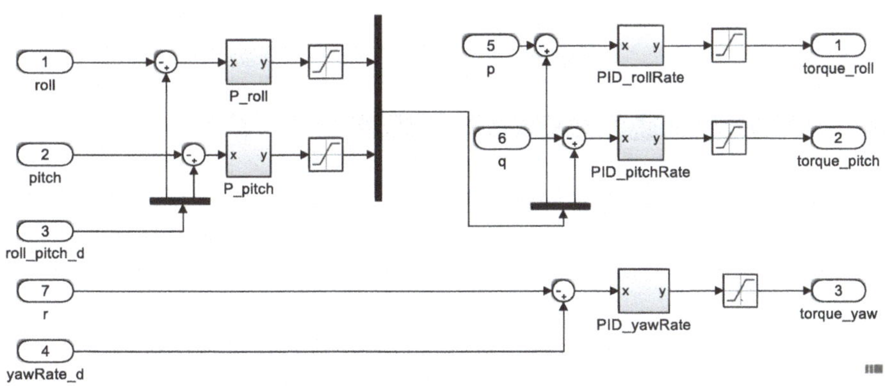

Fig. 9.10 Attitude controller, Simulink model "AttitudeControl_tune.slx"

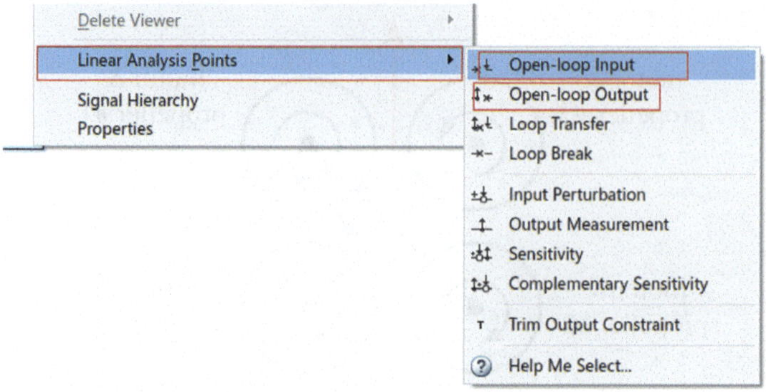

Fig. 9.11 Simulink menu for specifying signals as input and output

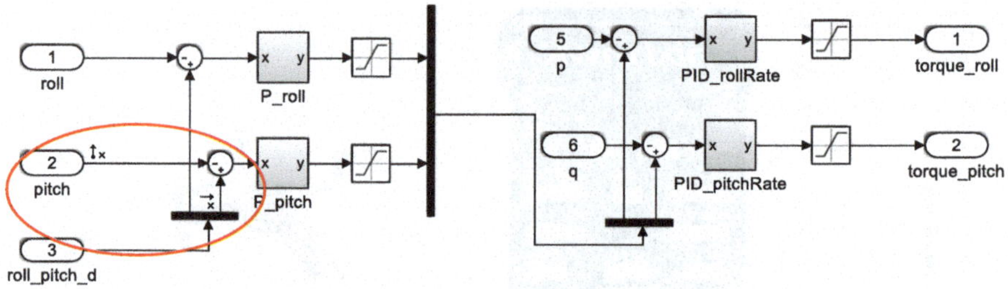

Fig. 9.12 Specified signals

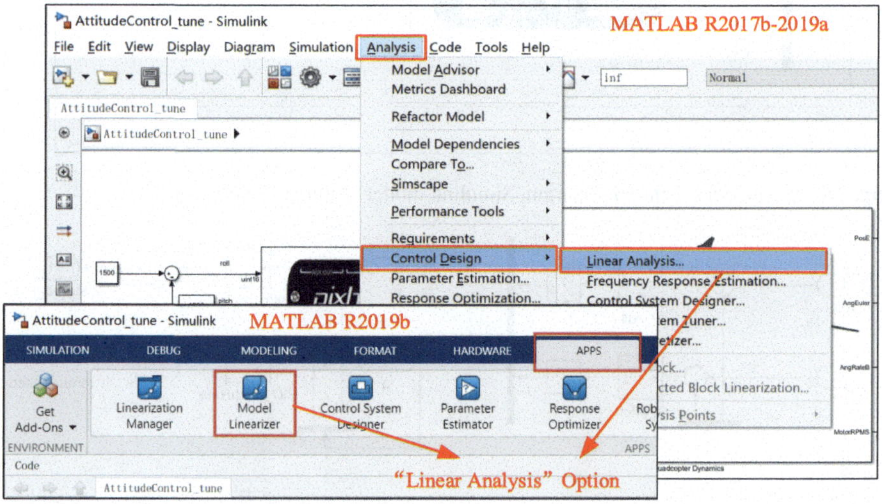

Fig. 9.13 Way to open “Linear Analysis”

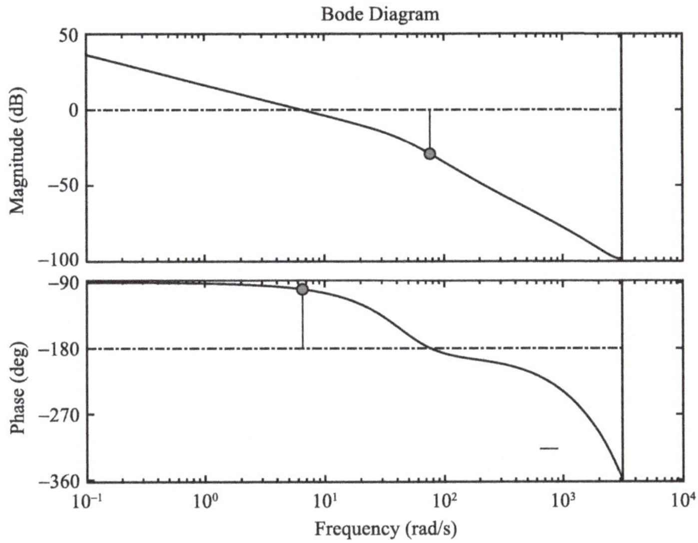

Fig. 9.14 Open-loop Bode plot of pitch angle loop

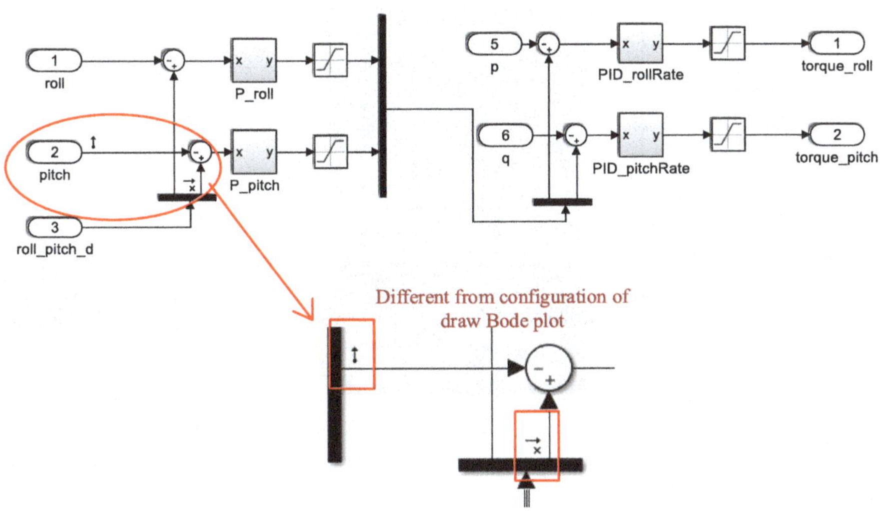

Fig. 9.15 Specified signals of SIL simulation

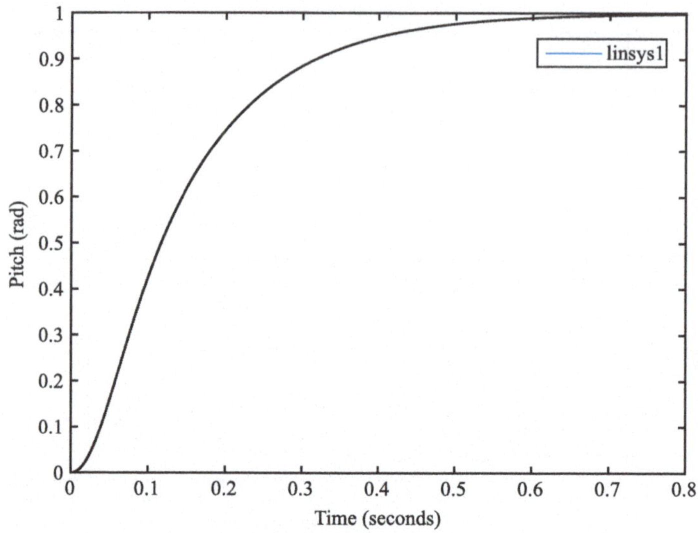

Fig. 9.16 Pitch angle step response

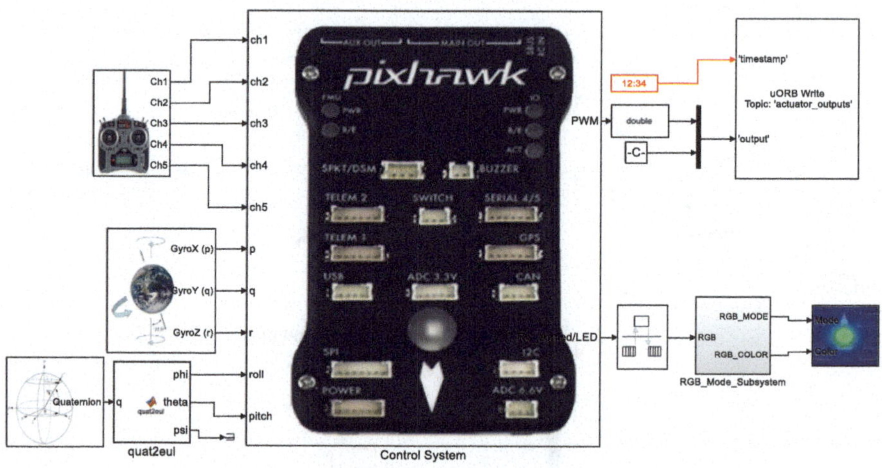

Fig. 9.17 HIL model to verify control allocation function, Simulink model "AttitudeControl-HIL.slx"

2) Obtain the pitch step response

Go to "Linear Analysis" in Simulink and click "Step" to get the step response, as shown in Fig. 9.16.

(4) **Step4: HIL simulation**

1) Open Simulink file for HIL

Open the Simulink file "e5\e5.1\HIL\AttitudeControl_HIL.slx", as shown in Fig. 9.17. It should be noted that "Control System" in the SIL simulation shown in Fig. 9.9 is the same as that in the HIL simulation in Fig. 9.17.

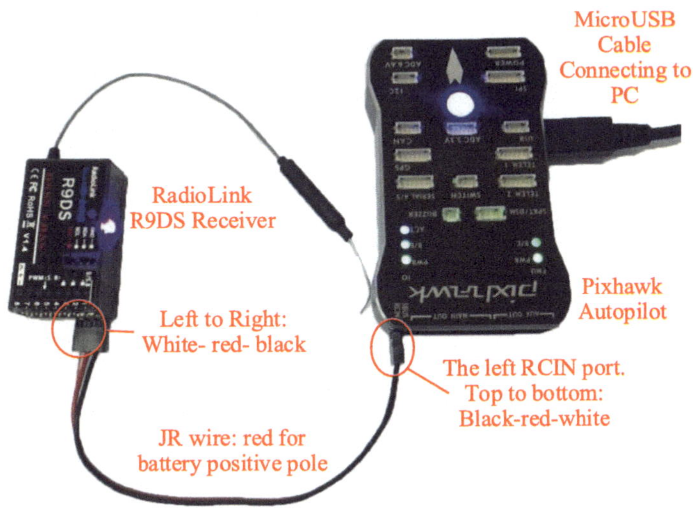

Fig. 9.18 Connection between Pixhawk hardware and RC receiver

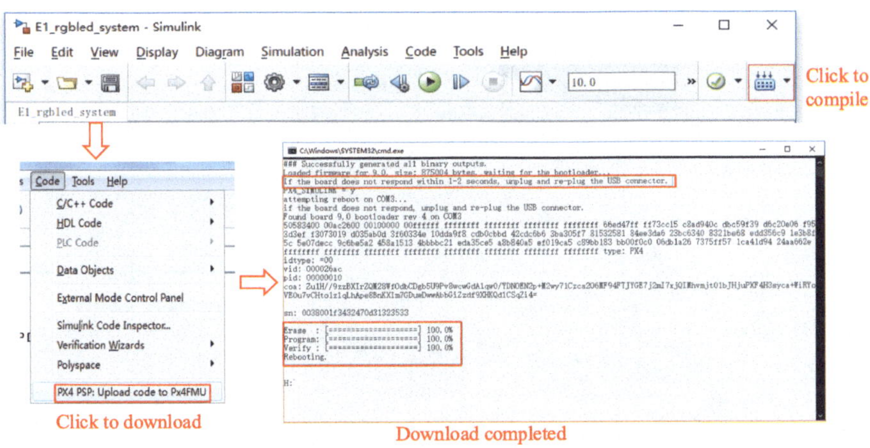

Fig. 9.19 Code compilation and upload process

2) Connect hardware
   Referring to the procedure in Sect. 2.3 in Chap. 2, the connection between an RC receiver and a Pixhawk autopilot can be set up as shown in Fig. 9.18. It should be noted that the airframe type "HIL Quadcopter X" should be selected in HIL simulations.
3) Compile and upload code
   Compile the HIL simulation model and upload the file to the given Pixhawk autopilot. Later, the designed attitude control program can be run on Pixhawk autopilot. The operation is shown in Fig. 9.19. For details, see Sect. 3.3.3 in Chap. 3.

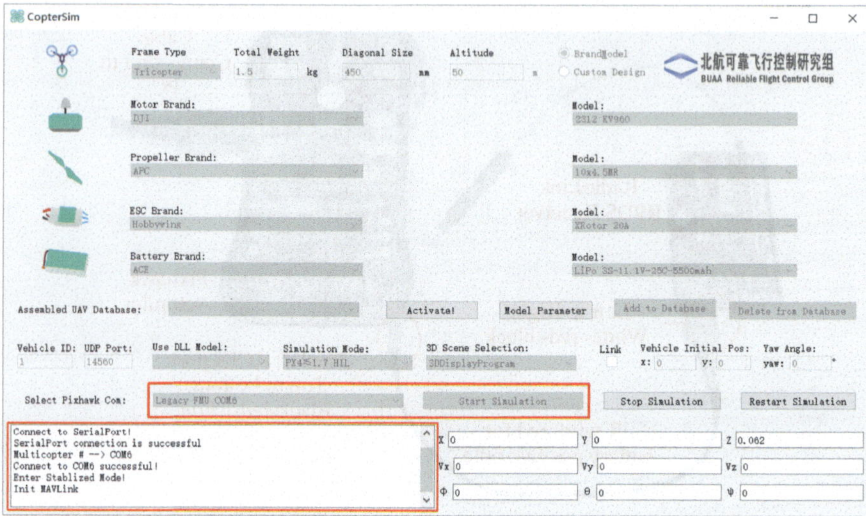

Fig. 9.20 User interface of CopterSim

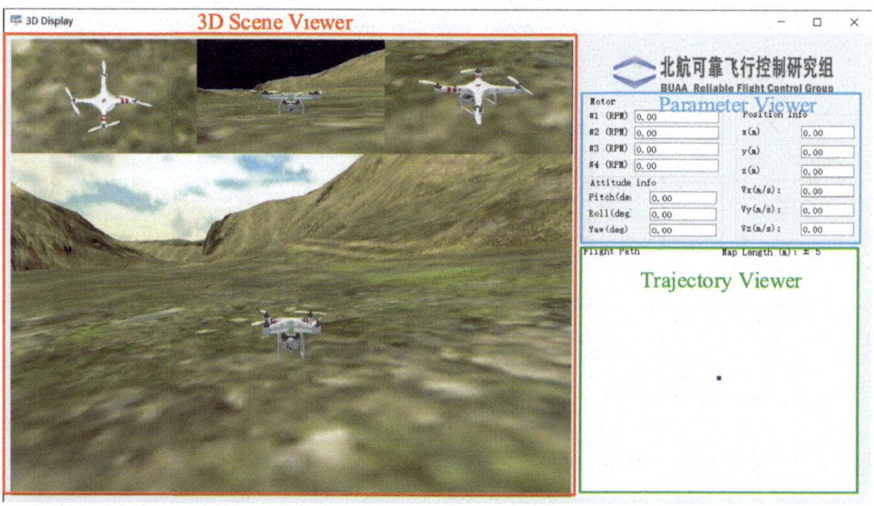

Fig. 9.21 User interface of 3DDisplay

4) Configure CopterSim
Double-click on the desktop shortcut CopterSim to open it. Readers can choose different propulsion systems using the following procedure. Click on "Model Parameters" to customize the model parameters and, then click on "Store and use the parameters" to make them available. The software will automatically match the serial port number. Readers

would click the "Run" button to enter the HIL simulation mode. After that, readers could see the message returned by the Pixhawk autopilot in the lower-left corner of the interface, as shown in Fig. 9.20.

5) Open 3DDisplay
Double-click on the desktop shortcut 3DDisplay to open it.
6) Simulation performance
Arm the quadcopter for manual control using the given RC transmitter.[8] As shown in Fig. 9.21, the quadcopter flight status is displayed on the left side of 3DDisplay's interface, the real-time flight data is plotted in the upper right corner, and the multicopter motion trajectory is plotted in the lower right corner.

9.3 Analysis Experiment

9.3.1 Experimental Objective

(1) Things to prepare
Software: MATLAB R2017b or above, Simulink-based Controller Design and Simulation Platform, Instructional Package "e5.2" (https://rflysim.com/course).
(2) Objectives

1) Adjust the PID controller parameters to improve the control performance, record the overshoot and settling time, and obtain a group of satisfied parameters;
2) Sweep the system to draw the Bode plot based on the satisfied parameters. Observe the system amplitude versus frequency response curve and the phase versus frequency response curve. Finally, analyze the stability margin.

9.3.2 Experimental Procedure

(1) **Step1: Initial model setup**
In this section, to adjust PID controller parameters, the Simulink-based controller design and simulation platform in the basic experiment should be simplified. The modified model is shown in the file "e5\e5.2\AttitudeControl_tune.slx". As the quadcopter is symmetrical, the PID parameters for the roll angle and pitch angle are the same. Hence, only one group of the roll or pitch angle parameters needs to be adjusted. The adjustment of PID parameters of the pitch angle is described below. First, let the quadcopter hover at an initial altitude of 100 m by setting the thrust value to 0.6085 (thrust percentage is 60.85%) and the initial speed

[8] For details, see Sect. 2.3.1.1 in Chap. 2.

of the motor to 557.1420 rad/s. Modify the corresponding parameters in the file "Init_control.m" as shown in Table 9.2.

(2) **Step2: Adjust the PID parameters for the angular velocity control loop**
In the file "AttitudeControl_tune.slx", replace the desired angular velocity with the step input and left-click on "q" signal line(corresponding to the angular velocity) and step input line, select "Enable Data Logging" to get the step response, as shown in Fig. 9.22. Modify the PID parameters of the angular velocity control loop in the file "Init_control.m". First, adjust the proportional term parameter and set the integral and derivative term parameters to 0. Later, run the file "Init_control.m" (this file must be run every time to update any altered parameters). Click on Simulink's "Run" button to view the input and output in the "Simulation Data Inspector". The proportional term parameter is gradually increased from a small value to a large value, which corresponds to increase "Kp_PITCH_AngleRate" variable in the file "Init_control.m". The step responses are shown in Fig. 9.23. Next, adjust the integral and derivative term parameters, i.e., "Ki_PITCH_AgngleRate", "Kd_PITCH_AgngleRate" variables in the file "Init_control.m". Finally, fine tune the proportional term parameter with the resulting response shown in Fig. 9.24.

(3) **Step3: Adjust the PID parameters for the angle control loop**
Adjust the proportional term for the angle control loop, corresponding to the variable "Kp_PITCH_ANGLE" in the file "Init_control.m". Based on the angular velocity control loop parameters obtained in Step 2, replace the desired pitch angle with a step input and set the "pitch" signal line and the step input as "Enable Data Logging" as shown in Fig. 9.25. Increase the proportional term parameter gradually and observe the step response in the "Simulation Data Inspector" shown in Fig. 9.26. The final selected parameters are:

```
Kp_PITCH_Angle=14
Kp_PITCH_AngleRate=0.10
Ki_PITCH_AngleRate=0.02
 Kd_PITCH_AngleRate=0.001
```

(4) **Step4: Sweep to get the Bode plot**
Specify signals as the input and the output for Bode plot. Specify the desired pitch angle input line as "Open-loop Input", and the actual output as "Open-loop Output". The Bode plot obtained using these values is shown in Fig. 9.27. Then, right-click on "Characteristics"—"Minimum Stability Margins", it can be observed that the gain margin is 22.6 dB, and the phase margin is 69.7°.

Table 9.2 Code in "Init_control.m"

```
ModelInitattitude=[0,0,-100];
ModelInitVelB=[0,0,0];
ModelInitAngEuler=[0,0,0];
ModelInitRateB=[0,0,0];
ModelInitRPM=557.1420;
```

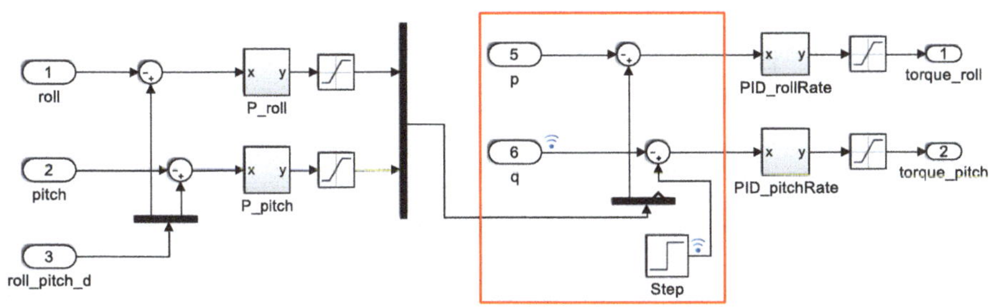

Fig. 9.22 Setting step response of angel velocity control loop

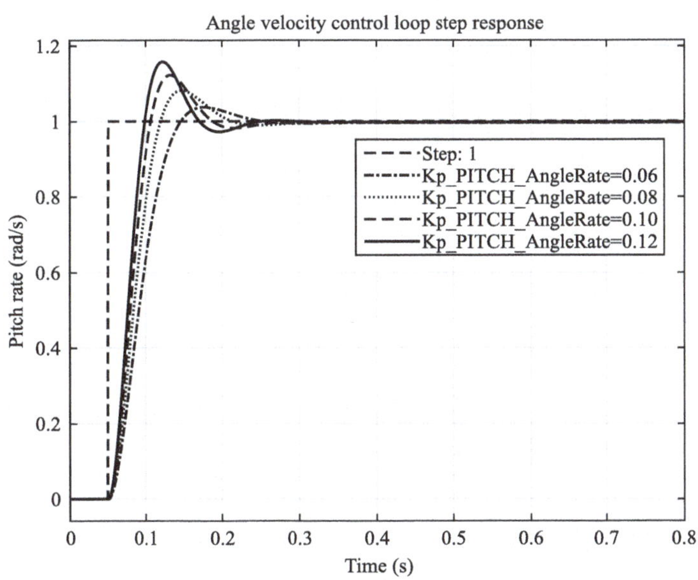

Fig. 9.23 Step response with different proportional term parameters

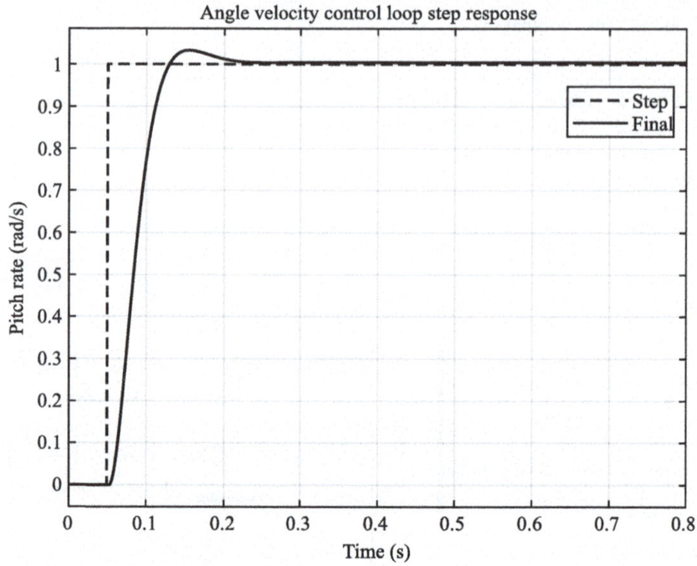

Fig. 9.24 Step response with a group of satisfied PID parameters

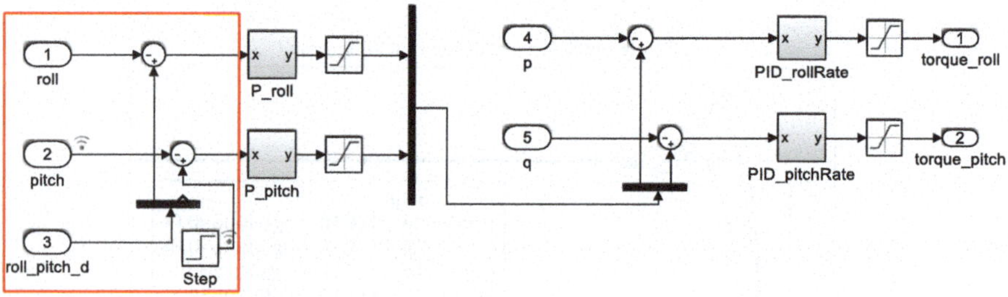

Fig. 9.25 Setting step response of angle control loop

9.4 Design Experiment

9.4.1 Experimental Objective

(1) Things to prepare

1) Hardware: Multicopter Hardware System, Pixhawk Autopilot System;
2) Software: MATLAB R2017b or above, Simulink-based Controller Design and Simulation Platform, HIL Simulation Platform, Instructional Package "e5.3" (https://rflysim.com/course).

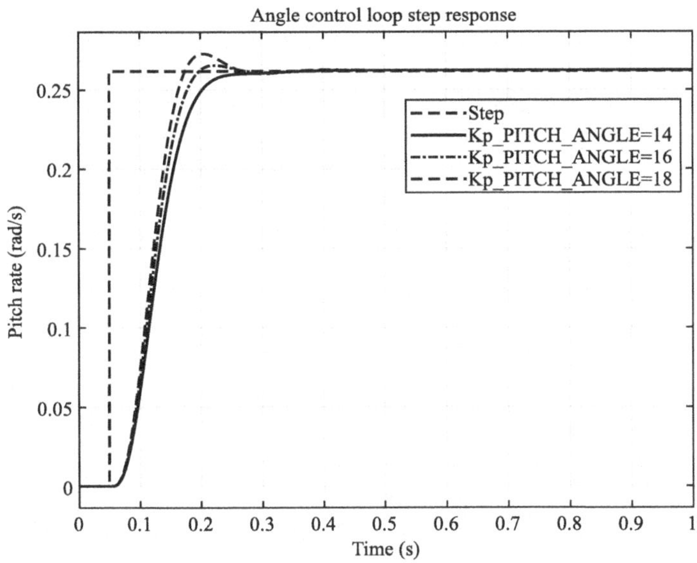

Fig. 9.26 Pitch angle step response with different parameters

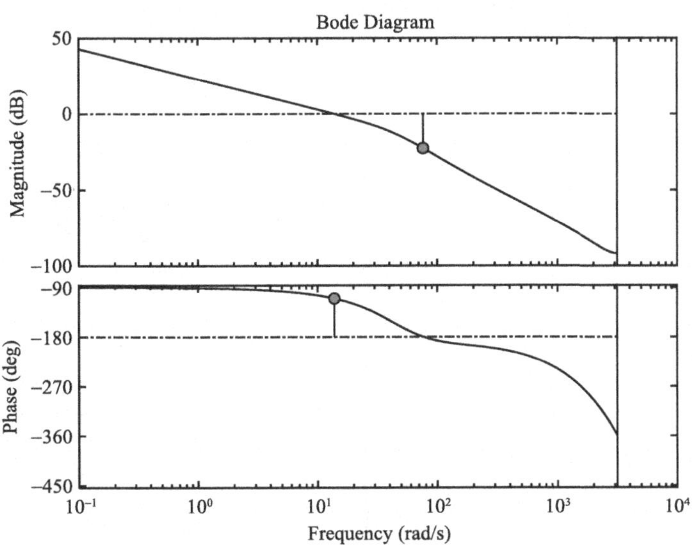

Fig. 9.27 Open-loop Bode plot of pitch angle control loop

(2) Objectives

1) Obtain the transfer function model of the attitude control loop, and then design a compensator for the attitude angular velocity control loop satisfying the following conditions: steady-state error $e_{rss} \leq 0.01$, phase margin >65° and cut-off frequency >10 rad/ s. The attitude angle control loop is satisfied when cut-off frequency >5 rad/s, phase margin >60°;
2) Complete the SIL simulation and HIL simulation experiments with the designed controller;
3) Use the designed controller to perform a flight test experiment.

9.4.2 *Experimental Design*

(1) **Step1: Simplify the overall structure**
Take the pitch channel for example. The simplified model is shown in Fig. 9.28.
(2) **Step2: Angle velocity-control loop analysis**
The input and output are the desired angular velocity and the angular velocity, respectively. Specify these signals as the input and output for the Bode plot, as shown in Fig. 9.29.

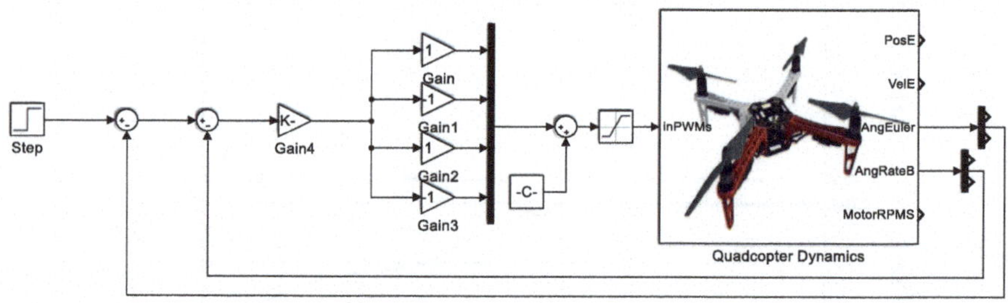

Fig. 9.28 Simplified model of pitch angle control loop, Simulink model "AttitudeControl_tune.slx"

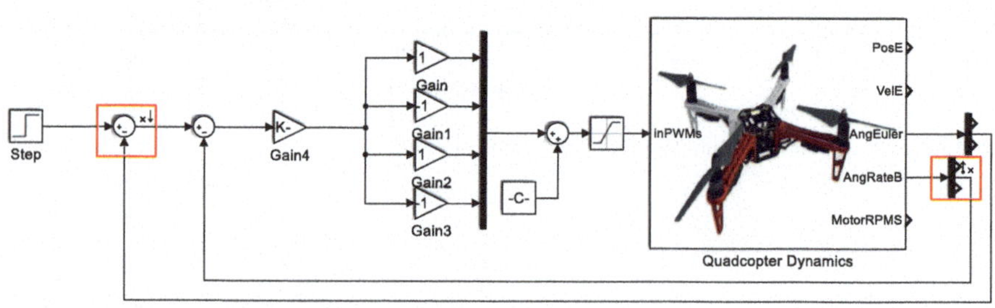

Fig. 9.29 Specifying signals as input and output

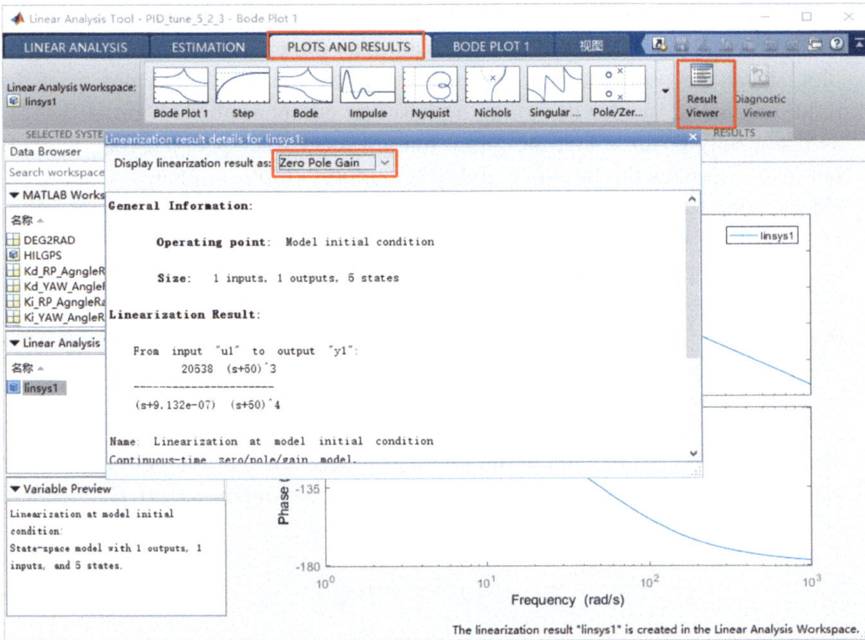

Fig. 9.30 Transfer function of angular velocity control loop

(3) **Step3: Obtain the transfer function model**
After the Bode plot is obtained, a variable, "linsys1" appears, in the "Linear Analysis Workspace". The transfer function model can be obtained using the operation shown in Fig. 9.30. The transfer function is

$$G(s) = \frac{20538}{(s + 9.132 \times 10^{-7})(s + 50)}$$

which is simplified as

$$G(s) = \frac{410.76}{s(0.02s + 1)}. \tag{9.14}$$

(4) **Step4: Adjust open-loop gain**
First, the open-loop gain is adjusted according to the steady-state error. When there is no compensator, the steady-state error of the system e_{rss} with an input $r(t) = t$ can be obtained by the final value theorem as

$$e_{rss} = \frac{1}{K}$$

where K is the open-loop gain. Because $e_{rss} \leq 0.01$, $K \geq 100$. As for the transfer function in Eq. (9.14), no adjustment is needed because its open-loop gain is 410.76.

(5) **Step5: Design a compensator for the angular velocity control loop**
First, draw a Bode plot of the initial system, as shown in Fig. 9.31. In the Bode plot, it can be observed that the phase margin $\gamma = 19.8°$ does not meet the requirements. Consider designing a phase-lag compensator [15, pp. 491–521] as it not only ensures that the phase margin meets the specified requirements but also improves the system's ability to suppress high-frequency noise. Consider setting the cut-off frequency at ω_c'=12.9rad/s, where the phase margin is 76° satisfying the requirements and the gain margin is 29.8 dB. At a frequency ω_c', the corresponding amplitude should be 0 dB after compensation. Based on the typical lag-lead compensation Bode plot, we can obtain the following relationship.

$$20\lg b + 20\lg \left|G(j\omega_c')\right| = 0.$$

Thus, $b = 10^{-1.49}$. To reduce the phase lag effect of the compensator on the curve at frequency ω_c', the cut-off frequency of the compensator $(bT)^{-1}$ should be located ten decades away from ω_c', i.e.,

$$(bT)^{-1} = 0.1\omega_c'.$$

Hence, T = 23.9558. The transfer function of the compensator is given as follows

$$G_c(s) = \frac{1 + bTs}{1 + Ts} = \frac{1 + 0.775194s}{1 + 23.955779s}.$$

(6) **Step6: Add the compensator to the angular velocity control loop**
The Bode plot obtained after adding the compensator, is shown in Fig. 9.32 where the phase margin is 70°.

(7) **Step7: Design a compensator for the angle control loop**

The angle control loop is also designed as a feedback closed loop, based on which, a compensator is designed. Specify signals as the input and the output, as shown in Fig. 9.33. The resultant open-loop Bode plot is obtained, as shown in Fig. 9.34.

It can be observed that the cut-off frequency is 1.04 rad/s at a phase margin of 88.1°. According to the design requirements of the attitude angle control loop, consider to increase the open-loop gain. According to Fig. 9.34, we hope to locate the new cut-off frequency at $\omega = 5.3$ rad/s, where the corresponding amplitude is -14 dB. This implies that after adding the compensator, the amplitude of Bode plot at the frequency $\omega = 5.3$ rad/s needs to be 0 dB. Then,

$$20\lg (K_2) = 14.$$

Thus, $K_2 = 5.0119$. The Bode plot obtained after adding the compensator is shown in Fig. 9.35, where the cut-off frequency is 5.3 rad/s and the amplitude margin is 67.1°.

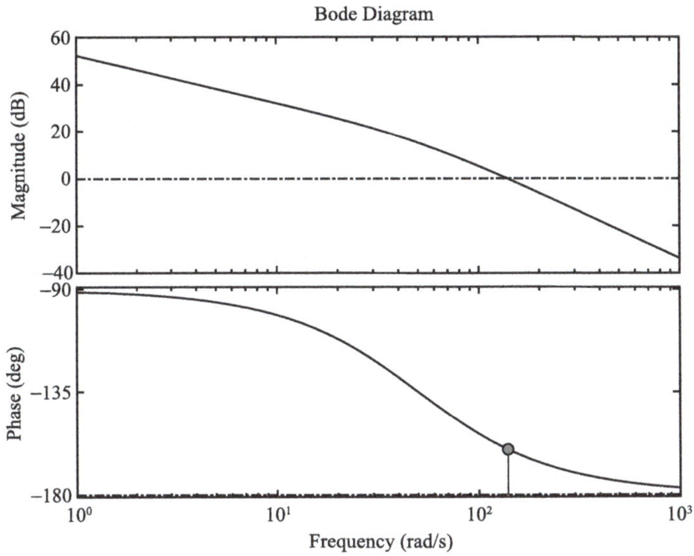

Fig. 9.31 Open-loop Bode plot of angle velocity control loop without compensation

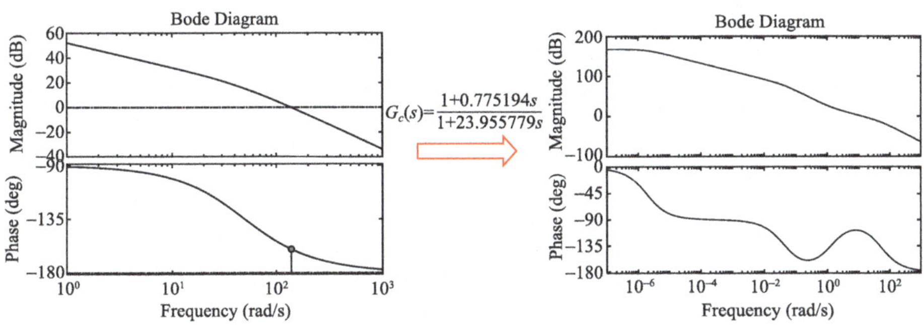

Fig. 9.32 Open-loop Bode plot of angle velocity control loop before and after compensation

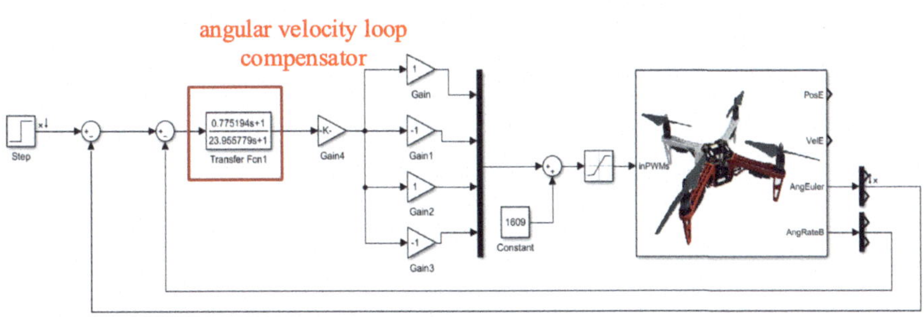

Fig. 9.33 Uncompensated model of the angle control loop, Simulink model "AttitudeControl_tune.slx"

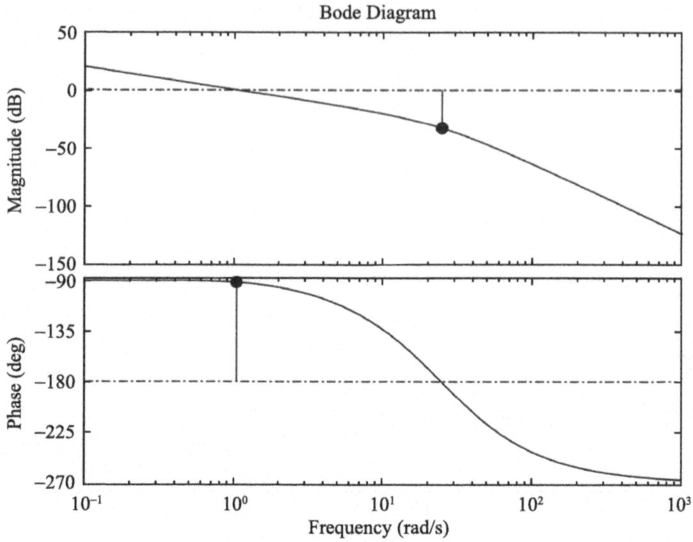

Fig. 9.34 Open-loop Bode plot of angle control loop without compensation

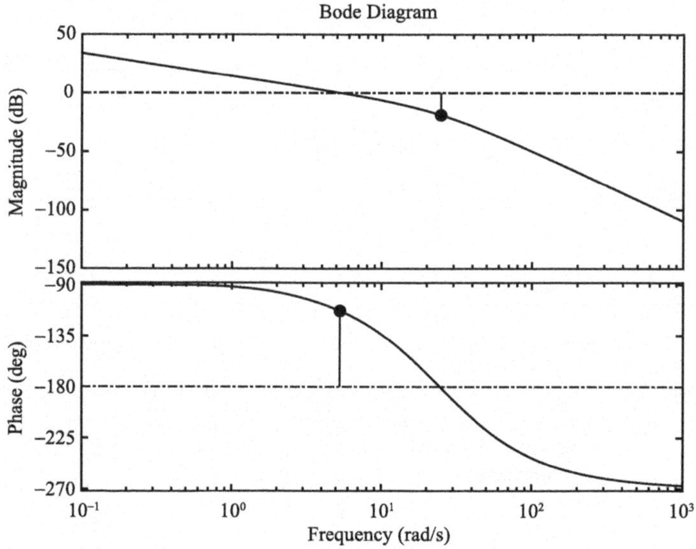

Fig. 9.35 Open-loop Bode plot of angle control loop after compensation

9.4.3 Simulation Procedure

(1) **Step1: Discretize the continuous-time compensator**
The designed compensator is an s transfer function, which has to be discretized so that it can be run on the Pixhawk autopilot, a digital computer. The "c2d" function in MATLAB is used as:

```
H = tf([num], [den])
Hd = c2d(H, Ts, 'zoh')
```

Here, "num" is the transfer function numerator coefficient vector, "den" is the transfer function denominator coefficient vector, and "Ts" is the sample time, "Ts = 0.004s". This MATLAB function discretizes the s transfer function using a zero-order hold on the input with a sample time "Ts". As the sample time of the Pixhawk autopilot for attitude control is 0.004s, the compensator is discretized using the same sample time 0.004s as well. The s transfer function is converted into a z transfer function as follows

$$G_c(s) = \frac{1 + 0.775194s}{1 + 23.955779s} \rightarrow G_c(z) = \frac{0.03236z - 0.03219}{z - 0.9998}.$$

(2) **Step2: Replace the control model**
Replace the continuous-time model with a discrete-time model in the Simulink-based controller design and simulation platform, as shown in Fig. 9.36.

(3) **Step3: HIL simulation**
It is observed that by releasing the control stick, the quadcopter can quickly return to its previous stable state without significant oscillations.

9.4.4 Flight Test Procedure

(1) **Step1: Quadcopter preparation**
For details, see Sect. 4.3.4 in Chap. 4.

(2) **Step2: Simulink model for flight test**
The model modified according to Step 1 is shown in Fig. 9.37. Unlike the HIL simulator in Fig. 9.17, the block in Fig. 9.37 is used to replace the PWM output shown in Fig. 9.17. Additionally, as shown in Fig. 9.36, the attitude controller is changed. A new data recording module is added to the model, as shown in Fig. 9.38. A "invalid.msg.specified" warning block appears automatically when the Simulink model is opened. The procedure to add the customized logging data module is as follows.

1) Customize the message file. Establish the file "costom_attctrl_e5.msg" as shown in Table. 9.3 (Note that a blank line must be left at the end of the .msg file, otherwise compilation errors will occur). Six 32-bit floating-point variables are defined, namely the roll angle, pitch angle, yaw rate, desired

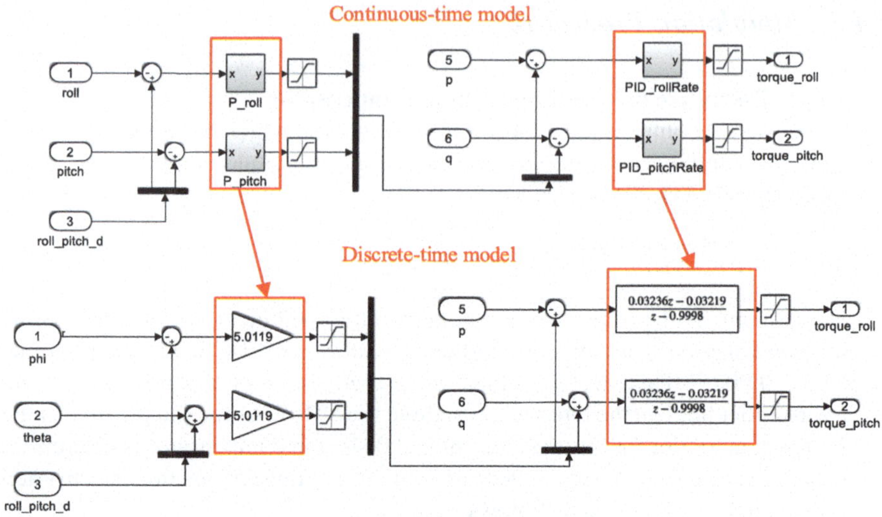

Fig. 9.36 PID controller discretized for the HIL simulation model

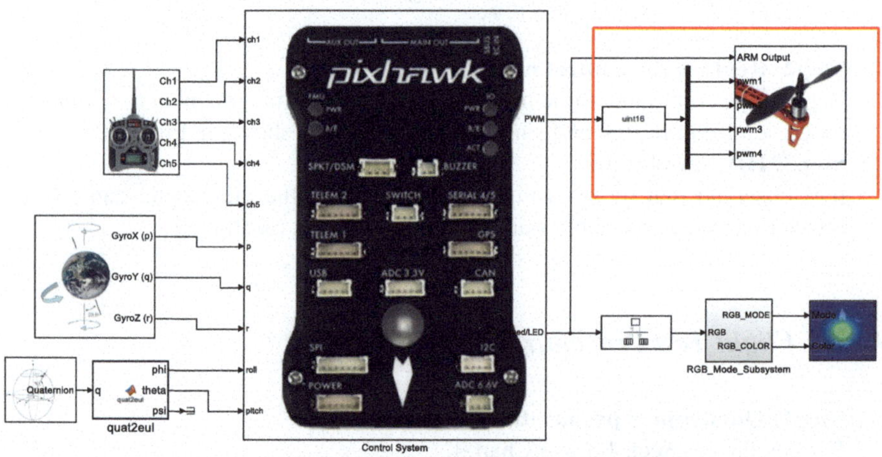

Fig. 9.37 Model for flight test, Simulink model "AttitudeControl_FLY.slx"

roll angle, desired pitch angle and desired yaw rate. Save the file to the folder "C:\PX4PSP\Firmware\ msg"("C:\PX4PSP" is the default software package installing directory).

2) Open the file "C:\PX4PSP\Firmware\msg\CmakeLists.txt" and add "costom_attctrl_e5.msg" to the bottom line of "set()".
3) Open the file "C:\PX4PSP\Firmware\src\modules\logger\logger.cpp" and add "add_topic("costom_attctrl_e5",4)" to the bottom line of "add_default_topic()", where "costom_attctrl_e5" represents the name of the message, and "4" means written cycle is 4 ms.

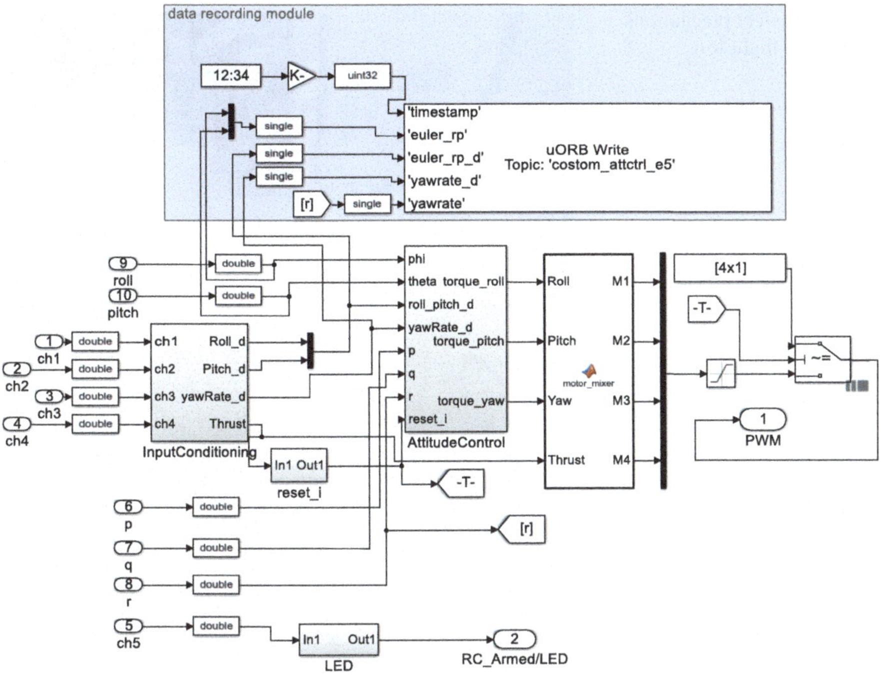

Fig. 9.38 Control system with a new data recording module

The modified file can be found in file “e5\e5.1\PSPfile\”. Subsequently, the warning block disappears.

(3) **Step3: Upload code**
This process is similar to that used for compiling and uploading the code in HIL simulations. The experimental flow is shown in Fig. 9.19. The detailed procedure is included in Sect. 3.3.3.

(4) **Step4: No propellers test**
The quadcopter configuration is shown in Fig. 9.8.

1) Thrust channel
   - Remove the propellers and, then connect the battery supply
   - Turn on the RC and switch the RC CH5 stick to arm the quadcopter
   - Pull up the left-hand stick on the RC transmitter (CH3) and observe if the motors rotate correctly.
2) Roll channel
   - Move the right-hand stick on the RC transmitter (CH1) to the left
   - Observe whether the speeds of the motors #1 and #4 increase or the speeds of motors #2 and #3 decrease.
3) Pitch channel
   - Pull up the right-hand stick on the RC transmitter (CH2)
   - Observe whether the speeds of motors #2 and #4 increase or the speeds of motors #1 and #3 decrease.

Fig. 9.39 Safety precautions for Outdoor flight test

Fig. 9.40 Pitch step response of a quadcopter on a test stand

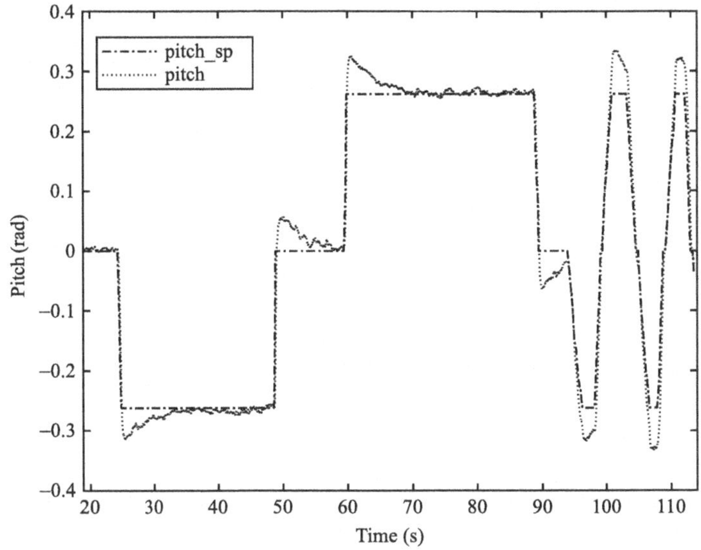

Fig. 9.41 Pitch step response of outdoor flight test

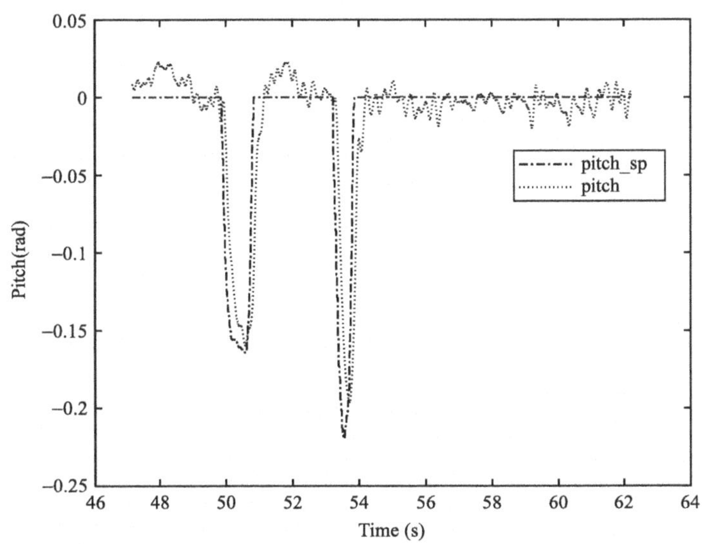

Fig. 9.42 A quadcopter on a test stand

4) Yaw channel
   - Move the left-hand stick on the RC transmitter (CH4) to the right
   - Observe whether the speeds of motors #1 and #2 increase or if the speeds of motors #3 and #4 decrease.
5) Automatic control test[9]
   - Keep the CH3 channel control stick in the middle and hold the quadcopter horizontally. In the next step, tilt the quadcopter to the left (rotating around the $o_b x_b$ axis). Observe whether the speeds of motors #2 and #3 increase or the speeds of motor #1 and #4 decrease.
   - Keep the quadcopter horizontal. Tilt it forward (rotating around the $o_b y_b$ axis). Observe whether the speeds of motors #1 and #3 increase or the speeds of motors #2 and #4 decrease.
   - Keep the quadcopter horizontal. Yaw the quadcopter to the right (rotating around the $o_b z_b$ axis) and observe if the speeds of motors #1 and #2 decrease or the speeds of motors #3 and #4 increase.

   If the quadcopter performs well in all the steps, then it is considered that the flight controller can accurately control the attitude of quadcopter.[10]

(5) **Step5: Indoor stand test**
Install the quadcopter on a test stand and install the propeller as shown in Fig. 9.42. Arm the quadcopter and test the flight.

(6) **Step6: Outdoor flight test**
To ensure safety, a rope is tethered to the quadcopter, and the other end is tethered to a heavy object. The remote pilot maintains a safe distance from the quadcopter during flight, as shown in Fig. 9.39. Notice that it may be difficult to control the altitude as there is no position feedback. Do not rotate the thrust control stick too fast. Furthermore, it is better to let the stick be near the middle position, and rotate the stick smoothly.

(7) **Step7. Analyze the data**
Take out the SD card in Pixhawd, read out the logger file "log001.ulg", and then copy the file to "e5\e5.4\". And use function "ulog2csv" to convert "ULG" file to "CSV" file for analysis. For example, run the function "ulog2csv('log001.ulg','log001')" can extract the contents of the file "log001.ulg" to the folder "log001". And the customized message data is in the file "log001_costom_attctrl_e5_0.csv". The stand test data is shown in Fig. 9.40. It can be observed that the pitch angle overshoots over the desired command, but can converge within a few seconds. The tracking performance is satisfactory. The outdoor flight data is shown in Fig. 9.41. It can be observed that the pitch angle of the quadcopter can track the given desired command.

---

[9]For details, see Chap. 11. Keep the control stick in the middle position, i.e., the desired attitude angle is 0. Tilt the quadcopter, feel whether the quadcopter has a moment to make its attitude turn zero.

[10]For details, see Sect. 3.5.3 in Chap. 3.

Table 9.3 Code in "Costom_attctrl_e5.msg"

```
#attitude data
float32[2] euler_rp
float32 yawrate
#desired attitude data
float32[2] euler_rp_d
float32 yawrate_d
```

9.5 Summary

(1) Based on the attitude model of the quadcopter, a widely-used PID control method is used, and the design of the attitude controller is completed in Simulink and MATLAB. The simulation performance is displayed in FlightGear.
(2) Use the PSP tool of Simulink to generate the embedded code and upload it to the Pixhawk autopilot for HIL simulations.
(3) Adjust the parameters of the PID controller, try to obtain the satisfied parameters, and use the "System Analysis Tool" in MATLAB to obtain the open-loop Bode plot of the entire system to observe the phase margin and gain margin of the corresponding closed-loop system.
(4) Lead-and-lag compensators are designed for both the angle control loop and angular velocity control loop. Furthermore, the design is verified by HIL simulations and flight test.

If you have any question, please go to https://rflysim.com for your information.

Chapter 10
Set-Point Controller Design Experiment

The position control described in this chapter is concerned with set-point control, which is the basic issue in the trajectory tracking and path following problems. The experimental flow of set-point position controller design is similar to that of the attitude controller design. This chapter introduces the design principles and methods of constructing a position controller for multicopters. It is hoped that readers can gradually master this part through three step-by-step experiments, namely basic, analysis and design experiments. In the basic experiment, readers will repeat the simulation of the position control of a multicopter to understand how a set-point position controller works. In the analysis experiment, readers will adjust the PID parameters of the set-point position controller so that the multicopter exhibits good control performance in the time domain. In the design experiment, readers will design compensators based on well-known principles of automatic control, allowing multicopters to achieve good control performance in the frequency domain.

10.1 Preliminary[1]

In order to make this chapter self-contained, the preliminary is borrowed from Chap. 11 of *Introduction to Multicopter Design and Control* [8].

10.1.1 Basic Concept

To make this section self-contained, several basic concepts are initially introduced. As shown in Fig. 10.1, depending on the given position $\mathbf{p}_{\mathrm{d}}$, position control is divided into three types, set-point control, trajectory tracking and path following.

[1]The model used here is established in Chap 6.

© Publishing House of Electronics Industry 2020, corrected publication 2021

Q. Quan et al., *Multicopter Design and Control Practice*,
https://doi.org/10.1007/978-981-15-3138-5_10

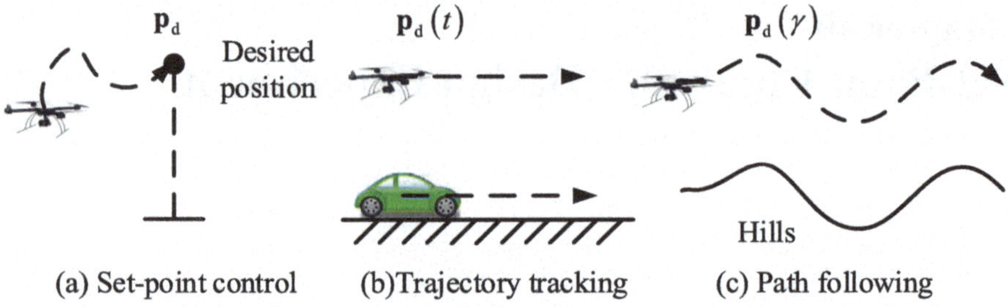

Fig. 10.1 Three types of position control

(1) Set-point control. The desired position $\mathbf{p}_d \in \mathbb{R}^3$ is a constant point. The aim of this part is to design a controller to make $\|\mathbf{p}(t) - \mathbf{p}_d\| \to 0$ or $\mathbf{p}(t) - \mathbf{p}_d \to \mathcal{B}(\mathbf{0}_{3\times1}, \delta)$[2] as $t \to \infty$, where $\delta \geq 0$. Here, it is assumed that the multicopter can fly to the given position, regardless of the trajectory. In practice, most applications fall into this category, as shown in Fig. 10.1a.
(2) Trajectory tracking. The desired trajectory $\mathbf{p}_d : [0, \infty) \to \mathbb{R}^3$ is a time-dependent trajectory. The aim of this step is to design a controller to yield $\|\mathbf{p}(t) - \mathbf{p}_d(t)\| \to 0$ or $\mathbf{p}(t) - \mathbf{p}_d(t) \to \mathcal{B}(\mathbf{0}_{3\times1}, \delta)$, as $t \to \infty$, where $\delta \geq 0$. This kind of application includes cases where a multicopter is required to track a given trajectory over a moving car, as shown in Fig. 10.1b.
(3) Path following.[3] The desired path $\mathbf{p}_d(\gamma(t)) \in \mathbb{R}^3$ is a path directly determined by the parameter γ rather than time t. The aim of this step to design a controller to ensure that $\|\mathbf{p}(t) - \mathbf{p}_d(\gamma(t))\| \to 0$ or $\mathbf{p}(t) - \mathbf{p}_d(\gamma(t)) \to \mathcal{B}(\mathbf{0}_{3\times1}, \delta)$, as $t \to \infty$, where $\delta \geq 0$. This kind of application includes multicopters that fly along with a profile over a hill, as shown in Fig. 10.1c.

Intuitively, the difference between a trajectory tracking problem and a path following problem lies in whether the curves describing the trajectories or the paths depend upon time. In a trajectory tracking problem, the desired trajectory is a time-dependent curve. However, the curve in a path following problem is independent of time directly. Path following is also called three-dimensional (3D) tracking, while trajectory tracking is called four-dimensional (4D) tracking with the time being added as a new dimension. When $\gamma(t) = t$, the path following problem degenerates into a trajectory tracking problem. Furthermore, if $\mathbf{p}_d(t) \equiv$ constant, the trajectory tracking problem degenerates into a set-point control problem. In other words, path following degenerates into trajectory tracking when a time constraint is imposed and trajectory tracking degenerates into set-point control when the desired trajectory is further restricted. Thus, a set-point control problem is a special case of the trajectory tracking problem. Meanwhile, a trajectory tracking problem is a special case of the path

[2]$\mathcal{B}(\mathbf{o}, \delta) \triangleq \{\xi \in \mathbb{R}^3 \,|\, \|\xi - \mathbf{o}\| \leq \delta\}$, and the notion $\mathbf{x}(t) \to \mathcal{B}(\mathbf{o}, \delta)$ means $\min_{\mathbf{y}\in\mathcal{B}(\mathbf{o},\delta)} \|\mathbf{x}(t) - \mathbf{y}\| \to 0$.

[3]Only parameterized path following is considered here.

following problem. This chapter mainly focuses on set-point control and trajectory tracking problems.

The desired attitude is set as the output of the position controllers. PID controllers are designed for the horizontal position channel and altitude channel according to the linear position model Eqs. (6.70)–(6.71), which generate the desired Euler angles ϕ_{d} and θ_{d} and thrust f_{d} as the output. In the following sections, traditional PID controllers are designed at the first. Later, PID controllers used in open source autopilots are introduced. Finally, PID controllers with saturation are designed.

10.1.2 Traditional PID Controller

(1) **Horizontal position channel**
According to the linear position model Eq. (6.70), the desired attitude angle $\boldsymbol{\Theta}_{\mathrm{hd}} = [\phi_{\mathrm{d}}\ \theta_{\mathrm{d}}]^{\mathrm{T}}$ is expected to be designed such that

$$\lim_{t\to\infty} \left\|\mathbf{e}_{\mathbf{p}_{\mathrm{h}}}(t)\right\| = 0$$

where $\mathbf{e}_{\mathbf{p}_{\mathrm{h}}} \triangleq \mathbf{p}_{\mathrm{h}} - \mathbf{p}_{\mathrm{hd}}$. First, the transient process is expected to follow:

$$\ddot{\mathbf{e}}_{\mathbf{p}_{\mathrm{h}}} = -\mathbf{K}_{\mathbf{p}_{\mathrm{h}}\mathrm{d}}\dot{\mathbf{e}}_{\mathbf{p}_{\mathrm{h}}} - \mathbf{K}_{\mathbf{p}_{\mathrm{h}}\mathrm{p}}\mathbf{e}_{\mathbf{p}_{\mathrm{h}}} \tag{10.1}$$

where $\mathbf{K}_{\mathbf{p}_{\mathrm{h}}\mathrm{d}}$ and $\mathbf{K}_{\mathbf{p}_{\mathrm{h}}\mathrm{p}} \in \mathbb{R}^{2\times 2}$ are diagonal positive definite matrices. Then, to realize the dynamics Eq. (10.1), acceleration $\ddot{\mathbf{p}}_{\mathrm{h}}$ is expected to satisfy the following condition.

$$\ddot{\mathbf{p}}_{\mathrm{h}} = \ddot{\mathbf{p}}_{\mathrm{hd}} - \mathbf{K}_{\mathbf{p}_{\mathrm{h}}\mathrm{d}}\dot{\mathbf{e}}_{\mathbf{p}_{\mathrm{h}}} - \mathbf{K}_{\mathbf{p}_{\mathrm{h}}\mathrm{p}}\mathbf{e}_{\mathbf{p}_{\mathrm{h}}}. \tag{10.2}$$

Combining Eq. (6.70) with Eq. (10.2) results in

$$-\,g\mathbf{A}_{\psi}\boldsymbol{\Theta}_{\mathrm{hd}} = \ddot{\mathbf{p}}_{\mathrm{hd}} - \mathbf{K}_{\mathbf{p}_{\mathrm{h}}\mathrm{d}}\dot{\mathbf{e}}_{\mathbf{p}_{\mathrm{h}}} - \mathbf{K}_{\mathbf{p}_{\mathrm{h}}\mathrm{p}}\mathbf{e}_{\mathbf{p}_{\mathrm{h}}}. \tag{10.3}$$

According to the above equation, $\boldsymbol{\Theta}_{\mathrm{hd}}$ can be written explicitly as follows:

$$\boldsymbol{\Theta}_{\mathrm{hd}} = -g^{-1}\mathbf{A}_{\psi}^{-1}\left(\ddot{\mathbf{p}}_{\mathrm{hd}} - \mathbf{K}_{\mathbf{p}_{\mathrm{h}}\mathrm{d}}\dot{\mathbf{e}}_{\mathbf{p}_{\mathrm{h}}} - \mathbf{K}_{\mathbf{p}_{\mathrm{h}}\mathrm{p}}\mathbf{e}_{\mathbf{p}_{\mathrm{h}}}\right). \tag{10.4}$$

This implies that if $\boldsymbol{\Theta}_{\mathrm{h}} = \boldsymbol{\Theta}_{\mathrm{hd}}$ in Eq. (6.70), then $\lim_{t\to\infty}\left\|\mathbf{e}_{\mathbf{p}_{\mathrm{h}}}(t)\right\| = 0$. In particular, when a set-point control problem is considered, namely $\dot{\mathbf{p}}_{\mathrm{hd}} = \ddot{\mathbf{p}}_{\mathrm{hd}} = \mathbf{0}_{2\times 1}$, Eq. (10.4) transforms into Eq. (10.5) as

$$\boldsymbol{\Theta}_{\mathrm{hd}} = -g^{-1}\mathbf{A}_{\psi}^{-1}\left(-\mathbf{K}_{\mathbf{p}_{\mathrm{h}}\mathrm{d}}\dot{\mathbf{p}}_{\mathrm{h}} - \mathbf{K}_{\mathbf{p}_{\mathrm{h}}\mathrm{p}}\left(\mathbf{p}_{\mathrm{h}} - \mathbf{p}_{\mathrm{hd}}\right)\right). \tag{10.5}$$

(2) **Altitude channel**
Similarly, according to the altitude channel model Eq. (6.71), the control objective is

$$\lim_{t\to\infty}\left|e_{p_z}(t)\right|=0$$

where $e_{p_z} \triangleq p_z - p_{z_d}$. First, the transient process is expected as follows:

$$\ddot{e}_{p_z} = -k_{p_z\mathrm{d}}\dot{e}_{p_z} - k_{p_z\mathrm{p}}e_{p_z} \tag{10.6}$$

where $k_{p_z\mathrm{d}}, k_{p_z\mathrm{p}} > 0$. To realize the dynamics Eq. (10.6), acceleration $\ddot{p}_z$ is expected to satisfy the condition.

$$\ddot{p}_z = \ddot{p}_{z_d} - k_{p_z\mathrm{d}}\dot{e}_{p_z} - k_{p_z\mathrm{p}}e_{p_z}. \tag{10.7}$$

Combining Eq. (6.71) with Eq. (10.7) results in

$$g - \frac{f_\mathrm{d}}{m} = \ddot{p}_{z_d} - k_{p_z\mathrm{d}}\dot{e}_{p_z} - k_{p_z\mathrm{p}}e_{p_z} \tag{10.8}$$

where f_d is the desired thrust. From the above equation, f_d can be written explicitly as follows:

$$f_\mathrm{d} = mg - m\left(\ddot{p}_{z_d} - k_{p_z\mathrm{d}}\dot{e}_{p_z} - k_{p_z\mathrm{p}}e_{p_z}\right). \tag{10.9}$$

This implies that if $f = f_\mathrm{d}$ in Eq. (6.71), then $\lim_{t\to\infty}\left|e_{p_z}(t)\right| = 0$. In particular, when a set-point control problem is considered, namely, $\dot{p}_{z_d} = \ddot{p}_{z_d} = 0$, Eq. (10.9) becomes Eq. (10.10).

$$f_\mathrm{d} = mg - m\left(-k_{p_z\mathrm{d}}\dot{p}_z - k_{p_z\mathrm{p}}\left(p_z - p_{z_d}\right)\right). \tag{10.10}$$

10.1.3 PID Controllers in Open-Source Autopilots

The design ideas of a position controller based on an open source autopilot are introduced in this section.

(1) **Horizontal position channel**

To satisfy $\lim_{t\to\infty}\left\|\mathbf{e}_{\mathbf{p}_\mathrm{h}}(t)\right\| = 0$, according to

$$\dot{\mathbf{p}}_\mathrm{h} = \mathbf{v}_\mathrm{h} \tag{10.11}$$

the desired value of $\mathbf{v}_\mathrm{h}$, namely $\mathbf{v}_\mathrm{hd}$, is designed as

$$\mathbf{v}_\mathrm{hd} = \mathbf{K}_{\mathbf{p}_\mathrm{h}}\left(\mathbf{p}_\mathrm{hd} - \mathbf{p}_\mathrm{h}\right) \tag{10.12}$$

where $\mathbf{K}_{\mathbf{p}_\mathrm{h}} \in \mathbb{R}^{2\times 2}$ is a diagonal positive definite matrix. Under the assumption that $\dot{\mathbf{p}}_\mathrm{hd} = \mathbf{0}_{2\times 1}$, if $\lim_{t\to\infty}\left\|\mathbf{e}_{\mathbf{v}_\mathrm{h}}(t)\right\| = 0$, then $\lim_{t\to\infty}\left\|\mathbf{e}_{\mathbf{p}_\mathrm{h}}(t)\right\| = 0$, where $\mathbf{e}_{\mathbf{v}_\mathrm{h}} \triangleq \mathbf{v}_\mathrm{h} - \mathbf{v}_\mathrm{hd}$. In fact, Eqs. (10.11) and (10.12) build the horizontal position control loop.

According to the following equation

$$\dot{\mathbf{v}}_{\mathrm{h}} = -g\mathbf{A}_{\psi}\boldsymbol{\Theta}_{\mathrm{h}} \tag{10.13}$$

the next task is to design the desired $\boldsymbol{\Theta}_{\mathrm{h}}$, namely $\boldsymbol{\Theta}_{\mathrm{hd}}$. Similar to Eq. (10.3), a PID controller is adopted as follows:

$$-g\mathbf{A}_{\psi}\boldsymbol{\Theta}_{\mathrm{hd}} = -\mathbf{K}_{\mathbf{v}_{\mathrm{h}}\mathrm{p}}\mathbf{e}_{\mathbf{v}_{\mathrm{h}}} - \mathbf{K}_{\mathbf{v}_{\mathrm{h}}\mathrm{i}}\int \mathbf{e}_{\mathbf{v}_{\mathrm{h}}} - \mathbf{K}_{\mathbf{v}_{\mathrm{h}}\mathrm{d}}\dot{\mathbf{e}}_{\mathbf{v}_{\mathrm{h}}} \tag{10.14}$$

where $\mathbf{K}_{\mathbf{v}_{\mathrm{h}}\mathrm{p}}$ and $\mathbf{K}_{\mathbf{v}_{\mathrm{h}}\mathrm{i}}, \mathbf{K}_{\mathbf{v}_{\mathrm{h}}\mathrm{d}} \in \mathbb{R}^{2\times 2}$. Under the assumption $\dot{\mathbf{v}}_{\mathrm{hd}} = \mathbf{0}_{2\times 1}$, if $\lim_{t\to\infty} \|\boldsymbol{\Theta}_{\mathrm{h}}(t) - \boldsymbol{\Theta}_{\mathrm{hd}}(t)\| = 0$, then $\lim_{t\to\infty} \left\|\mathbf{e}_{\mathbf{v}_{\mathrm{h}}}(t)\right\| = 0$. The desired attitude angle is derived from Eq. (10.14) as

$$\boldsymbol{\Theta}_{\mathrm{hd}} = g^{-1}\mathbf{A}_{\psi}^{-1}\left(\mathbf{K}_{\mathbf{v}_{\mathrm{h}}\mathrm{p}}\mathbf{e}_{\mathbf{v}_{\mathrm{h}}} + \mathbf{K}_{\mathbf{v}_{\mathrm{h}}\mathrm{i}}\int \mathbf{e}_{\mathbf{v}_{\mathrm{h}}} + \mathbf{K}_{\mathbf{v}_{\mathrm{h}}\mathrm{d}}\dot{\mathbf{e}}_{\mathbf{v}_{\mathrm{h}}}\right). \tag{10.15}$$

As a result, the designed attitude controller should ensure that $\lim_{t\to\infty} \|\boldsymbol{\Theta}_{\mathrm{h}}(t) - \boldsymbol{\Theta}_{\mathrm{hd}}(t)\| = 0$. Consequently, the horizontal position channel Eq. (6.70) can guarantee $\lim_{t\to\infty} \left\|\mathbf{e}_{\mathbf{p}_{\mathrm{h}}}(t)\right\| = 0$. In fact, Eqs. (10.13) and (10.15) build the horizontal velocity control loop. If the set-point control problem is considered, then $\dot{\mathbf{p}}_{\mathrm{hd}} = \ddot{\mathbf{p}}_{\mathrm{hd}} = \mathbf{0}_{2\times 1}$ in Eq. (10.15). To avoid noise caused by the differential, the differential term $\mathbf{K}_{\mathrm{hd}}\dot{\mathbf{e}}_{\mathbf{v}_{\mathrm{h}}}$ can be omitted. Thus, controller design of the horizontal position channel is composed of Eqs. (10.12) and (10.15).

A question often asked is that why is Eq. (10.12) written as a P controller, whereas Eq. (10.15) is in the PID form. The reason is that the channel (10.11) is a kinematic model, which is uncertainty-free. Therefore, the P controller Eq. (10.12) appropriate and enough. Because the channel Eq. (10.13) is a dynamic model subject to various uncertainties, the PID controller Eq. (10.15) is used to compensate for these uncertainties. A similar principle is applied to the altitude channel.

(2) **Altitude channel**

In order to satisfy $\lim_{t\to\infty} \left|e_{p_z}(t)\right| = 0$, according to

$$\dot{p}_z = v_z \tag{10.16}$$

the desired value of v_z, namely $v_{z\mathrm{d}}$, is written as

$$v_{z\mathrm{d}} = -k_{p_z}(p_z - p_{z\mathrm{d}}) \tag{10.17}$$

where $k_{p_z} > 0$. Under the assumption $\dot{p}_{z\mathrm{d}} = 0$, if $\lim_{t\to\infty} \left|e_{v_z}(t)\right| = 0$, then $\lim_{t\to\infty} \left|e_{v_z}(t)\right| = 0$, where $e_{v_z} \triangleq v_z - v_{z\mathrm{d}}$. In fact, Eqs. (10.16) and (10.17) build the altitude control loop. According to the following equation

$$\dot{v}_z = g - \frac{f}{m} \tag{10.18}$$

the next task is to design the desired thrust f_{d}. Similar to Eq. (10.8), a PID controller is adopted as follows:

$$g-\frac{f_{\mathrm{d}}}{m}=-k_{v_z\mathrm{p}}e_{v_z}-k_{v_z\mathrm{i}}\int e_{v_z}-k_{v_z\mathrm{d}}\dot{e}_{v_z} \tag{10.19}$$

where $k_{v_z\mathrm{p}},k_{v_z\mathrm{i}},k_{v_z\mathrm{d}}\in\mathbb{R}$. Under the assumption $\dot{v}_{z\mathrm{d}}=0$, if $\lim_{t\to\infty}|f(t)-f_{\mathrm{d}}(t)|=0$, then $\lim_{t\to\infty}\left|e_{v_z}(t)\right|=0$. The desired thrust f_{d} is derived from Eq. (10.19) as follows

$$f_{\mathrm{d}}=m\left(g+k_{v_z\mathrm{p}}e_{v_z}+k_{v_z\mathrm{i}}\int e_{v_z}+k_{v_z\mathrm{d}}\dot{e}_{v_z}\right). \tag{10.20}$$

In fact, Eqs. (10.18) and (10.20) build the altitude velocity control loop. If the set-point control problem is considered, then $\dot{p}_{z\mathrm{d}}=\ddot{p}_{z\mathrm{d}}=0$ in Eq. (10.20). To avoid the noise caused by the differential, the differential term $k_{v_z\mathrm{d}}\dot{e}_{v_z}$ can be omitted. So far, controller design of the altitude channel is completed, which is composed of Eqs. (10.17) and (10.20).

10.1.4 PID Controller with Saturation

Thus, the position control problem has been solved, i.e., the desired pitch angle and roll angle have been calculated. However, some other problems may arise in practical implementation, because $p_x-p_{x_\mathrm{d}}$ or $p_y-p_{y_\mathrm{d}}$ may be too large initially in practice. For example, $\left\|\mathbf{e}_{\mathbf{p}_\mathrm{h}}(0)\right\|=100$ km. This will lead to unacceptable θ_{d} and ϕ_{d}. As a result, the small-angle assumption is violated owing to which Eq. (6.70) is no longer valid. If the attitude controller uses these unacceptable desired pitch angle and roll angle values, then multicopters may crash. Therefore, PID controllers with saturation must be considered.

(1) **Horizontal position channel**
Upon considering saturation, the traditional PID controller Eq. (10.5) becomes

$$\mathbf{\Theta}_{\mathrm{hd}}=\mathrm{sat}_{\mathrm{gd}}\left(\mathrm{g}^{-1}\mathbf{A}_{\psi}^{-1}\left(\mathbf{K}_{\mathbf{p}_\mathrm{h}\mathrm{d}}\dot{\mathbf{p}}_\mathrm{h}+\mathbf{K}_{\mathbf{p}_\mathrm{h}\mathrm{p}}\left(\mathbf{p}_\mathrm{h}-\mathbf{p}_{\mathrm{hd}}\right)\right),a_0\right) \tag{10.21}$$

where $a_0>0$. Similarly, the PID controller used in open source autopilots is rewritten as

$$\begin{aligned}\mathbf{e}_{\mathbf{v}_\mathrm{h}}&=\mathrm{sat}_{\mathrm{gd}}\left(\mathbf{v}_\mathrm{h}-\mathbf{v}_{\mathrm{hd}},a_1\right)\\ \mathbf{\Theta}_{\mathrm{hd}}&=\mathrm{sat}_{\mathrm{gd}}\left(\mathrm{g}^{-1}\mathbf{A}_{\psi}^{-1}\left(\mathbf{K}_{\mathbf{v}_\mathrm{h}\mathrm{p}}\mathbf{e}_{\mathbf{v}_\mathrm{h}}+\mathbf{K}_{\mathbf{v}_\mathrm{h}\mathrm{i}}\int\mathbf{e}_{\mathbf{v}_\mathrm{h}}+\mathbf{K}_{\mathbf{v}_\mathrm{h}\mathrm{d}}\dot{\mathbf{e}}_{\mathbf{v}_\mathrm{h}}\right),a_2\right)\end{aligned} \tag{10.22}$$

where $a_1,a_2>0$. The parameters a_0, a_1 and a_2 related to saturation can be specified depending on the practical requirements. For example, if $\theta_{\mathrm{d}},\phi_{\mathrm{d}}\in[-\theta_{\max},\theta_{\max}]$, then the saturated horizontal position controller (10.22) becomes

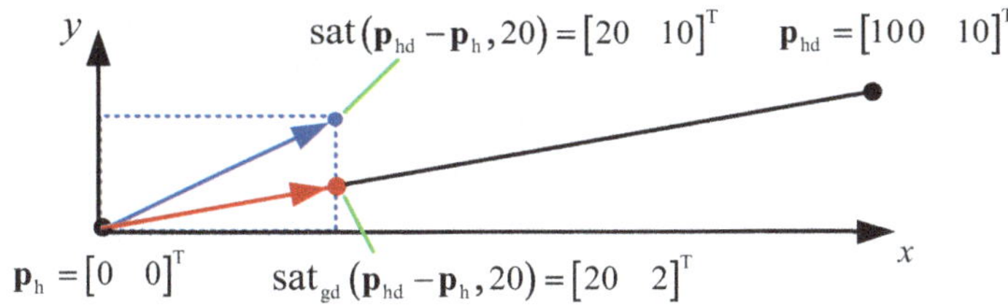

Fig. 10.2 Comparison of the results of two saturation functions

$$\boldsymbol{\Theta}_{hd} = \text{sat}_{gd}\left(g^{-1}\mathbf{A}_{\psi}^{-1}\left(\mathbf{K}_{\mathbf{v}_h p}\mathbf{e}_{\mathbf{v}_h} + \mathbf{K}_{\mathbf{v}_h i}\int \mathbf{e}_{\mathbf{v}_h} + \mathbf{K}_{\mathbf{v}_h d}\dot{\mathbf{e}}_{\mathbf{v}_h}\right), \theta_{max}\right). \tag{10.23}$$

Next, the difference between the direction-guaranteed saturation function $\text{sat}_{gd}(\mathbf{x}, a)$ and the traditional saturation function $\text{sat}(\mathbf{x}, a)$ will be discussed. The traditional saturation function is defined as

$$\text{sat}(\mathbf{x}, a) \triangleq \begin{bmatrix} \text{sat}(x_1, a) \\ \vdots \\ \text{sat}(x_n, a) \end{bmatrix}, \ \text{sat}(x_k, a) \triangleq \begin{cases} x_k & , |x_k| \le a \\ a \cdot \text{sign}(x_k) & , |x_k| > a \end{cases}$$

whose direction may be different from that of $\mathbf{x}$. Meanwhile, the direction-guaranteed saturation function $\text{sat}_{gd}(\mathbf{x}, a)$ can not only ensure that the absolute value of each element of the final vector is not greater than a, but also guarantee a direction same to that of $\mathbf{x}$. For example, as shown in Fig. 10.2, if $\mathbf{p}_h = [0\ 0]^T$ and $\mathbf{p}_{hd} = [100\ 10]^T$, then

$$\text{sat}_{gd}(\mathbf{p}_{hd} - \mathbf{p}_h, 20) = [20\ 2]^T$$

but

$$\text{sat}(\mathbf{p}_{hd} - \mathbf{p}_h, 20) = [20\ 10]^T.$$

Here $\text{sat}_{gd}(\mathbf{p}_{hd} - \mathbf{p}_h, 20)$ is still on the straight-line path from $\mathbf{p}_h$ to $\mathbf{p}_{hd}$, but sat $(\mathbf{p}_{hd} - \mathbf{p}_h, 20)$ undergoes deviation. Under these conditions, multicopters cannot fly along a straight line, and therefore more energy is consumed.

The following simulation is performed to study flight trajectories based on the two saturation functions. In the simulation, the initial value is set as $\mathbf{p}_h(0) = [0\ 0]^T$, and the desired position is set as $\mathbf{p}_{hd} = [4\ 6]^T$. The parameters related to saturation are $a_1 = 3$ and $a_2 = 12\pi/180$ in controller Eq. (10.22). The simulation results in Fig. 10.3 can be observed that the direction-guaranteed saturation function ensures that the multicopter flies along a straight line, while the traditional saturation function cannot.

(2) **Altitude channel**

To avoid the throttle command being out of range, saturation needs to be considered

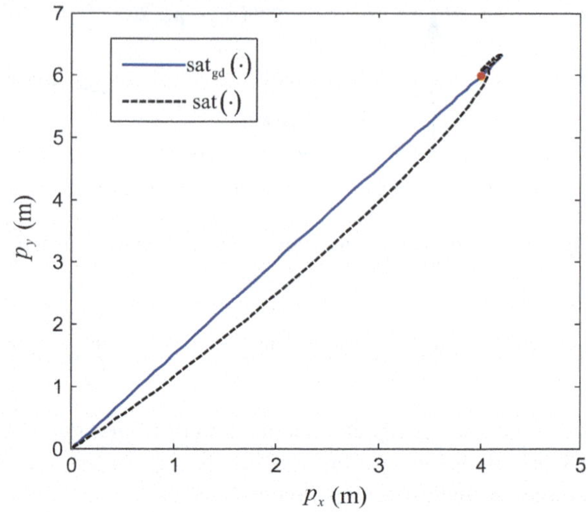

Fig. 10.3 Multicopter horizontal motion trajectory under action of PID controller based on two types of saturation functions

as well. Thus, the traditional PID controller Eq. (10.10) becomes

$$f_d = \text{sat}_{gd}\left(m\left(g + k_{p_z d}\dot{p}_z + k_{p_z p}\left(p_z - p_{z_d}\right)\right), a_3\right) \tag{10.24}$$

where $a_3 > 0$. Similarly, the PID controller design used in open source autopilots Eq. (10.20) is rewritten as

$$\begin{aligned} e_{v_z} &= \text{sat}_{gd}\left(v_z - v_{zd}, a_4\right) \\ f_d &= \text{sat}_{gd}\left(m\left(g + k_{v_z p}e_{v_z} + k_{v_z i}\int e_{v_z} + k_{v_z d}\dot{e}_{v_z}\right), a_5\right) \end{aligned} \tag{10.25}$$

where $a_4,\ a_5 > 0$. For a scale, the direction-guaranteed saturation function is the same as the traditional saturation function. Furthermore, if $f_d \in [f_{\min}, f_{\max}]$, then the saturated controller Eq. (10.25) becomes

$$\begin{aligned} f_d = {} & \text{sat}_{gd}\left(m\left(g + k_{v_z p}e_{v_z} + k_{v_z i}\int e_{v_z} + k_{v_z d}\dot{e}_{v_z}\right) - \frac{f_{\min}+f_{\max}}{2}, \frac{f_{\max}-f_{\min}}{2}\right) \\ & + \frac{f_{\min}+f_{\max}}{2}. \end{aligned} \tag{10.26}$$

10.2 Basic Experiment

10.2.1 Experimental Objective

(1) Things to prepare

1) Hardware: Multicopter Hardware System, Pixhawk Autopilot System;

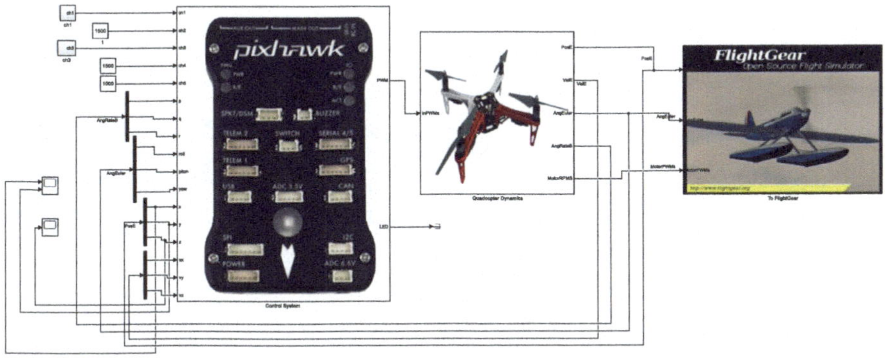

Fig. 10.4 SIL model for analyzing channel decoupling, Simulink model "PositionControl_Sim.slx"

2) Software: MATLAB R2017b or above, Simulink-based Controller Design and Simulation Platform, HIL simulation Platform, Instructional Package "e6.1" (https://rflysim.com/course).

(2) Objectives

1) Repeat the Simulink simulation of a quadcopter to analyze the channel decoupling between the control along $o_b x_b$ and $o_b y_b$ axes;
2) Sweep the open-loop position control system to obtain the Bode plot and further analyze the stability margin of the closed-loop position control system;
3) Perform the HIL simulation.

10.2.2 Experimental Procedure

(1) **Step1: SIL simulation—Channel decoupling**

1) Parameter Initialization
Run the file "e6\e6.1\Sim\Init_control.m" to initialize the parameters, after which the Simulink file "PosControl_Sim.slx" is opened automatically as shown in Fig. 10.4.
2) Run the simulation
Open the file "FlightGear-F450" and run the Simulink file "PosControl_Sim.slx". Later, the motion of the quadcopter can be observed in FlightGear, as shown in Fig. 10.5, where the quadcopter takes off, climbs up, and then flies along the $o_e y_e$ axis[4] and finally hovers.

[4]For details, see Sect. 6.1.1.2.

Fig. 10.5 Quadcopter in FlightGear

Fig. 10.6 Setting of "ch1" and "ch3"

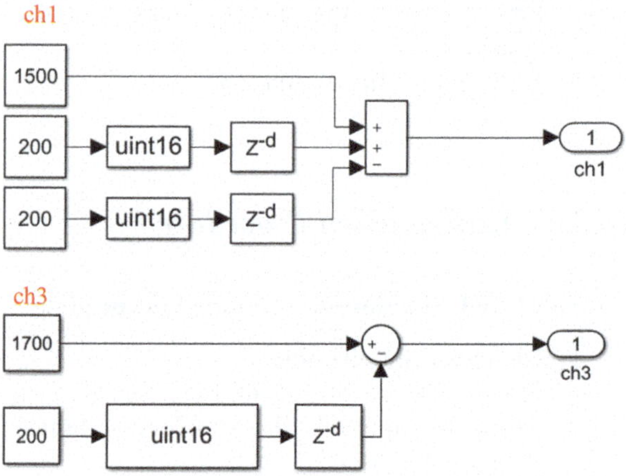

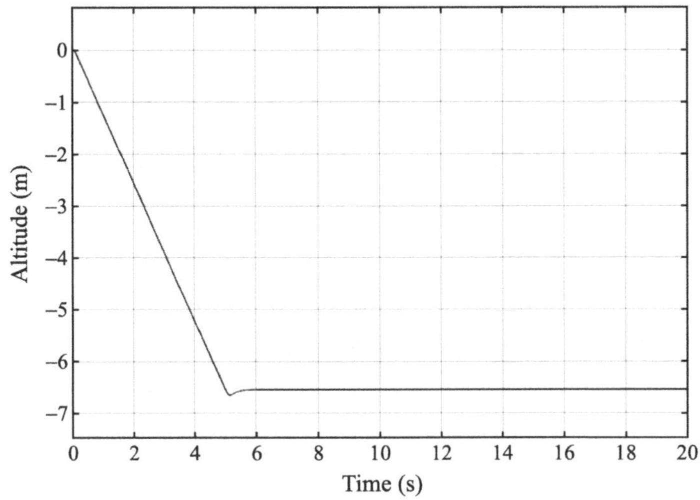

Fig. 10.7 Altitude response

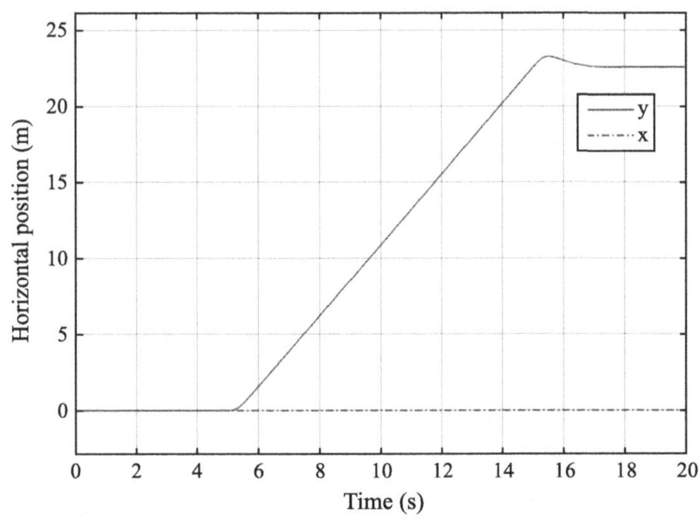

Fig. 10.8 Horizontal position response

3) Simulation results
   Regarding the inputs of the "Control System" module, i.e., "ch1–ch5"[5] in Fig. 10.4, "ch4" and "ch5" are similar to those used in Fig. 9.2 in Chap. 9 to control the yaw rate and LED, while "ch1", "ch2", and "ch3" are converted into velocity commands instead of the roll angle, pitch angle and thrust com-

[5]Corresponding to CH1–CH5 of the RC transmitter introduced in Sect. 2.3.1.1.

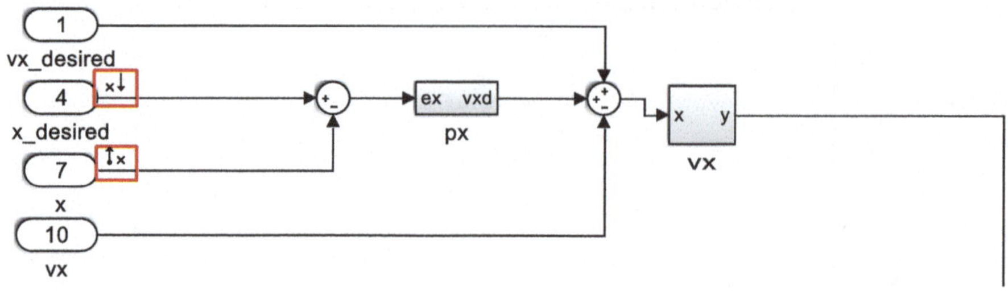

Fig. 10.9 Specifying signals as input and output

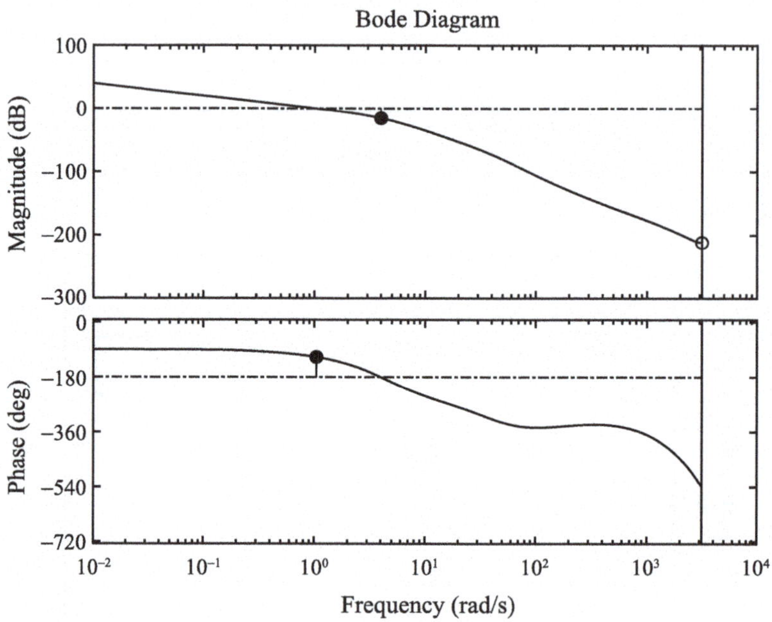

Fig. 10.10 Open-loop Bode plot of x-axis channel control system

mands. For example, the output of "ch1" is converted into the desired velocity along the $o_e y_e$ axis. For simplicity, the value of "ch4" (corresponding to the yaw channel) is set to 1500. The settings of "ch1" and "ch3" are shown in Fig. 10.6. The value of "ch3" is 1700 in the first 4.5 s and, then decreases to 1500. The value of "ch1" is 1500 in the first 5 s and, then increases to 1700; after 15 s, the value reverts to 1500 and remains constant. This means that in the first 4.5 s, there is only the desired velocity along the negative direction of the $o_e z_e$ axis, (note that $o_e z_e$ pointing toward the ground indicates a positive direction); and at 5 s, there is a desired velocity along the positive $o_e y_e$ axis, the desired velocity in the direction of $o_e z_e$ and $o_e x_e$ is 0; after 15 s, the desired

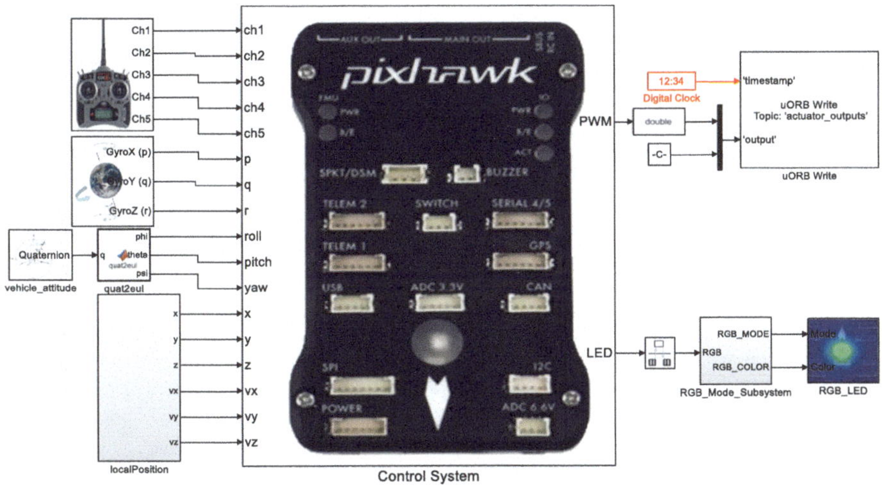

Fig. 10.11 HIL model to verify channel decoupling, Simulink model "AttitudeControlHIL.slx"

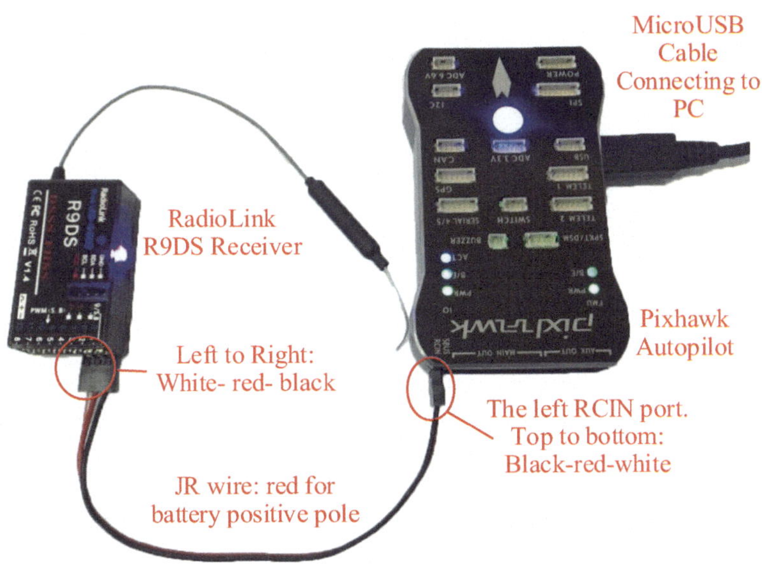

Fig. 10.12 Connection between Pixhawk hardware and RC receiver

velocity along the three directions is 0. Changes in the altitude change are shown in Fig. 10.7.

4) Channel decoupling analysis
According to the setting, the earth-fixed frame is aligned with the aircraft-

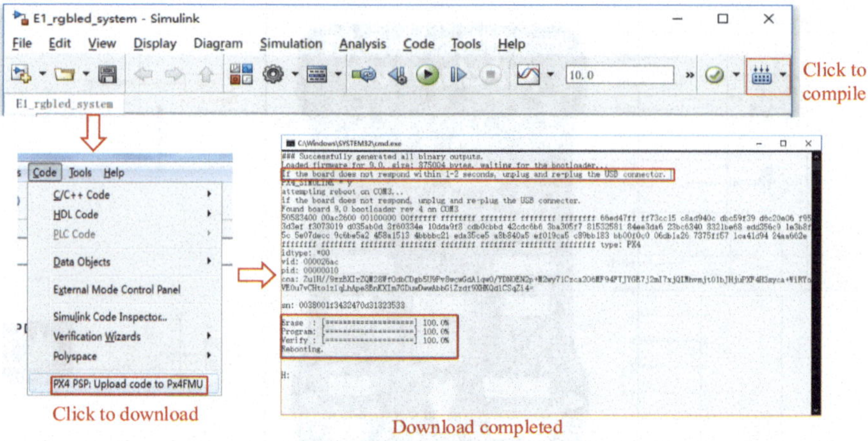

Fig. 10.13 Code compilation and upload process

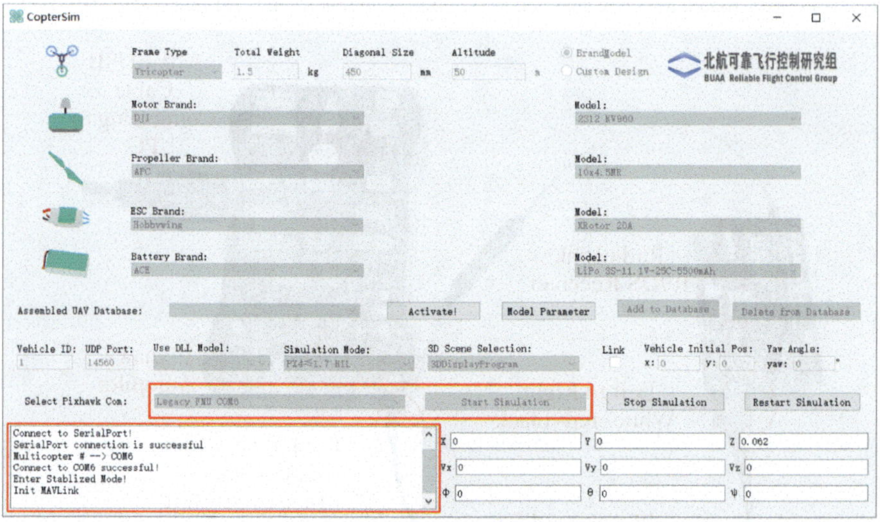

Fig. 10.14 User interface of CopterSim

body frame,[6] when the quadcopter hovers with yaw being zero. The resultant horizontal position of the quadcopter is shown in Fig. 10.8. It can be observed that position control along the $o_b y_b$ axis does not change the position along the $o_b x_b$ axis. This implies that control actions along the $o_b y_b$ and $o_b x_b$ axes have been decoupled.

[6]For details, see Sect. 6.2.2.

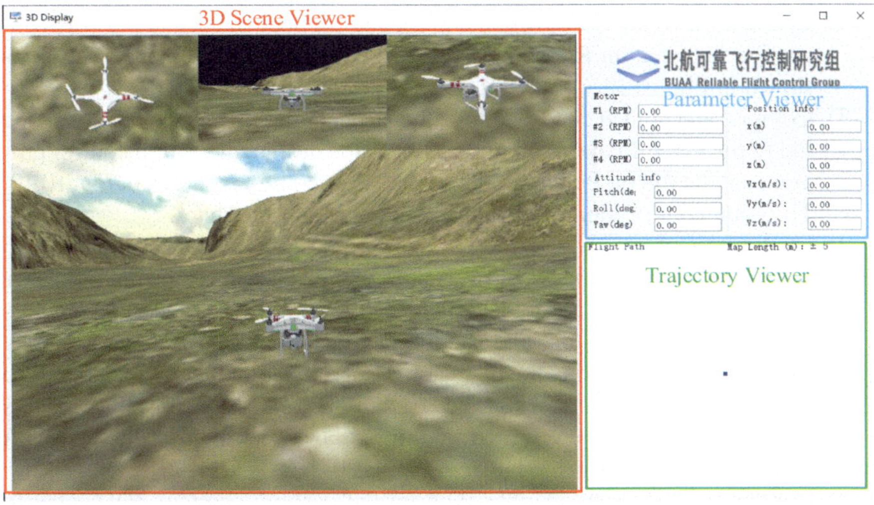

Fig. 10.15 User interface of 3DDisplay

(2) **Step2: SIL simulation—Stability margin**

1) Run the file "e6\e6.1\tune\Init_control.m" to initialize the parameters and, then the Simulink file "PosControl_tune.slx" is opened automatically. Open the model "Control System"–"position_control" of the Simulink file and specify the input and output signals for the Bode plot. Specify the desired x-axis channel (corresponding to the position of the quadcopter along the $o_e x_e$ axis) input as the "Open-loop Input" , and specify the actual output in the x-axis channel as "Open-loop Output". The specified signals are shown in Fig. 10.9.
2) Select "Analysis"—"Control Design"—"Linear Analysis" on the top menu bar.
3) From the context menu, select "Linear Analysis" and click "Bode" to get the Bode plot.
4) Right-click on the curve and select "Characteristics"—"All Stability Margins". It is observed that the gain margin is 15.3 dB at a frequency of 3.97 rad/s; the phase margin is 65.6° at a frequency of 1.04 rad/s, as shown in Fig. 10.10.

(3) **Step3: HIL simulation**

1) Open Simulink file for HIL. Run the file "e6\e6.1\HIL\Init_control.m" to initialize the parameters and, then the Simulink file "PosControl_HIL.slx" is opened automatically, as shown in Fig. 10.11. It should be noted that "Control System" in the SIL simulation shown in Fig. 10.4 is the same as that in the HIL simulation shown in Fig. 10.11.

2) Connect hardware. Referring the procedure in Sect. 2.3 in Chap. 2, the connection between an RC receiver and a Pixhawk autopilot can be set up as shown in Fig. 10.12.
3) Compile and upload code. Compile the HIL simulation model and upload the file to the given Pixhawk autopilot. Later, the designed position control program can be run on Pixhawk autopilot. The operation is shown in Fig. 10.13. The detailed procedure is introduced in Sect. 3.3.3 in Chap. 3.
4) Configure CopterSim. Double-click on the desktop shortcut CopterSim to open it. Readers can choose different propulsion systems using the following procedure. Click on "Model Parameters" to customize the model parameters and, then click on "Store and use the parameters" to make them available. The software will automatically match the serial port number. Readers would click the "Run" button to enter the HIL simulation mode. After that, readers could see the message returned by the Pixhawk autopilot in the lower-left corner of the interface, as shown in Fig. 10.14.
5) Configure 3DDisplay. Double-click on the desktop shortcut 3DDisplay to open it, as shown in Fig. 10.15.
6) Simulation performance. Arm the quadcopter for manual control using the given RC transmitter.[7] The quadcopter can hover and fly at a specified speed. When all control sticks are in the middle position, the quadcopter will keep hovering.

10.3 Analysis Experiment

10.3.1 Experimental Objective

(1) Things to prepare

1) Software: MATLAB R2017b or above, Simulink-based Controller Design and Simulation Platform, Instructional Package Package "e6.2" (https://rflysim.com/course).

(2) Objectives

1) Adjust PID controller parameters to improve its control performance and record the overshoot and the settling time, and then obtain a group of satisfied parameters;
2) Based on the obtained satisfied parameters, sweep the system to draw the Bode plot and analyze its stability margin.

[7] For details, see Sect. 3.5.3 in Chap. 3.

Table 10.1 Code corresponding to "Init_control.m"

```
ModelInit_PosE=[0,0,-100];
ModelInit_VelB=[0,0,0];
ModelInit_AngEuler=[0,0,0];
ModelInit_RateB=[0,0,0];
ModelInit_RPM=557.1420;
```

10.3.2 Experimental Procedure

(1) **Step1: Initial model setup**
The steps to adjust PID parameters are similar to those used for attitude control. First, adjust the altitude channel and then adjust the horizontal position channel. Furthermore, for each channel, first adjust the velocity control loop and then adjust the position loop. The necessary file can be found in the folder "e6\e6.2\tune". Because the PID adjusting methods for the channel control loops are similar, the PID parameters for velocity along the $o_e x_e$ axis are adjusted here as an example. First, let the quadcopter hover at the initial altitude of 100 m by setting the throttle value to 0.6085 and the initial speed of four motors to 557.1420 rad/s. Modify the corresponding parameters in the file "Init_control.m" as shown in Table 10.1.

(2) **Step2: Adjust the PID parameters of the velocity control loop**
Open the model "Control System"—"position_control" in the file "e6\e6.2\tune\PosControl_tune.slx". Replace "vx_desired" with the step input. Later, configure the step response and "vx" to "Enable Data logging" to get the step response of the velocity control loop, as shown in Fig. 10.16. Modify the PID parameters corresponding to the velocity control loop in the file "Init_control.m" . First, adjust the proportional term parameter and set the integral and derivative term

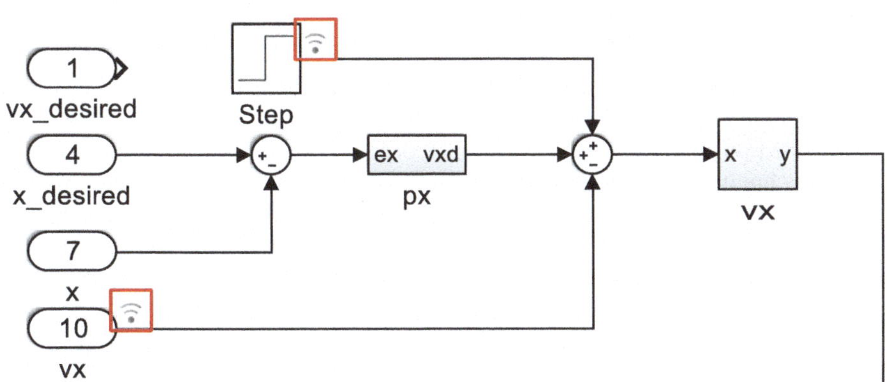

Fig. 10.16 Setting step response for velocity control loop

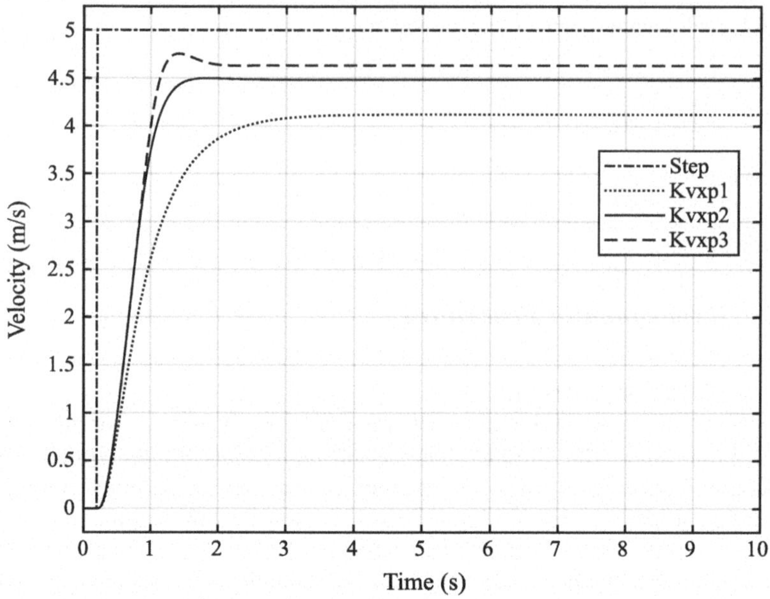

Fig. 10.17 Step response of velocity along $o_e x_e$ axis channel with different proportional term parameters

parameters to 0. After that, run the file "Init_control.m" (you must run the file each time to update any changes if you change them). Click on the Simulink "Run" button to view responses in the "Simulation Data Inspector". The proportional term parameter gradually increases from a small value to a large value, i.e., the variable "Kvxp" in the file "Init_control.m" increases. The obtained step responses are shown in Fig. 10.17. Later, adjust the integral and derivative term parameters, i.e., "Kvxi" and "Kvxd" in the file "Init_control.m". Finally, fine tune the proportional term parameter with the resulting response as shown in Fig. 10.18. The resulting PID parameters are "Kvxp = 2.5; Kvxi = 0.4; Kvxd = 0.01".

(3) **Step3: Adjust PID parameters for the position control loop**
Adjust the proportional term parameter for the position control loop. Using the obtained position control loop parameters in Step 2, replace "x_desired" with a step input, and set the step input and "x" in the "PosControl_tune.slx" file to "Enable Data Logging", as shown in Fig. 10.19. Increase the proportional term parameter "Kpxp" in the file "Init_control.m" gradually, and observe the step response in "Simulation Data Inspector". The obtained curve is shown in Fig. 10.20. The final selected parameters are "Kpxp = 1.0; Kvxp = 2.5; Kvxi = 0.4; Kvxd =0.01". The corresponding step response is shown in Fig. 10.21.

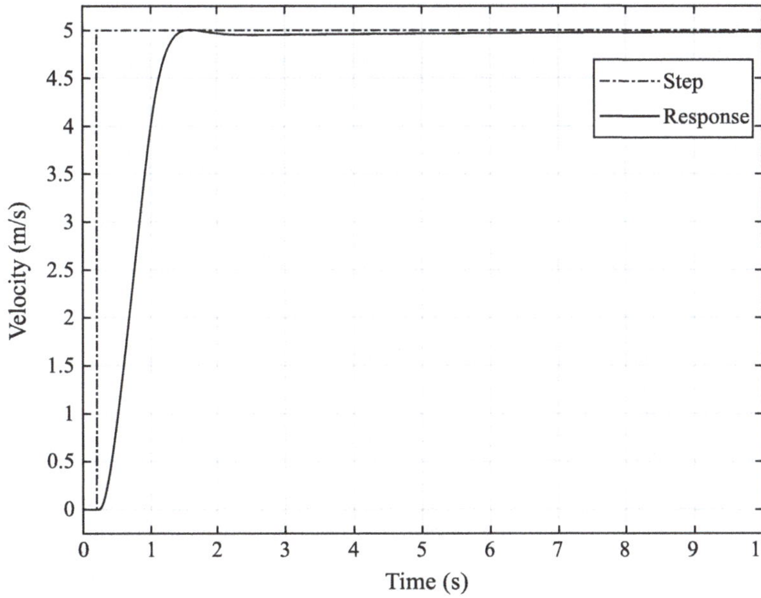

Fig. 10.18 Step response of velocity along $o_e x_e$ axis channel with a group of satisfied PID parameters

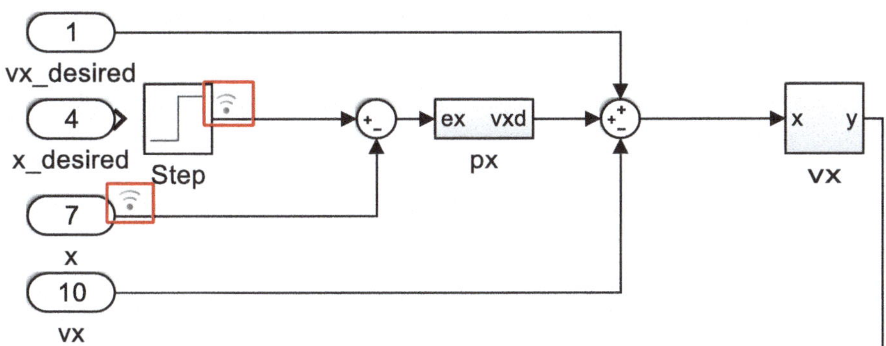

Fig. 10.19 Setting step response of Position control loop along the $o_e x_e$ axis

(4) **Step4: Sweep to get the Bode plot**
Specify input and output signals for the Bode plot. Specify "x_desired" as "Open-loop Input", and then specify "x" as "Open-loop Output". Using these parameters, the Bode plot can be drawn, as shown in Fig. 10.22.

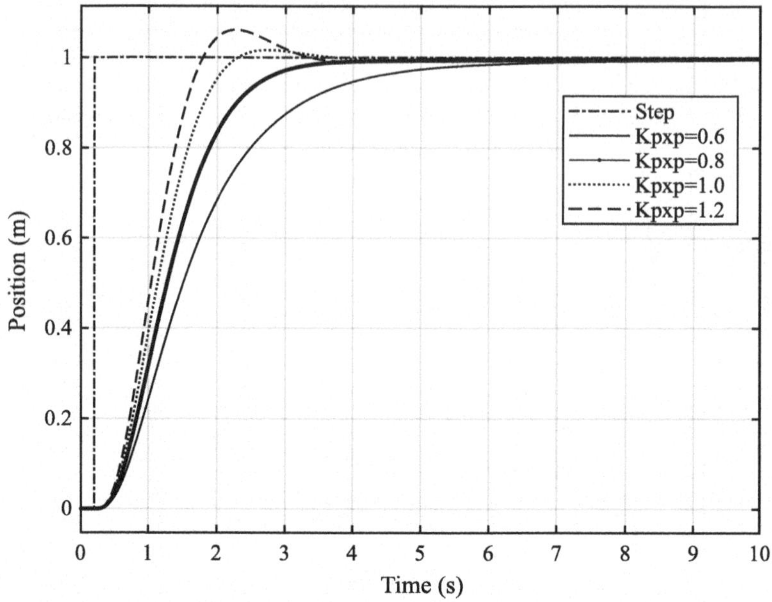

Fig. 10.20 Step response of position control loop along $o_e x_e$ axis with different proportional term parameters

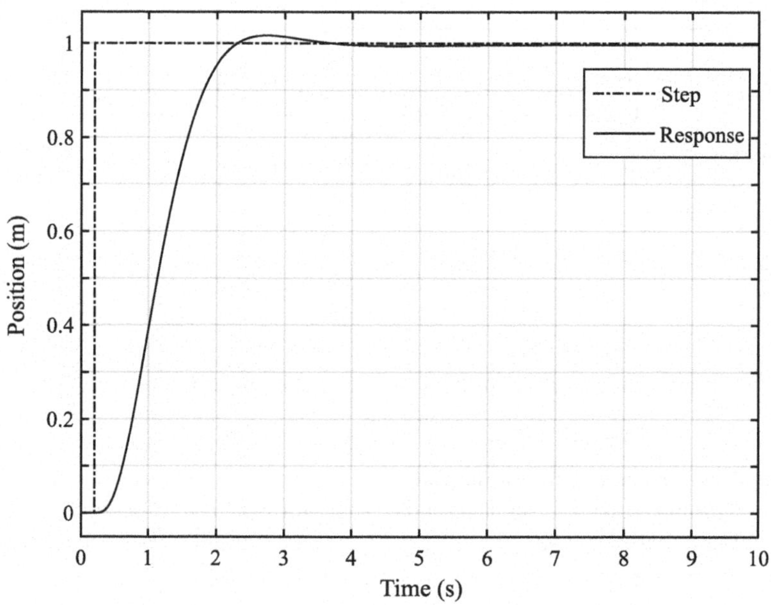

Fig. 10.21 Step response of position along $o_e x_e$ axis with a group of satisfied PID parameters

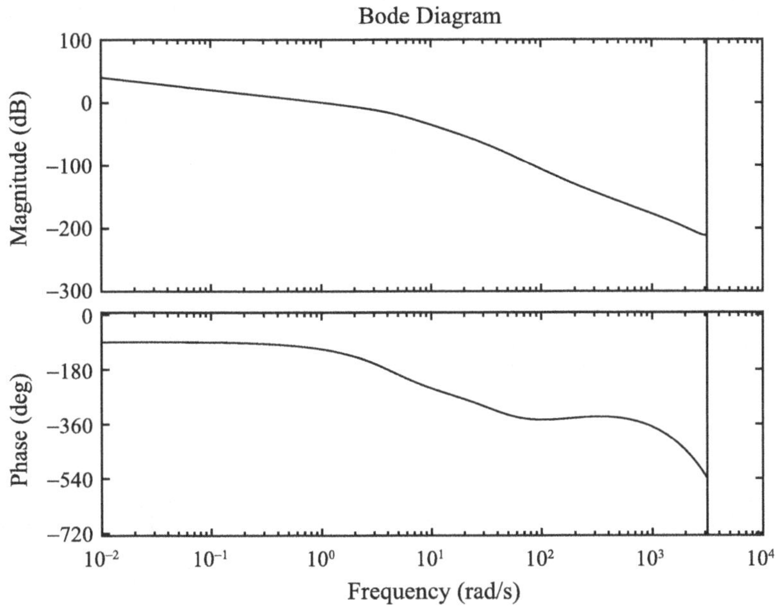

Fig. 10.22 Open-loop Bode plot of position control loop along the $o_e x_e$ axis

10.3.3 Remarks

(1) A time-invariant system is the premise of the analysis in the frequency domain. Hence, it is important to set the quadcopter at an equilibrium, such as hovering.
(2) Find out the input and output of the system, and then specify the input and output signals correctly. Select the output signal line as "Open-loop output" when testing the open-loop system, and "Output Measurement" when testing the closed-loop system. For details, please refer to the document https://ww2.mathworks.cn/help/slcontrol/ug/specify-portion-of-model-to-linearize-in-simulink-model.html.

10.4 Design Experiment

10.4.1 Experimental Objective

(1) Things to prepare
   1) Hardware: Multicopter Hardware System, Pixhawk Autopilot System;

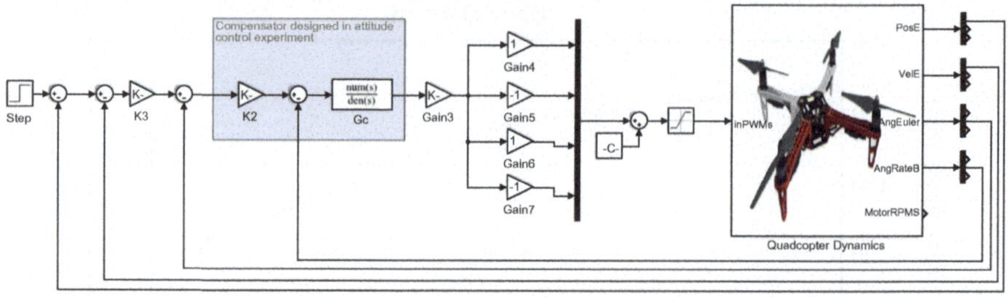

Fig. 10.23 Simplified model of position control loop along $o_e x_e$ axis channel, Simulink model "PosControl_tune.slx"

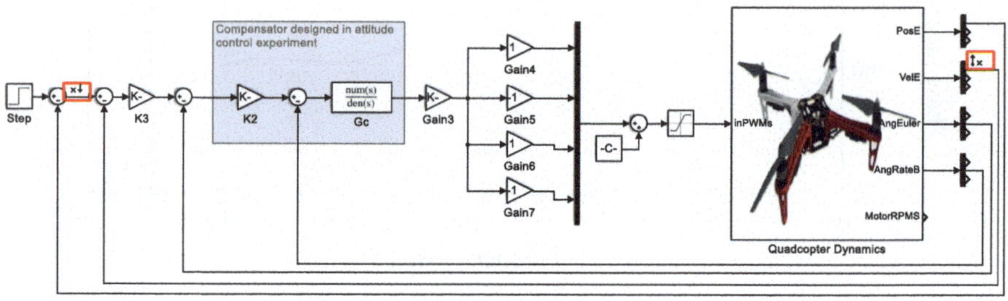

Fig. 10.24 Specifying signals as input and output

2) Software: MATLAB R2017b or above, Simulink-based Controller Design and Simulation Platform, HIL Simulation Platform, Instructional Package "e6.3" (https://rflysim.com/course).

(2) Objectives

1) Obtain the transfer function for the position control channel, and design a compensator for the existing controller using MATLAB "ControlSystemDesigner" in the velocity control loop to satisfy the conditions of step response steady-state error $e_{rss} \leq 0.01$, phase margin >75°, and cut-off frequency >2.0 rad/s. The position control loop satisfies that the cut-off frequency is >1 rad/s and phase margin is >60°.
2) Perform SIL simulations and HIL simulations with the designed controller.
3) Use the designed controller to perform outdoor flight test experiments.

10.4.2 Experimental Design

(1) **Step1: Simplify the overall structure**
Consider the x-axis channel for example. The simplified model is shown in Fig. 10.23.

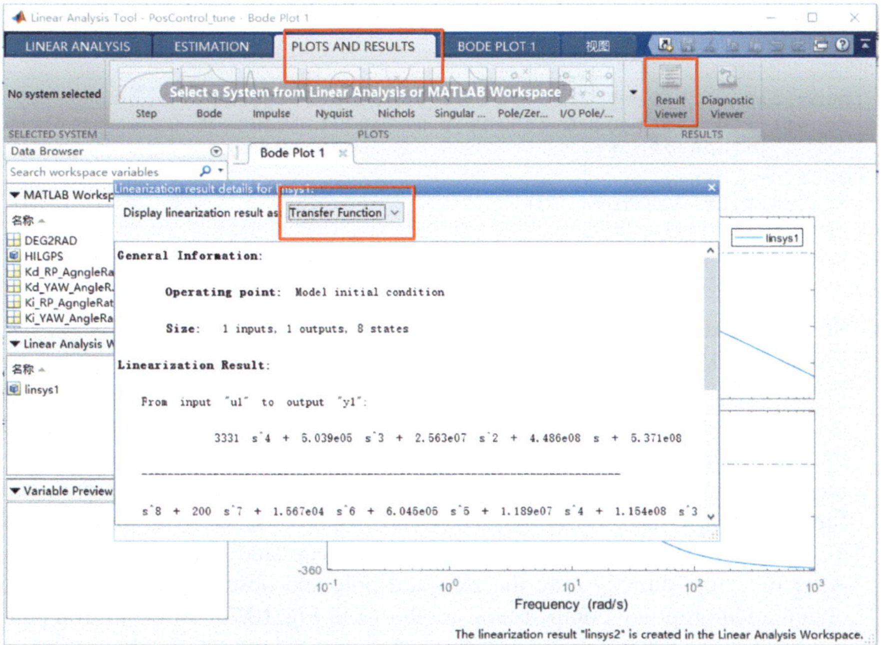

Fig. 10.25 Transfer function

(2) **Step2: Velocity control loop analysis**
The input is the desired velocity and the output is the velocity. By specifying signals as input and output, an open-loop Bode plot is obtained, as shown in Fig. 10.24.

(3) **Step3: Obtain the transfer function of the x-axis channel**
After the Bode plot is obtained, a variable, namely "linsys1", will appear in the "Linear Analysis Workspace". The transfer function can be obtained by the operation shown in Fig. 10.25 as

$$\frac{3330.9(s+1.29)(s+50)^3}{(s+1.964e-07)(s+1.253)(s+33.92)(s+50)^3(s^2+14.87s+101.1)}$$

which is simplified as follow

$$\frac{3330.9(s+1.29)}{s(s+1.253)(s+33.92)(s^2+14.87s+101.1)}.$$

According to this transfer function, write a file including code as shown in Table 10.2, and run it to put the information of the transfer function in the "Workspace" of MATLAB. After that, the control system toolbox can be used for compensator design, as shown in Fig. 10.26.

Table 10.2 Code in "v_tune.m"

```
num=[3331 5.039e05 2.563e07 4.486e08 5.371e08];
den=[1 200 1.567e04 6.045e05 1.189e07 1.154e08 5.557e08
    5.371e08 280.1];
G=tf(num,den);
controlSystemDesigner(G);
```

(4) **Step4: Use toolbox to design the compensator**
From Fig. 10.26, it can be observed that the step response is slow. Drag the curve in the Bode plot up to increase the open-loop gain so that the step response becomes fast. Later, as shown in Fig. 10.27, the response time reduces but an overshoot appears, and the phase margin is 50.1°. These do not meet the requirement. Consider adding a lead compensator to increase the phase margin, and further increase the cut-off frequency and response speed. The procedure is as follows. In the Bode plot, right-click and select "add Pole/Zero"—"Lead". After this step, directly drag the zero and pole and observe their response to obtain an appropriate compensator, as shown in Fig. 10.28. After the compensator is designed, right-click on "Edit Compensator" in the Bode plot, as shown in Fig. 10.29. The final compensator is obtained as

$$G_c = \frac{2.5(1 + 0.15s)}{1 + 0.013s}.$$

(5) **Step5: Design a compensator for the position control loop of the x-axis channel**
First, put the velocity control loop compensator designed in Step 4 into the model in Step 1, as shown in Fig. 10.23. Specify the input and output signals of the Bode plot, as shown in Fig. 10.30. The obtained Bode plot is shown in Fig. 10.31. It can be observed that the phase margin is 68.8°, and the cut-off frequency is 0.95 rad/s. These meet the requirements marginally. Gain should be increased slightly to increase the cut-off frequency. For example, choose the position loop gain to be 1.2 and redraw the Bode diagram. As a result, the phase margin is 65.3 °, and the cutoff frequency is 1.12. Therefore, the experimental requirement is satisfied.

10.4.3 Simulation Procedure

(1) **Step1: Discretize the continuous-time compensator**
The designed compensator is an s transfer function, which has to be discretized so that it can be run on the Pixhawk autopilot, a digital computer. The "c2d" function in MATLAB is adopted, as shown in Table 10.3, where "num" is the

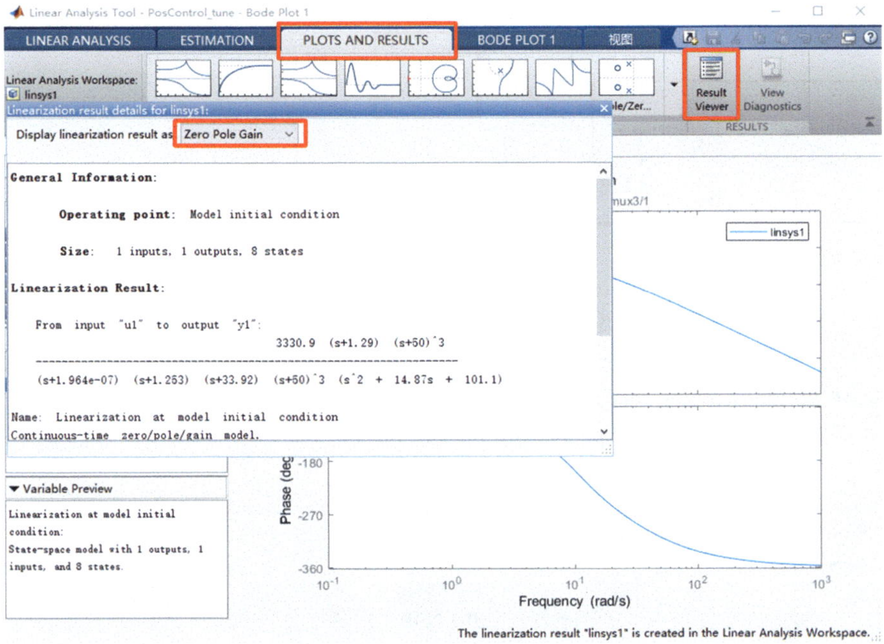

Fig. 10.26 Control system design based on Bode plot

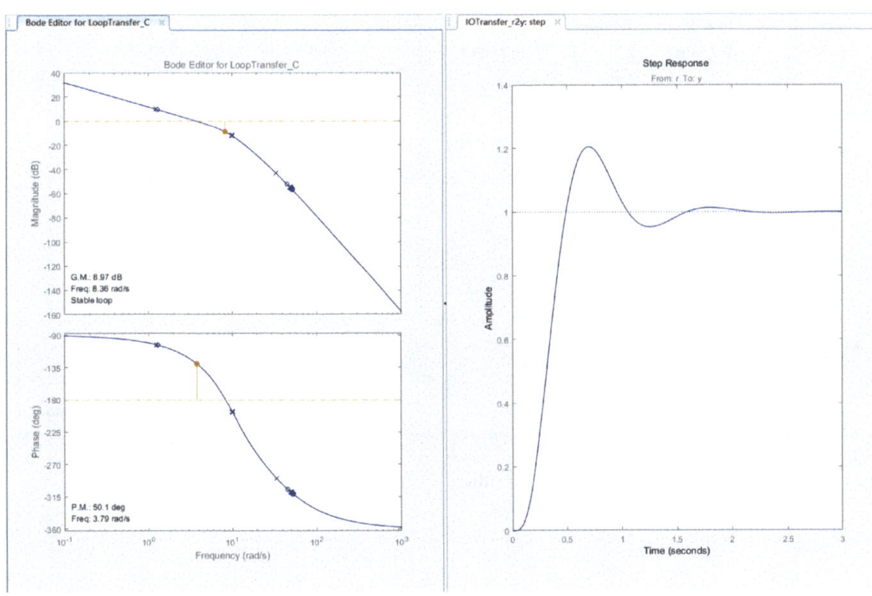

Fig. 10.27 Bode plot and step response after increasing the open-loop gain

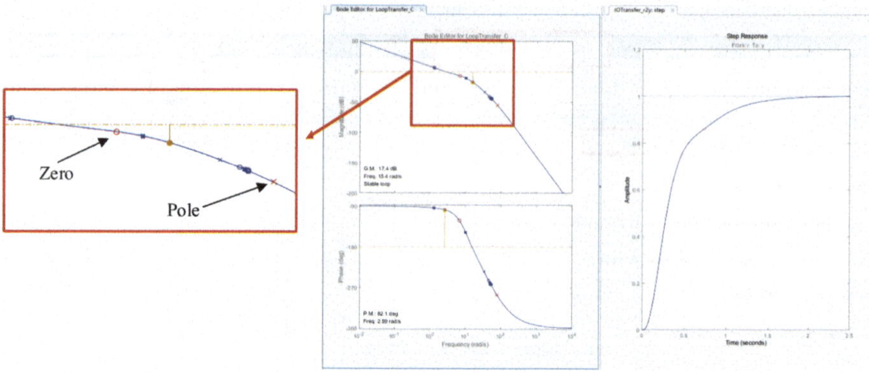

Fig. 10.28 Step response after adding a lead compensator

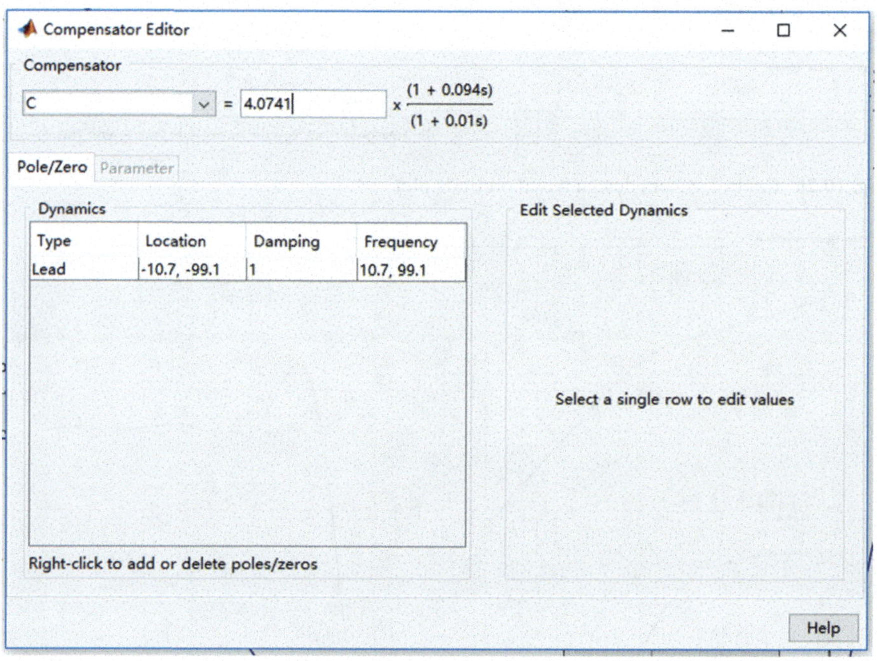

Fig. 10.29 Compensator obtained using toolbox

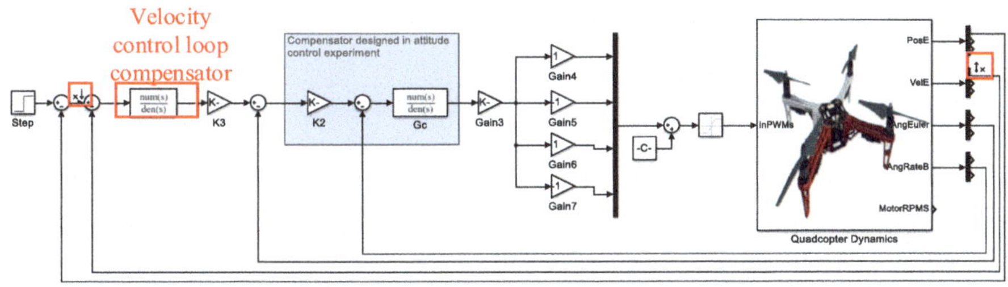

Fig. 10.30 Simplified model after adding velocity control loop compensator, Simulink model "PosControl_tune.slx"

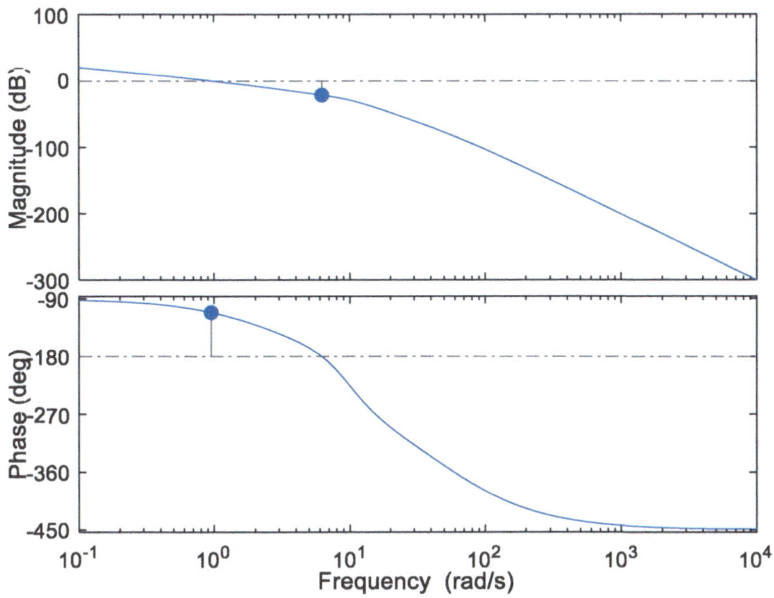

Fig. 10.31 Bode plot of the position control loop

transfer function numerator coefficient vector, "den" is the transfer function denominator coefficient vector, and "Ts" is sample time. Because the sample time of the Pixhawk autopilot for position control is 0.01s, the compensator is discretized by a sample time of 0.01s. This MATLAB function discretizes the s transfer function using the zero-order hold on the input with a sample time "Ts = 0.01" seconds. The transfer function after discretizing is as follows

$$G_c(s) = \frac{2.5(1 + 0.15s)}{1 + 0.013s} \rightarrow G_c(z) = \frac{28.85z - 27.5}{z - 0.4634}.$$

Table 10.3 Code in file "c2d.m"

```
H = tf([num], [den])
Hd = c2d(H, Ts, 'zoh')
```

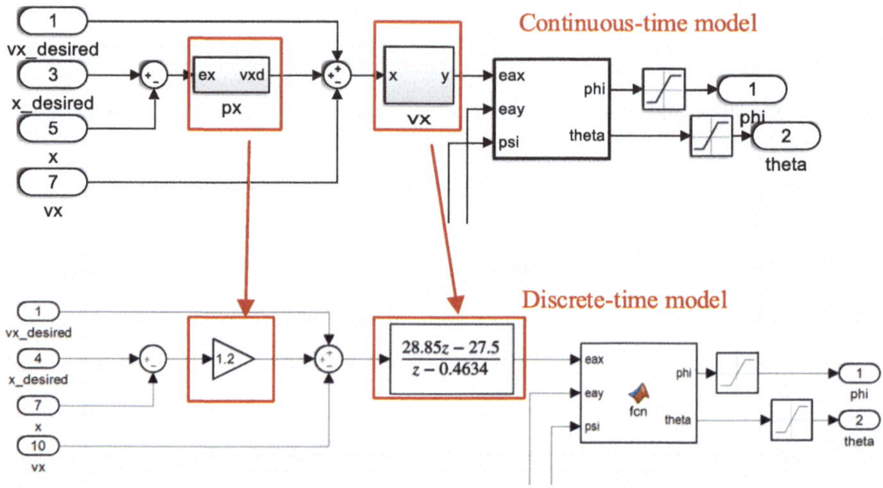

Fig. 10.32 Controller discretized for HIL simulation

(2) **Step2: Replace the controller**
Replace the continuous-time PID controller with the discrete-time model in Simulink model "PosControl_HIL.slx", as shown in Fig. 10.32.

(3) **Step3: HIL simulation**
The quadcopter can fly along a straight line and hover, as shown in Fig. 10.33.

10.4.4 Flight Test Procedure

(1) **Step1: Quadcopter preparation**
For details, see Sect. 4.3.4 in Chap. 4.

(2) **Step2: Simulink model for flight test**
The model modified according to Step 1 is shown in Fig. 10.34. Unlike the HIL simulation shown in Fig. 10.11, the block in Fig. 10.34 is used to replace the PWM output part in Fig. 10.11. Additionally, the compensator designed in Sect. 10.4.3 is added in the model in Fig. 10.34. In order to record flight data, a new data recording module is also used, the detailed procedure can be found in Sect. 9.4.4.

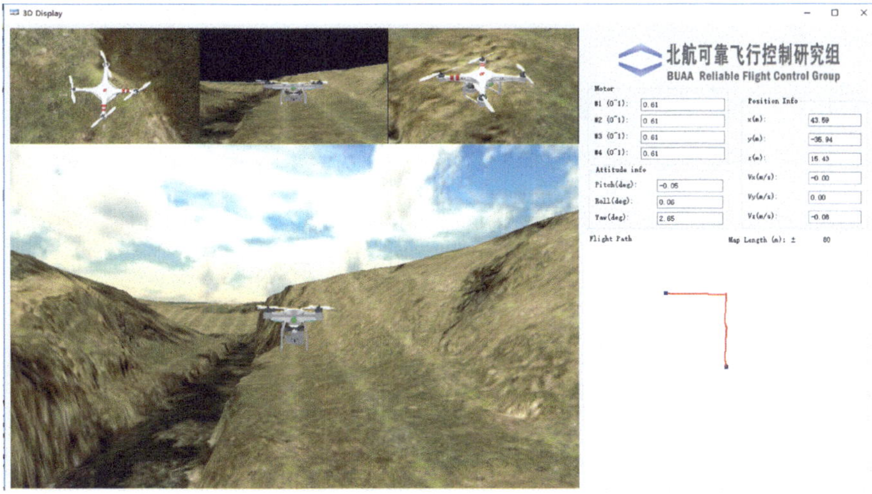

Fig. 10.33 HIL simulation shown in 3DDisplay

(3) **Step3. Upload code**
This process is similar to that used for compiling and uploading the code in HIL simulations. The experimental flow is shown in Fig. 10.13. The detailed procedure is included in Sect. 3.3.3.

(4) **Step4. Outdoor flight test**
To ensure safety, a rope is tethered to the quadcopter, and the other end is tethered to a heavy object. The remote pilot maintains a safe distance from the quadcopter during flight, as shown in Fig. 9.39.

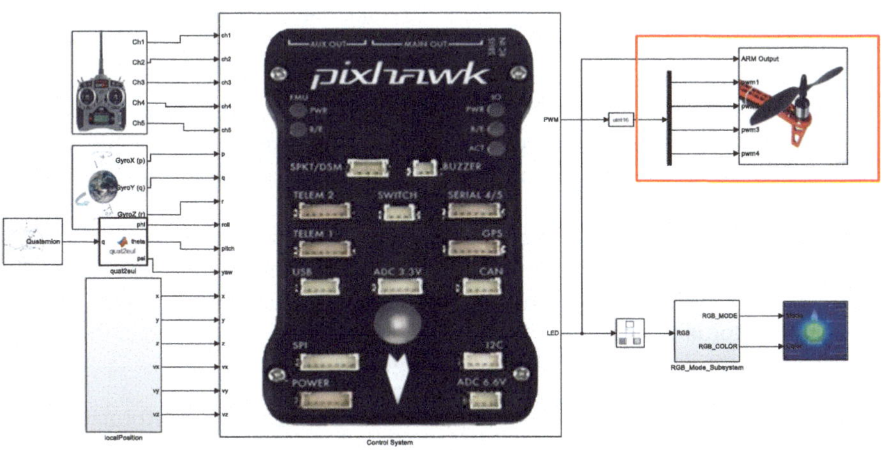

Fig. 10.34 Model for flight test, Simulink model "PosControl_FLY.slx"

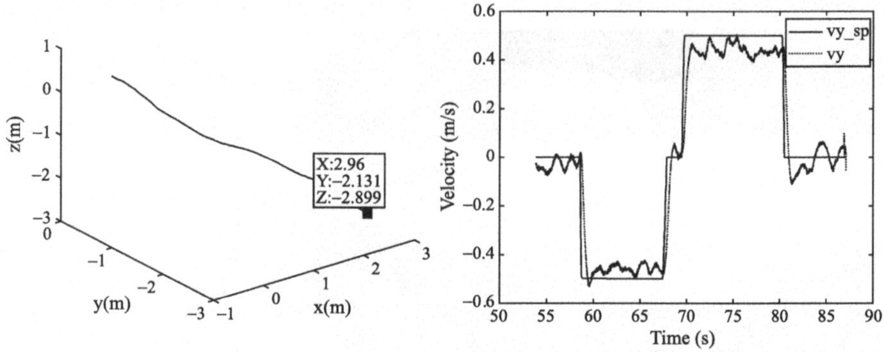

Fig. 10.35 Position and velocity along the $o_e y_e$ axis

(5) **Step5. Analyze data**
The actual flight data is shown in Fig. 10.35. In the left-hand plot, the quadcopter reaches the specified position from 0. The designed set-point controller functions well. The right-hand plot represents the velocity control response and it can be observed that the quadcopter flies at the specified speed with a fast velocity response.

10.5 Summary

(1) Based on the position control model of a quadcopter, a widely-used PID control method is developed, and the design of the position controller is completed in Simulink and MATLAB. The simulation performance is displayed in FlightGear.
(2) The PSP tool of Simulink was used to generate the embedded code which was then uploaded to the Pixhawk autopilot for HIL simulation and flight test.
(3) The parameters of the PID controller are adjusted to get the satisfied parameters. The system analysis tool in MATLAB/Simulink is adopted to obtain Bode plots corresponding to the open-loop position control system and velocity control system to observe the phase margin and gain margin of the corresponding closed-loop systems.
(4) In order to satisfy the given requirements, the system compensation method is adopted. Unlike in the attitude control design experiment, this chapter directly relies on the MATLAB toolbox to facilitate the compensator design. A lead compensator is designed for the velocity control loop. Furthermore, the design is verified through HIL simulation and flight test.

If you have any question, please go to https://rflysim.com for your information.

Chapter 11
Semi-autonomous Control Mode Design Experiment

In Semi-Autonomous Control (SAC), a multicopter's attitude stabilization or hover is realized by the autopilot, while position control is often manually realized by a remote pilot with a Radio Control (RC) transmitter. The SAC is based on the attitude controller and position controller designed in the previous chapters. This chapter will involve three modes (stabilize mode, altitude hold mode, loiter mode) of SAC of multicopters. Readers will gradually master this part of the knowledge through three step-by-step experiments, namely, basic, analysis and design experiments from shallow to deep. In the basic experiment, readers will repeat the simulation of the stabilize mode and observe its control performance. In the analysis experiment, readers will change the stabilize mode to altitude hold mode and then analyze their difference. Finally, in the design experiment, readers will further realize the loiter mode and design a state machine to switch among the three modes.

11.1 Preliminary

In order to make this chapter self-contained, the preliminary is from Chap. 13 of *Introduction to Multicopter Design and Control* [8].

11.1.1 Semi-autonomous Control

In the SAC, a multicopter's attitude stabilization or hover is realized by autopilot, while the position control is often manually realized by a remote pilot with an RC transmitter. In the SAC manner, a Ground Control Station (GCS) is optional. Thus, the control command is a sum of two control inputs, one of which is generated by the RC transmitter and the other from the Automatic Controller (AC) of the autopilot. In fact, the SAC consists of AC and RC, and a multicopter in the SAC mode is controlled by either AC or RC. Generally, according to the degree of the AC, a multicopter in

© Publishing House of Electronics Industry 2020, corrected publication 2021

Q. Quan et al., *Multicopter Design and Control Practice*,
https://doi.org/10.1007/978-981-15-3138-5_11

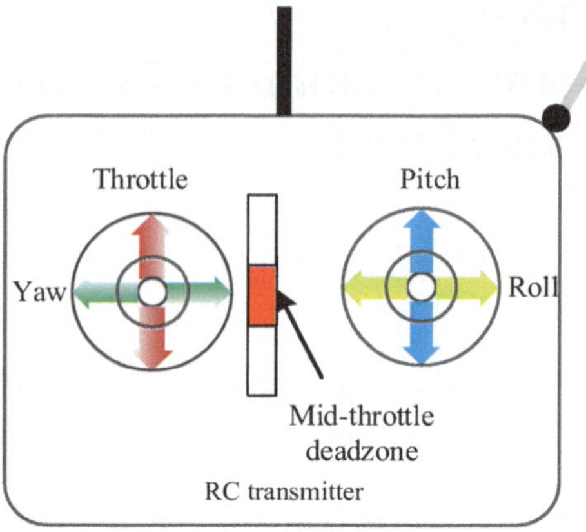

Fig. 11.1 Stick function and dead zone in RC transmitter

SAC can be in one of three main modes, i.e., "stabilize mode", "altitude hold mode", and "loiter mode".

11.1.1.1 Stabilize Mode

The stabilize mode allows a remote pilot to fly a multicopter manually, but itself-levels the roll and pitch axis. Under RC, an RC transmitter is used to control its roll/pitch and then drive the multicopter to tilt toward the desired direction. When the remote pilot releases the roll and pitch control sticks, the multicopter automatically switches itself to AC. Then, its attitude will be stabilized, but the position drift will occur. During this process, the remote pilot will need to give roll, pitch, and throttle commands regularly to keep the multicopter in place as it is pushed around by the wind. The throttle command controls the average motor speed to maintain the altitude. If the remote pilot puts the throttle control stick completely down, then the motors will operate at their minimum rate, and if the multicopter is in the air, it will lose altitude control and tumble. In addition, when the remote pilot releases the yaw control stick, the multicopter will maintain its current heading.

11.1.1.2 Altitude Hold Mode

In the altitude hold mode, a multicopter maintains a consistent altitude while allowing roll, pitch, and yaw to be controlled normally. When the throttle control stick is in the mid-throttle dead zone (40–60%), the multicopter automatically switches itself to AC (see Fig. 11.1). Then, the throttle command is automatically given to maintain the current altitude, but the horizontal position drift will occur. AC can not only stabilize the attitude, but also keep the altitude. The remote pilot will need to give roll and pitch commands regularly to keep the multicopter in place. Going outside of the mid-throttle dead zone (i.e., below 40% or above 60% for example), the multicopter will enter RC. That is, the multicopter will descend or climb depending upon the deflection of the throttle control stick. When the throttle control stick is completely down, the multicopter will descend at the maximum permissible speed and if it is at the very top, it will climb at a maximum permissible speed. The altitude hold mode needs the support of height sensors, such as barometers or ultrasonic range finders.

11.1.1.3 Loiter Mode

The loiter mode automatically attempts to maintain the current location, heading, and altitude. When the remote pilot releases the roll, pitch, and yaw control sticks and pushes the throttle control stick to the mid-throttle dead zone, the multicopter automatically switches itself to AC and maintain the current location, heading and altitude. Precise GPS position, low magnetic interference on the compass, and low vibrations are all important in achieving a good hovering performance. The remote pilot can control the multicopter's position once by pushing the control sticks out of the midpoints. Horizontal location can be adjusted by the roll and pitch control sticks. When the remote pilot releases the control sticks, the multicopter will slow to a stop. Altitude can be controlled by the throttle control stick just as in the altitude hold mode. The heading can be set with the yaw control stick. The loiter mode needs the support from both height sensors and position sensors, such as GPS receiver and cameras. From the analysis above, in the SAC, a multicopter is switched automatically between AC and RC. It is worth noting that RC has a higher priority than AC. This implies that if RC is active, AC will be disconnected from the multicopter. Only when a control stick is close to its midpoint, the AC of the corresponding channel then takes over the control of the multicopter.

11.1.2 Radio Control

Define $(\cdot)_{\mathrm{d}}$, $(\cdot)_{\mathrm{drc}}$ to be desired values and RC values, respectively. To make things easy, let $\theta_{\mathrm{drc}} = u_{\theta}$, $\phi_{\mathrm{drc}} = u_{\phi}$, $\dot{\psi}_{\mathrm{drc}} = u_{\omega_z}$ and $f_{\mathrm{drc}} = u_T$. In the SAC, the throttle/ yaw control sticks and roll/pitch control sticks are manipulated to specify the desired total thrust $f_{\mathrm{d}} = f_{\mathrm{drc}}$, the desired yaw rate $\dot{\psi}_{\mathrm{d}} = \dot{\psi}_{\mathrm{drc}}$, the desired pitch angle $\theta_{\mathrm{d}} = \theta_{\mathrm{drc}}$, and the desired roll angle $\phi_{\mathrm{d}} = \phi_{\mathrm{drc}}$, respectively. For a direct-type RC transmitter, the throttle control stick can stop at any position when it is not pushed, and the total thrust is proportional to the deflection of throttle control stick. For the throttle control stick, the relationship between the deflection and the total thrust is shown in Fig. 11.2. Here, $\sigma_{\mathrm{drc}} \in [0, 1]$ is the deflection of throttle control stick or throttle command, where $\sigma_{\mathrm{drc}} = 0.5$ represents the midpoint of the throttle control stick. If $\sigma_{\mathrm{drc}} \in [0.4, 0.6]$, then $f_{\mathrm{drc}}(\sigma_{\mathrm{drc}}) = 0$. Then, AC starts to introduce the feedback in altitude for altitude hold. Here, it is supposed that the feedforward mg has been given by AC, so f_{drc} starts from $-mg$ as shown in Fig. 11.2. The dead zone is used

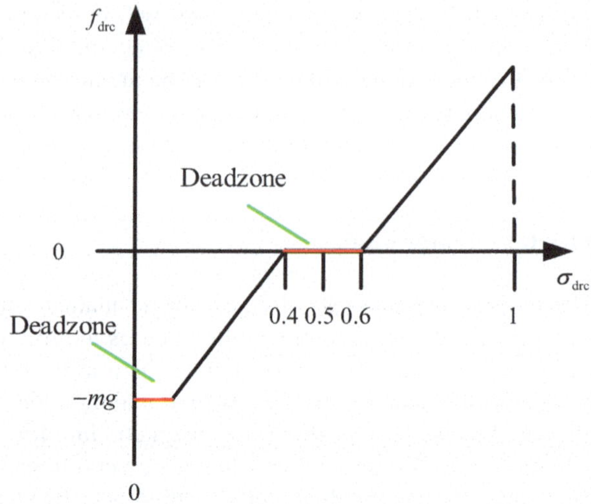

Fig. 11.2 Relationship between the deflection and the total thrust

to reduce the effect by a slight change of the throttle control stick. Similarly, other control sticks also have dead zones.

In addition, some multicopters, such as the Phantom quadcopter launched by DJI, use the increment type throttle to control the total thrust. The increment-type throttle has two features. One is that the control stick can turn back to the midpoint automatically if it is released. The other is that the throttle control stick position is proportional to the desired vertical speed v_{zd} or thrust rate instead of the desired total thrust. For simplicity, this chapter just considers the direct-type throttle RC transmitter because this kind of RC transmitter is adopted by most open-source autopilots.

11.1.3 Automatic Control

When all control sticks are close to their midpoints, as shown in Fig. 11.1, the AC takes over the control of the multicopter. In fact, regardless of which mode of SAC is active, the decision-making module of an autopilot will produce the desired hover position $\mathbf{p}_{\mathrm{dac}}$ and yaw ψ_{dac}. On the other hand, the state estimation module of the autopilot will produce the position estimate $\hat{\mathbf{p}} = [\,\hat{p}_x\ \hat{p}_y\ \hat{p}_z\,]^{\mathrm{T}}$ and attitude estimate $\hat{\boldsymbol{\Theta}} = [\,\hat{\phi}\ \hat{\theta}\ \hat{\psi}\,]^{\mathrm{T}}$. The AC aims to drive the multicopter so that $\lim_{t\to\infty} \left\|\hat{\mathbf{p}}(t) - \mathbf{p}_{\mathrm{dac}}(t)\right\| = 0$ and $\lim_{t\to\infty} \left\|\hat{\psi}(t) - \psi_{\mathrm{dac}}(t)\right\| = 0$ are satisfied. Thus, AC structures for the three modes of SAC can be the same. However, because different sensors are used, the estimation accuracy is different for the three modes, namely the stabilize, altitude hold, and loiter modes. In the following subsection,

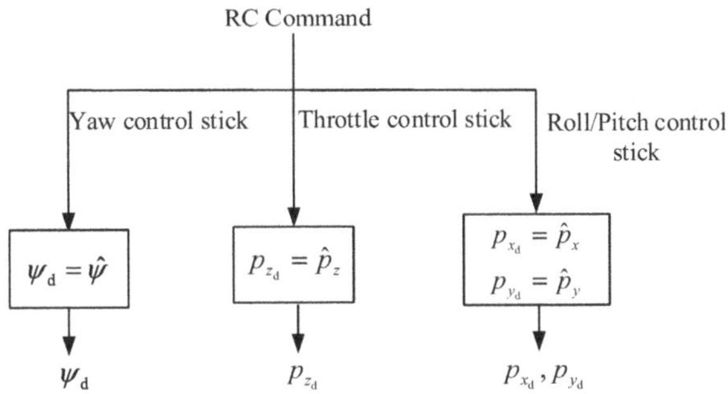

Fig. 11.3 Principle of producing the desired position and yaw angle in stabilize mode

the principle of the three modes is investigated. It should be noted that the proposed method is only one way to realize the three modes.

11.1.3.1 Stabilize Mode

The stabilize mode produces the desired thrust and moments according to the desired position $\mathbf{p}_d = \hat{\mathbf{p}}$ and desired yaw $\psi_d = \hat{\psi}$. As shown in Fig. 11.3, the autopilot can produce the desired position $\mathbf{p}_d$, and yaw angle ψ_d according to the RC commands.

By recalling the controllers proposed in Eqs. (10.5) and (10.10) in Chap. 10, the horizontal position controller is

$$\boldsymbol{\Theta}_{hd} = -g^{-1}\mathbf{A}_{\psi}^{-1}\left(-\mathbf{K}_{\mathbf{p}hd}\dot{\hat{\mathbf{p}}}_h - \mathbf{K}_{\mathbf{p}hp}\left(\hat{\mathbf{p}}_h - \mathbf{p}_{hd}\right)\right)$$

and the altitude controller is

$$f_d = mg - m\left(-k_{p_z d}\dot{\hat{p}}_z - k_{p_z p}\left(\hat{p}_z - p_{z_d}\right)\right).$$

Since $\mathbf{p}_d = \hat{\mathbf{p}}$, the horizontal position controller and altitude controller become

$$\begin{aligned}\boldsymbol{\Theta}_{hd} &= g^{-1}\mathbf{A}_{\psi}^{-1}\mathbf{K}_{\mathbf{p}hd}\dot{\hat{\mathbf{p}}}_h \\ f_d &= mg + mk_{p_z d}\dot{\hat{p}}_z.\end{aligned}$$

Generally, let $\boldsymbol{\Theta}_{hd} = \left[\,\phi_d\ \theta_d\,\right]^{\mathrm{T}} = \mathbf{0}_{2\times1}$, because $\dot{\hat{\mathbf{p}}}_h$ may be unavailable or inaccurate, or the magnetometer is unavailable. This implies that AC can make the multicopter self-level. Moreover, AC cannot hold the altitude anymore. A simple yaw controller has the following form

$$\tau_z = -k_{\psi}\left(\hat{\psi} - \psi_d\right) - k_{\dot{\psi}}\dot{\psi}.$$

The yaw angle estimate $\hat{\psi}$ may be inaccurate and may drift slowly in a range because the magnetometer is sensitive to the environment. Therefore, if the estimate to the

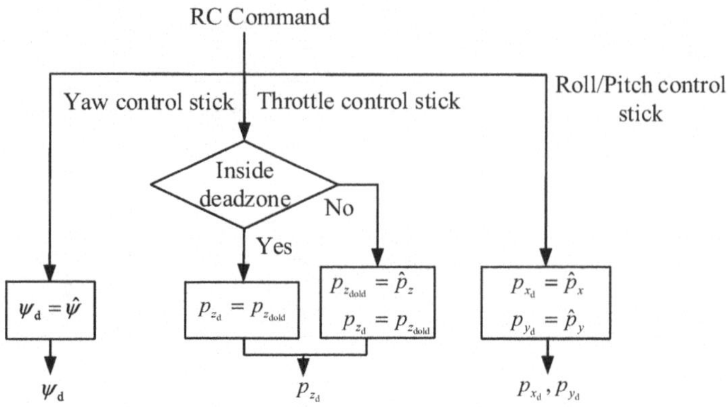

Fig. 11.4 Principle of producing the desired position and yaw angle in altitude hold mode

yaw angle is used as feedback, the heading may oscillate. On the other hand, the yaw rate measured by gyroscopes is fairly accurate. So a simple way to keep the heading is to make the yaw rate be zero. This is realized by setting $\psi_d = \hat{\psi}$. As a result, the yaw controller becomes

$$\tau_z = -k_{\dot{\psi}}\dot{\psi}.$$

Stabilize mode cannot make the multicopter hover steadily due to the lack of feedback of the horizontal position signal. This mode is often used when a multicopter is operated without a GPS receiver and height sensors or they fail.

11.1.3.2 Altitude Hold Mode

The altitude hold mode produces the desired thrust and moment according to the desired altitude $p_{z_d} = p_{z_{dold}}$, horizontal position $\mathbf{p}_{hd} = \hat{\mathbf{p}}_h$ and yaw angle $\psi_d = \hat{\psi}$, where $\mathbf{p}_{hd} = \hat{\mathbf{p}}_h$ implies $\theta_d = \phi_d = 0$.

As shown in Fig. 11.4, the time, denoted by t_{z_d}, is recorded when the throttle control stick turns back to the midpoint, and then, the altitude estimate $\hat{p}_z\left(t_{z_d}\right)$ is saved as $p_{z_{dold}} = \hat{p}_z\left(t_{z_d}\right)$. At the same time, the altitude hold mode starts to hold the multicopter's altitude at $p_{z_d} = p_{z_{dold}}$. Just like the stabilize mode, the altitude hold mode cannot make the multicopter hover for lack of feedback in the horizontal position. The altitude hold mode is often used when the height sensors are available while position sensors or electronic compasses are unavailable.

11.1.3.3 Loiter Mode

As shown in Fig. 11.5, the loiter mode produces the desired thrust and moment according to the desired position $\mathbf{p}_d = \mathbf{p}_{dold}$ and yaw angle $\psi_d = \psi_{dold}$.

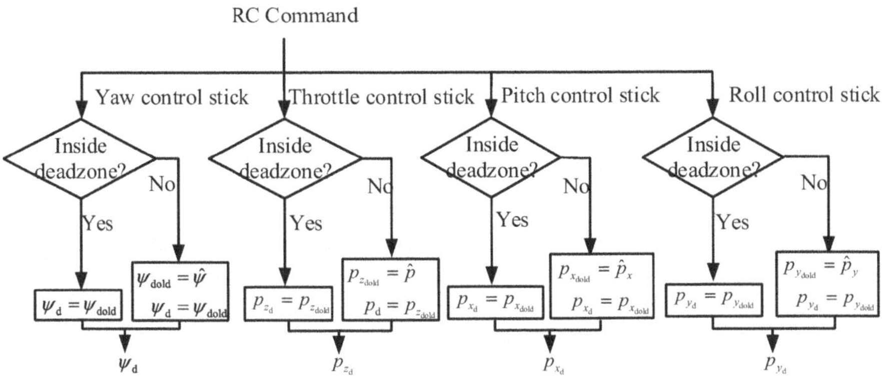

Fig. 11.5 Principle of producing the desired position and yaw angle in loiter mode

The time, denoted by $t_{\psi_d}, t_{z_d}, t_{x_d}, t_{y_d}$, is recorded when the four control sticks turn back to the midpoint, respectively. The estimates are saved to be desired ones that

$$\begin{aligned}\psi_{\text{dold}} &= \hat{\psi}\left(t_{\psi_d}\right)\\ p_{z_{\text{dold}}} &= \hat{p}_z\left(t_{z_d}\right)\\ p_{x_{\text{dold}}} &= \hat{p}_x\left(t_{x_d}\right)\\ p_{y_{\text{dold}}} &= \hat{p}_y\left(t_{y_d}\right).\end{aligned}$$

At the same time, the loiter mode starts to control the multicopter to hover at $\mathbf{p}_d = \mathbf{p}_{\text{dold}}$ and $\psi_d = \psi_{\text{dold}}$. The loiter mode is often used when the height sensors, position sensors, and electronic compasses are all available.

11.1.4 Switching Logic Between RC and AC

The closed-loop control block diagram under the SAC mode is shown in Fig. 11.6. Based on Fig. 11.6, the switching logic between RC and AC is discussed.

11.1.4.1 Yaw Command Switching Logic

In the SAC, as shown in Fig. 11.6, the total yaw command is

$$\psi_d = \psi_{\text{dac}} + \psi_{\text{drc}}$$

where ψ_{dac} is produced by the AC in autopilot, and ψ_{drc} represents the command from the RC transmitter. As described in the previous section, $\psi_{\text{dac}} = \hat{\psi}$ or $\psi_{\text{dac}} = \psi_{\text{dold}}$

depends on which mode (stabilize mode, altitude hold mode, or loiter mode) the multicopter is in. RC command ψ_{drc} is presented as

$$\psi_{\text{drc}} = \dot{\psi}_{\text{drc}} \Delta t$$

where $\dot{\psi}_{\text{drc}}$ is the yaw control command, and Δt is a command period or can be considered as an adjustable gain.

When the yaw control stick is at its midpoint, then $\psi_{\text{drc}} = 0$, and the autopilot will control the yaw to ψ_{dac} depending on which mode the autopilot is in. If the yaw control stick leaves its midpoint, then $\psi_{\text{dac}} = \hat{\psi}$; namely, the RC will take over the control of the yaw channel completely.

11.1.4.2 Throttle Command Switching Logic

As shown in Fig. 11.6, the total throttle command is expressed as

$$f_{\text{d}} = f_{\text{dac}} + f_{\text{drc}} \left(\sigma_{\text{drc}} \right)$$

where f_{dac} is produced by AC in autopilot, and $f_{\text{drc}} \left(\sigma_{\text{drc}} \right)$ represents the command from the RC transmitter. More precisely, the waypoint producing module produces the desired position according to the current mode (stabilize mode, altitude hold mode, or loiter mode) the multicopter is in f_{dac}, and then, the position controller in autopilot produces the f_{dac}. The RC command function $f_{\text{drc}} \left(\sigma_{\text{drc}} \right)$ is presented in Fig. 11.2.

When the throttle control stick is at its midpoint, $f_{\text{drc}} \left(\sigma_{\text{drc}} \right) = 0$, the autopilot will control the position according to the total throttle command f_{dac}. Once the throttle

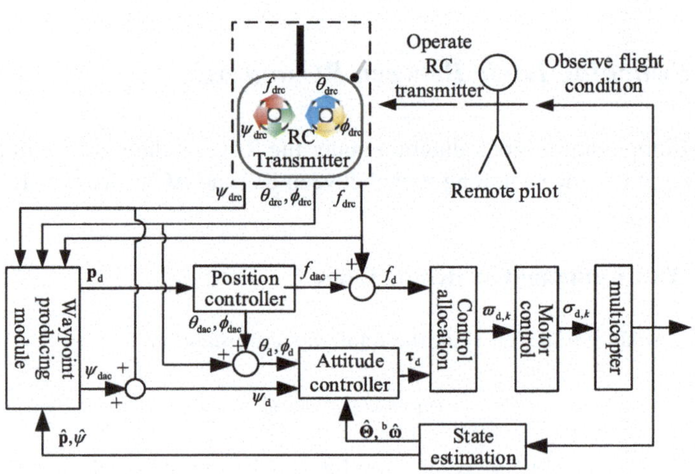

Fig. 11.6 Closed-loop control block diagram under the SAC mode

control stick leaves its midpoint, $p_{z_\text{d}} = \hat{p}_z$, AC does not offer the feedback in altitude except for velocity feedback, and RC will take over the control of the altitude channel completely.

11.1.4.3 Roll/Pitch Command Switching Logic

As shown in Fig. 11.6, the total roll/pitch command is

$$\begin{aligned}\phi_\text{d} &= \phi_\text{dac} + \phi_\text{drc}\\ \theta_\text{d} &= \theta_\text{dac} + \theta_\text{drc}\end{aligned}$$

where ϕ_dac and θ_dac are produced by AC in autopilot; ϕ_drc and θ_drc represent the command from the RC transmitter. Similar to the throttle command f_dac, the roll/pitch commands ϕ_dac and θ_dac are also produced by the position controller in autopilot.

Take the pitch control stick as an example. When the pitch control stick is at its midpoint, $\theta_\text{drc} = 0$, the autopilot will control the pitch to θ_dac depending on which mode the autopilot is in. Once the pitch control stick leaves its midpoint, $p_{x_\text{d}} = \hat{p}_x$. Then, the AC does not offer the feedback in the horizontal location except for velocity feedback. This implies that RC will take over the control of the horizontal channel completely.

11.2 Basic Experiment

11.2.1 Experimental Objective

(1) Things to prepare

1) Hardware: Multicopter System, Pixhawk Autopilot System;
2) Software: MATLAB R2017b or above, Simulink-based Controller Design and Simulation Platform, HIL Simulation Platform, Instructional Package "e7.1" (https://rflysim.com/course).

(2) Objectives

1) On the Simulink-based controller design and simulation platform, repeat the given code and compare the desired attitude with the attitude response during flight in the stabilize mode; then, record the horizontal position response when the desired attitude is set to 0; finally, record the vertical position response when the throttle stick is returned to the middle position;
2) Perform the HIL simulation.

11.2.2 Experimental Procedure

(1) **Step1: SIL simulation**

1) Open the Simulink-based controller design and simulation platform
Run the file "e7\e7.1\Sim\Init_control.m" to initialize the parameters. Next, the file "StabilizeControl_Sim.slx" will pop up automatically as shown in Fig. 11.7. To simulate some uncertainties, a constant disturbance has been added to the output of the attitude angle, shown in Fig. 11.7.
2) Run the simulation and analyze the recorded experimental results. First, run "StabilizeControl_Sim.slx". Then, observe and record the response and analyze the experimental results. The desired pitch angle and the roll angle are 0, and the responses of the roll angle and pitch angle are shown in Fig. 11.8. Because of the added constant disturbance, the initial roll angle is not 0. However, using the attitude controller, the roll/pitch angle approaches the desired roll/pitch angle. During the process, a non-zero velocity is generated. It can be observed that the horizontal position drifts because of the disturbance shown in Fig. 11.9. In the stabilize mode, the quadcopter can only be stable in attitude and cannot maintain its horizontal position. As for the altitude, it cannot be held either (see Fig. 11.10) due to lack of feedback for the altitude.

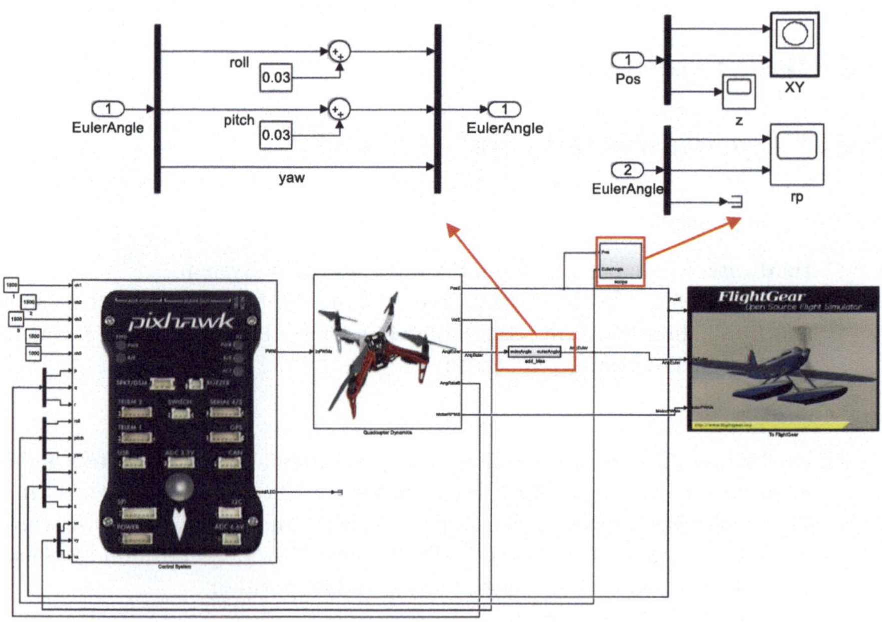

Fig. 11.7 SIL model for stabilize mode, Simulink model "StabilizeControlSim.slx"

(2) **Step2: HIL simulation**

1) Open the Simulink file "e7\e7.1\HIL\StabilizeControl_HIL.slx", as shown in Fig. 11.11. It should be noted that "Control System" in the SIL simulation shown in Fig. 11.7 is the same as that in the HIL simulation shown in Fig. 11.11.
2) Referring the procedure in Sect. 2.3 in Chap. 2, the connection between an RC receiver and a Pixhawk autopilot can be set up as shown in Fig. 11.12.
3) Compile and upload the code
Compile the HIL simulation model and upload the file to the given Pixhawk autopilot. Later, the designed Semi-autonomous control program can be run on Pixhawk autopilot. The operation is shown in Fig. 11.13. The detailed procedure is introduced in Sect. 3.3.3.
4) Configure CopterSim
Double-click on the desktop shortcut CopterSim to open it. Readers can choose different propulsion systems using the following procedure. Click on "Model Parameters" to customize the model parameters and, then click on "Store and use the parameters" to make them available. The software will automatically match the serial port number. Readers would click the "Run" button to enter the HIL simulation mode. After that, readers could see the message returned by the Pixhawk autopilot in the lower-left corner of the interface, as shown in Fig. 11.14.
5) Configure 3DDisplay
Double-click on the desktop shortcut 3DDisplay to open it.
6) Simulation performance
Arm the quadcopter for manual control using the given RC transmitter.[1] As shown in Fig. 11.15, given a desired attitude by the remote control, the quadcopter can quickly track the desired attitude. When all the sticks are in the middle, the attitude is basically horizontal, but the position in the lower right corner of the 3DDisplay software interface is still moving, indicating that the quadcopter is drifting.

11.3 Analysis Experiment

11.3.1 Experimental Objective

(1) Things to prepare

1) Hardware: Multicopter Hardware System, Pixhawk Autopilot System;
2) Software: MATLAB R2017b or above, Simulink-based Controller Design and Simulation Platform, HIL Simulation Platform, Instructional Package "e7.2" (https://rflysim.com/course).

[1] For details, see Sect. 2.3.1.1 in Chap. 2.

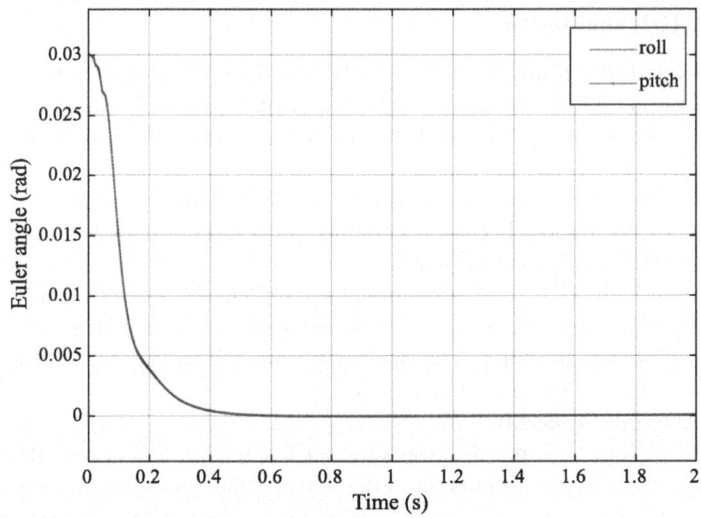

Fig. 11.8 Response of roll angle and pitch angle

Fig. 11.9 Horizontal position drifting when control stick in the middle

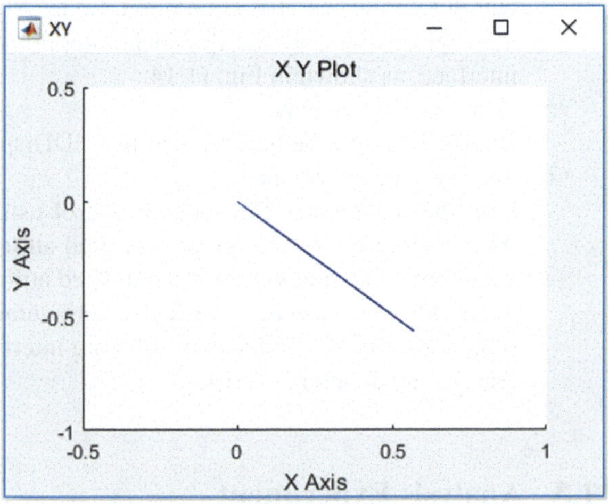

(2) Objectives

1) Design the altitude hold mode based on the stabilize mode. Through the experimental data, compare the attitude and position in the altitude hold mode with those in the stabilize mode.
2) Perform the HIL simulation.

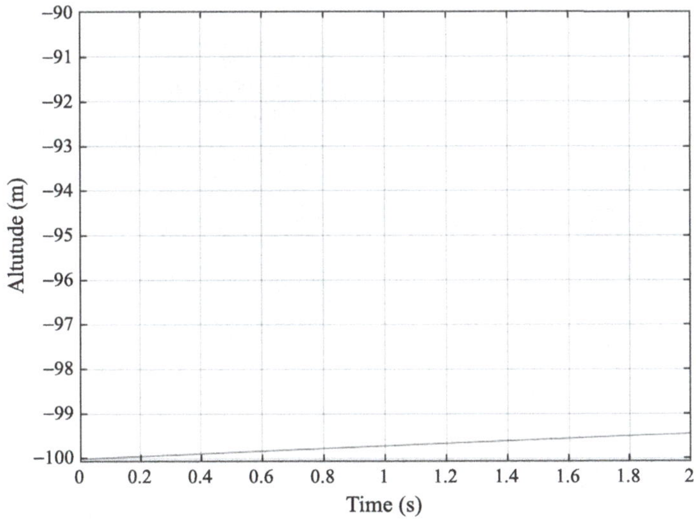

Fig. 11.10 Altitude response when throttle stick in the middle

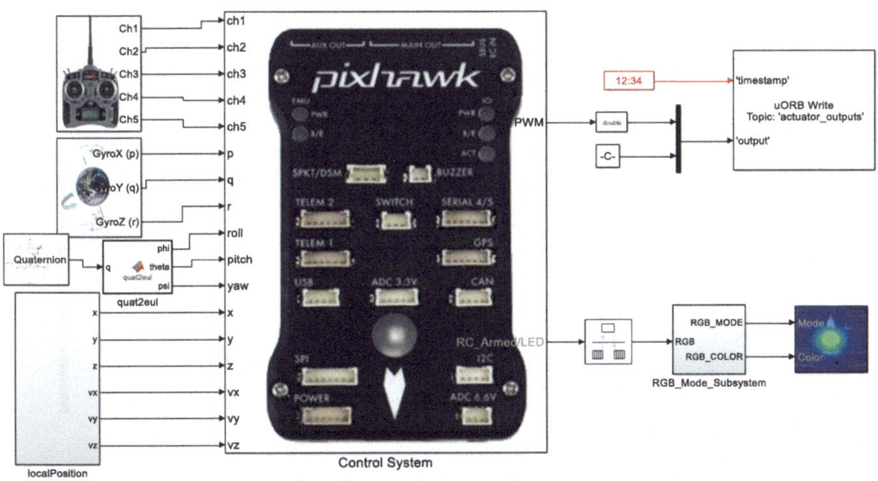

Fig. 11.11 HIL model for stabilize mode, Simulink model "StabilizeControl_HIL.slx"

11.3.2 Experimental Analysis

To realize altitude control, the input of RC transmitter in CH3 channel[2] by remote pilots is converted into v_{zdrc}, i.e., the velocity along the $o_e Z_e$ axis, and then generated the desired throttle f_{drc}. The implementation logic of the stabilize mode is shown in Fig. 11.16. Through "Switch1", θ_{drc},ϕ_{drc} are selected as the desired attitude angles

[2]For details, see Sect. 2.3.1.1 in Chap. 2.

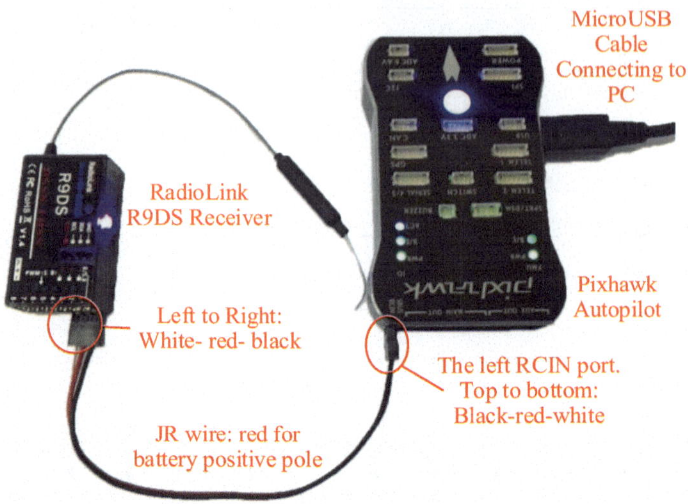

Fig. 11.12 Connection between the Pixhawk hardware and the RC receiver

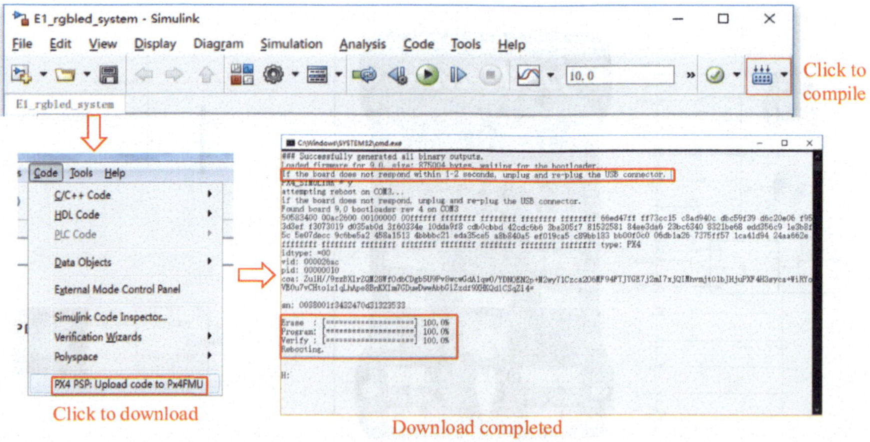

Fig. 11.13 Code compilation and upload process

(the expected yaw angle is the same in all three modes); through "Switch2", f_{drc} is selected as the desired throttle. In the altitude hold mode, through "Switch1", θ_{drc}, ϕ_{drc} are selected as the desired attitude angles; and through "Switch2", f_{dap} is selected as the desired throttle, as shown in Fig. 11.17.

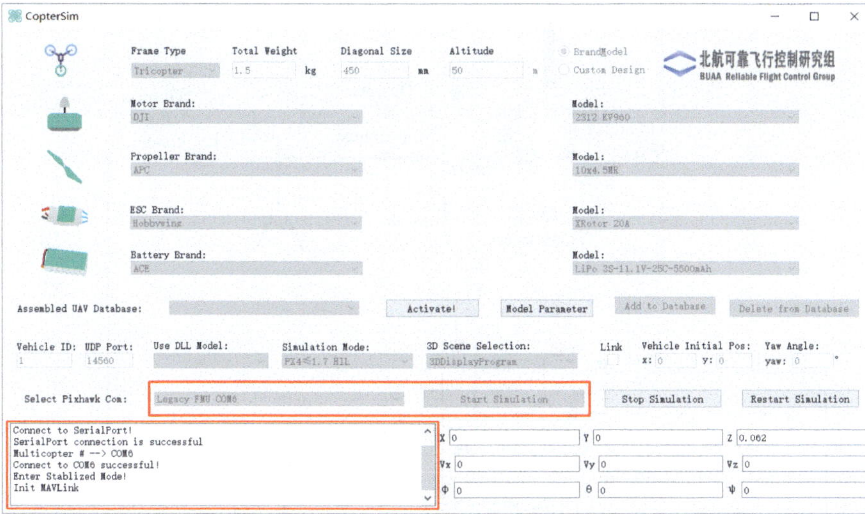

Fig. 11.14 User interface of CopterSim

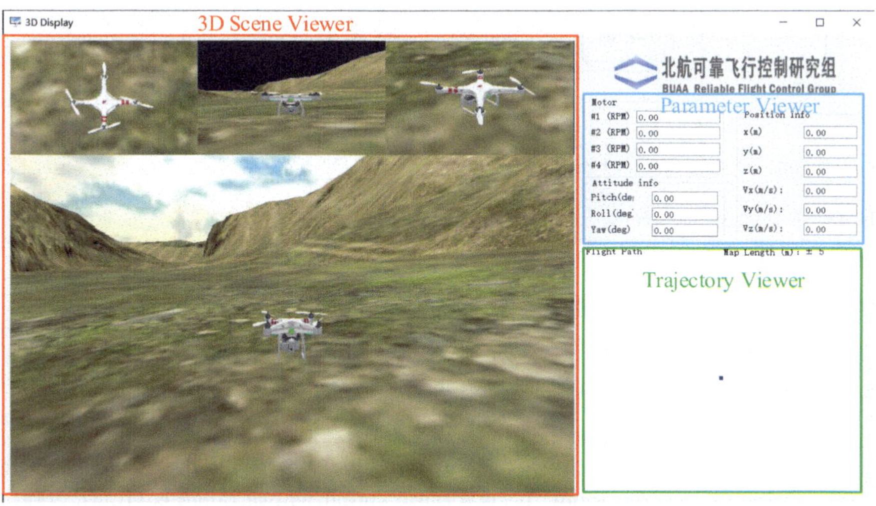

Fig. 11.15 User interface of 3DDisplay

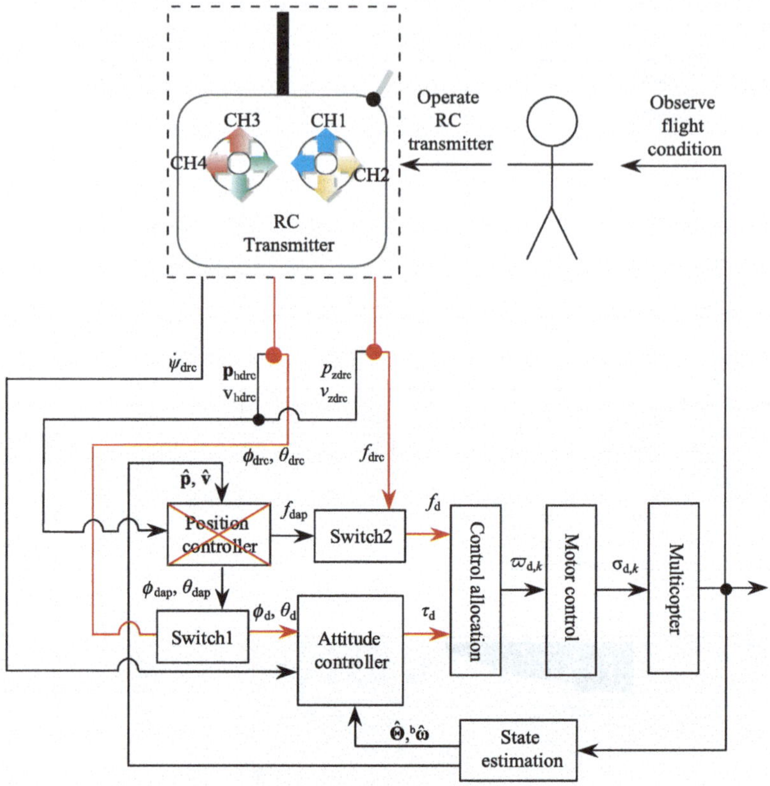

Fig. 11.16 Implementation logic of stabilize mode

11.3.3 Experimental Procedure

(1) **Step1: SIL simulation**

1) Add the dead zone to the input signal of RC transmitter
The dead zone module is shown in Fig. 11.18 with the code given in Table 11.1. If the input "u" is a ramp signal whose range is [1000, 2000], then the corresponding output is shown in Fig. 11.19. After normalization, the range of the output signal amplitude is [−1, 1].

2) Determine the desired position
When the throttle control stick deviates from the middle position, i.e., out of the dead zone range, the desired altitude is changed to be the current altitude. With such a desired altitude, the altitude feedback does not work anymore because the altitude error is always zero. Only altitude velocity feedback works for the altitude controller. On the other hand, when the throttle control stick is within the middle dead zone range, i.e., within the dead zone, the desired altitude is the altitude at the moment when the control

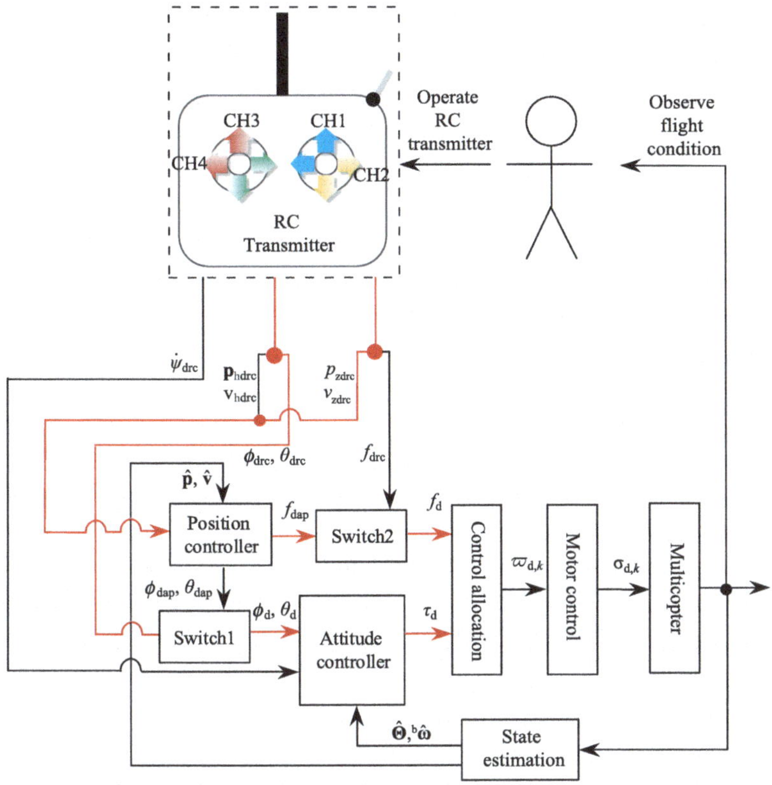

Fig. 11.17 Implementation logic of altitude hold mode

stick just returned back to the dead zone. If the control stick is always in the middle position, the desired altitude remains the same. The design is shown in Table 11.2.

3) Realize the controller for the altitude hold mode
Add the dead zone module and the desired position module designed in the previous two steps to the controller in Fig. 11.7. Besides, change the input of mode type from 0 to 1, indicating that the altitude hold mode is available, as shown in Fig. 11.20.
4) Run the simulation and analyze the test results
Run the file "e7\e7.2\Sim\HeightControl_Sim.slx". Readers can find that the attitude and horizontal position response are the same as that in the stabilize mode which means that the attitude can remain stable, whereas the horizontal position cannot remain stable, as shown in Fig. 11.21. When the altitude input is between 1460 and 1540, meaning the throttle control stick is within middle dead zone range, the altitude change is very small with an error of ±0.002m as shown in Fig. 11.22. When the throttle exceeds the dead

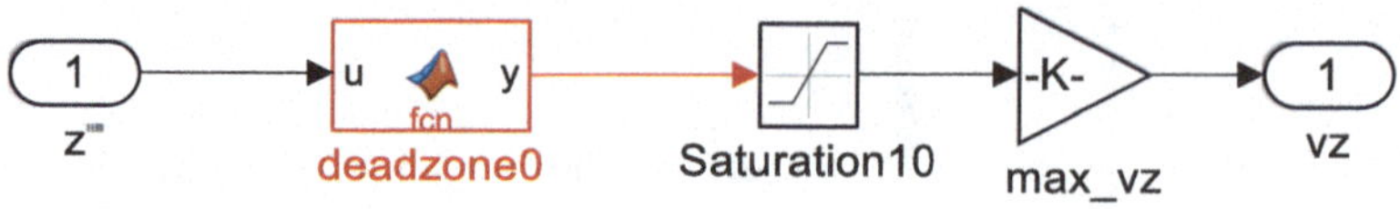

Fig. 11.18 Dead zone model in Simulink

zone, such as the throttle input is 1600, as shown in Fig. 11.24, the altitude velocity follows the desired velocity steadily while the altitude continues to increase, as shown in Fig. 11.23.

(2) **Step2: HIL simulation**

1) Open the Simulink file for HIL
Open the file "e7\e7.2\HIL\HeightControlHIL.slx", as shown in Fig. 11.25. It should be noted that "Control System" in the attitude hold mode of the SIL simulation is the same as that in the HIL simulation.
2) Connect the hardware
Referring the procedure in Sect. 2.3 in Chap. 2, the connection between an RC receiver and a Pixhawk autopilot can be set up as shown in Fig. 11.12.
3) Compile and upload the code
Compile the HIL simulation model and upload the file to the given Pixhawk autopilot. Later, the designed attitude control program can be run on Pixhawk autopilot. The operation is shown in Fig. 11.13. The detailed procedure is introduced in Sect. 3.3.3.
4) Configure CopterSim
Double-click on the desktop shortcut CopterSim to open it. Readers can choose different propulsion systems using the following procedure. Click on "Model Parameters" to customize the model parameters and, then click on "Store and use the parameters" to make them available. The software will automatically match the serial port number. Readers would click the "Run" button to enter the HIL simulation mode. After that, readers could see the message returned by the Pixhawk autopilot in the lower-left corner of the interface, as shown in Fig. 11.14.
5) Configure 3DDisplay
Double-click on the desktop shortcut 3DDisplay to open it.
6) Simulation performance
Arm the quadcopter for manual control.[3] As shown in Fig. 11.15, when controlling the quadcopter, the attitude and horizontal positions of the quadcopter are the same as that in the stabilize mode. When the throttle stick is back to the middle, the altitude of the quadcopter can keep stable.

[3]For details, see Sect. 3.5.3 in Chap. 3.

Table 11.1 Code in the dead zone module

```
function y = fcn(u)
%Function description:
%  Control the range of the remote control input from RCMin to RCMax, which can be obtained from QGC.And set a dead zone with a dead zone size of dead Zone,the output y is normalized to 0~1.
%Input:
%  u:PWM input.
%Output:
%  y:normonalized output after passing the dead zone,range from 0 to 1.
RCMin = 1100;
RCMax = 1900;
RCMid = (RCMin + RCMax)/2;
deadZoneRate = 0.05;
deadZone = deadZoneRate*(RCMax - RCMin);
k = 1/(RCMax - RCMid - deadZone);
%Limiting
if(u < RCMin)
    u = RCMin;
elseif(u > RCMax)
    u = RCMax;
end
%Dead zone and normalization
if(u > RCMid + deadZone)
    y = (u-RCMid - deadZone)*k;
elseif(u < RCMid - deadZone)
    y = (u - RCMid + deadZone)*k;
else
    y = 0;
end

end
```

11.4 Design Experiment

11.4.1 Experimental Objective

(1) Things to prepare

1) Hardware: Multicopter System, Pixhawk Autopilot System;
2) Software: MATLAB R2017b or above, Simulink-based Controller Design and Simulation Platform, HIL Simulation Platform, Experiment Instruction Package “e7.3” (https://rflysim.com/course).

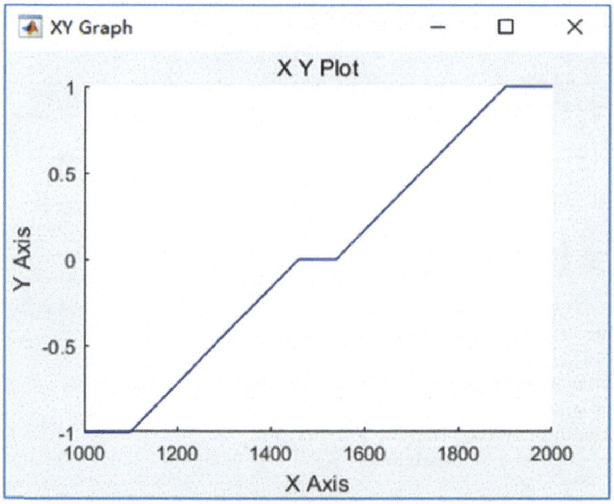

Fig. 11.19 Relationship of RC transmitter's input and output

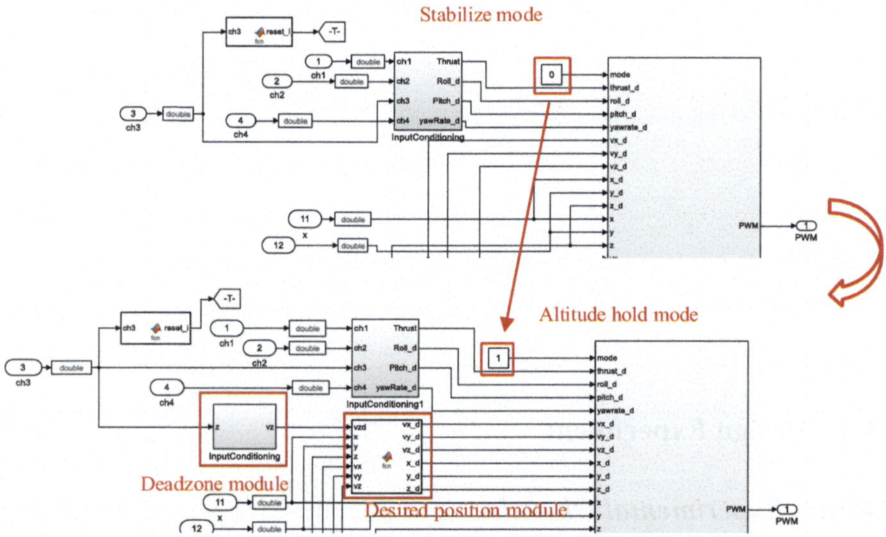

Fig. 11.20 Difference between stabilize mode and altitude hold mode

(2) Objectives

1) Design and realize the loiter mode based on the stabilize mode. Through the experimental data, compare the attitude and position in the loiter mode with those in the stabilize mode.
2) Realize switching among three modes by the three-position switch. Then, perform the HIL simulation and flight test.

Table 11.2 A main function for altitude hold mode

```
function [vx_d, vy_d, vz_d, x_d, y_d, z_d] = fcn(vzd, x, y, z, vx, vy, vz)
%Function description:
%Altitude control logic implementation.
%When the desired speed (specified by the remote controller) is < 0.001, and
    the current speed is <6 m/s, the altitude hold is turned on, and the
    desired altitude is that at which the above condition is satisfied, the
    altitude loop and the velocity loop work simultaneously;
%when the condition is not satisfied, only the speed loop control works.
%Input:
%  vzd: desired speed specified by the remote control
%  x, y, z: current position
%  vx, vy,vz: current position
%Output:
%  vx_d, vy_d, vz_d, x_d, y_d, z_d:desired velocity and position
persistent z1;
if isempty(z1)
    z1= z;
end
persistent hold_z_flag;
if isempty(hold_z_flag)
    hold_z_flag = 0;
end
if abs(vzd) < 0.001 && abs(vz) < 6
    hold_z = 1;
else
    hold_z = 0;
    hold_z_flag = 0;
end
% Throttle in the middle position, hold the height
if (hold_z > 0.5) && (hold_z_flag < 0.5)
    z1 = z;
    hold_z_flag = 1;
end
%The throttle is not within the dead zone range, and the desired altitude is
    the
%current altitude. Altitude controller only works with speed loop
if hold_z < 0.5
    z1 = z;
    hold_z_flag = 0;
end
vx_d = vx;
vy_d = vy;
vz_d = vzd;
x_d = x;
y_d = y;
z_d = z1;
```

11.4.2 Experimental Design

(1) **Step1: Design the loiter mode controller**
In the analysis experiment, to realize altitude control, the input of the RC transmitter in CH3 channel is converted into the altitude velocity along $o_e z_e$ axis v_{zdrc} which is the input of position control. Similarly, to design the loiter mode

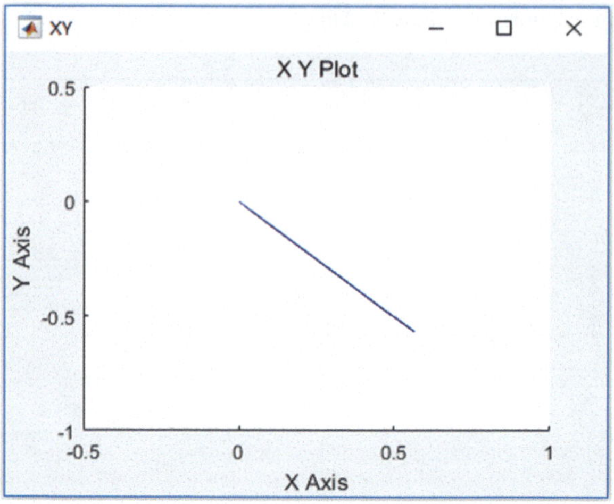

Fig. 11.21 Position response in altitude hold mode

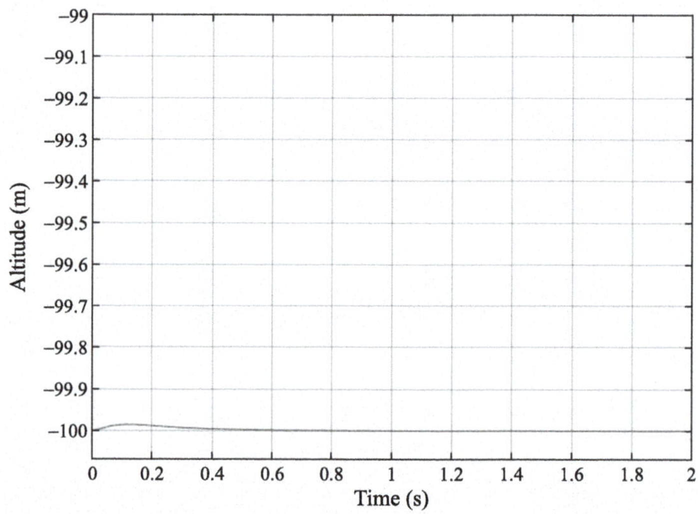

Fig. 11.22 Altitude response when throttle control stick is in the dead zone

controller, the inputs of the RC transmitter in CH1 channel and CH2 channel are converted into the desired velocities along the $o_e x_e$ and $o_e y_e$ axes, as shown in Fig. 11.26. In the loiter mode, θ_{dap},ϕ_{dap} are selected as the desired attitude angle through "Switch1"; and f_{dap} is selected as the desired throttle through "Switch2", as shown in Fig. 11.27.

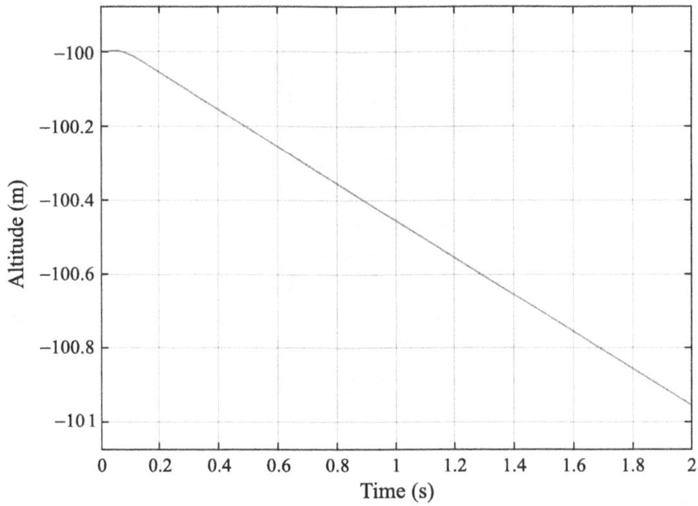

Fig. 11.23 Altitude response when the PWM value of "ch3" is 1600

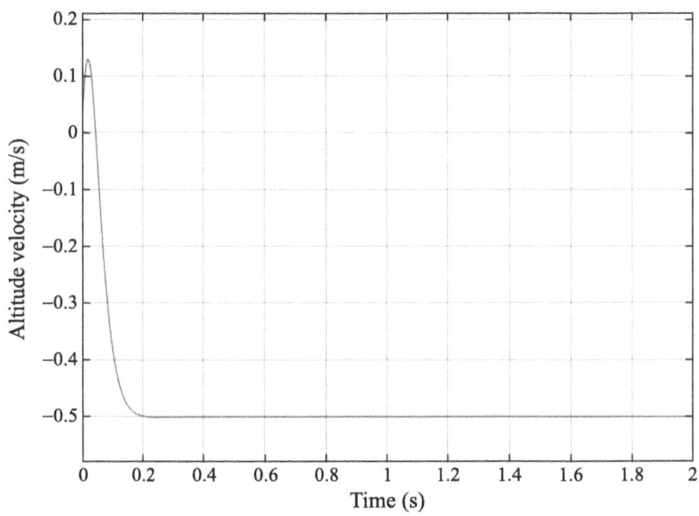

Fig. 11.24 Altitude velocity response when PWM value of "ch3" is 1600

(2) **Step2: Design dead zone for loiter mode**
The dead zone configuration for the loiter mode is the same as that for the altitude hold mode, as shown in Fig. 11.28. If the input of "u" is a ramp signal, whose range is within [1000, 2000], the corresponding response is shown in Fig. 11.19. Besides, while the input signal is normalized at the same time, the range of the output signal amplitude is within [−1, 1].

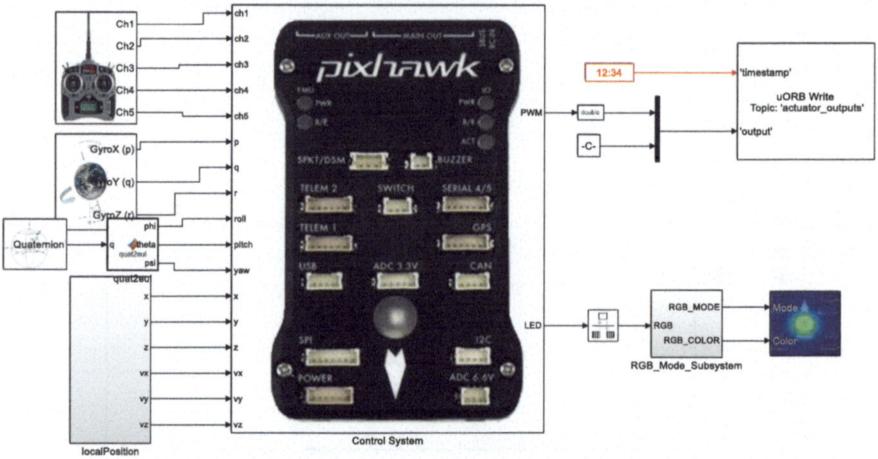

Fig. 11.25 HIL model for altitude hold mode, Simulink model "HeightControlHIL.slx"

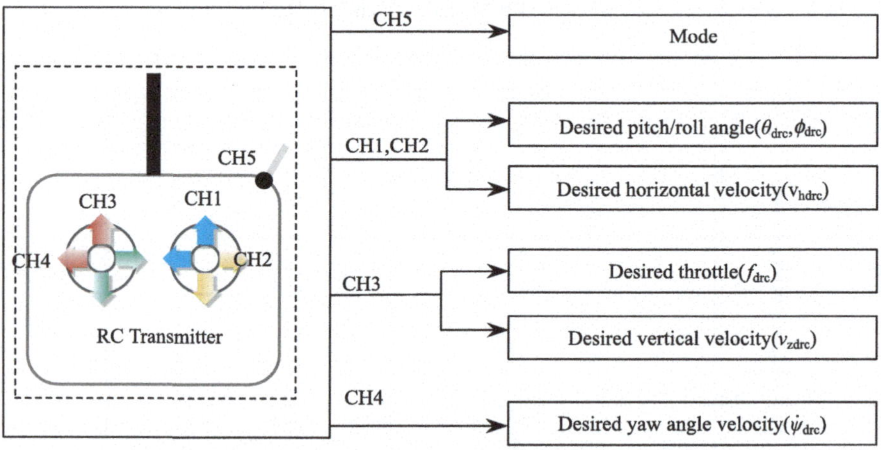

Fig. 11.26 RC transmitter input mapping

(3) **Step3: Determine the desired horizontal position**
When the RC sticks (CH1, CH2 or CH3) deviate from the middle dead zone range, the desired position for AC is always the current position. As a result, the position feedback of AC does not work. There is only the desired velocity by RC, which is calculated and then transmitted to the AC controller. When the RC stick (CH1, CH2 or CH3) is in the middle, the desired position is the position at the moment when the stick comes back to the middle. If the control stick is always in the middle, the desired position remains the same. The specified design is given in Table 11.3.

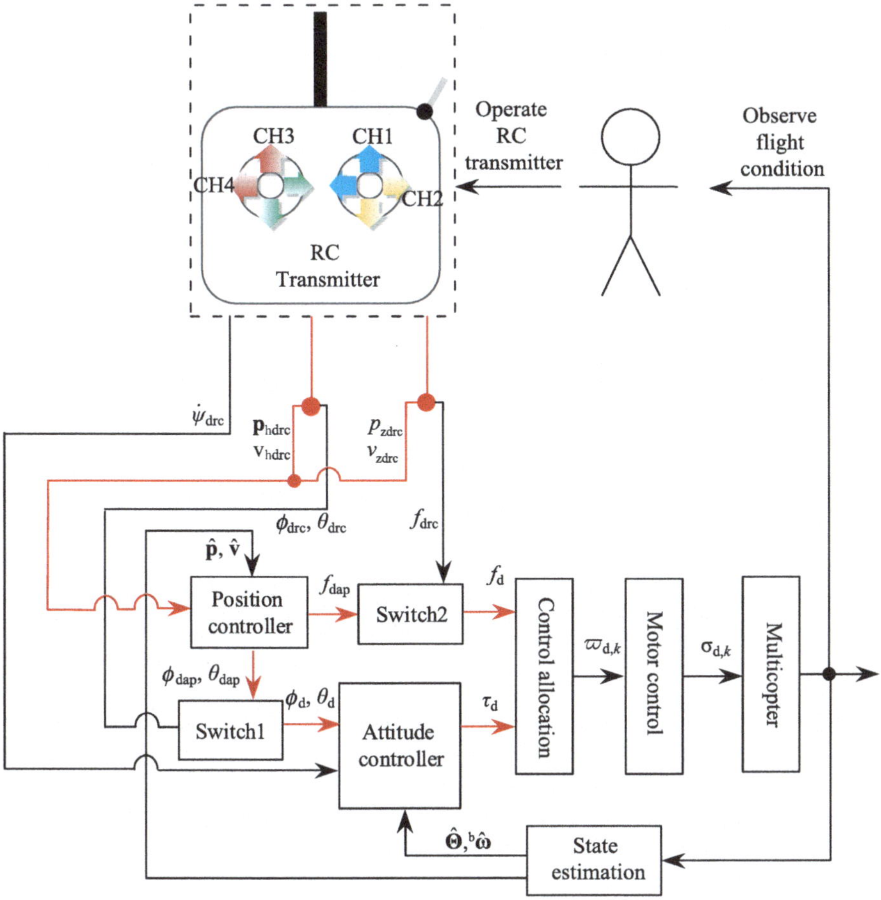

Fig. 11.27 Implementation logic of loiter mode

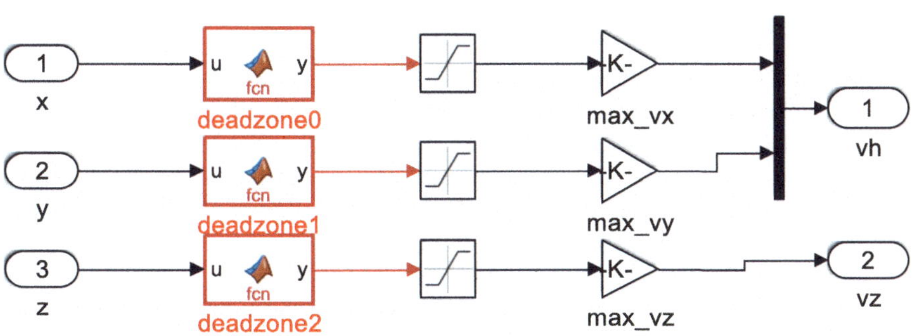

Fig. 11.28 Dead zone setting for loiter mode

(4) **Step4: Design the mode switching controller**
Specify a three-position switch of the RC transmitter as the mode switch, here CH5 is chosen. As for CH5, when the PWM value is within [1000, 1400), let the multicopter be in the stabilize mode; when the PWM value is within [1400,1600), let the multicopter be in the altitude hold mode; when the PWM value is within [1600, 2000), let the multicopter be in the loiter mode, as shown in Table 11.4.

11.4.3 Simulation Procedure

(1) **Step1: SIL simulation**

1) Write a position control model
Add the dead zone and the position controller designed in Sect. 11.4.2 to the stabilize mode model shown in Sect. 11.2.2. Then, change the input mode from 0 to 2, indicating that the loiter model is available, as shown in Fig. 11.29.
2) Run the simulation and analyze test results
Run the file "Init_control.m" to initialize the parameters. Next, the Simulink file "PosControl_Sim.slx" will open automatically. It can be observed from Fig. 11.30 that the altitude response is the same as that in the altitude hold mode, which means that the altitude can remain stable. When the values of the "ch1" and "ch2" (corresponding to CH1 and CH2 of the RC transmitter) are between 1460 and 1540 for the roll and pitch channels, the horizontal position response is shown in Fig. 11.31. It can be observed that, in the presence of the constant disturbance on the roll and pitch channels, the disturbance is well rejected in the loiter mode. When the value of "ch2" is 1600 for the pitch channel, the observed velocity along the $o_e x_e$ axis is shown in Fig. 11.32. The velocity can follow the desired velocity steadily. At the same time, the horizontal position is shown in Fig. 11.33.

(2) **Step2: HIL simulation**

1) Build the HIL simulation model
Replace the "Control System" module in the "StabilizeControl_HIL.slx" model with the "Control System" module in the Simulink model of the position control mode, and add the CH5 channel for mode switch shown in Table 11.4. Then you can get the HIL simulation model "e7\e7.3\HIL\ModeSwitch_HIL.slx", as shown in Fig. 11.34.
2) Connect hardware
Referring the procedure in Sect. 2.3 in Chap. 2, the connection between an RC receiver and a Pixhawk autopilot can be set up as shown in Fig. 11.12.
3) Compile and upload code
Compile the HIL simulation model and upload the file to the given Pixhawk autopilot. Later, the designed semi-autonomous control program can be run on Pixhawk autopilot. The operation is shown in Fig. 11.13. The detailed procedure is introduced in Sect. 3.3.3.

Table 11.3 A main function for the loiter mode

```
function [vx_d, vy_d, vz_d, x_d, y_d, z_d] = fcn(vxd, vyd, vzd, x, y, z, vx, vy, vz)
%Function description:
%Position control logic implementation.
%When the desired speed (specified by the remote controller) is < 0.001, and
%the current speed is <6 m/s, the position hold is turned on, and
%the desired altitude is the position at which the above condition
%is satisfied, the altitude loop and the velocity loop act simultaneously;
%when the condition is not satisfied, Only the velocity loop control works.
%Input:
%  vxd, vyd, vzd: desired speed specified by the remote control
%  x, y, z: current position
%  vx, vy,vz: current position
%Output:
%  vx_d, vy_d, vz_d, x_d, y_d, z_d:desired velocity and position
persistent x1;
if isempty(x1)
    x1 = 0;
end
persistent y1;
if isempty(y1)
    y1 = 0;
end
persistent z1;
if isempty(z1)
    z1 = z;
end
persistent hold_x_flag;
if isempty(hold_x_flag)
    hold_x_flag = 0;
end
persistent hold_y_flag;
if isempty(hold_y_flag)
    hold_y_flag = 0;
end
persistent hold_z_flag;
if isempty(hold_z_flag)
    hold_z_flag = 0;
end
if abs(vxd) < 0.001 && abs(vx) < 6
    hold_x = 1;
else
    hold_x = 0;
    hold_x_flag = 0;
end
if abs(vyd) < 0.001 && abs(vy) < 6
    hold_y = 1;
else
    hold_y = 0;
    hold_y_flag = 0;
end
if abs(vzd) < 0.001 && abs(vz) < 6
    hold_z = 1;
else
    hold_z = 0;
    hold_z_flag = 0;
end
if (hold_x > 0.5) && (hold_x_flag < 0.5)
    x1 = x;
    hold_x_flag = 1;
end
```

(continued)

Table 11.3 (continued)

```
if (hold_y > 0.5) && (hold_y_flag < 0.5)
    y1 = y;
    hold_y_flag = 1;
end
% throttle in the middle position, hold the height
if (hold_z > 0.5) && (hold_z_flag < 0.5) z1 = z; hold_z_flag = 1;
end
%The rocker is not in the middle position and the desired position is the current
    position.
%The position controller only works with the speed loop
if hold_x < 0.5
    x1 = x;
    hold_x_flag = 0;
end
if hold_y < 0.5
    y1 = y;
    hold_y_flag = 0;
end
if hold_z < 0.5
    z1 = z;
    hold_z_flag = 0;
end
vx_d = vxd; vy_d = vyd; vz_d = vzd; x_d = x1; y_d = y1; z_d = z1;
```

Table 11.4 Mode switch

```
function control_mode = fcn(ch5)
%0:Stabilize mode
%1: Altitude hold mode
%2: loiter mlde
if ch5<1400
control_mode=0;
elseif ch5<1600
control_mode=1;
else
control_mode=2;
end
```

4) Configure CopterSim
Double-click on the desktop shortcut CopterSim to open it. Readers can choose different propulsion systems using the following procedure. Click on "Model Parameters" to customize the model parameters and, then click on "Store and use the parameters" to make them available. The software will automatically match the serial port number. Readers would click the "Run" button to enter the HIL simulation mode. After that, readers could see the message returned by the Pixhawk autopilot in the lower-left corner of the interface, as shown in Fig. 11.14.

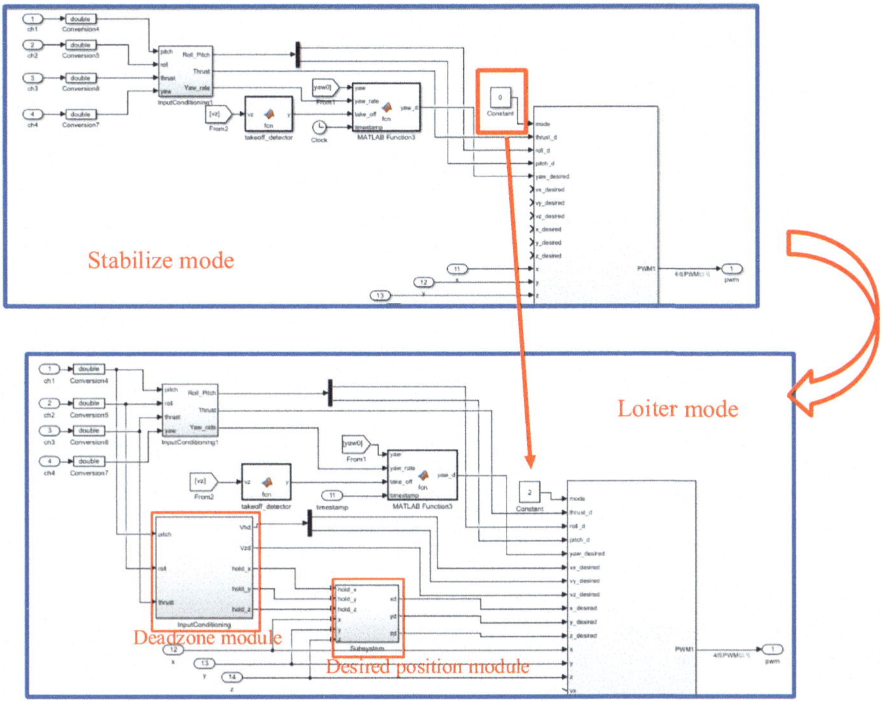

Fig. 11.29 Difference between stabilize mode and loiter mode

5) Configure 3DDisplay
   Double-click on the desktop shortcut 3DDisplay to open it.
6) Simulation performance
   Arm the quadcopter for manual control.[4] Rotate the left-upper switch corresponding to CH5 for mode switching. When the quadcopter is in the stabilize mode, its response is the same as that in the basic experiment; when it is switched to the altitude hold mode, the performance is the same as that in the analysis experiment; when it is switched to the loiter mode and all the sticks return to the center, the quadcopter stays in the air after adjusting.

[4]For details, see Sect. 2.3.1.1 in Chap. 2.

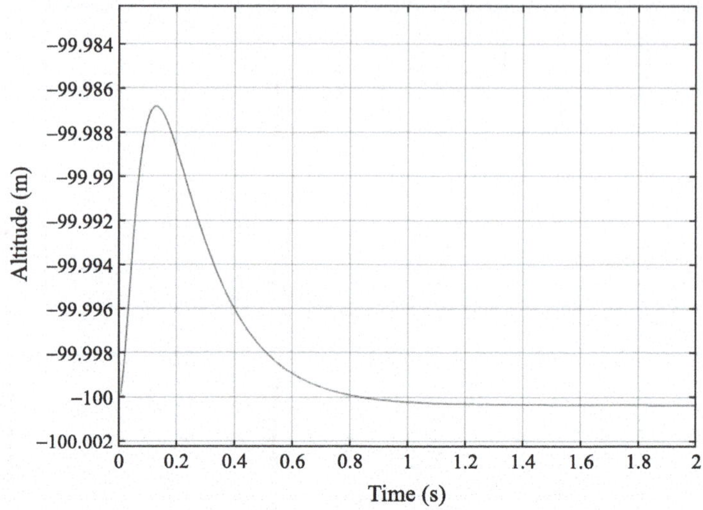

Fig. 11.30 Altitude response when the throttle control stick is in the middle

Fig. 11.31 Horizontal position response when pitch and roll sticks are in the middle

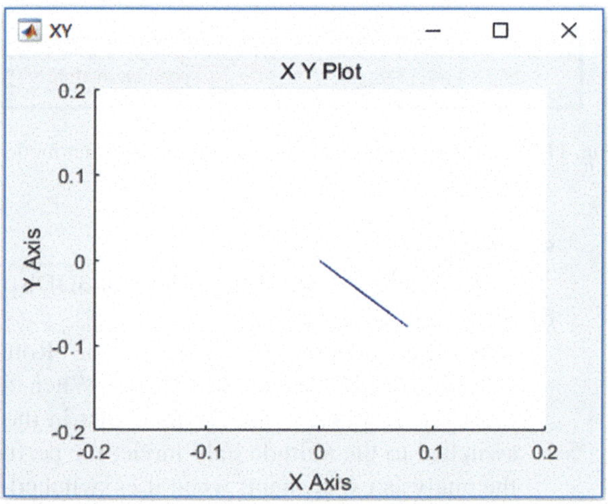

11.4.4 Flight Test Procedure

(1) **Step1: Quadcopter preparation**
For details, see Sect. 4.3.4 in Chap. 4.
(2) **Step2: Simulink model for flight test**
The model is modified according to the configuration of Step 1 shown in Fig. 11.35. Replace the PWM output part in HIL Simulink module with the block in Fig. 11.35. In order to record flight data, a new data recording method is also used, and the detailed procedure can be found in Sect. 9.4.4.

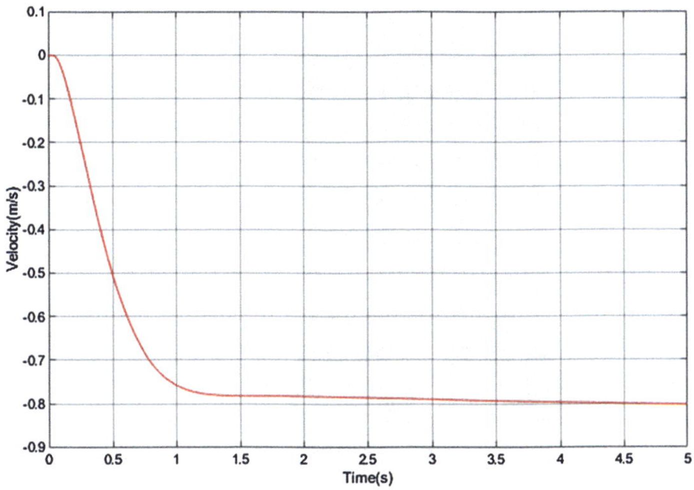

Fig. 11.32 Velocity response along the $o_e x_e$ axis when PWM value of "ch2" is 1600

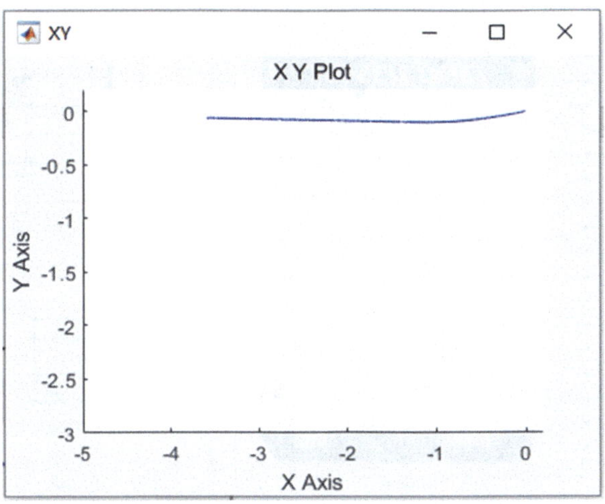

Fig. 11.33 Horizontal position response when PWM value of "ch2" is 1600

(3) **Step3: Upload code**
This process is similar to that used for compiling and uploading the code in HIL simulations. The experimental flow is shown in Fig. 11.13. The detailed procedure is introduced in Sect. 3.3.3.

(4) **Step4: Outdoor flight test**
To ensure safety, a rope is tethered to the quadcopter, and the other end is tethered to a heavy object. The remote pilot maintains a safe distance from the quadcopter during flight, as shown in Fig. 9.39.

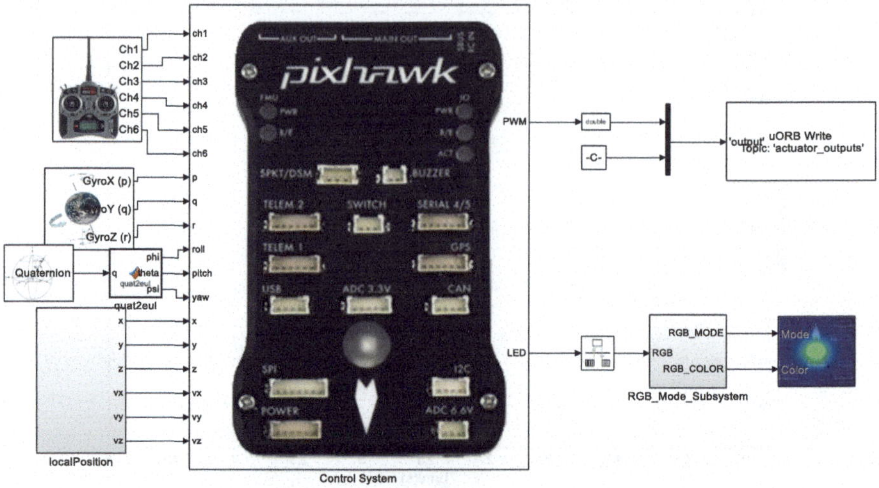

Fig. 11.34 HIL model for mode switching, Simulink model "ModeSwitch_HIL.slx"

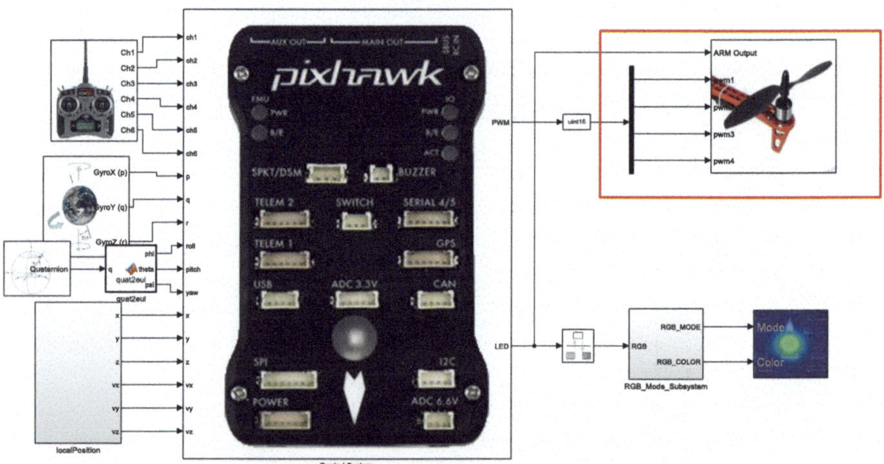

Fig. 11.35 Simulink model for flight test, Simulink model "ModeSwitch_FLY.slx"

(5) **Step5: Analyze data**
The flight test data is shown in Fig. 11.36. It can be observed that in the first 60s, the quadcopter is in the stabilize mode, corresponding the phase "a", and then shifts to the altitude hold mode, corresponding to the phase "b". Here, the throttle stick is at the middle and the altitude remains the same. Later, it enters the loiter mode where the position remains the same, corresponding to the phase "c". At the same time, the pitch and roll sticks are at the middle and the throttle stick is centered. The flight performance and data show that the three modes of the quadcopter can be switched correctly.

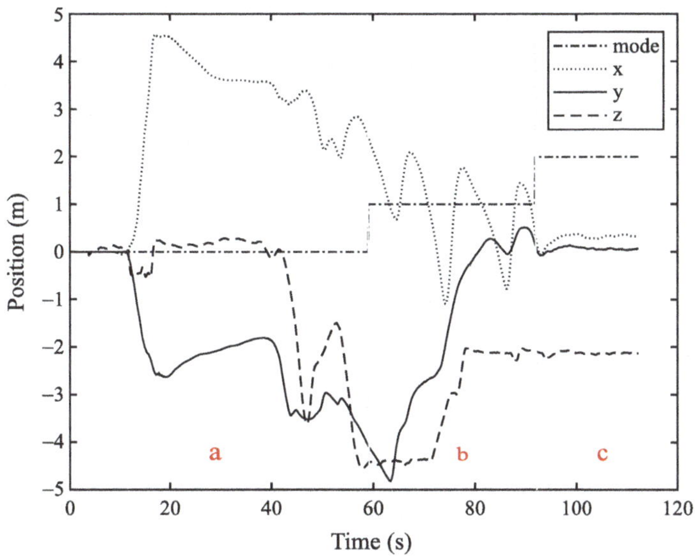

Fig. 11.36 Flight test data

11.5 Summary

(1) In the basic experiment, the quadcopter maintains the desired attitude and position in the "stabilize" mode under ideal conditions, i.e., when no uncertainties exist. However, due to environmental disturbances and measurement errors (similar to disturbances given in the SIL simulation), the position of quadcopter will drift.

(2) In the design of the "altitude" mode, the PWM inputs of the RC transmitter are converted into desired angles. The key point of the altitude hold mode design is the control logic related to the dead zone. In the dead zone, the altitude feedback of AC works to ensure the altitude remains constant. When the control stick is out of the dead zone, only the velocity feedback of the AC works for tracking the command from RC in the altitude channel.

(3) In the design experiment, based on the design of the altitude hold mode, the design of the loiter mode is realized; in this mode, the command for the horizontal position is from the roll and pitch control stick. For mode switching, using the three-position switch in the RC transmitter, readers can convert the input of the RC transmitter into the corresponding signal to trigger the desired mode.

If you have any question, please go to https://rflysim.com for your information.

Chapter 12
Failsafe Logic Design Experiment

For multicopters, several types of failure cannot be avoided, including communication breakdown, sensor failure, and propulsion system anomalies. These failures may cause the abortion of missions, multicopter crashes, and injuries or even death. To guarantee safety, the multicopter's decision-making module should prevent or mitigate unsafe consequences of system failures. For this purpose, some flight modes are defined and switched according to the health evaluation results of each component. This chapter will introduce the principle and method of a multicopter failsafe system and its design; readers will gradually master this part of the knowledge through three step-by-step experiments: basic, analysis, and design from shallow to deep. In the basic experiment, readers will repeat the Stateflow simulation to understand basic mode switching between the manual flight mode (it can be stabilize mode, altitude-hold mode, or loiter mode) and Return-to-Launch (RTL) as well as switching between the manual flight mode and the auto-landing mode. In the analysis experiment, readers will design a new logic to realize a multicopter that can switch between the RTL mode and auto-landing mode manually. In the design experiment, readers will design a new logic and controllers to realize a multicopter that can return and land automatically under conditions in which the connection to the Radio Control (RC) transmitter is abnormal.

12.1 Preliminary

To make this chapter self-contained, the preliminary is from Chap. 14 of *Introduction to Multicopter Design and Control* [8].

© Publishing House of Electronics Industry 2020, corrected publication 2021

Q. Quan et al., *Multicopter Design and Control Practice*,
https://doi.org/10.1007/978-981-15-3138-5_12

12.1.1 Safety Issues

First, the major types of failures that may cause accidents will be introduced. Here, the following three types of failures are mainly considered: communication breakdown, sensor failure, and propulsion system anomaly.

12.1.1.1 Communication Breakdown

Communication breakdown mainly refers to a contact anomaly between the RC transmitter and the multicopter, or between the Ground Control Station (GCS) and the multicopter. Such failures can be categorized as follows.

(1) RC transmitter not calibrated. An RC transmitter without calibration implies that the remote pilot does not calibrate the RC transmitter before the first flight of the multicopter. As a result, the flight control system cannot recognize the user instructions given by the sticks of the RC transmitter. This will lead to flight accidents due to the misinterpretation of the user instructions.
(2) Loss of RC. Loss of RC implies that the RC transmitter is unable to communicate with the corresponding RC receiver onboard before the multicopter takes off or during flight. The loss of RC will result in the multicopter going out of control, which will lead to an accident. The causes of RC loss include the following: 1) the remote pilot turns off the RC transmitter; 2) the multicopter flies outside of the control range of the RC transmitter; 3) the RC transmitter or the receiver onboard loses power or fails; 4) the wires connecting the onboard RC receiver to the autopilot are broken.
(3) Loss of GCS. Loss of GCS implies that the GCS is unable to communicate with the corresponding multicopter before the multicopter takes off or during flight. The loss of GCS will cause the multicopter to fail to reach the mission position, and then the task will fail. The causes of GCS loss include but are not limited to the following: 1) the remote pilot turns off the GCS; 2) the multicopter flies outside of the control range of the GCS; 3) the GCS or the corresponding signal receiver onboard loses power or fails; 4) the wires connecting the radio telemetry receiver for the GCS to the autopilot are broken.

12.1.1.2 Sensor Failure

Sensor failure mainly implies that a sensor on the multicopter cannot measure accurately or cannot work properly. Such failures can be categorized as follows:

(1) Barometer failure. Barometer failure will cause a multicopter to fail to measure the flight altitude accurately. The causes of barometer failure include the following: 1) barometer hardware failure; 2) height measurement results from

barometers and other height measurement sensors (ultrasonic range finder, etc.) are inconsistent.

(2) Compass failure. Compass failure will result in a multicopters orientation going out of control, meaning the yaw channel cannot be controlled effectively. The causes of compass failure include the following: 1) compass hardware failure; 2) compass not calibrated; 3) compass offset too high, which is often caused by metal objects being placed too close to the compass; 4) regional magnetic field too high or too low (for example, it is 35% above or below the expected value); 5) the internal and external compasses are pointing to different directions (for example, the difference is greater than 45°, which is generally caused by the external compass orientation being set incorrectly).

(3) GPS failure. GPS failure implies that a GPS receiver cannot measure location information accurately. In this case, the multicopter cannot hover or complete the pre-programmed route. After losing the location information from the GPS receiver, the position estimation within several seconds is only acceptable with dead reckoning.

(4) Inertial Navigation System (INS) failure. An INS is a navigation aid that uses a computer, motion sensors (accelerometers), and rotation sensors (gyroscopes) to continuously calculate via dead reckoning the position, orientation, and velocity (direction and speed of movement) of a moving object. In this chapter, INS failure mainly indicates anomalies in accelerometers and gyroscopes, which implies that the system cannot correctly measure attitude angle and attitude angular velocity. The causes of INS failure include but are not limited to: 1) The INS is not calibrated. If the accelerometers are not calibrated, then the measurements from accelerometers may be inconsistent. 2) Accelerometer or gyroscope hardware failures. These may cause accelerometer or gyroscope measurements to be inconsistent. 3) Gyroscopes calibrated unsuccessfully. It mainly implies that the gyroscope calibration fails to capture offsets. This is mostly caused by a multicopter being moved during the gyroscope calibration. Hardware failures of sensors can also cause such a failure. This may cause gyroscope measurement to be inconsistent.

12.1.1.3 Propulsion System Anomaly

A propulsion system anomaly mainly refers to either battery failure, or hardware failure of propulsors of the flight control system caused by Electronic Speed Controllers (ESCs), motors, or propellers.

(1) Battery failure. This usually refers to a lack of power caused by low battery capacity or a degradation in the battery life and is mainly reflected in the following three aspects. 1) The capacity is low and then the output voltage is low, which implies that the motor cannot be driven properly. 2) The increase of battery internal resistance will shorten the battery life. 3) Over-charging, over-discharging,

discharging under low battery voltage, and low temperature condition will decrease the battery's capacity.

(2) ESC failure. This is mainly reflected in the following two aspects. 1) An ESC cannot correctly recognize the Pulse Width Modulation (PWM) instructions given by the autopilot or RC transmitter. 2) An ESC cannot drive the motor as expected.
(3) Motor failure. This mainly means that the motor cannot work correctly under the given ESC control signal.
(4) Propeller failure. This is mainly caused by worn and broken blades, or a loose blade from the propeller shaft, etc. Such a failure often occurs in the case that the motor and propeller are damaged due to a strong collision caused by the improper operation of remote pilots. These crashes will further cause the poor contact in the wires connecting the motor to ESC. Secondly, due to an aggressive maneuver or a motor rotation jam, the working current may be too high so that it damages these electronic components and related solder joints. Thirdly, these components have reached their life span. For motors, the phenomenon of demagnetization may occur under working condition with high temperature. This will also lead to motor failure during flight.

12.1.2 Failsafe Suggestions

Health check of key components of a multicopter will be carried out in the pre-flight process. If the components do not pass the check, then the remote pilot cannot arm the multicopter. During flight, failures will make the multicopter go out of control. Therefore, failsafe for key components will be considered in the autopilot design.

12.1.2.1 Communication Failsafe

The communication failsafe features include: the RC system failsafe and the GCS failsafe. When a multicopter is in flight, it is recommended to perform the following protective measures if RC or GCS is lost:

(1) Do nothing if the multicopter is already disarmed.
(2) The multicopter will be immediately disarmed if it has landed or the remote pilot's throttle control stick is at the bottom.[1]
(3) RTL. if the multicopter has a GPS lock and the straight-line distance from the home position is greater than the threshold (2 m for example).[2]

[1]This implies that the pilot has given up the control.

[2]This indicates that the multicopter does not arrive at the HOME point. Thus, RTL is executed.

(4) Immediately land if the multicopter has no GPS lock or the straight-line distance from the home position is less than the set threshold (2 m for example).[3]
(5) In addition, if the contact between the RC transmitter and the onboard RC receiver is reestablished (or the contact between the GCS and the onboard signal receiver is reestablished), the multicopter should still stay in the current mode rather than being switched back to resume the task it left off before the failsafe was triggered.[4] However, if remote pilots manually manipulate the flight mode switch, they can retake the control of multicopters.

12.1.2.2 Sensor Failsafe

The sensor failsafes include barometer failsafe, compass failsafe, GPS failsafe, and INS failsafe.

(1) The barometer failsafe. In the SAC mode, according to the user setting, it is suggested that the multicopter be switched from the loiter mode or the altitude hold mode to the stabilize mode,[5] or land directly. On the other hand, in the Fully-Autonomous Control (FAC) mode, it is suggested that the multicopter land immediately.
(2) The compass failsafe. In the SAC mode, according to the user setting, it is suggested that the multicopter be switched from the loiter mode to the altitude hold mode, or land directly. On the other hand, in the FAC mode, it is suggested that the multicopter land immediately.
(3) The GPS failsafe. In the SAC mode, according to the user setting, it is suggested that the multicopter be switched from the loiter mode to the altitude hold mode, or land directly. On the other hand, in the FAC mode, it is suggested that the multicopter land immediately.
(4) The INS failsafe. In both SAC and FAC modes, it is suggested that the multicopter land urgently by gradually reducing the throttle command.

12.1.2.3 Propulsion System Failsafe

If the propulsion system of a multicopter is evaluated to be abnormal, then the following actions should be taken:

(1) Do nothing if the multicopter is already disarmed.

---

[3]This indicates that the multicopter has already arrived at the HOME point. Thus, landing is executed.

[4]One of the reasons is that the remote pilot may put the throttle and other control sticks at improper positions after the RC is out of contact. In particular, beginners may be in panic facing the RC out of contact. Consequently, the improper throttle position will cause a sudden change in the multicopter speed. That may in turn cause an accident.

[5]The loiter, altitude hold and stabilize modes have been explained in Sect. 11.1 in Chap. 11.

(2) The multicopter will be immediately disarmed if it has landed or if the remote pilot's throttle control stick is at the bottom for a given time.
(3) RTL if the multicopter has a GPS receiver and the straight-line distance from the home position is greater than a set threshold (2 m for example). Moreover, the battery voltage is required to be higher than a set threshold (or other indices to show the battery with a sufficient capacity to support RTL).
(4) In other cases, it is suggested that the multicopter land directly. If a multicopter has one propulsor (including a propeller, a motor, and an ESC) that has failed, it may lose control in the hover state. In this case, the multicopter should adopt a limited control scheme immediately to land urgently by giving up the yaw or follow methods given in. If the multicopter is still controllable in the hover state, then control reallocation is often adopted or robust stabilizing control is used by regarding the damage as a disturbance.

12.1.3 A Safe Semi-autonomous Autopilot Logic Design

In this section, for simplicity, a Semi-Autonomous Autopilot (SAA) logic is realized by using an Extended Finite-State Machine (EFSM). State automaton is a mathematical model to describe a discrete-event system or a hybrid system. Generally, the following conditions are assumed to be true: (1) the system has a finite number of states; (2) system behavior in a specific state should remain the same; (3) the system always stays in a certain mode for a certain period of time; (4) the number of conditions for state switching is finite; (5) a switch of the system state is a response to a set of events; (6) the time of state switch is negligible. To apply the EFSM, it is necessary to define the multicopter states, flight states, and events that may occur. Based on these, the state transition conditions are constructed to guarantee safety. It should be pointed out that the EFSM not only realizes the failsafe, but also allows the user observe and understand the decision-making process clearly.

12.1.3.1 Requirement Description

(1) System composition
Here, SAA design is considered. The main sensors on an autopilot include the GPS receiver, compass, INS, and barometer. Here, the RC transmitter is only used without considering the GCS.
(2) Functional requirements
    1) The user can arm the multicopter by the RC transmitter and then allow it to take off. The user can manually control the multicopter to land and disarm it.
    2) The user can control the multicopter to fly, return, and land automatically through the RC transmitter.

3) The multicopter can realize spot hover (LOITER MODE in autopilots) if the GPS receiver, compass and barometer are all healthy.
4) The multicopter can realize altitude hold hover (ALTITUDE HOLD MODE in autopilots) if a GPS receiver or compass is not installed or is unhealthy.
5) The multicopter can realize attitude self-stabilization (STABILIZE MODE in autopilots), if the barometer is unhealthy.

(3) Safety requirements
This part mainly refers to the failsafe in Sect. 12.1.2 by using the SAC.

12.1.3.2 Multicopter State and Flight Mode Definition

The whole process from takeoff to landing of the multicopter is divided into three multicopter states and three flight modes. They form the basis of the logic design. First, three multicopter states are defined as follows:

(1) POWER OFF STATE. This state implies that a multicopter is out of power. In this state, the user can disassemble, modify, and replace the hardware of a multicopter.
(2) STANDBY STATE. When a multicopter is connected to the power module, it enters the preflight state immediately. In this state, the multicopter is not armed, and users can arm the multicopter manually. Subsequently, the multicopter will perform a safety check and then can transit into the next state according to results of the safety check.
(3) GROUND_ERROR STATE. This state indicates that the multicopter has a safety problem. In this state, the buzzer will turn on an alarm to alert the user that there are errors in the system. Furthermore, three kinds of flight modes are defined.
(4) MANUAL FLIGHT MODE. This mode allows a remote pilot to control a multicopter manually. It further contains three submodes: LOITER MODE, ALTITUDE HOLD MODE, and STABILIZE MODE. Under normal circumstances, if the multicopter is in MANUAL FLIGHT MODE, then the default submode is in LOITER MODE. If the multicopter does not install the "GPS + compass", or "GPS + compass" are unhealthy, then the flight mode is degraded to ALTITUDE HOLD MODE. Furthermore, if the barometer is unhealthy, then the flight mode is further degraded to STABILIZE MODE.
(5) RTL MODE. Under this mode, the multicopter will return to the home location from the current position and hover there. In the process, if the current relative altitude comparison with that of the home location is higher than a set value (the default relative height is 15 m in the autopilot), the multicopter will maintain the current altitude and return home. Otherwise, it will first ascend to the preset height before returning home.

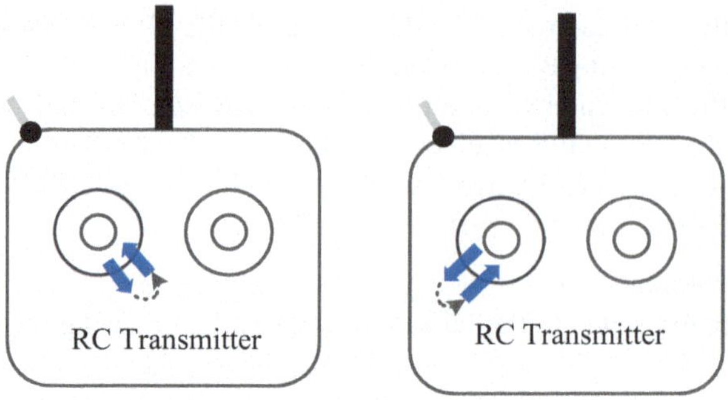

(a) Arm instruction (b) Disarm instruction

Fig. 12.1 Arm and disarm instruction

(6) AUTO-LANDING MODE. In this mode, the multicopter realizes the automatic landing by adjusting the throttle command according to the estimated height.[6]

12.1.3.3 Event Definition

Here, two kinds of events are defined: Manual Input Events (MIEs) and Automatic Trigger Events (ATEs). These events will cause the state or mode transition.

(1) MIEs
MIEs are instructions from remote pilots sent through the RC transmitter, as shown in Fig. 12.1.

MIE2: Manual operation instruction (1: Switch to MANUAL FLIGHT MODE; 2: Switch to RTL MODE; 3: Switch to AUTO-LANDING MODE). This instruction is realized by the flight mode switch (a three-position switch) as shown in Fig. 12.2. When the multicopter is powered on, $MIE2 = 0$ is temporarily set as 0. Then, according to the position of the flight mode switch, the switch will correspondingly happen.

MIE3: Turn on or turn off the multicopter. (1: turn on; 0: turn off).

MIE4: Power cutoff for maintenance. (1: repaired; 0: repairing).

(2) ATEs
ATEs are independent of the remote pilot's operations, but mainly generated by the status of components on board and status of the multicopter.

ATE1: Health status of INS and status of multicopter (1: healthy; 0: unhealthy)

ATE2: Health status of GPS (1: healthy; 0: unhealthy)

ATE3: Health status of barometer (1: healthy; 0: unhealthy)

ATE4: Health status of compass (1: healthy; 0: unhealthy)

[6]Even if the barometer fails, the height estimation is acceptable within a short time. Similarly, the other estimates generated by filters will continue to be used for a short time, even if related sensors fail.

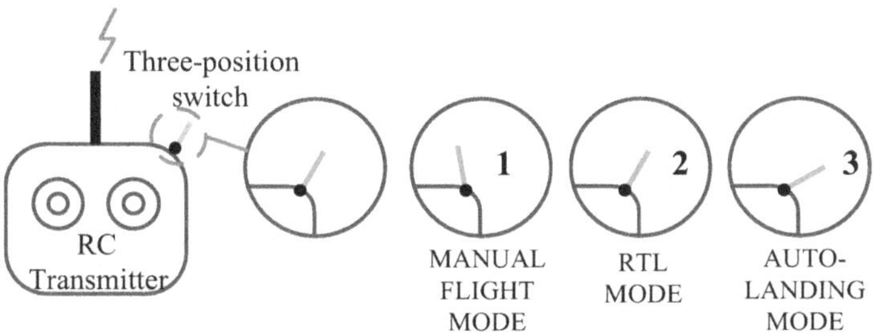

Fig. 12.2 Manual operation instruction

ATE5: Health status of propulsion system (1: healthy; 0: unhealthy)

ATE6: Status of connections of RC (1: normal; 0: abnormal)

ATE7: The status of battery's capacity (1: adequate; 0: inadequate, able to perform RTL; -1: inadequate, unable to perform RTL)

ATE8: Comparison of the multicopter's altitude and a specified threshold, (1: the multicopter's altitude is lower than the specified threshold; 0: the multicopter's altitude is not lower than the specified threshold.)

ATE9: Comparison of the multicopter's throttle command and a specified threshold over a time horizon, (1: the multicopter's throttle command is less than the specified threshold, as $\sigma_{\mathrm{drc}} < \sigma_{\mathrm{drcT}}$ for $t > t_{\mathrm{T}}$; 0: otherwise)

ATE10: Comparison of the multicopter's distance from HOME point and a specified threshold, (1: the multicopter's distance from HOME point is greater than the specified threshold, as $d > d_{\mathrm{T}}$; 0: the multicopter's distance from HOME point is not greater than the specified threshold, as $d \leqslant d_{\mathrm{T}}$.)

12.1.3.4 Autopilot Logic Design

Based on the multicopters' state, flight modes, and events, the autopilot logic is designed in the form of an EFSM as shown in Fig. 12.3, where C_i represents the transition condition, $i = 1, \ldots, 21$.

C1: $MIE3 = 1$

C2: $MIE3 = 0$

$$\text{C3: } (MIE1 = 1)\,\&\,(MIE2 = 1)\,\&\,(ATE1 = 1)\,\&\,(ATE5 = 1)\,\&\,(ATE6 = 1)\,\&\,(ATE7 = 1)$$

This condition implies a successful arm operation. This condition is true, when (1) the remote pilot arms the multicopter ($MIE1 = 1$); (2) the multicopter passes the check that the INS and propulsion system are both healthy ($ATE1 = 1\&ATE5 = 1$); (3) the connection to the RC transmitter is normal ($ATE6 = 1$); (4) the battery's capacity is adequate ($ATE7 = 1$); and (5) the flight mode switch to MANUAL

FLIGHT MODE happens ($MIE2 = 1$). Then, the multicopter is armed and switched from STANDBY STATE to MANUAL FLIGHT MODE.

C4: $[(MIE1 = 0 \& ATE8 = 1) | (ATE9 = 1)] \& (ATE1 = 1) \& (ATE5 = 1) \& (ATE6 = 1)$

This condition describes the disarm operation, including manual and automatic disarming. The manual disarm operation requires that the remote pilot disarm the multicopter ($MIE1 = 0$), and the multicopter be on the ground or lower than a

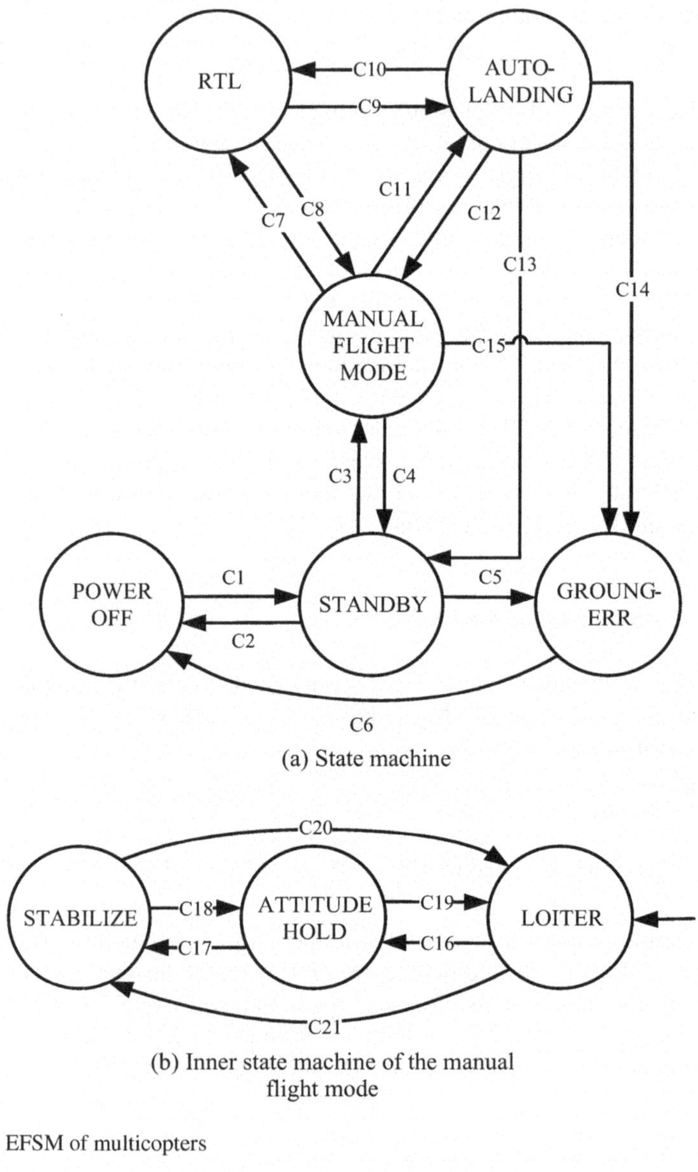

Fig. 12.3 EFSM of multicopters

given height ($ATE8 = 1$). The automatic disarm operation is to disarm the multicopter automatically when the throttle command is less than a given value beyond a given time threshold ($ATE9 = 1$). In this situation, it is required that the INS and propulsion system both be healthy ($ATE1 = 1\&ATE5 = 1$), and the connection to the RC transmitter must be normal ($ATE6 = 1$). Then, the multicopter is switched from MANUAL FLIGHT MODE to STANDBY STATE. If there is an unhealthy INS, or propulsion system, or the connection to the RC transmitter is abnormal, then the multicopter is switched from MANUAL FLIGHT MODE to GROUND_ERROR STATE, for which condition C15 gives more details.

C5: $(MIE1 = 1)\,\&[(ATE1 = 0)|(ATE5 = 0)|(ATE6 = 0)|(ATE7 = 0|ATE7 = -1)]$

This condition implies an unsuccessful arm operation. When the remote pilot tries to arm the multicopter ($MIE1 = 1$), the multicopter will check itself. If the INS is unhealthy ($ATE1 = 0$), the propulsion system is unhealthy ($ATE5 = 0$), the connection to the RC is abnormal ($ATE6 = 0$), or the battery's capacity is inadequate ($ATE7 = 0|ATE7 = -1$), then the multicopter cannot be armed successfully and will be switched from STANDBY STATE to GROUND_ERROR STATE.

$$\text{C6: } MIE4 = 1$$

This condition means that the power is cut off for maintenance so that unhealthy components can be replaced or a manual health check can be conducted.

$$\begin{aligned}\text{C7: } &(ATE1 = 1\&ATE2 = 1\&ATE3 = 1\&ATE4 = 1\&ATE5 = 1\&ATE10 = 1)\&\\ &[(MIE2 = 2\&ATE7 \geq 0)|(ATE6 = 0\&ATE7 \geq 0)|(ATE7 = 0)]\end{aligned}$$

This condition implies a switch from MANUAL FLIGHT MODE to RTL. Such a switch can take place in one of the following three cases: (1) the flight mode switch to RTL happens ($MIE2 = 2$), where the battery capacity must be adequate ($ATE7 \geq 0$); or (2) the connection to the RC transmitter is abnormal ($ATE6 = 0$), where the battery capacity is adequate ($ATE7 \geq 0$); or (3) the battery capacity is inadequate, but the multicopter is able to perform RTL ($ATE7 \geq 0$). Furthermore, the INS, GPS, barometer, compass, and propulsion system must be healthy ($ATE1 = 1\&ATE2 = 1\&ATE3 = 1\&ATE4 = 1\&ATE5 = 1$), and the distance from the multicopter to the predefined HOME point is required to be greater than a given threshold ($ATE10 = 1$).

$$\text{C8: } (MIE2 = 1)\&(ATE1 = 1\&ATE5 = 1)\&(ATE6 = 1)\&(ATE7 = 1)$$

This condition means that the flight mode switch to MANUAL FLIGHT MODE happens ($MIE2 = 1$) during the RTL process. In this case, the INS and propulsion system must be healthy($ATE1 = 1\&ATE5 = 1$), the connection to the RC transmitter must be normal($ATE6 = 1$), and the battery capacity must be adequate ($ATE7 = 1$). Here, note that when the contact between the RC transmitter and onboard RC receiver is reestablished, the multicopter will stay in RTL. Only if the remote pilot manually switches the flight mode switch to another position, and then back to MANUAL FLIGHT MODE again, the control of the multicopter can be retaken.

$$\begin{aligned}\text{C9: } &(MIE2 = 3)|(ATE1 = 0|ATE2 = 0|ATE3 = 0|\\ &ATE4 = 0|ATE5 = 0)|(ATE7 = -1)|(ATE10 = 0)\end{aligned}$$

This condition implies a switch from RTL to AUTO-LANDING. This switch will occur in any of the following four cases: (1) the flight mode switch to AUTO-LANDING happens ($MIE2 = 3$); or (2) there is a health problem in the INS, GPS, barometer, compass, or propulsion system ($ATE1 = 0|ATE2 = 0|ATE3 = 0|ATE4 = 0|ATE5 = 0$); or (3) the battery's capacity is inadequate, and the multicopter is unable to perform RTL ($ATE7 = -1$); or (4) the distance from the multicopter to the HOME point is required to not greater than a given threshold beyond a given time threshold ($ATE10 = 0$).

$$\begin{aligned}\text{C10: } &(MIE2 = 2)\&(ATE1 = 1\&ATE2 = 1\&ATE3 = 1\&\\ &ATE4 = 1\&ATE5 = 1)\&(ATE7 \geq 0)\&(ATE10 = 1)\end{aligned}$$

This condition indicates that the flight mode switch to RTL happens ($MIE2 = 2$). The following requirements must be met (1) the INS, GPS, barometer, compass, and propulsion system be all healthy ($ATE1 = 1\&ATE2 = 1\&ATE3 = 1\&ATE4 = 1\&ATE5 = 1$); (2) the battery capacity be adequate ($ATE7 \geq 0$); and (3) the distance from the multicopter to the HOME point be greater than a given threshold ($ATE10 = 1$).

$$\begin{aligned}\text{C11: } &(MIE2 = 3)|(ATE7 = -1)|(ATE1 = 0|ATE5 = 0)|\\ &[(ATE6 = 0)\&(ATE7 \geq 0)\&(ATE2 = 0|ATE4 = 0|ATE10 = 0)]\end{aligned}$$

This condition implies a switch from MANUAL FLIGHT MODE to AUTO-LANDING, including four situations: (1) the flight mode switch to AUTO-LANDING happens ($MIE2 = 3$); or (2) the battery's capacity is inadequate, and the multicopter is unable to perform RTL ($ATE7 = -1$); or (3) there is a health problem in the INS, or propulsion system ($ATE1 = 0|ATE5 = 0$); or (4) the connection to the RC transmitter is abnormal ($ATE6 = 0$), where the battery capacity must be adequate ($ATE7 \geq 0$), and the distance from the multicopter to the HOME point is not greater than a given threshold ($ATE10 = 0$), or the GPS or compass is unhealthy ($ATE2 = 0|ATE4 = 0$).

$$\text{C12: } (MIE2 = 1)\&(ATE1 = 1\&ATE5 = 1)\&(ATE6 = 1)\&(ATE7 = 1)$$

This condition indicates that the flight mode switch to MANUAL FLIGHT MODE happens during the landing process ($MIE2 = 1$), where the INS and propulsion system must be healthy ($ATE1 = 1 \& ATE5 = 1$), the connection to the RC transmitter is normal ($ATE6 = 1$), and the battery capacity must be adequate ($ATE7 = 1$). Here, it should be noted that when the contact between the RC transmitter and the onboard RC receiver is reestablished, the multicopter will stay in AUTO-LANDING MODE. Only if the remote pilot manually changes, the flight mode switch to another position, and then back to MANUAL FLIGHT MODE, control of the multicopter can be retaken.

$$\text{C13: } (ATE8 = 1 | ATE9 = 1) \& (ATE1 = 1 \& ATE5 = 1 \& ATE6 = 1)$$

This condition indicates that during the landing process, the multicopter is considered to have landed successfully and be disarmed, when (1) the height is less than a predetermined threshold ($ATE8 = 1$); or (2) the throttle command is less than a given value beyond a given time threshold ($ATE9 = 1$). Furthermore, it is required that the INS and propulsion system must be both healthy ($ATE1 = 1 \& ATE5 = 1$), and the connection to the RC transmitter must be normal ($ATE6 = 1$). In this case, the multicopter is switched from AUTO-LANDING to STANDBY STATE.

$$\text{C14: } (ATE8 = 1 | ATE9 = 1) \& (ATE1 = 0 | ATE5 = 0 | ATE6 = 0)$$

This condition indicates that during the landing process, the multicopter is considered to have landed and be disarmed. Its state is switched from AUTO-LANDING to GROUND_ERROR STATE, when (1) the height is lower than a predetermined threshold ($ATE8 = 1$); or (2) the throttle command is less than a given value beyond a given time threshold ($ATE9 = 1$). Furthermore, the INS or propulsion system must be unhealthy, or the connection to the RC transmitter be abnormal ($ATE1 = 0 | ATE5 = 0 | ATE6 = 0$).

$$\begin{aligned} \text{C15: } &[(MIE1 = 0 \& ATE8 = 1) | \\ &(ATE9 = 1)] \& (ATE1 = 0 | ATE5 = 0 | ATE6 = 0) \end{aligned}$$

This condition describes the disarm operation, including manual and automatic disarming. The manual disarm operation requires that the remote pilot disarm the multicopter ($MIE1 = 0$), and the multicopter be on the ground or lower than a given height ($ATE8 = 1$). The automatic disarm operation is to disarm the multicopter automatically when the throttle command is less than a given value beyond a given time threshold ($ATE9 = 1$). In this situation, if the INS or propulsion system is unhealthy ($ATE1 = 0 | ATE5 = 0$), or the connection to the RC transmitter is abnormal ($ATE6 = 0$), then the multicopter is switched from MANUAL

FLIGHT MODE to GROUND_ERROR STATE. For the three submodes in MANUAL FLIGHT MODE, the switch conditions are given as follows:

C16: $ATE2 = 0|ATE4 = 0$

This condition indicates that if the GPS or compass is unhealthy $(ATE2 = 0|ATE4 = 0)$, then the flight mode is switched from LOITER MODE to ALTITUDE HOLD MODE.

C17: $ATE3 = 0$

This condition indicates that if the barometer is unhealthy $(ATE3 = 0)$, then the flight mode is switched from ALTITUDE HOLD MODE to STABILIZE MODE.

C18: $ATE3 = 1\&(ATE2 = 0|ATE4 = 0)$

This condition indicates that if the barometer is healthy $(ATE3 = 1)$, and the GPS or compass is unhealthy $(ATE2 = 0|ATE4 = 0)$, then the flight mode is switched from STABILIZE MODE to ALTITUDE HOLD MODE.

C19: $ATE2 = 1\&ATE4 = 1$

This condition indicates that if the GPS and compass are healthy $(ATE2 = 1\&ATE4 = 1)$, then the flight mode is switched from ALTITUDE HOLD MODE to LOITER MODE.

C20: $ATE2 = 1\&ATE3 = 1\&ATE4 = 1$

This condition indicates that if the GPS, compass, and barometer are all healthy $(ATE2 = 1\&ATE3 = 1\&ATE4 = 1)$, then the flight mode is switched from STABILIZE MODE to LOITER MODE.

C21: $ATE3 = 0$

This condition indicates that if the barometer is unhealthy $(ATE3 = 0)$, then the flight mode is switched from LOITER MODE to STABILIZE MODE.

12.2 Basic Experiment

12.2.1 Experimental Objective

(1) Things to prepare

1) Hardware: Multicopter Hardware System, Pixhawk Autopilot System;
2) Software: MATLAB R2017b or above, Simulink-based Controller Design and Simulation Platform, HIL Simulation Platform, Instructional Package Package "e8.1" (https://rflysim.com/course).

(2) Objectives

1) In SIL simulation, repeat the given code to realize RTL MODE, AUTO-LANDING MODE and MANUAL FLIGHT MODE[7];
2) Perform the HIL simulation.

[7]The definitions can be found in Sect. 12.1.3.2.

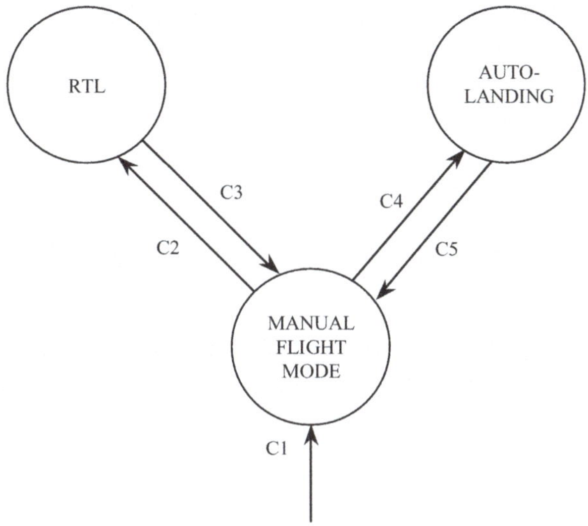

Fig. 12.4 EFSM of quadcopter in basic experiment

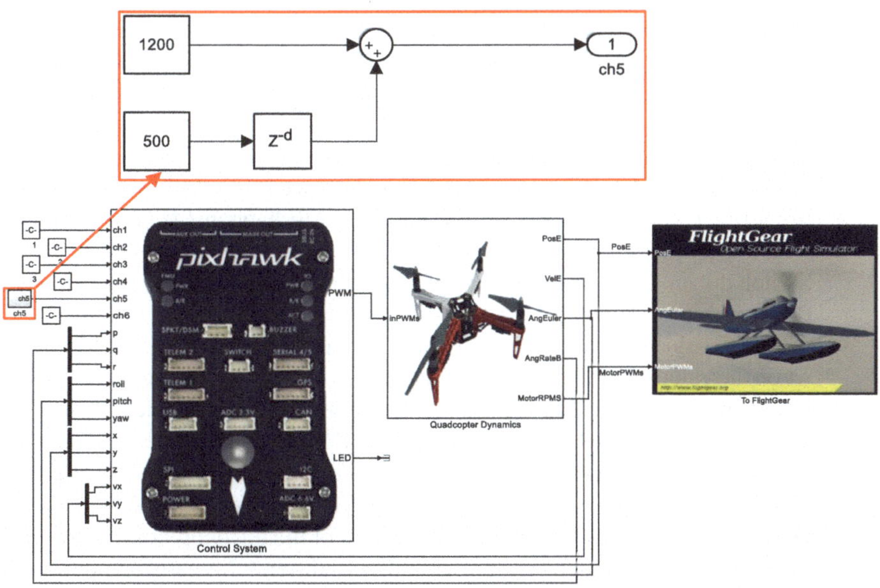

Fig. 12.5 SIL model for basic experiment, Simulink model “e8_1sim.slx”

12.2.2 Experimental Procedure

(1) **Step1: SIL simulation**

1) Parameter Initialization
Run the file "e8\e8.1\Sim\Init_control.m" to initialize the parameters, then the Simulink file "e8_1_sim.slx" will open automatically as shown in Fig. 12.5. The EFSM of the basic experiment is shown in Fig. 12.4, where "C1, C2, C3, C4, C5" represent the transition condition in the following.
• C1:$MIE1 = 1$
This condition implies a successful arm operation. This condition is true when the remote pilot arms the multicopter ($MIE1 = 1$).
• C2: $MIE2 = 2$
This condition implies a switch from MANUAL FLIGHT MODE to RTL MODE. Such a switch can take place in the following case: the flight mode switch to RTL MODE happens ($MIE2 = 2$).
• C3, C5:$MIE2 = 1$
These conditions imply a switch from RTL MODE to MANUAL FLIGHT MODE and a switch from AUTO-LANDING MODE to MANUAL FLIGHT MODE. Such a switch can take place in the following case: the flight mode switch to MANUAL FLIGHT MODE happens ($MIE2 = 1$);
• C4:$MIE2 = 3$
This condition implies a switch from MANUAL FLIGHT MODE to AUTO-LANDING MODE. Such a switch can take place in the following case: the flight mode switch to AUTO-LANDING MODE happens($MIE2 = 3$).
2) RTL simulation
Switch the mode by changing the value of "ch5"(corresponding to CH5 of the RC transmitter). When the PWM value of "ch5" is less than 1400, $MIE2 = 1$; when the value of "ch5" is greater than 1400 and less than 1600, $MIE2 = 2$; when the value of "ch5" is greater than 1600, $MIE2 = 3$. The setting of "ch5" is shown in Fig. 12.5. In the first 10 s, the input value of the "ch5" channel is 1200, corresponding to MANUAL FLIGHT MODE, $MIE2 = 1$; after 10 s, the value is set to 1500, corresponding to RTL MODE, $MIE2 = 2$. The simulation performance can be observed in Fig. 12.6. In 0–10 s, the quadcopter is in MANUAL FLIGHT MODE, and the state of the quadcopter is determined by the values of "ch1–ch4". After 10 s, the condition C2 is satisfied, the quadcopter enters RTL MODE. Then, the altitude remains unchanged while the horizontal position gradually returns to 0.
3) Auto-Landing simulation
Here, the PWM value of the "ch5" is modified as follows. In the first 10 s, the input value is 1200, corresponding to MANUAL FLIGHT MODE, $MIE2 = 1$; after 10 s, the input value is 1700, corresponding to the AUTO-LANDING MODE, $MIE2 = 3$. The simulation is observed in Fig. 12.7. In 0–10 s, the quadcopter takes off and climbs up under the given command in MANUAL FLIGHT MODE. After 10 s, the condition C4 is satisfied and the quadcopter enters AUTO-LANDING MODE. Then, the horizontal position remains unchanged, while the altitude gradually reduces to 0.

(2) **Step2: HIL simulation**

1) Parameters initialization
Open the Simulink file "e8\e8.1\HIL\e8_1_HIL.slx" and run the file "Init_control.m" to initialize the parameters in the same folder, as shown in Fig. 12.8. It should be noted that "Control System" in the SIL simulation shown in Fig. 12.5 is the same as that in the HIL simulation shown in Fig. 12.8.
2) Connect hardware
Referring the procedure in Sect. 2.3 in Chap. 2, the connection between an RC receiver and a Pixhawk autopilot can be set up as shown in Fig. 12.9.
3) Compile and download code
Compile the HIL simulation model and upload the file to the given Pixhawk autopilot. Later, the designed failsafe program can be run on Pixhawk autopilot. The operation is shown in Fig. 12.10. The detailed procedure is introduced in Sect. 3.3.3.
4) Configure CopterSim
Double-click on the desktop shortcut CopterSim to open it. Readers can choose different propulsion systems using the following procedure. Click on "Model Parameters" to customize the model parameters and, then click on "Store and use the parameters" to make them available. The software will automatically match the serial port number. Readers would click the "Run" button to enter the HIL simulation mode. After that, readers could see the message returned by the Pixhawk autopilot in the lower-left corner of the interface, as shown in Fig. 12.11.
5) Configure 3DDisplay
Double-click on the desktop shortcut 3DDisplay to open it.
6) Simulation performance
Arm the quadcopter for manual control, and fly the quadcopter to a certain altitude. Then push CH5 from the farthest position to the middle position to realize the RTL. Futher, push CH5 to the closest position, and the quadcopter is still in the RTL MODE. This is because there is no switch between RTL MODE and AUTO-LANDING MODE in the state transition conditions. Then push CH5 to the farthest position, the quadcopter re-enters the MANUAL FLIGHT MODE, quickly push the CH5 to the closest position, and the quadcopter enters the AUTO-LANDING MODE (Fig. 12.12).

12.3 Analysis Experiment

12.3.1 Experimental Objective

(1) Things to prepare

1) Hardware: Multicopter Hardware System, Pixhawk Autopilot System;
2) Software: MATLAB R2017b or above, Simulink-based Controller Design and Simulation Platform, HIL Simulation Platform, Instructional Package Package "e8.2" (https://rflysim.com/course).

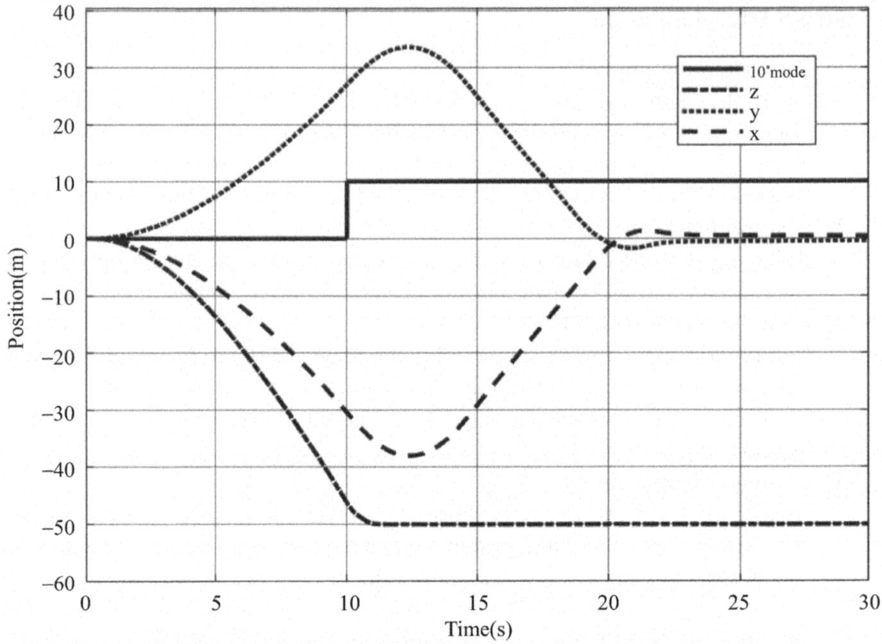

Fig. 12.6 Position response with RTL MODE

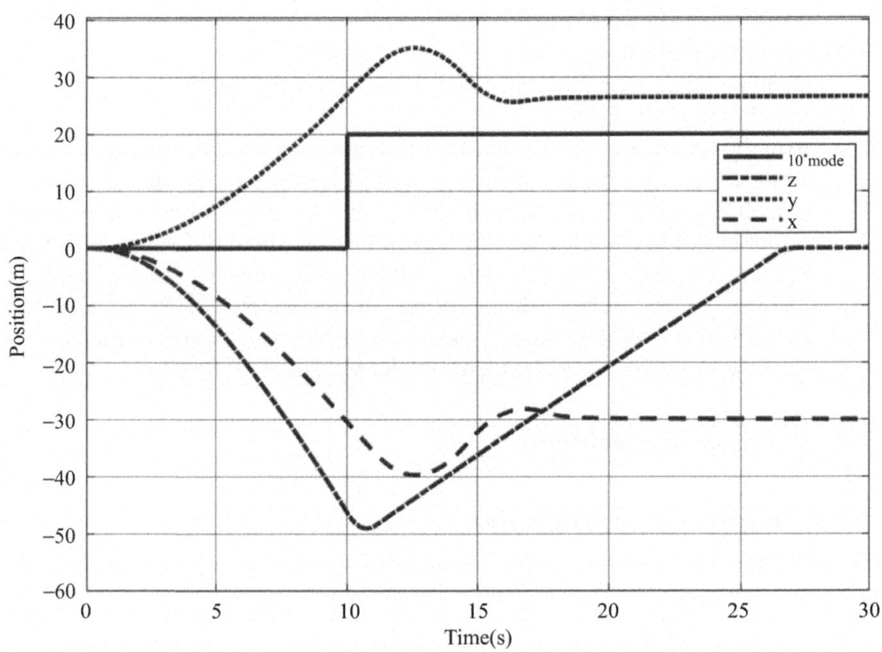

Fig. 12.7 Position response with AUTO-LANDING MODE

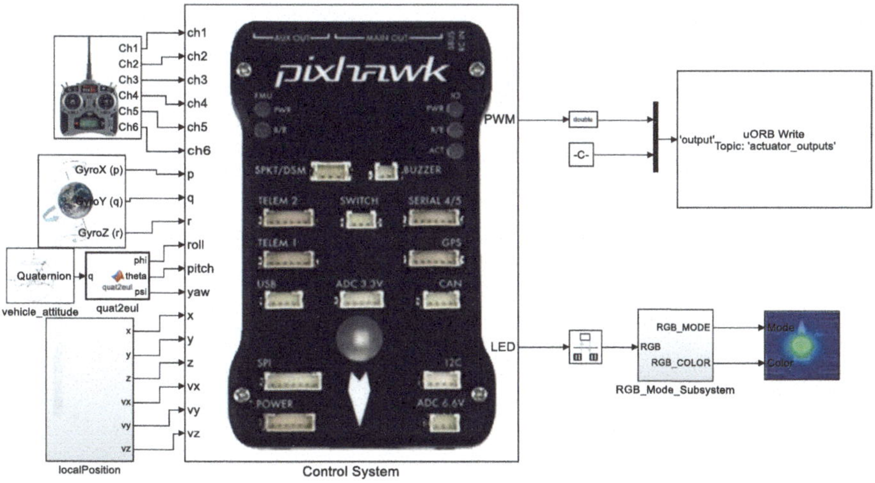

Fig. 12.8 HIL model for basic experiment, Simulink model "e8_1_HIL.slx"

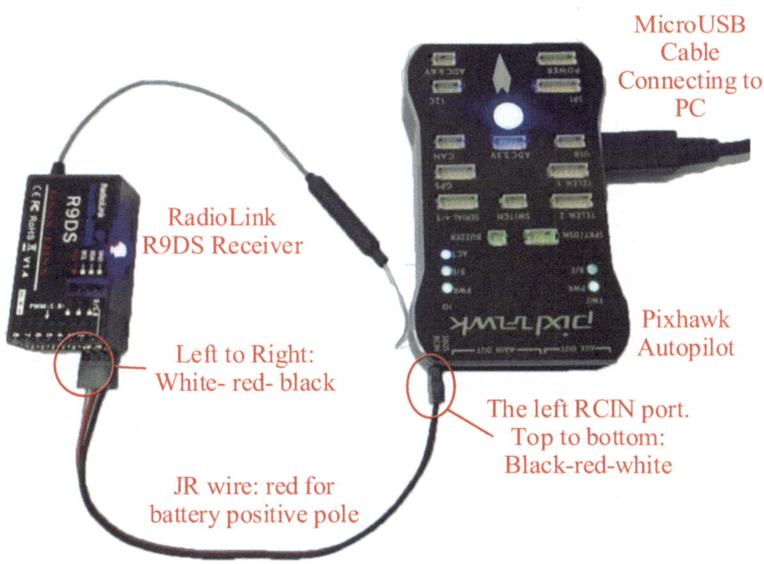

Fig. 12.9 Connection between Pixhawk hardware and RC receiver

(2) Objectives

1) Based on the basic experiment, design a new logic to realize that the quadcopter can switch between RTL MODE and AUTO-LANDING MODE manually.
2) Perform the HIL simulation.

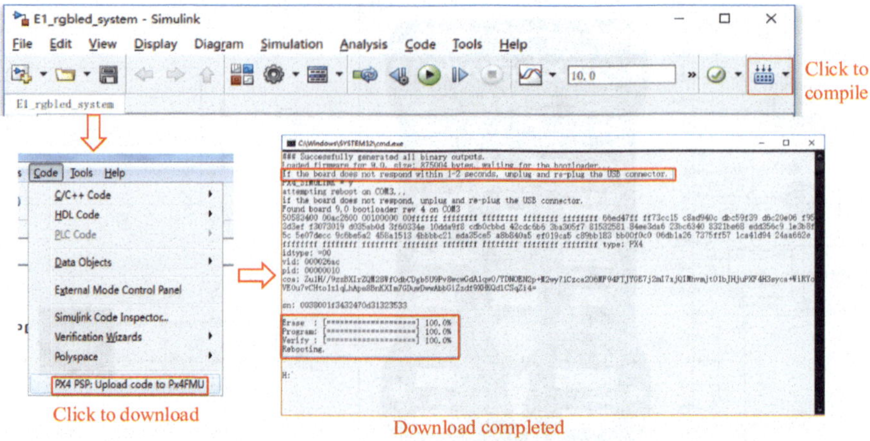

Fig. 12.10 Code compilation and upload process

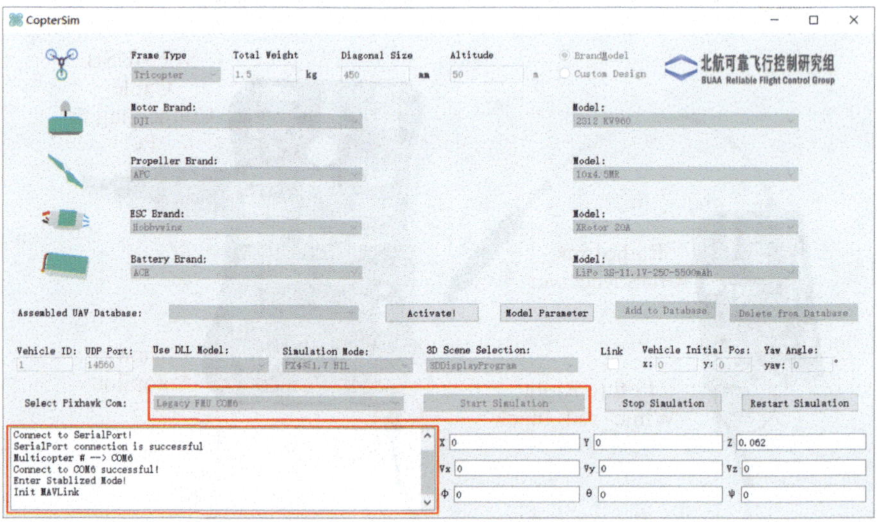

Fig. 12.11 User interface of CopterSim

12.3.2 *Experimental Analysis*

This experiment adds a dual switching between RTL MODE and AUTO-LANDING MODE based on the basic experiment. The corresponding EFSM is shown in Fig. 12.13.

The new transition conditions are presented below.

- C6: $MIE2 = 3$

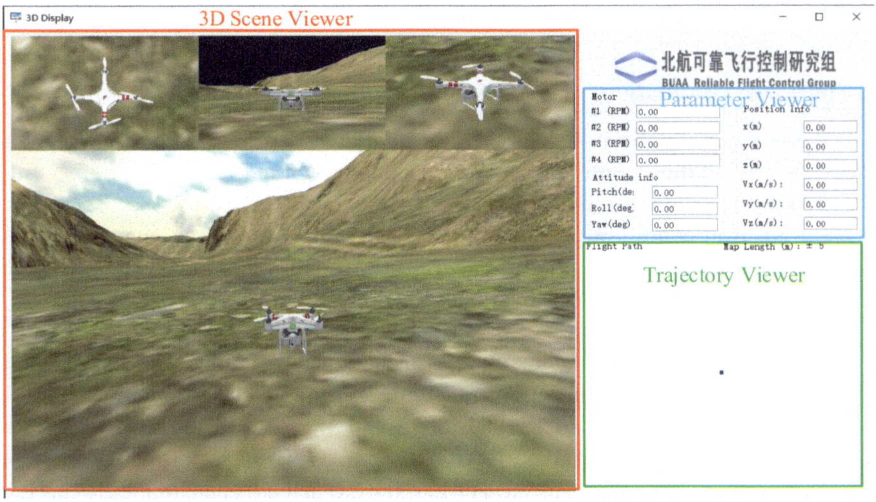

Fig. 12.12 User interface of 3DDisplay

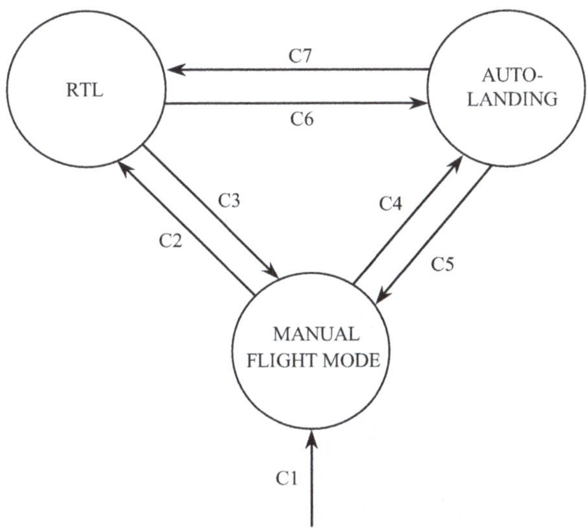

Fig. 12.13 EFSM of quadcopter in analysis experiment

This condition implies a switch from RTL MODE to AUTO-LANDING MODE. This switch will occur in the following case: the flight mode switching to AUTO-LANDING MODE happens ($MIE2 = 3$).

- C7: $MIE2 = 2$

This condition implies a switch from AUTO-LANDING MODE to RTL MODE. This switch will occur in the following case: any flight mode switching to AUTO-LANDING MODE happens ($MIE2 = 2$).

12.3.3 Experimental Procedure

(1) **Step1: SIL simulation**

1) Copy the model from the basic experiment and find the state machine named "mode switch" in the "Control System model" as shown in Fig. 12.14.

2) Modify the state machine

The transition condition of the state machine is determined by events, such as $MIE2$, which is the input condition. The transition conditions between RTL MODE and AUTO-LANDING MODE are added, as shown in Fig. 12.15.

3) Modify the model input

Change the input value of the "ch5" in the model to simulate the PWM value from three-position switch of an RC transmitter. During 0–10 s, the value is 1200, corresponding to MANUAL FLIGHT MODE and $MIE2 = 1$; during 10–30 s, the value is 1500, corresponding to RTL MODE and $MIE2 = 2$; during 30–50 s, the value is 1800, corresponding to AUTO-LANDING MODE and $MIE2 = 3$. The entire process simulates the RC transmitter switching from MANUAL FLIGHT MODE to RTL MODE and then to AUTO-LANDING MODE.

4) Save the model and initialize the parameters

Save the model to the file "e8\e8.2\Sim\e8_2_sim.slx" and run the file "e8\e8.2\Sim\Init_control.m" to initialize the parameters.

5) Simulation performance

The simulation results are directly shown in Fig. 12.16. It can be observed that during 0–10 s, the quadcopter flies freely in the air; then, during 10–30 s, it enters

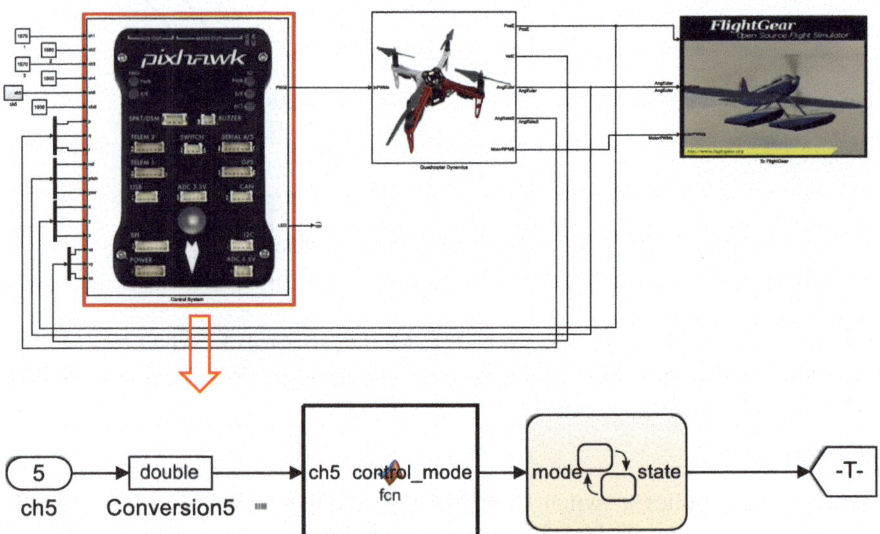

Fig. 12.14 Mode switch model in "Control System" submodule

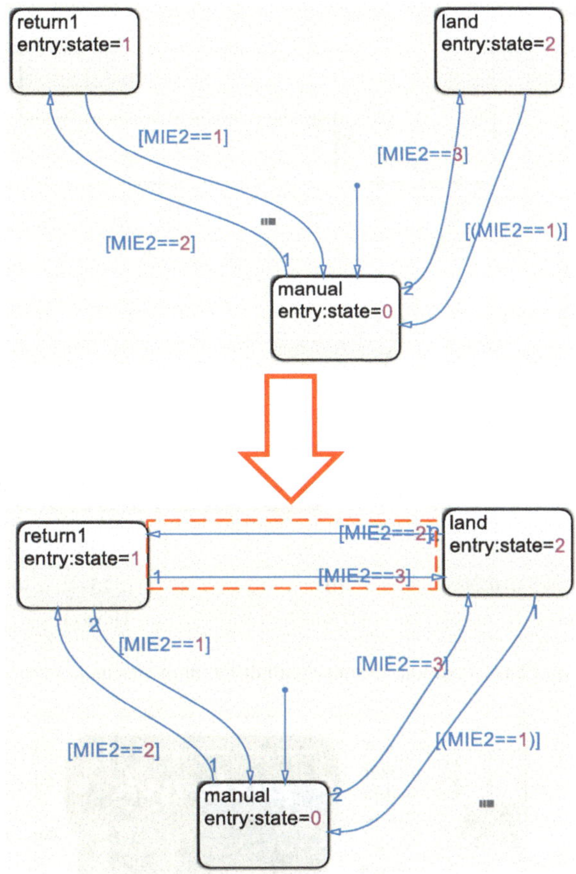

Fig. 12.15 Modify state machine for analysis experiment

RTL MODE, the altitude remains unchanged while the horizontal position goes to (0, 0); after 30 s, it enters AUTO-LANDING MODE, where the horizontal position remains (0, 0) and the altitude drops down to zero.

(2) **Step2: HIL simulation**

1) Open the Simulink file for HIL

Open the Simulink file "e8\e8.2\HIL\e8_2_HIL.slx", as shown in Fig. 12.17. It should be noted that "Control System" in the SIL simulation shown in Fig. 12.14 is the same as that in the HIL simulation shown in Fig. 12.17.

2) Connect hardware

Referring the procedure in Sect. 2.3 in Chap. 2, the connection between an RC receiver and a Pixhawk autopilot can be set up as shown in Fig. 12.9.

3) Compile and upload the code

Compile the HIL simulation model and upload the file to the given Pixhawk autopilot. Later, the designed failsafe program can be run on Pixhawk autopilot. The operation is shown in Fig. 12.10. The detailed procedure is introduced in Sect. 3.3.3.

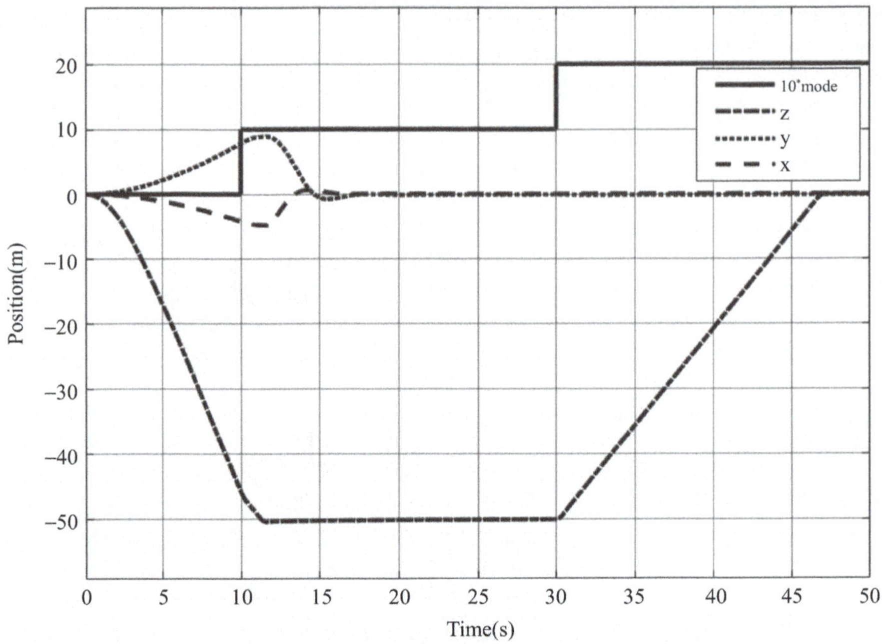

Fig. 12.16 Position response with different modes in analysis experiment

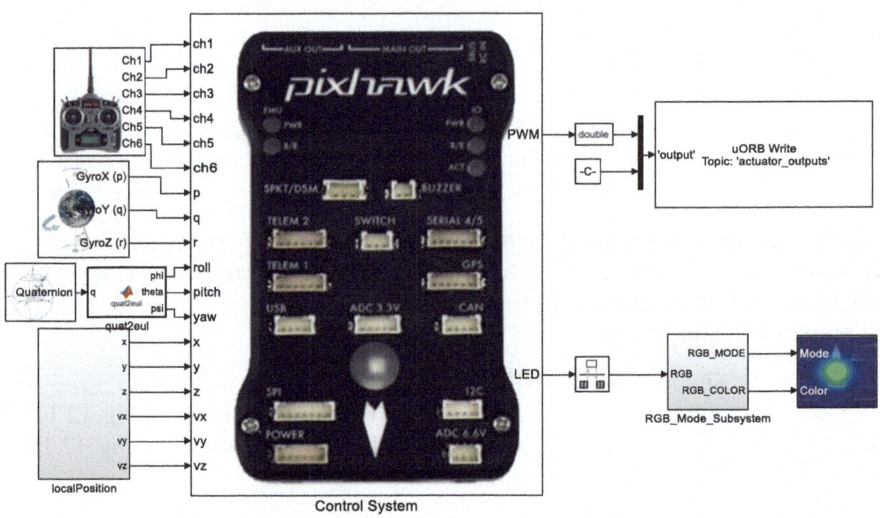

Fig. 12.17 HIL model for analysis experiment, Simulink model "e8_2_HIL.slx"

4) Configure CopterSim

Double-click on the desktop shortcut CopterSim to open it. Readers can choose different propulsion systems using the following procedure. Click on "Model Parameters" to customize the model parameters and, then click on "Store and use the parameters" to make them available. The software will automatically match the serial port number. Readers would click the "Run" button to enter the HIL simulation mode. After that, readers could see the message returned by the Pixhawk autopilot in the lower-left corner of the interface, as shown in Fig. 12.11.

5) Configure 3DDisplay

Double-click on the desktop shortcut 3DDisplay to open it.

6) Simulation performance

Arm the quadcopter and fly it to a certain altitude. Then push the three-position switch CH5 back to realize RTL MODE. After RTL is completed, then push CH5 back for auto-landing. This is different from the basic experiment, because of no switch between returning and landing in the basic experiment.

12.3.4 Remark

The quadcopter needs some time in the process of returning and landing. During the simulation process, sufficient time should be left for the whole simulation.

12.4 Design Experiment

12.4.1 Experimental Objective

(1) Things to prepare

1) Hardware: Multicopter Hardware System, Pixhawk Autopilot System;
2) Software: MATLAB R2017b or above, Simulink-based Controller Design and Simulation Platform, HIL Simulation Platform, Instructional Package Package "e8.3" (https://rflysim.com/course).

(2) Objectives

1) Based on the analysis experiment, a new state machine should be designed to deal with the case in which the RC transmitter is power-off. In addition, a requirement should be added that: if the quadcopter is close to HOME point, it will land directly; and if the quadcopter is at a certain distance from HOME point, it will first return and then land.
2) Perform the HIL simulation and flight test.

12.4.2 Experimental Design

(1) **Step1: Flight modes**
To simplify the logic design, the whole process from takeoff to landing of the multicopter is divided into five flight modes, where two additional modes for the case of failure are added.

1) MANUAL FLIGHT MODE
2) RTL MODE
3) AUTO-LANDING MODE
4) FAIL LANDING MODE: the quadcopter automatically lands when the RC transmitter is power-off.
5) FAIL RTL MODE: the quadcopter automatically returns to the initial horizontal position when the RC transmitter is power-off.

(2) **Step2: Event definition**

1) MIEs are instructions from remote pilots sent through the RC transmitter, including the following:
   - $MIE1$: Arm and disarm instructions.
   - $MIE2$: Manual operation instruction.
2) ATEs are independent of the remote pilot's operations, but mainly generated by the status of components on board and status of the multicopter.
   - $ATE1$: Status of connections of RC. $ATE1 = 1$: normal, $ATE1 = 0$: abnormal.
   - $ATE3$: Multicopter's distance from HOME point. $ATE3 = 1$: quadcopter's distance from HOME point is not greater than a specified threshold, $ATE3 = 0$: quadcopter's distance from HOME point is greater than the specified threshold.

(3) **Step3. Design the State Machine**
The state machine is shown in Fig. 12.18, where Ci represents the corresponding transition condition.

- C1:$MIE1 = 1$

This condition implies a successful arm operation. This condition is true, when the remote pilot arms the quadcopter ($MIE1 = 1$);

- C2, C7, C10, C15: $ATE1 = 1 \& MIE2 = 2$

These conditions describe the transition conditions of the quadcopter from MANUAL FLIGHT MODE(C2), AUTO-LANDING MODE(C7), FAIL LANDING MODE(C15), FAIL RTL MODE(C10) to RTL MODE, respectively. The quadcopter must meet the following requirements: the RC must be connected normally ($ATE1 = 1$) and a manual operation instruction to RTL MODE is performed ($MIE2 = 2$);

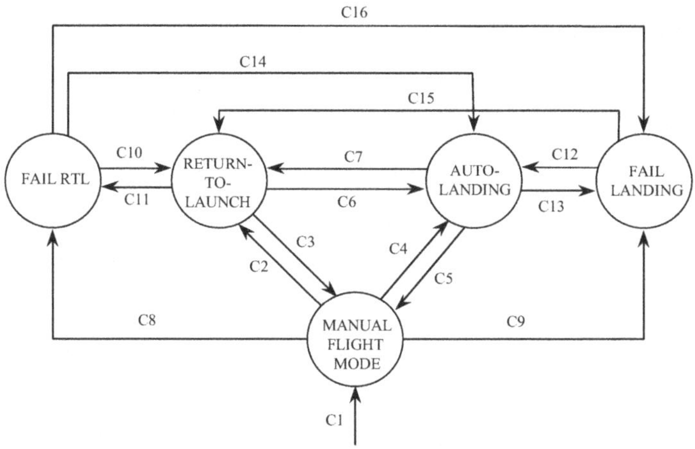

Fig. 12.18 EFSM of design experiment

- C3, C5: $ATE1 = 1\&MIE2 = 1$

There are conditions for switching of a quadcopter from RTL MODE(C3) and AUTO-LANDING MODE(C5) to MANUAL FLIGHT MODE, respectively. The quadcopter must meet the following requirements: the RC is connected normally ($ATE1 = 1$) and manual operation instruction to MANUAL FLIGHT MODE is performed ($MIE2 = 1$);

- C4, C6, C12, C14:$ATE1 = 1\&MIE2 = 3$

These conditions describe the conditions for a quadcopter from MANUAL FLIGHT MODE(C4), FAIL LANDING MODE(C6), FAIL LANDING MODE(C12) and FAIL RTL MODE(C14) to AUTO-LANDING MODE, respectively. The quadcopter must meet the following requirements: the RC must be connected properly ($ATE1 = 1$) and manual operation instruction to AUTO-LANDING MODE ($MIE2 = 3$);

- C8: $ATE1 = 0\&ATE3 = 0$

This condition implies a switch from MANUAL FLIGHT MODE to FAIL RTL MODE. Such a switch will take place in the following case: the connection to the RC transmitter is abnormal ($ATE1 = 0$) and the quadcopter's distance from HOME point is greater than the specified threshold ($ATE3 = 0$);

- C9: $ATE1 = 0\&ATE3=1$

This condition implies a switch from MANUAL FLIGHT MODE to FAIL LANDING MODE. Such a switch will take place in the following case: the connection to the RC transmitter is abnormal ($ATE1 = 0$) and the quadcopter's distance from the HOME point is less than the specified threshold ($ATE3 = 1$);

- C11, C13: $ATE1 = 0$

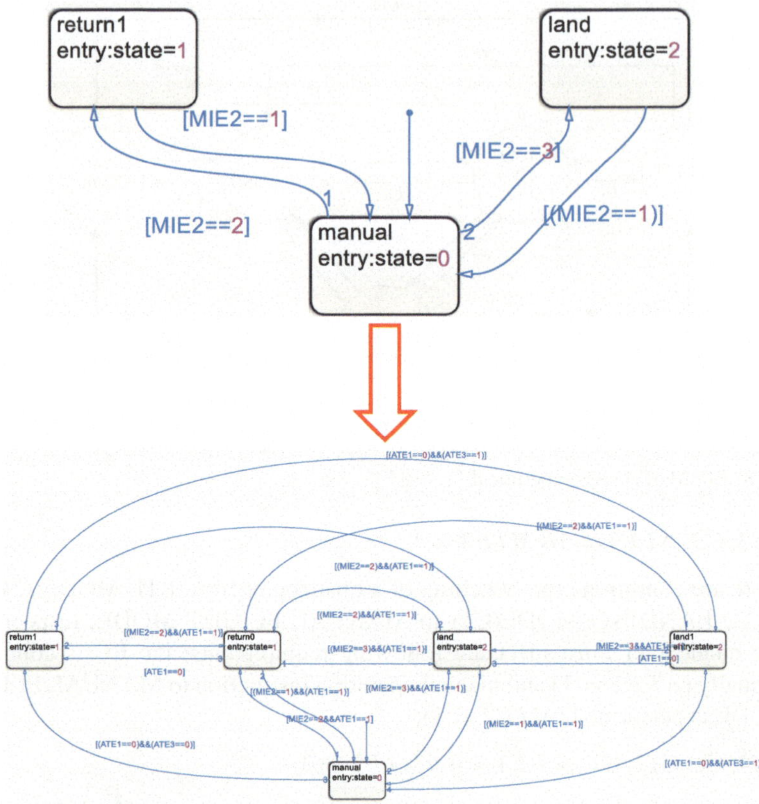

Fig. 12.19 State machine designed in design experiment

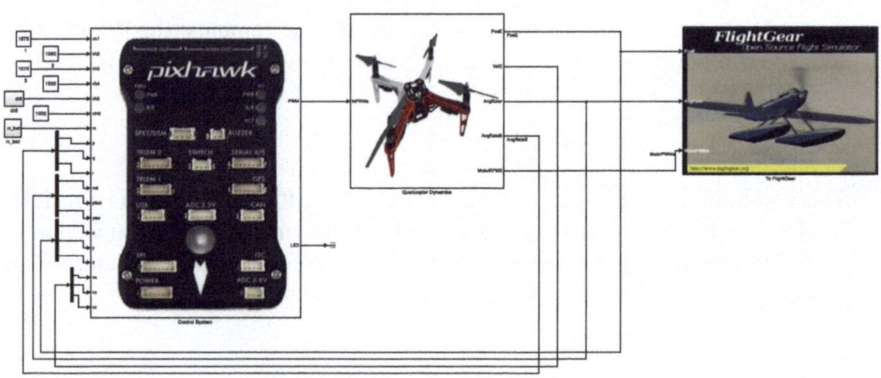

Fig. 12.20 SIL model for design experiment, Simulink model "e8_3.slx"

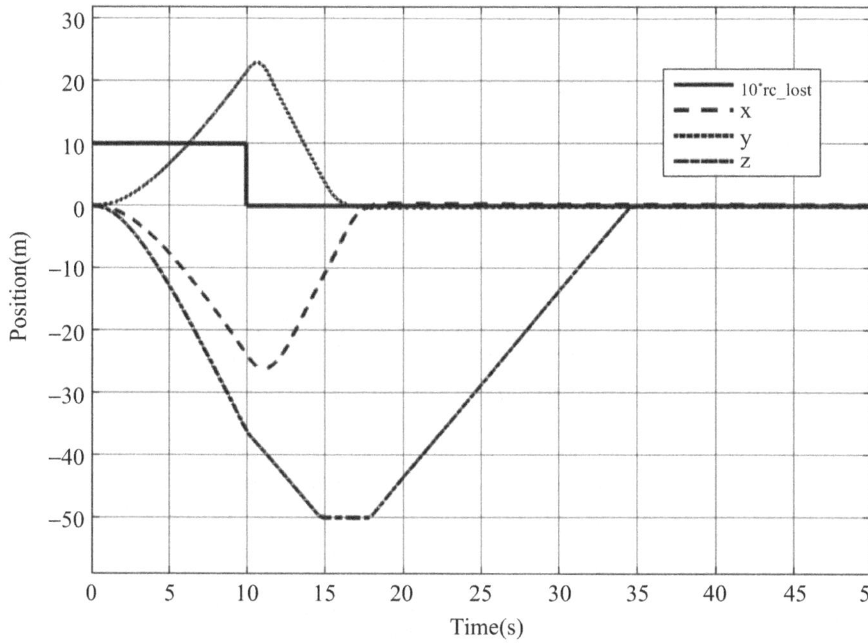

Fig. 12.21 Position response when RC transmitter is power-off

These conditions imply switches from RTL MODE(C11) and LANDING MODE(C13) to FAIL LANDING MODE, respectively. The conditions are triggered by the connection to the RC transmitter being abnormal ($ATE1 = 0$).

- C16: $ATE1 = 0\&ATE3 = 1$

This condition implies a switch from FAIL RTL MODE to FAIL LANDING MODE. Such a switch will take place in the following case: the connection to the RC transmitter is abnormal ($ATE1 = 0$) and the quadcopter's distance from HOME point is not greater than the specified threshold ($ATE3 = 1$). In this case, the quadcopter can start landing.

(4) **Step4: Modify the Simulink model**

1) Modify the state machine model
According to Steps 1–3, the state machine model is modified and refined based on that in the analysis experiment, as shown in Fig. 12.19.
2) Modify the RC input
Add an RC channel to simulate the RC transmitter power failure event, as shown in Fig. 12.20, the input of which is 0 in the first 10 s, corresponding to the normal connection of RC transmitter. It becomes 1 after 10 s, corresponding to the connection lost.

3) Save model
Save the model to "e8\e8.3\Sim\e8_3_sim.slx".

12.4.3 Simulation Procedure

(1) **Step1: SIL simulation**
Run the file "Init_control.m" to initialize the parameters. Click on the "Run" button to run the simulation, the result is shown in Fig. 12.21. It can be observed that, during 0–10 s, the quadcopter is in MANUAL FLIGHT MODE and it can be piloted freely. After 10 s, the RC is out of contact because of RC transmitter being power-off. At this time, the horizontal position of the quadcopter is far from HOME point and the altitude is less than the threshold. The quadcopter first climbs up to the safe altitude we set, then enters RTL MODE. The return process is completed at 18s with the horizontal position being 0. Then, the quadcopter enters AUTO-LANDING MODE and finally completes landing at 35s.

(2) **Step2: HIL simulation**

1) Open the Simulink file for HIL
Open the Simulink file "e8\e8.3\HIL\e8_3_HIL.slx" as shown in Fig. 12.22, and run the file "Init_control.m" to initialize the parameters. It should be noted that "Control System" in the SIL simulation shown in Fig. 12.20 is the same as that in the HIL simulation shown in Fig. 12.22.
2) Connect the hardware
Referring the procedure in Sect. 2.3 in Chap. 2, the connection between an RC receiver and a Pixhawk autopilot can be set up as shown in Fig. 12.9.

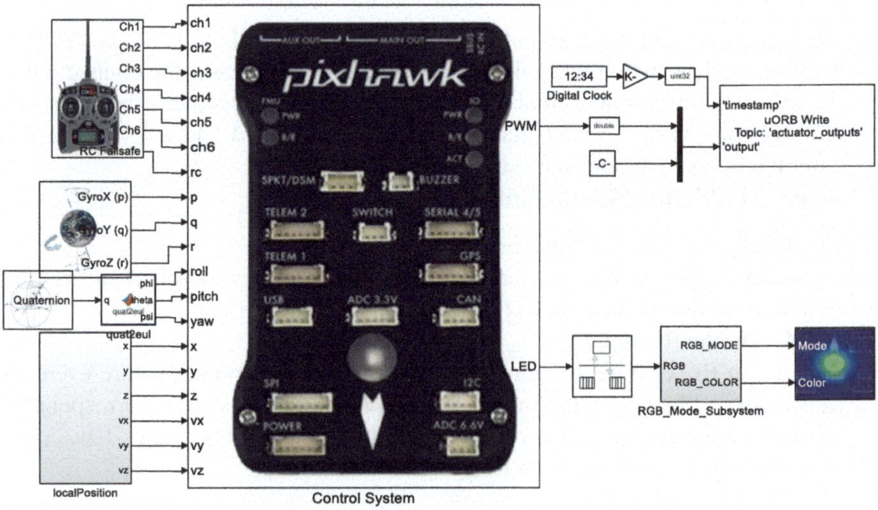

Fig. 12.22 HIL model for design experiment, Simulink model "e8_3_HIL.slx"

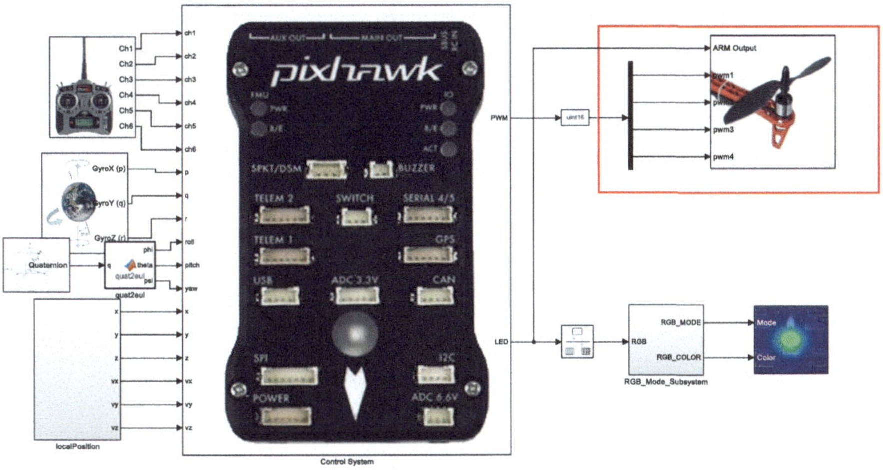

Fig. 12.23 Model for design experiment flight test, Simulink model "e8_4_FLY.slx"

3) Compile and upload the code
   Compile the HIL simulation model and upload the file to the given Pixhawk autopilot. Later, the designed failsafe program can be run on Pixhawk autopilot. The operation is shown in Fig. 12.10. The detailed procedure is introduced in Sect. 3.3.3.
4) Configure CopterSim
   Double-click on the desktop shortcut CopterSim to open it. Readers can choose different propulsion systems using the following procedure. Click on "Model Parameters" to customize the model parameters and, then click on "Store and use the parameters" to make them available. The software will automatically match the serial port number. Readers would click the "Run" button to enter the HIL simulation mode. After that, readers could see the message returned by the Pixhawk autopilot in the lower-left corner of the interface, as shown in Fig. 12.11.
5) Configure 3DDisplay
   Double-click on the desktop shortcut 3DDisplay to open it.
6) Simulation performance
   Arm the quadcopter which is first in MANUAL FLIGHT MODE for a while. Then, turn off the power of the RC transmitter. In this case, the RC is out of contact. By the logic designed, the quadcopter returns to HOME point and land automatically.

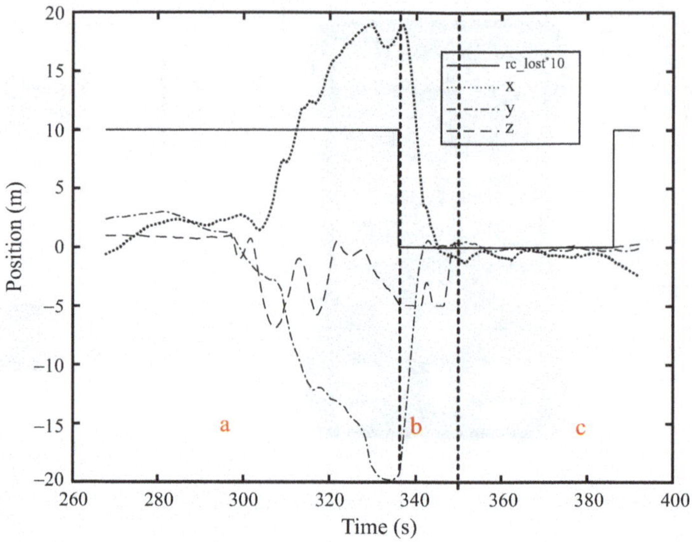

Fig. 12.24 Position response with RC transmitter being power-off in flight test

12.4.4 Flight Test Procedure

(1) **Step1: Quadcopter Preparation**
For details, see Sect. 4.3.4 in Chap. 4.

(2) **Step2: Simulink model for flight test**
The modified model according to Step 1 is shown in Fig. 12.23. The HIL simulator is shown in Fig. 12.22, where the PWM output part in Fig. 12.22 is replaced with the block in Fig. 12.23. Another change is the safe altitude set to be 5 m. In order to record flight data, a new data recording method is also used, the detailed procedure can be found in Sect. 9.4.4.

(3) **Step3: Upload code**
This process is similar to that used for compiling and uploading the code in HIL simulations. The experimental flow is shown in Fig. 12.10. The detailed procedure is introduced in Sect. 3.3.3.

(4) **Step4: Outdoor flight test**
To ensure safety, a rope is tethered to the quadcopter, and the other end is tethered to a heavy object. The remote pilot maintains a safe distance from the quadcopter during flight, as shown in Fig. 9.39.

(5) **Step5: Analyze data**
The log data of the Pixhawk Autopilot System is shown in Fig. 12.24. During the first 335 s, the quadcopter is in the stabilizing mode, corresponding the phase "a", during which it flies freely under the control of the flight controller. At about 335 s, the RC transmitter loses contact, corresponding the phase "b". During this phase, the altitude of the quadcopter rises to 5 m, and simultaneously the

horizontal position returns to the HOME point. Finally, the quadcopter makes landing, corresponding the phase "c". During this phase, the quadcopter's horizontal position stays in the HOME point, while the altitude is decreased to 0. The actual flight results and experimental data show that the quadcopter can realize the designed failsafe.

12.5 Summary

(1) According to the given requirement, a state machine with several modes and events can be designed to simplify flight control.
(2) It is important to understand the three flight modes. RTL MODE is about returning to base and then hovering at a certain altitude. AUTO-LANDING MODE is about landing but the horizontal position keeps the same. In the two modes, the automatic control of the autopilot is performed. In MANUAL FLIGHT MODE, both the RC and automatic control work.
(3) In this chapter, the main considerations are the communication failure as well as the altitude and horizontal position constraints. In an actual situation, it is necessary to consider other failures with the health evaluation (including preflight and in flight), making the design more practical.

If you have any question, please go to https://rflysim.com for your information.

Correction to: Multicopter Design and Control Practice

Correction to:
Q. Quan et al., *Multicopter Design and Control Practice*, https://doi.org/10.1007/978-981-15-3138-5

The original version of the book has been revised. The erratum book has now been updated with the below-listed corrections.

Category	Occurnces
Textual (replacement)	143
Textual (deletion)	28
URL corrections	16
Equation	18
Figure	9
Table	10

The updated version of the book can be found at https://doi.org/10.1007/978-981-15-3138-5

© Publishing House of Electronics Industry 2021

Q. Quan et al., *Multicopter Design and Control Practice*,
https://doi.org/10.1007/978-981-15-3138-5_13

Appendix A
RflySim Advanced Functions

The RflySim platform released with this book only opens up some basic functions required for the course requirements. This platform also supports some advanced practical functions, such as swarm simulation, high-fidelity UE4 scenes, and HIL simulations for other types of vehicles (fixed-wing aircraft, small cars, etc). This chapter introduces the installation method of RflySim platform and the advanced functions of the experimental platform after registration.

A.1 Installation Method

The registration method is presented as follows.

(1) As shown in Fig. A.1, visit https://rflysim.com to acquire the installation package and activation key, and then enter the package path and run the command "OnekeyScript" in MATLAB.
(2) Paste the activation key in the input box shown in Fig. A.1, and Click "OK" to install the whole platform automatically.

As shown in Fig. A.2, the complete UI of CopterSim after registration is mainly divided into five parts. Figure A.2a is the interface for configuring the multicopter model and flight environment parameters, where readers can select propulsion system components to assemble different types of multicopters for subsequent simulations. Figure A.2b is the model parameter calculation and database management interface, which is mainly used to calculate the model parameters of the assembled multicopter and store the results into the database for subsequent use with convenient database management functions. Figure A.2c is the advanced function area, including swarm simulation, UE4 scene selection, and other functions, which will be described in

© Publishing House of Electronics Industry 2020, corrected publication 2021

Q. Quan et al., *Multicopter Design and Control Practice*,
https://doi.org/10.1007/978-981-15-3138-5

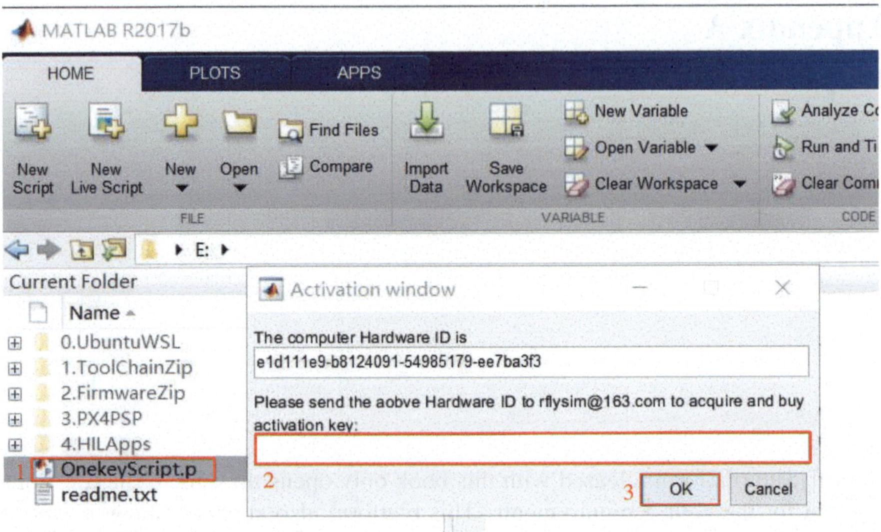

Fig. A.1 CopterSim registration dialog

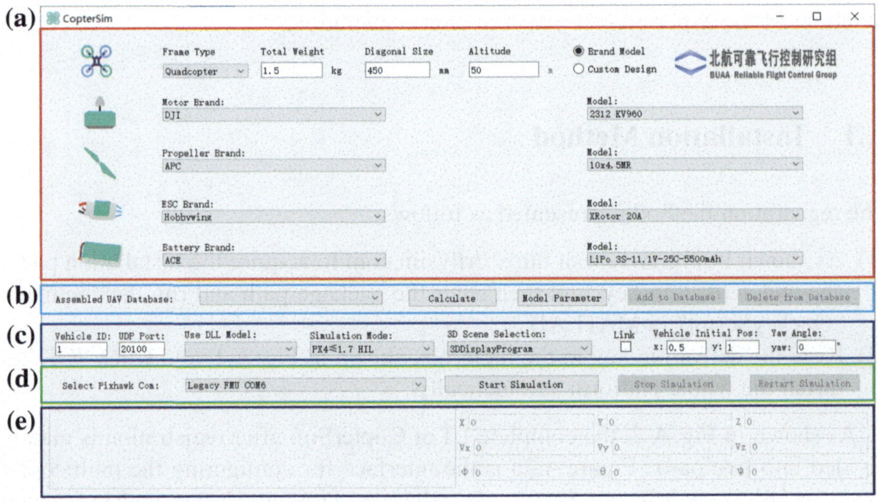

Fig. A.2 CopterSim interface after registration

detail later. Figure A.2d is mainly used to connect the Pixhawk autopilot and control the start and stop of the simulation. Figure A.2e is used to display the real-time message received from the Pixhawk autopilot as well as the position and attitude information of the simulated vehicle model.

A.2 Custom Multicopter Model

In the first line of the model configuration interface shown in Fig. A.2a, the configurable parameters include basic information such as the total weight, diagonal size, and flight altitude. In the second to the fifth rows of Fig. A.2a, readers can select the propulsion system components or parameters for the multicopter components, such as motors, propellers, ESCs, and batteries. CopterSim offers two options for selecting propulsion components. As shown in Fig. A.3, the first is to assemble a multicopter directly from the propulsion system branded model database which covers many common products on the market.

Since the propulsion system product database is difficult to cover all of the products in the world, CopterSim also provides a function to customize the component parameters which enables a more flexible way to configure a multicopter model. As shown in Fig. A.3a, click on the "Custom Design" option at the end of the first line to enter the component custom parameter interface shown in Fig. A.3b. In this interface, readers can customize the detailed parameters of motors (KV value, no-load voltage, internal resistance), ESCs, batteries, and propellers. In theory, by using the "Custom Design" function, readers can simulate any propulsion system components, even products that do not exist on the market.

After a multicopter model is configured, clicking on the "Calculate" button in Fig. A.2b will automatically analyze the feasibility of the inputted multicopter configuration. Unpractical multicopter configurations may cause simulation failures, such as insufficient thrust to take off, ESC and motor overheat due to excessive battery voltage, or the propeller size is too large for the fuselage. As shown in Fig. A.4, after checking the inputted multicopter configuration, CopterSim will prompt a warning dialog if the configuration is unpractical, and the reader needs to re-select a practicable multicopter configuration.

For beginners or readers who are not familiar with multicopter design, it is a difficult task to configure a multicopter that can work. So CopterSim also provides a convenient function to directly select vehicle configurations through a prepared model database. As shown in Fig. A.5, in the drop-down list of the "Model Database" item, the reader can select a usable multicopter configuration (or modify parameters based on it) to quickly obtain the multicopter model for simulations.

After configuring a multicopter through the three methods mentioned above, clicking the "Add to Database" button shown in Fig. A.5 will store the new model into the model database for use in the future. Readers can also delete the current model from the database by clicking the "Delete from Database" button shown in Fig. A.5.

A.3 Set Initial Position

As shown in Fig. A.2c, CopterSim provides an interface to modify the multicopter initial position and altitude. As shown in Fig. A.6, readers can enter the "x" and "y" coordinates (unit: m) of the multicopter and the value of the yaw angle (unit: degree),

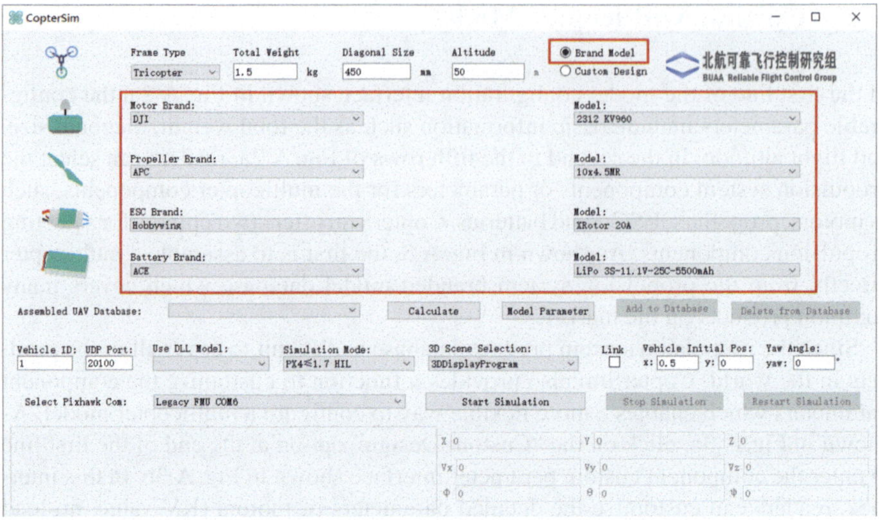

(a) Configure multicopter model by selecting propulsion components

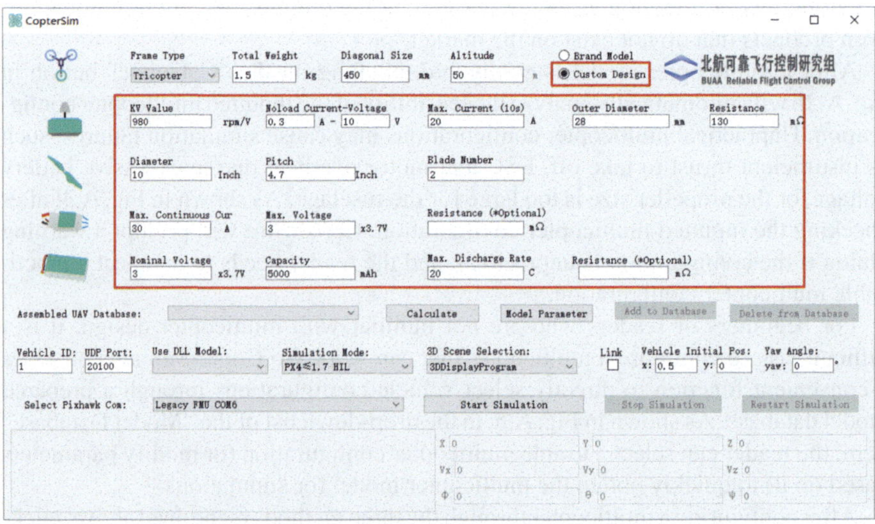

(b) Configure multicopter model by customizing propulsion parameters

Fig. A.3 Custom model parameter input interface

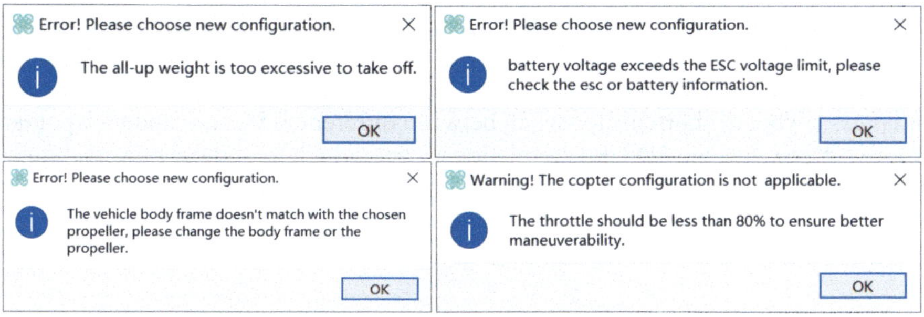

Fig. A.4 Model configuration feasibility detection warning dialog

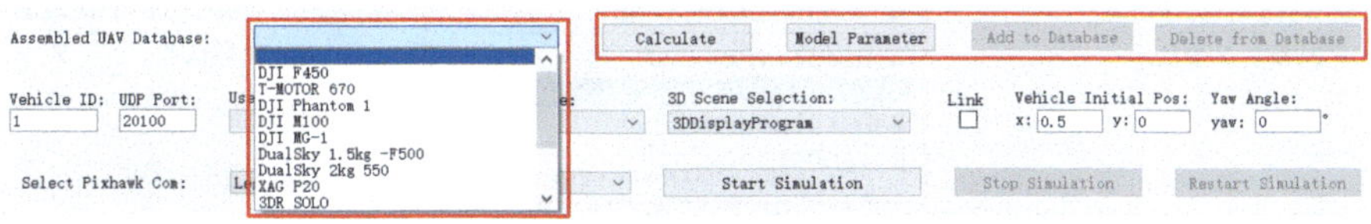

Fig. A.5 Multicopter model selection from model database

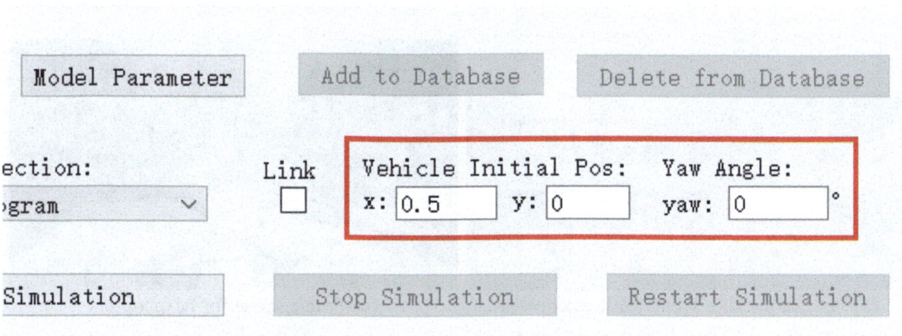

Fig. A.6 Setting initial simulation position

where the x-direction and y-direction point to the front and right of the multicopter, and the yaw angle is positive when rotating to right side.

A.4 3D Scenes Based on UE4

As shown in Fig. A.2c, CopterSim provides an ability to select different display scenes. The default scene is "3DDisplayprogram", which is based on a lightweight 3D engine—Ogre. It has low requirements for computer configurations, but the loading speed is slow, and the display performance is not realistic enough. As shown in Fig. A.7, click the drop-down box of "3D Scene Selection" to check all other available 3D scenes in the RflySim Advanced Platform, including "MountainTerrain" and "NeighborhoodPark", which are more realistic 3D scenes based on UE4 3D engine. After selecting the corresponding scene, manually open the RflySim3D software and press keyboard shortcuts to switch between different 3D scenes and viewpoints. Figure A.8 presents the 3D UE4 simulation scenes, which have more powerful functions and more convenient extensibility.

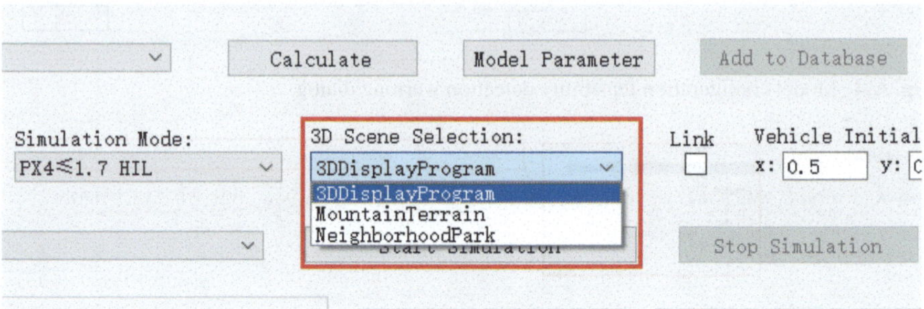

Fig. A.7 3D scene selection

(a) Plateau ice lake scene for quadcopters

(b) Forest scene for hexacopters

(c) Neighborhood street scene for cars

(d) Mountain scene for fixed-wing aircraft

Fig. A.8 3D simulation scene based on UE4

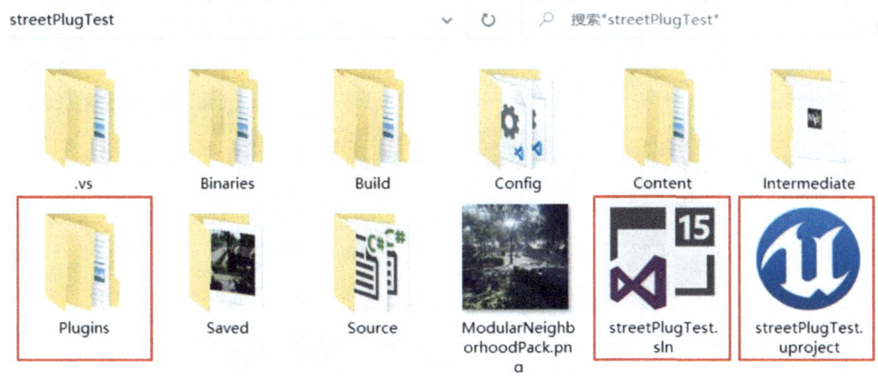

Fig. A.9 Custom UE4 3D scene project folder

A.5 Developing 3D Scenes Through UE4

Through the UE4 plugin, readers can quickly build their own flight scenarios in UE4. The specific procedure is listed as follows.

(1) Create a new UE4 scene project. As shown in Fig. A.9, readers can purchase or develop a 3D scene project based on UE4 (engine version should be 4.22). The detailed process will be not introduced here because it can be found in the official tutorial.[1]
(2) Package the UE4 project to "Windows (64-bit)" files and copy the obtained scene content folder to RflySim3D folder (Fig. A.10).
(3) Export the map terrain data. Select the terrain file in UE4, exported to a ".png" format file, and store three key position points to a ".txt" file. Then, copy the resulting ".png" and ".txt" files to the "CopterSim\external\map" directory (see Fig. A.11). Finally, after restarting CopterSim, readers can select the new developed 3D scene from the drop-down menu shown in Fig. A.7.

[1] http://api.unrealengine.com/CHN/.

Fig. A.10 UE4 editor

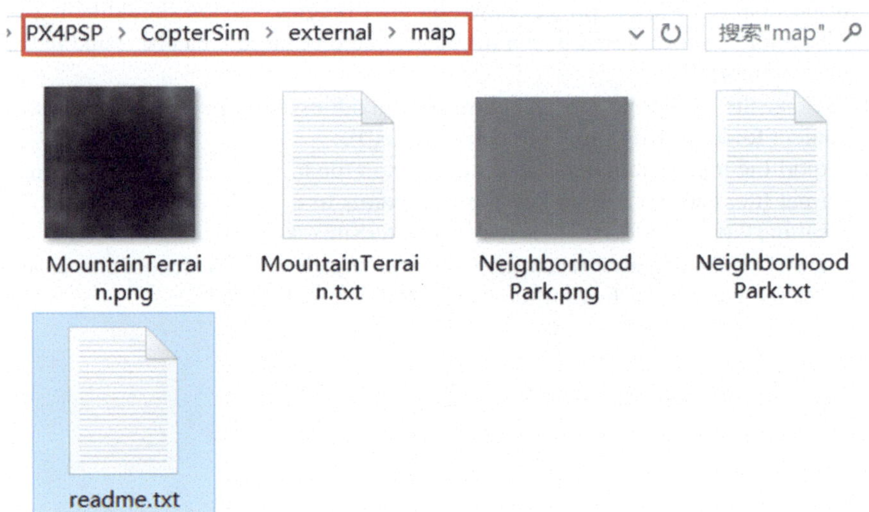

Fig. A.11 Importing terrain mesh data into CopterSim

A.6 HIL Simulation for Other Types of Multicopters

In the previous courses, readers need to set up the Pixhawk autopilot through QGC to enter the "HIL Quadcopter X" airframe mode. This mode only supports the quadcopter X configuration, which has strong limitations in practical usage. In addition to quadcopters, CopterSim can be used to all multicopter types supported by the PX4 autopilot software. The specific steps are presented as follows.

(1) Select a multicopter airframe in QGC

1) Correctly connect the Pixhawk autopilot with the QGC.
2) Select a required multicopter airframe in the "Airframe" tab (see Fig. A.12), such as hexacopters, octocopters, coaxial multicopters, etc.
3) Select a corresponding airframe size from the drop-down list of the airframe (e.g., F450, and 3DR DIY Quad).
4) Confirm that the Pixhawk is in the selected airframe mode applying the airframe and restarting.

Through the above steps, readers can set an airframe type that can be used for the actual flight with controller parameters matching the airframe size.

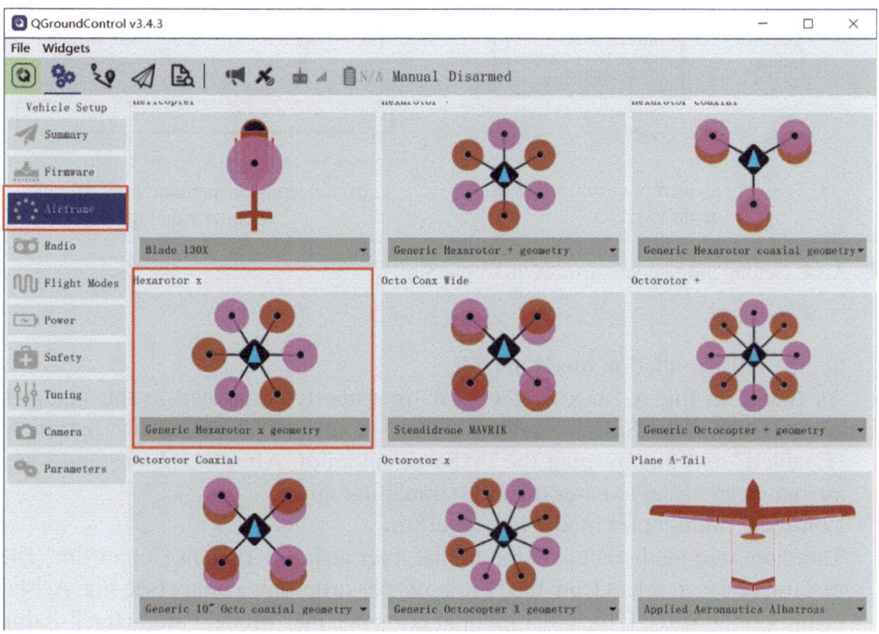

Fig. A.12 Choosing a required multicopter airframe in QGC

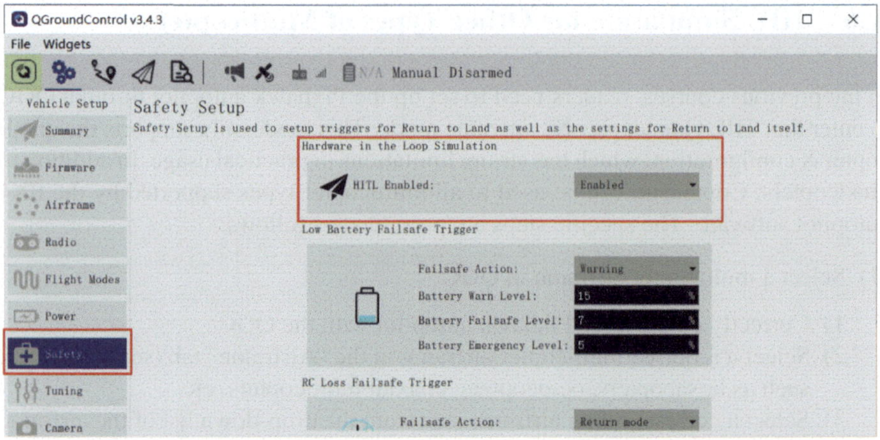

Fig. A.13 Setting "HITL Enabled" option to "Enabled"

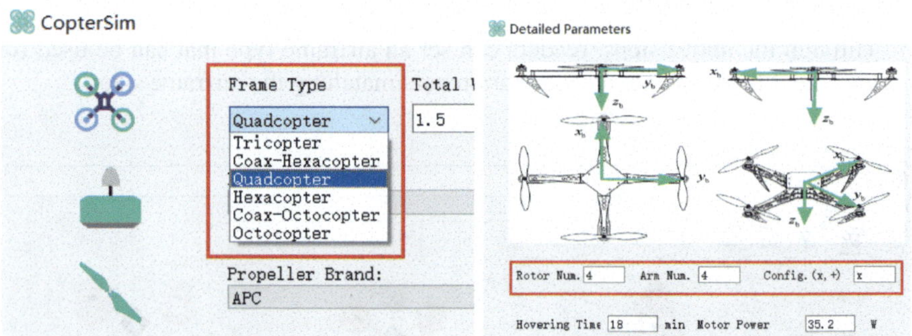

(a) Selecting an airframe from CopterSim main interface

(b) Selecting an airframe from Model Parameter dialog

Fig. A.14 Setting multicopter airframe in CopterSim

(2) Set the HIL simulation mode in QGC
As shown in Fig. A.13, after the QGC is properly connected to the Pixhawk autopilot, in the "Safety" tab of QGC, set the "HITL Enabled" option to "Enabled" and then re-plug the Pixhawk autopilot. After the above steps, any airframe can be set to enter the HIL simulation mode.

(3) Configure multicopter model in CopterSim
There are two methods to set the multicopter airframe type in CopterSim. The first method is to select from the "Frame Type" drop-down menu (see Fig. A.14a) on the CopterSim UI; the other method is to open the "Model Parameters" dialog to set the number of arms, the number of rotors, and the head orientation.

(4) Start HIL simulation
Taking a hexacopter as an example, set the Pixhawk airframe type to a general hexacopter in QGC (see Fig. A.12); enable the HIL mode in QGC (see Fig. A.13);

then, configure hexacopter parameters in CopterSim, as shown in Fig. A.14b. Next, insert the Pixhawk autopilot into the computer, select the Pixhawk serial port in CopterSim, and click "Start Simulation" button to start the HIL simulation for the hexacopter. If the UE4 scenes shown in Fig. A.7 are used, a hexacopter will be displayed automatically, as shown in Fig. A.8b.

A.7 HIL Simulations of Other Models

As shown in Fig. A.12, a Pixhawk autopilot supports not only multicopters, but also fixed-wing aircraft, vertical take-off and landing aircraft, rovers, boats, and other types of vehicle. To support all vehicle types supported by Pixhawk autopilots, CopterSim also provides an interface to import the Simulink model as DLL model files to perform HIL simulations.

(1) Use DLL model files in CopterSim
As shown in Fig. A.15a, copy the DLL model file generated by Simulink to the folder "CopterSim\external\model". After re-opening the CopterSim, readers can see all available DLL model files in the "Use DLL Model" drop-down list in Fig. A.15b. These DLL model files can be used to simulate any aircraft or vehicles. Select a DLL model, configure the Pixhawk autopilot to the desired airframe in QGC, and finally start HIL simulation in CopterSim. Readers can also add a corresponding 3D vehicle model in the UE4 editor shown in Fig. A.10 to form a closed-loop HIL simulation system. For example, Fig. A.8 shows the HIL simulation scene of the multicopter, fixed-wing aircraft, and trolley.

(2) Method to generate DLL model file in Simulink
As shown in Fig. A.16, open the Simulink source code folder "Multicopter-Model", and open the desired ".slx" Simulink file (multicopter, aircraft, etc.) in it. Then, modify the model parameters or replace some of the modules in Simulink to meet the simulation model requirements. Next, click the "compile" button in Simulink to generate C/C++ code. Finally, run the "GenarateMod-

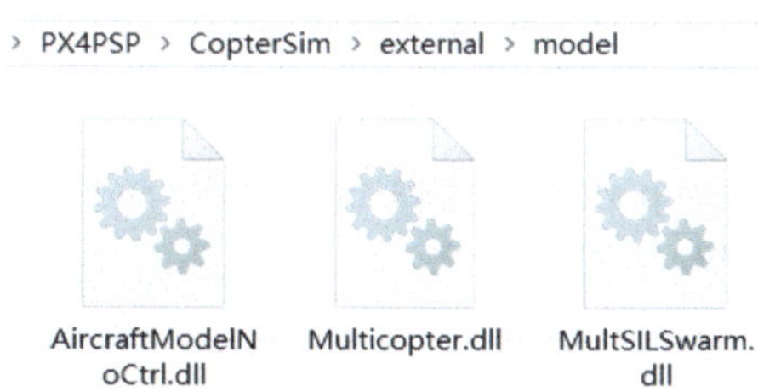

(a) The DLL files in CopeterSim folder

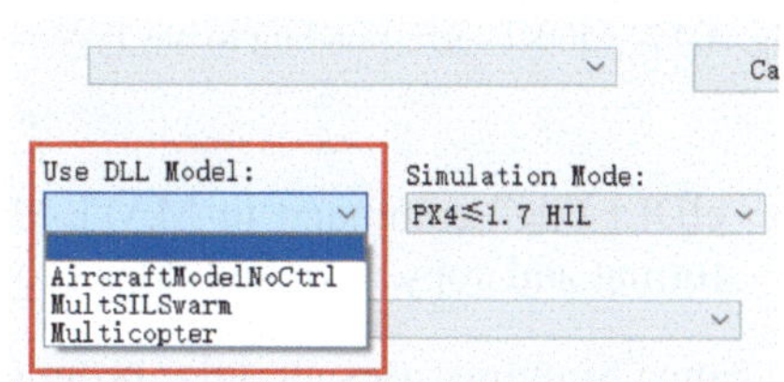

(b) DLL Model options in CopeterSim main interface

Fig. A.15 Select DLL model for simulation

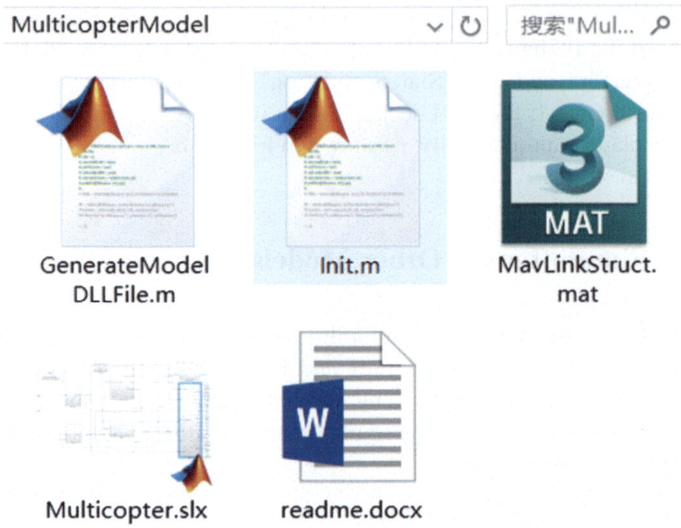

Fig. A.16 Simulink source code to generate DLL model file

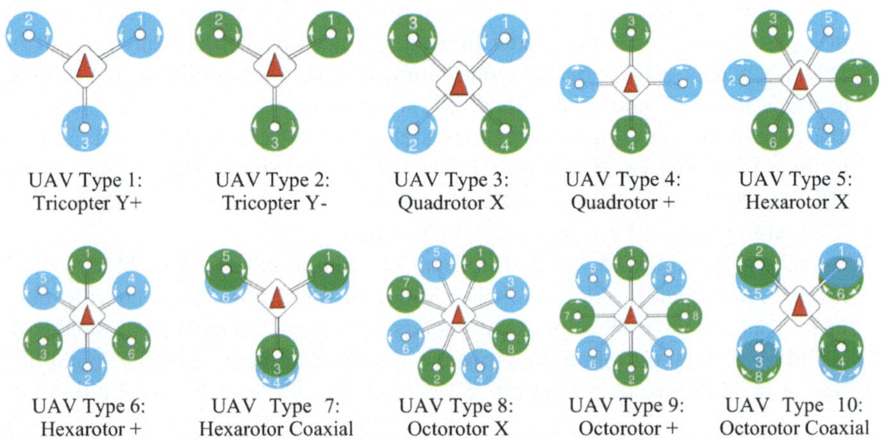

Fig. A.17 Model corresponding to uavType parameter in Simulink model

elDLLFile" command in MATLAB to generate the DLL model file in ".dll" format and copy it to the folder shown in Fig. A.15.

Since Simulink uses modular programming methods, in the provided ".slx" file, it is easy to obtain different vehicle configurations by changing parameters or some modules. For example, the multicopter types presented in Fig. A.16 can be easily applied by modifying the airframe type parameter ModelParam_uavType in the "Init.m" initialization script (see Fig. A.17) to a corresponding value. The default multicopter configuration of the previous experiment was uavType = 3, which is the conventional X-shaped quadcopter.[2]

A.8 Swarm Simulation

CopterSim also supports swarm simulations. As shown in Fig. A.18, open multiple CopterSim programs and set the ID value of each aircraft to the desired value. Each CopterSim needs to connect to a Pixhawk autopilot, and then click the "Start Simulation" button to start the HIL simulation separately. The 3D display programs (only supports the vision developed by UE4) will automatically receive all vehicle data from the local network and display them in the same 3D display software.

The current distributed simulation architecture theoretically supports HIL simulation for any number of vehicles. The limitation mainly depends on the following factors: the number of computer USB interfaces (each Pixhawk will occupy a USB port), the memory size of the computer (each CopterSim will occupy a certain amount of memory space), the computer processor speed (need to ensure that each CopterSim can run in real-time).

Currently, CopterSim supports two methods to start the swarm simulation:

(1) The first method is to open multiple CopterSims manually. Repeat to click the CopterSim.exe file to open an expected number of CopterSim programs one by one, each of them assigned with a unique ID automatically . Then, connect the serial ports and click to start the simulation.
(2) The second method is to open an expected number of CopterSim programs together through a "bat" script, with automatically completing software configuration and serial port selection. Open the "CopterSim\ AutoStartScriptTemp.bat" script with a text editor, and the following content will be observed.

    - start CopterSim.exe 1 1 20100 0 0 1 0 235 0 0 4
    - start CopterSim.exe 1 2 20102 0 0 1 0 235 0.5 0 5
    - start CopterSim.exe 1 3 20104 0 0 1 0 235.5 0 0 6
    - start CopterSim.exe 1 4 20106 0 0 1 0 235.5 1 0 7

    The above code shows a Windows batch bat script that opens four CopterSim programs at a time. As corresponding with Fig. A.19, the 0th number followed by CopterSim.exe means "whether the simulation starts automatically"; the 1st to 9th numbers correspond to the options of the advanced functions in the CopterSim UI shown in Fig. A.2c; the 10th number corresponds to the Pixhawk USB serial port number.

[2]These configurations correspond to the multicopter models supported by the PX4 autopilot. Readers can refer to the official website: http://dev.px4.io/master/en/airframes/airframe_reference.html.

Fig. A.18 CopterSim swarm simulation

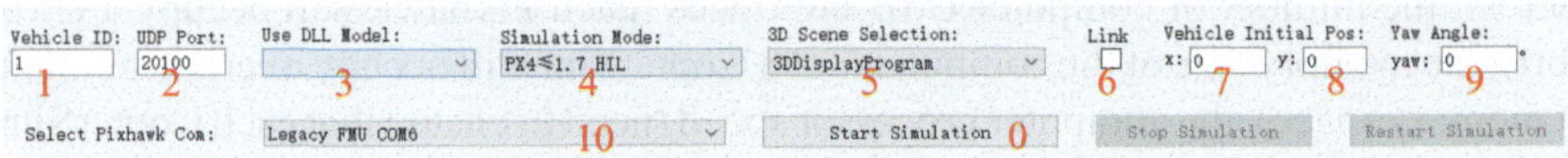

Fig. A.19 Corresponding relation between CopterSim's parameters and script's numbers

A.9 Simulation Mode Settings

In addition to HIL simulation, CopterSim also supports other simulation modes to improve test efficiency. As shown in Fig. A.20, there are several types of simulation modes supported by CopterSim.

(1) Change Simulation Modes. In RflySim platform, you can use a Pixhawk hardware to perform HIL simulations, or run PX4 SITL controller in WSL environment to perform SIL simulations without Pixhawk hardware.
(2) SIL simulation with Simulink. The swarm simulation speed is very slow if all full dynamics models are running directly by Simulink. As a result, it is not convenient to observe the simulation results in real-time. Therefore, readers can convert the Simulink model into a DLL model file which can be imported to CopterSim by the code generation method mentioned in the previous section. Then, Simulink controller can exchange sensor data and control signals with CopterSim programs through the UDP to form a closed-loop swarm SIL simulation platform. In swarm simulations, the UDP port number in Fig. A.20 needs to be different among CopterSim programs, and then Simulink can communicate with each CopterSim program through the corresponding port number. It should be noted that the DLL model source code presented in Fig. A.15 also supports to simulate a complete closed-loop system with both model and controller. This makes it possible to directly send the speed commands to DLL models, and perform the SIL simulations to test the top-level decision-making algorithms.

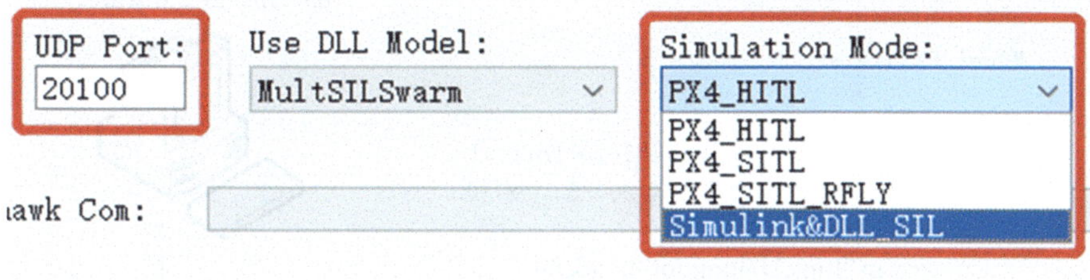

Fig. A.20 CopterSim simulation mode setting

(3) Connecting with the PX4 autopilot software (without Pixhawk hardware) to perform SIL simulations on a computer. The PX4 source code also supports to be compiled into a firmware that can be run in a computer. Run *N* PX4 autopilot software on a computer, and then open *N* CopterSim programs to connect with them correspondingly. Then, it is possible to perform swarm SIL simulations with PX4 autopilot software on the same computer.

A.10 Multi-computer Distributed Simulation

As mentioned in the previous section, the computing performance and memory space of a single computer are limited, so it is impossible to support swarm simulation for an unlimited amount of vehicles. Therefore, CopterSim also provides the function to perform distributed simulation through Local Area Network (LAN) by using multiple computers. As shown in Fig. A.21, enable the "Link" option on the UI of CopterSim to start the network broadcast function. The flight data of each CopterSim will be received by all computers via the LAN. Therefore, with enough computers connecting to the LAN, and a certain amount of CopterSim programs (in HIL or SIL simulation modes) are run on each computer, the platform can realize swarm distributed simulation for any number of vehicles. At the same time, each computer can open multiple 3D scene programs to observe all vehicles from multiple perspectives. Note that if the "Link" checkbox is unchecked, the flight information of CopterSim can only be received by this computer, so it will not affect other computers, which is more suitable for the use in experimental courses.

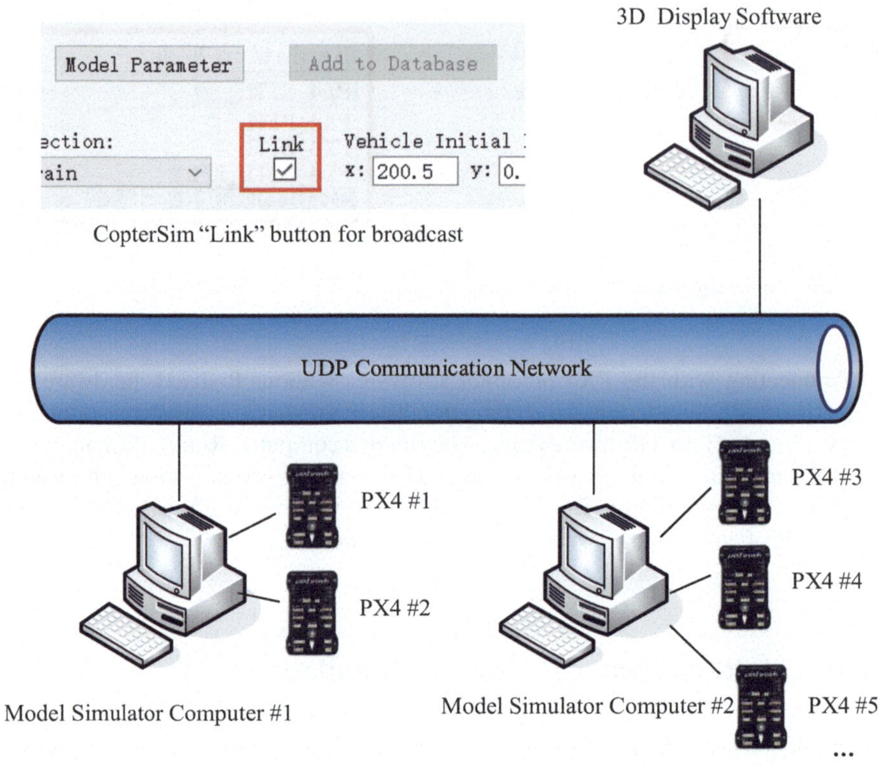

Fig. A.21 CopterSim distributed simulation architecture

A.11 Swarm Control in Simulink

For swarm simulation, communication and real-time control is a difficult task. In response to these needs, we have developed tutorials with source codes (see Fig. A.22) based on the advanced functions of this experimental platform. The main implemented features are shown in the following.

(1) Open-loop Swarm SIL simulation. The advanced platform can send swarm data to the UE4 scene display software with the Simulink UDP interface to preview the trajectories of all vehicles. With this function, on the one hand, it is convenient to generate trajectories for static or moving obstacles, so complex swarm flight scenes can be visualized as shown in Fig. A.23. On the other hand, it provides the opportunity to preview the swarm flight performance, which helps developers to improve the designed trajectories.

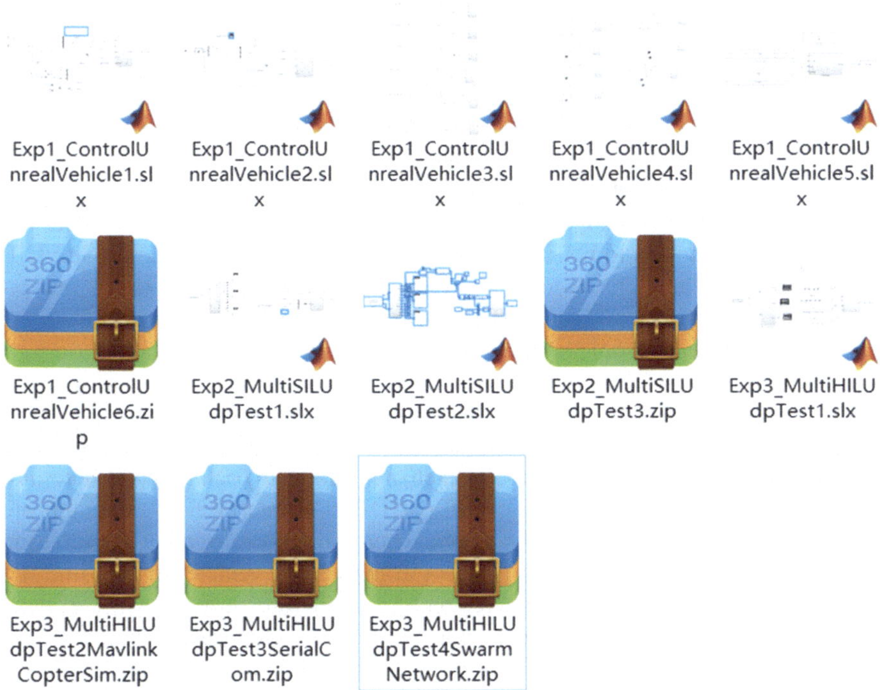

Fig. A.22 Source code for Simulink swarm control

Fig. A.23 Example for heterogeneous unmanned systems in UE4

(2) Closed-loop Swarm SIL simulation. The advanced platform provides a Simulink UDP interface to exchange motion states and control signals from all CopterSim SIL control models to realize the closed-loop swarm SIL simulation. The CopterSim SIL control model can be 1) a DLL model file including both a multicopter motion model and a velocity-loop controller; or 2) CopterSim motion model controlled by a PX4 SITL controller.
(3) Closed-loop Swarm HIL simulation. The advanced platform provides a Simulink UDP interface to exchange motion states and control signals from all CopterSim HIL control models to realize the swarm HIL simulation. The CopterSim HIL control model is a closed-loop control system involving real Pixhawk autopilots. CopterSim can automatically forward Mavlink message packets from the Pixhawk autopilot serial port to Simulink. Furthermore, interfaces are also developed to directly encode and decode Mavlink message in Simulink. As a result, users can control each Pixhawk+CopterSim HIL system by Simulink.
(4) Closed-loop swarm flight by Simulink. The advanced platform provides the serial communication interface to directly exchange data with each Pixhawk autopilot via wireless radio telemetry, as well as UDP/ROS interfaces to communicate with vehicles in the same WIFI network. Therefore, it is convenient to realize the functions of GCSs to receive flight data from multiple vehicles in real-time and send back control signals. Compared with GCS software developed by C/C++, it is more convenient to tune the controller parameters on line and observe the real-time control performance in Simulink. Besides, the Simulink swarm controller can also be converted into C/C++ to improve the running speed.

Through the above procedure, developers can perform the development, simulations and flight tests in a step-by-step process, which may significantly improve the development efficiency for swarm control systems.

A.12 Vision-based Control Interface

As shown in Fig. A.24, the provided UE4 3D display software also supports a view switching function to easily acquire image data from multiple viewing angles. The image data can be acquired and processed in real-time through Simulink, Python, C/C++, and other code platforms through shared memory inter-process communication or capture screen images from UE4 software. The obtained vision data can be returned to CopterSim or Simulink control through the UDP network to form a HIL simulation loop with vision feedback.

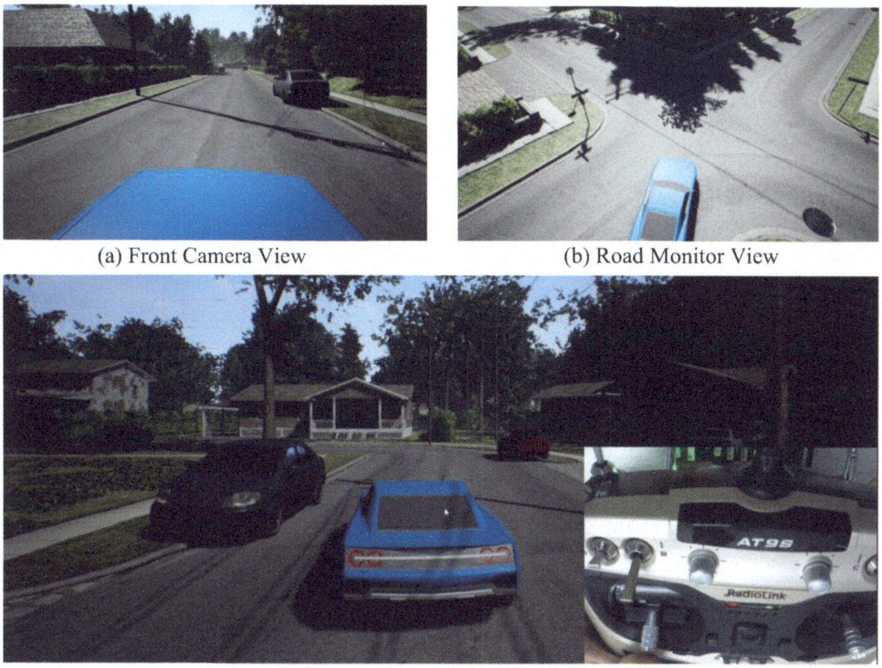

(a) Front Camera View

(b) Road Monitor View

(c) Chase View for Real-time HIL Simulation

Fig. A.24 Different views of a car

Appendix B
How Teachers Use This Book

This book provides eight experiments, each of which includes a basic experiment, an analysis experiment, and a design experiment. The code for the basic and analysis experiments will be open on the designated website while the code for design experiments will only be open to the teachers who open the course. Students can first learn experimental principles and basic code through basic experiments and analysis experiments, based on which they can then carry out design experiments. To ensure that different students have different design and experimental goals, there are two solutions given as follows: modifying the goals in the propulsion system design and modeling experiments for different students, or opening new experiments. A detailed introduction is in the following.

B.1 Modify the Goals in the Propulsion System Design and Modeling Experiments for Different Students

To ensure that the design goals of different students are different, a solution is given below.

B.1.1 Modify the Design Experiment of the Multicopter Propulsion System

(1) Things to prepare
https://flyeval.com/paper/.
(2) Objectives

1) Design a multicopter. The altitude is 0m, the local temperature is 25 °C, the weight of the airframe, autopilot, and accessories is 1 kg totally, circumferential circle radius is smaller than 59.23 inches, the total weight is lighter

© Publishing House of Electronics Industry 2020, corrected publication 2021

Q. Quan et al., *Multicopter Design and Control Practice*,
https://doi.org/10.1007/978-981-15-3138-5

than (7+Y) kg $\times$ 9.8 m/s^2, hover endurance is longer than 15 min, the hover throttle is less than 65% of the full throttle.

2) List flight parameters and basic flight performance parameters of the multicopters and compare the calculated results with that generated by https://flyeval.com/paper/.

Here, for students who select this course, we can modify the goals. For example, dividing the last digit of a student ID by 4 obtains the remainder X = {0: hexacopter, 1: coaxial hexacopter, 2: octocopter, 3: coaxial octocopter} corresponding to different types of multicopters. What is more, dividing the second-last digit of the student ID by 4 obtains the remainder Y, which corresponds to modifying the overall weight.

B.1.2 Modify the Design Experiment for the Multicopter Modeling Experiment

(1) Things to prepare
The well-designed multicopter configuration by "Multicopter propulsion system design experiment"—"Design Experiment" and model parameters provided by https://flyeval.com/paper/.

(2) Objectives
Obtain multicopter mathematical models, and build a complete multicopter model on MATLAB/Simulink, and add a 3D multicopter model to FlightGear. In terms of the attitude model, the quaternion model, the rotation matrix model, or the Euler angle model can be used for students who select this course. Dividing the third-last digit of their student ID obtains the remainder Z = {0: quaternion model, 1: rotation matrix model, 2: Euler angle model}.

With the two modified experiments above, it is difficult for two students to have the same task. Based on the results from the two modified experiments, the following attitude controller design experiment and set-point controller design experiment are different correspondingly. With the number of students increased, new types of multicopters can be added, such as a quadcopter, pentacopter, heptcopter, and so on. If facing a large number of students, it is recommended that students perform HIL simulation and indoor flight experiments, while only teachers show the process of outdoor flight tests by taking the students' safety into consideration.

B.2 Opening New Experiments

(1) Opening new experiments on filter design
At present, this book only provides complementary filter and Kalman filter experiments, and only for attitude estimation. As for the estimation method, there are

many new methods, such as unscented Kalman filter, particle filter, and so on. In the aspect of the model and measurement available, position estimation with GPS can be adopted as a new experiment, for example.

(2) Opening new experiments on multicopter attitude control design and multicopter position control design
At present, this book only provides control experiments for general readers with the classical automatic control background. For advanced and practical control methods, such as Active Disturbance Rejection Control (ADRC) [16] and model predictive control [17], students can try in experiments, especially in postgraduate related experimental courses.

(3) Other open experiments
In addition to the existing experiments, teachers can also open other courses:

(1) Multicopter tracking controller design. The simplest task is to design a controller to track a given 4D trajectory (including time and position), such as a circle (the multicopter' head needs to point to the center of the circle) or a number "8".
(2) Multicopter path following controller design. The simplest task is to design path-following controllers for straight lines, circles, and the number "8". Unlike tracking, the multicopter does not need to fly to a given point on a path at a given time, as long as its entire path follows the given path.
(3) Multicopter obstacle avoidance controller design. Control a multicopter to avoid a stationary spherical or stationary cylindrical obstacle, or several obstacles. Besides, teachers can also design an experiment which requires students to design a controller to avoid a geofence.
(4) Multicopter area coverage decision-making design. Refer to the tasks in Sect. 13.1.12 of book [8]. Consider multicopter recharging, design an experiment which asks students for designing a controllers with its flight routes covering a whole rectangular farmland.

References

1. A. Terry Bahill, S. J. Henderson, Requirements development, verification, and validation exhibited in famous failures. Syst. Eng. **8**(1), 1–14 (2005)
2. T.A. Talay, Introduction to the aerodynamics of flight, Scientific and Technical Information Office, NASA, USA, Tech. Rep. SP-367 (1975)
3. Namebay, F3-radio control soaring (2019). [Online]. https://www.fai.org/page/f3-radio-control-soaring
4. D. Palmer, The FAA's interpretation of the special rule for model aircraft. J. Air Law Commer. **80**, 567 (2015)
5. D.M. Henderson, Euler angles, quaternions, and transformation matrices for space shuttle analysis. NASA; United States, Tech. Rep. NASA-CR-151435 (1977)
6. Q. Zeng, W. Zhang, Z. Huang, R. Dong, Improving aerospace engineering students' achievements by an open aero control experiment apparatus. IEEE Trans. Educ. **57**(4), 229–234 (2014)
7. H. Hanrahan, The washington accord: history, development, status and trajectory, in *7th ASEE Global Colloquium on Engineering Education* (2008), pp. 19–23
8. Q. Quan, *Introduction to Multicopter Design and Control* (Springer, Singapore, 2017)
9. B.L. Stevens, F.L. Lewis, *Aircraft Control and Simulation, 2end* (Wiley, New Jersey, 2004)
10. P. Aggarwal, *MEMS-Based Integrated Navigation* (Artech House, 2010)
11. K. Madsen, H.B. Nielsen, O. Tingleff, Methods for non-linear least squares problems (2nd ed.), Richard Petersens Plads, Building 321, DK-2800 Kgs. Lyngby (2004) p. 60
12. S. Poddar, V. Kumar, A. Kumar, A comprehensive overview of inertial sensor calibration techniques. J. Dyn. Syst. Measurement Control **139**(1), 011006 (2017)
13. D. Simon, *Optimal State Estimation: Kalman, H Infinity, and Nonlinear Approaches* (Wiley, 2006)
14. C.W. Kang, C.G. Park, Attitude estimation with accelerometers and gyros using fuzzy tuned kalman filter, in *European Control Conference (ECC) 2009*, pp. 3713–3718
15. K. Ogata, Y. Yang, *Modern Control Engineering* (Pearson Upper Saddle River, NJ, 2010)
16. J. Han, From PID to active disturbance rejection control. IEEE Trans. Ind. Electron. **56**(3), 900–906 (2009)
17. L. Grüne, J. Pannek, *Nonlinear Model Predictive Control: Theory and Algorithms* (Springer-Verlag, London, 2017)

© Publishing House of Electronics Industry 2020, corrected publication 2021
Q. Quan et al., *Multicopter Design and Control Practice*,
https://doi.org/10.1007/978-981-15-3138-5

The manufacturer's authorised representative in the EU is Springer Nature Customer Service Centre GmbH, Europaplatz 3, 69115 Heidelberg, Germany. If you have any concerns regarding our products, please contact ProductSafety@springernature.com

Printed and bound by CPI Group (UK) Ltd, Croydon, CR0 4YY
30/09/2026
02225556-0013